SCHAUM'S OUTLINE OF

THEORY AND PROBLEMS

OF

FINITE MATHEMATICS

•

BY

SEYMOUR LIPSCHUTZ, Ph.D.
Associate Professor of Mathematics
Temple University

•

SCHAUM'S OUTLINE SERIES
McGRAW-HILL, INC.
New York St. Louis San Francisco Auckland Bogotá
Caracas Lisbon London Madrid Mexico Milan
Montreal New Delhi Paris San Juan Singapore
Sydney Tokyo Toronto

ISBN 07-037987-4

26 27 28 29 30 SH SH 9 8 7 6 5 4 3 2 1

Cover design by Amy E. Becker.

Preface

Finite mathematics has in recent years become an integral part of the mathematical background necessary for such diverse fields as biology, chemistry, economics, psychology, sociology, education, political science, business and engineering. This book, in presenting the more essential material, is designed for use as a supplement to all current standard texts or as a textbook for a formal course in finite mathematics.

The material has been divided into twenty-five chapters, since the logical arrangement is thereby not disturbed while the usefulness as a text and reference book on any of several levels is greatly increased. The basic areas covered are: logic; set theory; vectors and matrices; counting – permutations, combinations and partitions; probability and Markov chains; linear programming and game theory. The area on vectors and matrices includes a chapter on systems of linear equations; it is in this context that the important concept of linear dependence and independence is introduced. The area on linear programming and game theory includes a chapter on inequalities and one on points, lines and hyperplanes; this is done to make this section self-contained. Furthermore, the simplex method is given for solving linear programming problems with more than two unknowns and for solving relatively large games. In using the book it is possible to change the order of many later chapters or even to omit certain chapters without difficulty and without loss of continuity.

Each chapter begins with a clear statement of pertinent definitions, principles and theorems together with illustrative and other descriptive material. This is followed by graded sets of solved and supplementary problems. The solved problems serve to illustrate and amplify the theory, bring into sharp focus those fine points without which the student continually feels himself on unsafe ground, and provide the repetition of basic principles so vital to effective learning. Proofs of theorems and derivations of basic results are included among the solved problems. The supplementary problems serve as a complete review of the material in each chapter.

More material has been included here than can be covered in most first courses. This has been done to make the book more flexible, to provide a more useful book of reference and to stimulate further interest in the topics.

I wish to thank many of my friends and colleagues, especially P. Hagis, J. Landman, B. Lide and T. Slook, for invaluable suggestions and critical review of the manuscript. I also wish to express my gratitude to the staff of the Schaum Publishing Company, particularly to N. Monti, for their unfailing cooperation.

S. LIPSCHUTZ

Temple University
June, 1966

CONTENTS

CONTENTS

Chapter 1

Propositions and Truth Tables

STATEMENTS

A *statement* (or *verbal assertion*) is any collection of symbols (or sounds) which is either *true* or *false*, but not both. Statements will usually be denoted by the letters

$$p, q, r, \ldots$$

The truth or falsity of a statement is called its *truth value*.

Example 1.1: Consider the following expressions:

(i) Paris is in England. (iii) Where are you going?

(ii) $2+2=4$ (iv) Put the homework on the blackboard.

The expressions (i) and (ii) are statements; the first is false and the second is true. The expressions (iii) and (iv) are not statements since neither is either true or false.

COMPOUND STATEMENTS

Some statements are composite, that is, composed of substatements and various logical connectives which we discuss subsequently. Such composite statements are called *compound statements*.

Example 2.1: "Roses are red and violets are blue" is a compound statement with substatements "Roses are red" and "Violets are blue".

Example 2.2: "He is intelligent or studies every night" is, implicitly, a compound statement with substatements "He is intelligent" and "He studies every night".

The fundamental property of a compound statement is that its truth value is completely determined by the truth values of its substatements together with the way in which they are connected to form the compound statement. We begin with a study of some of these connectives.

CONJUNCTION, $p \wedge q$

Any two statements can be combined by the word "and" to form a compound statement called the *conjunction* of the original statements. Symbolically,

$$p \wedge q$$

denotes the conjunction of the statements p and q, read "p and q".

Example 3.1: Let p be "It is raining" and let q be "The sun is shining". Then $p \wedge q$ denotes the statement "It is raining and the sun is shining".

The truth value of the compound statement $p \wedge q$ satisfies the following property:

[T_1] If p is true and q is true, then $p \wedge q$ is true; otherwise, $p \wedge q$ is false.

In other words, the conjunction of two statements is true only in the case when each substatement is true.

Example 3.2: Consider the following four statements:

(i) Paris is in France and $2+2 = 4$.

(ii) Paris is in France and $2+2 = 5$.

(iii) Paris is in England and $2+2 = 4$.

(iv) Paris is in England and $2+2 = 5$.

By property $[T_1]$, only the first statement is true. Each of the other statements is false since at least one of its substatements is false.

A convenient way to state property $[T_1]$ is by means of a table as follows:

p	q	$p \wedge q$
T	T	T
T	F	F
F	T	F
F	F	F

Here, the first line is a short way of saying that if p is true and q is true then $p \wedge q$ is true. The other lines have analogous meaning. We regard this table as defining precisely the truth value of the compound statement $p \wedge q$ as a function of the truth values of p and of q.

DISJUNCTION, $p \vee q$

Any two statements can be combined by the word "or" (in the sense of "and/or") to form a new statement which is called the *disjunction* of the original two statements. Symbolically,

$$p \vee q$$

denotes the disjunction of the statements p and q and is read "p or q".

Example 4.1: Let p be "Marc studied French at the university", and let q be "Marc lived in France". Then $p \vee q$ is the statement "Marc studied French at the university or (Marc) lived in France".

The truth value of the compound statement $p \vee q$ satisfies the following property:

$[T_2]$ If p is true or q is true or both p and q are true, then $p \vee q$ is true; otherwise $p \vee q$ is false.

Accordingly, the disjunction of two statements is false only when both substatements are false. The property $[T_2]$ can also be written in the form of the table below, which we regard as defining $p \vee q$:

p	q	$p \vee q$
T	T	T
T	F	T
F	T	T
F	F	F

Example 4.2: Consider the following four statements:

(i) Paris is in France or $2+2 = 4$.

(ii) Paris is in France or $2+2 = 5$.

(iii) Paris is in England or $2+2 = 4$.

(iv) Paris is in England or $2+2 = 5$.

By property $[T_2]$, only (iv) is fai Each of the other statements is true since at least one of its substatements is tı

Remark: The English word "or" is commonly used in two distinct ways. Sometimes it is used in the sense of "p or q or both", i.e. at least one of the two alternates occurs, as above, and sometimes it is used in the sense of "p or q but not both", i.e. exactly one of the two alternatives occurs. For example, the sentence "He will go to Harvard or to Yale" uses "or" in the latter sense, called the *exclusive disjunction.* Unless otherwise stated, "or" shall be used in the former sense. This discussion points out the precision we gain from our symbolic language: $p \vee q$ is defined by its truth table and *always* means "p and/or q".

NEGATION, $\sim p$

Given any statement p, another statement, called the *negation* of p, can be formed by writing "It is false that..." before p or, if possible, by inserting in p the word "not". Symbolically,

$$\sim p$$

denotes the negation of p (read "not p").

Example 5.1: Consider the following three statements:

(i) Paris is in France.

(ii) It is false that Paris is in France.

(iii) Paris is not in France.

Then (ii) and (iii) are each the negation of (i).

Example 5.2: Consider the following statements:

(i) $2+2 = 5$

(ii) It is false that $2+2 = 5$.

(iii) $2+2 \neq 5$

Then (ii) and (iii) are each the negation of (i).

The truth value of the negation of a statement satisfies the following property:

[$\mathbf{T_3}$] If p is true, then $\sim p$ is false; if p is false, then $\sim p$ is true.

Thus the truth value of the negation of any statement is always the opposite of the truth value of the original statement. The defining property **[$\mathbf{T_3}$]** of the connective can also be written in the form of a table:

p	$\sim p$
T	F
F	T

Example 5.3: Consider the statements in Example 5.1. Observe that (i) is true and (ii) and (iii), each its negation, are false.

Example 5.4: Consider the statements in Example 5.2. Observe that (i) is false and (ii) and (iii), each its negation, are true.

PROPOSITIONS AND TRUTH TABLES

By repetitive use of the logical connectives ($\wedge$, $\vee$, $\sim$ and others discussed subsequently), we can construct compound statements that are more involved. In the case where the substatements $p, q, \ldots$ of a compound statement $P(p, q, \ldots)$ are variables, we call the compound statement a *proposition.*

Now the truth value of a proposition depends exclusively upon the truth values of its variables, that is, the truth value of a proposition is known once the truth values of its variables are known. A simple concise way to show this relationship is through a *truth table.* The truth table, for example, of the proposition $\sim(p \wedge \sim q)$ is constructed as follows:

p	q	$\sim q$	$p \wedge \sim q$	$\sim(p \wedge \sim q)$
T	T	F	F	T
T	F	T	T	F
F	T	F	F	T
F	F	T	F	T

Observe that the first columns of the table are for the variables $p, q, \ldots$ and that there are enough rows in the table to allow for all possible combinations of T and F for these variables. (For 2 variables, as above, 4 rows are necessary; for 3 variables, 8 rows are necessary; and, in general, for n variables, 2^n rows are required.) There is then a column for each "elementary" stage of the construction of the proposition, the truth value at each step being determined from the previous stages by the definitions of the connectives $\wedge, \vee, \sim$. Finally we obtain the truth value of the proposition, which appears in the last column.

Remark: The truth table of the above proposition consists precisely of the columns under the variables and the column under the proposition:

p	q	$\sim(p \wedge \sim q)$
T	T	T
T	F	F
F	T	T
F	F	T

The other columns were merely used in the construction of the truth table.

Another way to construct the above truth table for $\sim(p \wedge \sim q)$ is as follows. First construct the following table:

p	q	$\sim$	$(p$	$\wedge$	$\sim$	$q)$
T	T					
T	F					
F	T					
F	F					
Step						

Observe that the proposition is written on the top row to the right of its variables, and that there is a column under each variable or connective in the proposition. Truth values are then entered into the truth table in various steps as follows:

p	q	$\sim$	$(p$	$\wedge$	$\sim$	$q)$
T	T		T			T
T	F		T			F
F	T		F			T
F	F		F			F
Step			1			1

(a)

p	q	$\sim$	$(p$	$\wedge$	$\sim$	$q)$
T	T		T		F	T
T	F		T		T	F
F	T		F		F	T
F	F		F		T	F
Step			1		2	1

(b)

p	q	$\sim$	$(p$	$\wedge$	$\sim$	$q)$
T	T		T	F	F	T
T	F		T	T	T	F
F	T		F	F	F	T
F	F		F	F	T	F
Step			1	3	2	1

(c)

p	q	$\sim$	$(p$	$\wedge$	$\sim$	$q)$
T	T	T	T	F	F	T
T	F	F	T	T	T	F
F	T	T	F	F	F	T
F	F	T	F	F	T	F
Step		4	1	3	2	1

(d)

The truth table of the proposition then consists of the original columns under the variables and the last column entered into the table, i.e. the last step.

Solved Problems

STATEMENTS

1.1. Let p be "It is cold" and let q be "It is raining." Give a simple verbal sentence which describes each of the following statements:
(1) $\sim p$, (2) $p \wedge q$, (3) $p \vee q$, (4) $q \vee \sim p$, (5) $\sim p \wedge \sim q$, (6) $\sim\sim q$.

In each case, translate $\wedge$, $\vee$ and $\sim$ to read "and", "or" and "It is false that" or "not", respectively, and then simplify the English sentence.

(1) It is not cold.
(2) It is cold and raining.
(3) It is cold or it is raining.
(4) It is raining or it is not cold.
(5) It is not cold and it is not raining.
(6) It is not true that it is not raining.

1.2. Let p be "He is tall" and let q be "He is handsome." Write each of the following statements in symbolic form using p and q.

(1) He is tall and handsome.
(2) He is tall but not handsome.
(3) It is false that he is short or handsome.
(4) He is neither tall nor handsome.
(5) He is tall, or he is short and handsome.
(6) It is not true that he is short or not handsome.

(Assume that "He is short" means "He is not tall", i.e. $\sim p$.)

(1) $p \wedge q$
(2) $p \wedge \sim q$
(3) $\sim(\sim p \vee q)$
(4) $\sim p \wedge \sim q$
(5) $p \vee (\sim p \wedge q)$
(6) $\sim(\sim p \vee \sim q)$

TRUTH VALUES OF COMPOUND STATEMENTS

1.3. Determine the truth value of each of the following statements.
(i) $3+2 = 7$ and $4+4 = 8$. (ii) $2+1 = 3$ and $7+2 = 9$. (iii) $6+4 = 10$ and $1+1 = 3$.

By property $[T_1]$, the compound statement "p and q" is true only when p and q are both true. Hence: (i) False, (ii) True, (iii) False.

1.4. Determine the truth value of each of the following statements.

(i) Paris is in England or $3+4 = 7$.
(ii) Paris is in France or $2+1 = 6$.
(iii) London is in France or $5+2 = 3$.

By property $[T_2]$, the compound statement "p or q" is false only when p and q are both false. Hence: (i) True, (ii) True, (iii) False.

1.5. Determine the truth value of each of the following statements.

(i) It is not true that London is in France.
(ii) It is not true that London is in England.

By property $[T_3]$, the truth value of the negation of p is the opposite of the truth value of p. Hence: (i) True, (ii) False.

1.6. Determine the truth value of each of the following statements.

(i) It is false that $2+2 = 4$ and $1+1 = 5$.

(ii) Copenhagen is in Denmark, and $1+1 = 5$ or $2+2 = 4$.

(iii) It is false that $2+2 = 4$ or London is in France.

(i) The conjunctive statement "$2+2 = 4$ and $1+1 = 5$" is false since one of its substatements "$1+1 = 5$" is false. Accordingly its negation, the given statement, is true.

(ii) The disjunctive statement "$1+1 = 5$ or $2+2 = 4$" is true since one of its substatements "$2+2 = 4$" is true. Hence the given statement is true since it is the conjunction of two true statements, "Copenhagen is in Denmark" and "$1+1 = 5$ or $2+2 = 4$".

(iii) The disjunctive statement "$2+2 = 4$ or London is in France" is true since one of its substatements "$2+2 = 4$" is true. Accordingly its negation, the given statement, is false.

TRUTH TABLES OF PROPOSITIONS

1.7. Find the truth table of $\sim p \wedge q$.

p	q	$\sim p$	$\sim p \wedge q$
T	T	F	F
T	F	F	F
F	T	T	T
F	F	T	F

Method 1

p	q	$\sim$	p	$\wedge$	q
T	T	F	T	F	T
T	F	F	T	F	F
F	T	T	F	T	T
F	F	T	F	F	F
Step		2	1	3	1

Method 2

1.8. Find the truth table of $\sim(p \vee q)$.

p	q	$p \vee q$	$\sim(p \vee q)$
T	T	T	F
T	F	T	F
F	T	T	F
F	F	F	T

Method 1

p	q	$\sim$	$(p$	$\vee$	$q)$
T	T	F	T	T	T
T	F	F	T	T	F
F	T	F	F	T	T
F	F	T	F	F	F
Step		3	1	2	1

Method 2

1.9. Find the truth table of $\sim(p \vee \sim q)$.

p	q	$\sim q$	$p \vee \sim q$	$\sim(p \vee \sim q)$
T	T	F	T	F
T	F	T	T	F
F	T	F	F	T
F	F	T	T	F

Method 1

p	q	$\sim$	$(p$	$\vee$	$\sim$	$q)$
T	T	F	T	T	F	T
T	F	F	T	T	T	F
F	T	T	F	F	F	T
F	F	F	F	T	T	F
Step		4	1	3	2	1

Method 2

1.10. Find the truth table of the following: (i) $p \wedge (q \vee r)$, (ii) $(p \wedge q) \vee (p \wedge r)$.

Since there are three variables, we will need $2^3 = 8$ rows in the truth table.

p	q	r	$q \vee r$	$p \wedge (q \vee r)$
T	T	T	T	T
T	T	F	T	T
T	F	T	T	T
T	F	F	F	F
F	T	T	T	F
F	T	F	T	F
F	F	T	T	F
F	F	F	F	F

(i)

p	q	r	$p \wedge q$	$p \wedge r$	$(p \wedge q) \vee (p \wedge r)$
T	T	T	T	T	T
T	T	F	T	F	T
T	F	T	F	T	T
T	F	F	F	F	F
F	T	T	F	F	F
F	T	F	F	F	F
F	F	T	F	F	F
F	F	F	F	F	F

(ii)

Observe that both propositions have the same truth table.

MISCELLANEOUS PROBLEMS

1.11. Let Apq denote $p \wedge q$ and let Np denote $\sim p$. Rewrite the following propositions using A and N instead of $\wedge$ and $\sim$.

(i) $p \wedge \sim q$ (iii) $\sim p \wedge (\sim q \wedge r)$

(ii) $\sim(\sim p \wedge q)$ (iv) $\sim(p \wedge \sim q) \wedge (\sim q \wedge \sim r)$

(i) $p \wedge \sim q = p \wedge Nq = ApNq$

(ii) $\sim(\sim p \wedge q) = \sim(Np \wedge q) = \sim(ANpq) = NANpq$

(iii) $\sim p \wedge (\sim q \wedge r) = Np \wedge (Nq \wedge r) = Np \wedge (ANqr) = ANpANqr$

(iv) $\sim(p \wedge \sim q) \wedge (\sim q \wedge \sim r) = \sim(ApNq) \wedge (ANqNr) = (NApNq) \wedge (ANqNr) = ANApNqANqNr$

Observe that there are no parentheses in the final answer when A and N are used instead of $\wedge$ and $\sim$. In fact, it has been proved that parentheses are never needed in any proposition using A and N.

1.12. Rewrite the following propositions using $\wedge$ and $\sim$ instead of A and N.

(i) $NApq$ (iii) $ApNq$ (v) $NAANpqr$

(ii) $ANpq$ (iv) $ApAqr$ (vi) $ANpAqNr$

(i) $NApq = N(p \wedge q) = \sim(p \wedge q)$ (iii) $ApNq = Ap(\sim q) = p \wedge \sim q$

(ii) $ANpq = A(\sim p)q = \sim p \wedge q$ (iv) $ApAqr = Ap(q \wedge r) = p \wedge (q \wedge r)$

(v) $NAANpqr = NAA(\sim p)qr = NA(\sim p \wedge q)r = N[(\sim p \wedge q) \wedge r] = \sim[(\sim p \wedge q) \wedge r]$

(vi) $ANpAqNr = ANpAq(\sim r) = ANp(q \wedge \sim r) = A(\sim p)(q \wedge \sim r) = \sim p \wedge (q \wedge \sim r)$

Notice that the propositions involving A and N are unraveled from right to left.

Supplementary Problems

STATEMENTS

1.13. Let p be "Marc is rich" and let q be "Marc is happy". Write each of the following in symbolic form.

(i) Marc is poor but happy.

(ii) Marc is neither rich nor happy.

(iii) Marc is either rich or unhappy.

(iv) Marc is poor or else he is both rich and unhappy.

1.14. Let p be "Erik reads *Newsweek*", let q be "Erik reads *Life*" and let r be "Erik reads *Time*". Write each of the following in symbolic form.

(i) Erik reads *Newsweek* or *Life*, but not *Time*.

(ii) Erik reads *Newsweek* and *Life*, or he does not read *Newsweek* and *Time*.

(iii) It is not true that Erik reads *Newsweek* but not *Time*.

(iv) It is not true that Erik reads *Time* or *Life* but not *Newsweek*.

1.15. Let p be "Audrey speaks French" and let q be "Audrey speaks Danish". Give a simple verbal sentence which describes each of the following.

(i) $p \vee q$ (iii) $p \wedge \sim q$ (v) $\sim\sim p$

(ii) $p \wedge q$ (iv) $\sim p \vee \sim q$ (vi) $\sim(\sim p \wedge \sim q)$

1.16. Determine the truth value of each of the following statements.

(i) $3+3 = 6$ and $1+2 = 5$.

(ii) It is not true that $3+3 = 6$ or $1+2 = 3$.

(iii) It is true that $2+2 \neq 4$ and $1+2 = 3$.

(iv) It is not true that $3+3 \neq 6$ or $1+2 \neq 5$.

TRUTH TABLES OF PROPOSITIONS

1.17. Find the truth table of each of the following.

(i) $p \vee \sim q$, (ii) $\sim p \wedge \sim q$, (iii) $\sim(\sim p \wedge q)$, (iv) $\sim(\sim p \vee \sim q)$.

1.18. Find the truth table of each of the following.

(i) $(p \wedge \sim q) \vee r$, (ii) $\sim p \vee (q \wedge \sim r)$, (iii) $(p \vee \sim r) \wedge (q \vee \sim r)$, (iv) $\sim(p \vee \sim q) \wedge (\sim p \vee r)$.

MISCELLANEOUS PROBLEMS

1.19. Let Apq denote $p \wedge q$ and let Np denote $\sim p$. (See Problem 1.11.) Rewrite the following propositions using A and N instead of $\wedge$ and $\sim$.

(i) $\sim p \wedge q$, (ii) $\sim p \wedge \sim q$, (iii) $\sim(p \wedge \sim q)$, (iv) $(\sim p \wedge q) \wedge \sim r$.

1.20. Rewrite the following propositions using $\wedge$ and $\sim$ instead of A and N.

(i) $NApNq$, (ii) $ANApqNr$, (iii) $AApNrAqNp$, (iv) $ANANpANqrNp$.

Answers to Supplementary Problems

1.13. (i) $\sim p \wedge q$, (ii) $\sim p \wedge \sim q$, (iii) $p \vee \sim q$, (iv) $\sim p \vee (p \wedge \sim q)$

1.14. (i) $(p \vee q) \wedge \sim r$, (ii) $(p \wedge q) \vee \sim(p \wedge r)$, (iii) $\sim(p \wedge \sim r)$, (iv) $\sim[(r \vee q) \wedge \sim p]$

1.15. (i) Audrey speaks French or Danish.

(ii) Audrey speaks French and Danish.

(iii) Audrey speaks French but not Danish.

(iv) Audrey does not speak French or she does not speak Danish.

(v) It is not true that Audrey does not speak French.

(vi) It is not true that Audrey speaks neither French nor Danish.

1.16. (i) F, (ii) F, (iii) F, (iv) F

1.17.

p	q	$p \vee \sim q$	$\sim p \wedge \sim q$	$\sim(\sim p \wedge q)$	$\sim(\sim p \vee \sim q)$
T	T	T	F	T	T
T	F	T	F	T	F
F	T	F	F	F	F
F	F	T	T	T	F

1.18.

p	q	r	(i)	(ii)	(iii)	(iv)
T	T	T	T	F	T	F
T	T	F	F	T	T	F
T	F	T	T	F	F	F
T	F	F	T	F	T	F
F	T	T	T	T	F	T
F	T	F	F	T	T	T
F	F	T	T	T	F	F
F	F	F	F	T	T	F

1.19. (i) $ANpq$, (ii) $ANpNq$, (iii) $NApNq$, (iv) $AANpqNr$

1.20. (i) $\sim(p \wedge \sim q)$, (ii) $\sim(p \wedge q) \wedge \sim r$, (iii) $(p \wedge \sim r) \wedge (q \wedge \sim p)$, (iv) $\sim[\sim p \wedge (\sim q \wedge r)] \wedge \sim p$

Chapter 2

Algebra of Propositions

TAUTOLOGIES AND CONTRADICTIONS

Some propositions $P(p, q, \ldots)$ contain only T in the last column of their truth tables, i.e. are true for any truth values of their variables. Such propositions are called *tautologies*. Similarly, a proposition $P(p, q, \ldots)$ is called a *contradiction* if it contains only F in the last column of its truth table, i.e. is false for any truth values of its variables.

Example 1.1: The proposition "p or not p", i.e. $p \vee \sim p$, is a tautology and the proposition "p and not p", i.e. $p \wedge \sim p$, is a contradiction. This is verified by constructing their truth tables:

p	$\sim p$	$p \vee \sim p$
T	F	T
F	T	T

p	$\sim p$	$p \wedge \sim p$
T	F	F
F	T	F

Since a tautology is always true, the negation of a tautology is always false, i.e. is a contradiction, and vice versa. That is,

Theorem 2.1: If $P(p, q, \ldots)$ is a tautology then $\sim P(p, q, \ldots)$ is a contradiction, and conversely.

Now let $P(p, q, \ldots)$ be a tautology, and let $P_1(p, q, \ldots)$, $P_2(p, q, \ldots)$, $\ldots$ be any propositions. Since $P(p, q, \ldots)$ does not depend upon the particular truth values of its variables $p, q, \ldots$, we can substitute P_1 for p, P_2 for q, $\ldots$ in the tautology $P(p, q, \ldots)$ and still have a tautology. In other words:

Theorem 2.2 (Principle of Substitution): If $P(p, q, \ldots)$ is a tautology, then $P(P_1, P_2, \ldots)$ is a tautology for any propositions $P_1, P_2, \ldots$.

LOGICAL EQUIVALENCE

Two propositions $P(p, q, \ldots)$ and $Q(p, q, \ldots)$ are said to be *logically equivalent*, or simply *equivalent* or *equal*, denoted by

$$P(p, q, \ldots) \equiv Q(p, q, \ldots)$$

if they have identical truth tables.

Example 2.1: The truth tables of $\sim(p \wedge q)$ and $\sim p \vee \sim q$ follow:

p	q	$p \wedge q$	$\sim(p \wedge q)$
T	T	T	F
T	F	F	T
F	T	F	T
F	F	F	T

p	q	$\sim p$	$\sim q$	$\sim p \vee \sim q$
T	T	F	F	F
T	F	F	T	T
F	T	T	F	T
F	F	T	T	T

Accordingly, the propositions $\sim(p \wedge q)$ and $\sim p \vee \sim q$ are logically equivalent:

$$\sim(p \wedge q) \equiv \sim p \vee \sim q$$

Example 2.2: The statement

"It is false that roses are red and violets are blue"

can be written in the form $\sim(p \wedge q)$ where p is "Roses are red" and q is "Violets are blue". By the preceding example, $\sim(p \wedge q)$ is logically equivalent to $\sim p \vee \sim q$; that is, the given statement is equivalent to the statement

"Either roses are not red or violets are not blue."

ALGEBRA OF PROPOSITIONS

Propositions, under the relation of logical equivalence, satisfy various laws or identities which are listed in Table 2.1 below. In fact, we formally state:

Theorem 2.3: Propositions satisfy the laws of Table 2.1.

LAWS OF THE ALGEBRA OF PROPOSITIONS	
Idempotent Laws	
1*a*. $p \vee p \equiv p$	1*b*. $p \wedge p \equiv p$
Associative Laws	
2*a*. $(p \vee q) \vee r \equiv p \vee (q \vee r)$	2*b*. $(p \wedge q) \wedge r \equiv p \wedge (q \wedge r)$
Commutative Laws	
3*a*. $p \vee q \equiv q \vee p$	3*b*. $p \wedge q \equiv q \wedge p$
Distributive Laws	
4*a*. $p \vee (q \wedge r) \equiv (p \vee q) \wedge (p \vee r)$	4*b*. $p \wedge (q \vee r) \equiv (p \wedge q) \vee (p \wedge r)$
Identity Laws	
5*a*. $p \vee f \equiv p$	5*b*. $p \wedge t \equiv p$
6*a*. $p \vee t \equiv t$	6*b*. $p \wedge f \equiv f$
Complement Laws	
7*a*. $p \vee \sim p \equiv t$	7*b*. $p \wedge \sim p \equiv f$
8*a*. $\sim\sim p \equiv p$	8*b*. $\sim t \equiv f$, $\sim f \equiv t$
De Morgan's Laws	
9*a*. $\sim(p \vee q) \equiv \sim p \wedge \sim q$	9*b*. $\sim(p \wedge q) \equiv \sim p \vee \sim q$

Table 2.1

In the above table, t and f denote variables which are restricted to the truth values true and false, respectively.

Solved Problems

TAUTOLOGIES AND CONTRADICTIONS

2.1. Verify that the proposition $p \vee \sim(p \wedge q)$ is a tautology.

Construct the truth table of $p \vee \sim(p \wedge q)$:

p	q	$p \wedge q$	$\sim(p \wedge q)$	$p \vee \sim(p \wedge q)$
T	T	T	F	T
T	F	F	T	T
F	T	F	T	T
F	F	F	T	T

Since the truth value of $p \vee \sim(p \wedge q)$ is T for all values of p and q, it is a tautology.

2.2. Verify that the proposition $(p \wedge q) \wedge \sim(p \vee q)$ is a contradiction.

Construct the truth table of $(p \wedge q) \wedge \sim(p \vee q)$:

p	q	$p \wedge q$	$p \vee q$	$\sim(p \vee q)$	$(p \wedge q) \wedge \sim(p \vee q)$
T	T	T	T	F	F
T	F	F	T	F	F
F	T	F	T	F	F
F	F	F	F	T	F

Since the truth value of $(p \wedge q) \wedge \sim(p \vee q)$ is F for all values of p and q, it is a contradiction.

LOGICAL EQUIVALENCE

2.3. Prove the Associative Law: $(p \wedge q) \wedge r \equiv p \wedge (q \wedge r)$.

Construct the required truth tables:

p	q	r	$p \wedge q$	$(p \wedge q) \wedge r$	$q \wedge r$	$p \wedge (q \wedge r)$
T	T	T	T	T	T	T
T	T	F	T	F	F	F
T	F	T	F	F	F	F
T	F	F	F	F	F	F
F	T	T	F	F	T	F
F	T	F	F	F	F	F
F	F	T	F	F	F	F
F	F	F	F	F	F	F

Since the truth tables are identical, the propositions are equivalent.

2.4. Prove that disjunction distributes over conjunction; that is, prove the Distributive Law: $p \vee (q \wedge r) \equiv (p \vee q) \wedge (p \vee r)$.

Construct the required truth tables:

p	q	r	$q \wedge r$	$p \vee (q \wedge r)$	$p \vee q$	$p \vee r$	$(p \vee q) \wedge (p \vee r)$
T	T	T	T	T	T	T	T
T	T	F	F	T	T	T	T
T	F	T	F	T	T	T	T
T	F	F	F	T	T	T	T
F	T	T	T	T	T	T	T
F	T	F	F	F	T	F	F
F	F	T	F	F	F	T	F
F	F	F	F	F	F	F	F

Since the truth tables are identical, the propositions are equivalent.

2.5. Prove that the operation of disjunction can be written in terms of the operations of conjunction and negation. Specifically, $p \vee q \equiv \sim(\sim p \wedge \sim q)$.

Construct the required truth tables:

p	q	$p \vee q$	$\sim p$	$\sim q$	$\sim p \wedge \sim q$	$\sim(\sim p \wedge \sim q)$
T	T	T	F	F	F	T
T	F	T	F	T	F	T
F	T	T	T	F	F	T
F	F	F	T	T	T	F

Since the truth tables are identical, the propositions are equivalent.

2.6. There are exactly four non-equivalent propositions of one variable; the truth tables of such propositions follow:

p	$P_1(p)$	$P_2(p)$	$P_3(p)$	$P_4(p)$
T	T	T	F	F
F	T	F	T	F

Find four such propositions.

Observe that

p	$\sim p$	$p \vee \sim p$	$p \wedge \sim p$
T	F	T	F
F	T	T	F

Hence $P_1(p) \equiv p \vee \sim p$, $P_2(p) \equiv p$, $P_3(p) \equiv \sim p$, $P_4(p) \equiv p \wedge \sim p$.

2.7. Determine the number of non-equivalent propositions of two variables p and q.

The truth table of a proposition $P(p, q)$ will contain $2^2 = 4$ lines. In each line T or F can appear as follows:

p	q	P_1	P_2	P_3	P_4	P_5	P_6	P_7	P_8	P_9	P_{10}	P_{11}	P_{12}	P_{13}	P_{14}	P_{15}	P_{16}
T	T	T	T	T	T	T	T	T	T	F	F	F	F	F	F	F	F
T	F	T	T	T	T	F	F	F	F	T	T	T	T	F	F	F	F
F	T	T	T	F	F	T	T	F	F	T	T	F	F	T	T	F	F
F	F	T	F	T	F	T	F	T	F	T	F	T	F	T	F	T	F

In other words, there are $2^4 = 16$ non-equivalent propositions of two variables p and q.

2.8. Determine the number of non-equivalent propositions of: (i) three variables p, q and r; (ii) n variables $p_1, p_2, \ldots, p_n$.

(i) The truth table of a proposition $P(p, q, r)$ will contain $2^3 = 8$ lines. Since in each line T or F can appear, there are $2^8 = 256$ non-equivalent propositions of three variables.

(ii) The truth table of a proposition $P(p_1, \ldots, p_n)$ will contain 2^n lines; hence, as above, there are 2^{2^n} non-equivalent propositions of n variables.

NEGATION

2.9. Prove De Morgan's Laws: (i) $\sim(p \wedge q) \equiv \sim p \vee \sim q$; (ii) $\sim(p \vee q) \equiv \sim p \wedge \sim q$.

In each case construct the required truth tables.

(i)

p	q	$p \wedge q$	$\sim(p \wedge q)$	$\sim p$	$\sim q$	$\sim p \vee \sim q$
T	T	T	F	F	F	F
T	F	F	T	F	T	T
F	T	F	T	T	F	T
F	F	F	T	T	T	T

(ii)

p	q	$p \vee q$	$\sim(p \vee q)$	$\sim p$	$\sim q$	$\sim p \wedge \sim q$
T	T	T	F	F	F	F
T	F	T	F	F	T	F
F	T	T	F	T	F	F
F	F	F	T	T	T	T

2.10. Verify: $\sim\sim p \equiv p$.

p	$\sim p$	$\sim\sim p$
T	F	T
F	T	F

2.11. Use the results of the preceding problems to simplify each of the following propositions:

(i) $\sim(p \vee \sim q)$, (ii) $\sim(\sim p \wedge q)$, (iii) $\sim(\sim p \vee \sim q)$.

(i) $\sim(p \vee \sim q) \equiv \sim p \wedge \sim\sim q \equiv \sim p \wedge q$

(ii) $\sim(\sim p \wedge q) \equiv \sim\sim p \vee \sim q \equiv p \vee \sim q$

(iii) $\sim(\sim p \vee \sim q) \equiv \sim\sim p \wedge \sim\sim q \equiv p \wedge q$

2.12. Simplify each of the following statements.

(i) It is not true that his mother is English or his father is French.

(ii) It is not true that he studies physics but not mathematics.

(iii) It is not true that sales are decreasing and prices are rising.

(iv) It is not true that it is not cold or it is raining.

(i) Let p denote "His mother is English" and let q denote "His father is French". Then the given statement is $\sim(p \vee q)$. But $\sim(p \vee q) \equiv \sim p \wedge \sim q$. Hence the given statement is logically equivalent to the statement "His mother is not English and his father is not French".

(ii) Let p denote "He studies physics" and let q denote "He studies mathematics". Then the given statement is $\sim(p \wedge \sim q)$. But $\sim(p \wedge \sim q) \equiv \sim p \vee \sim\sim q \equiv \sim p \vee q$. Hence the given statement is logically equivalent to the statement "He does not study physics or he studies mathematics".

(iii) Since $\sim(p \wedge q) \equiv \sim p \vee \sim q$, the given statement is logically equivalent to the statement "Sales are increasing or prices are falling".

(iv) Since $\sim(\sim p \vee q) \equiv p \wedge \sim q$, the given statement is logically equivalent to the statement "It is cold and it is not raining".

ALGEBRA OF PROPOSITIONS

2.13. Simplify the proposition $(p \vee q) \wedge \sim p$ by using the laws of the algebra of propositions listed on Page 11.

Statement		Reason
(1)	$(p \vee q) \wedge \sim p \equiv \sim p \wedge (p \vee q)$	(1) Commutative law
(2)	$\equiv (\sim p \wedge p) \vee (\sim p \wedge q)$	(2) Distributive law
(3)	$\equiv f \vee (\sim p \wedge q)$	(3) Complement law
(4)	$\equiv \sim p \wedge q$	(4) Identity law

2.14. Simplify the proposition $p \vee (p \wedge q)$ by using the laws of the algebra of propositions listed on Page 11.

Statement		Reason
(1)	$p \vee (p \wedge q) \equiv (p \wedge t) \vee (p \wedge q)$	(1) Identity law
(2)	$\equiv p \wedge (t \vee q)$	(2) Distributive law
(3)	$\equiv p \wedge t$	(3) Identity law
(4)	$\equiv p$	(4) Identity law

2.15. Simplify the proposition $\sim(p \vee q) \vee (\sim p \wedge q)$ by using the laws of the algebra of propositions listed on Page 11.

Statement		Reason
(1)	$\sim(p \vee q) \vee (\sim p \wedge q) \equiv (\sim p \wedge \sim q) \vee (\sim p \wedge q)$	(1) De Morgan's law
(2)	$\equiv \sim p \wedge (\sim q \vee q)$	(2) Distributive law
(3)	$\equiv \sim p \wedge t$	(3) Complement law
(4)	$\equiv \sim p$	(4) Identity law

MISCELLANEOUS PROBLEMS

2.16. The propositional connective $\veebar$ is called the *exclusive disjunction*; $p \veebar q$ is read "p or q but not both".

(i) Construct a truth table for $p \veebar q$.

(ii) Prove: $p \veebar q \equiv (p \vee q) \wedge \sim(p \wedge q)$. Accordingly $\veebar$ can be written in terms of the original three connectives $\wedge$, $\vee$ and $\sim$.

(i) Now $p \veebar q$ is true if p is true or if q is true but not if both are true; hence the truth table of $p \veebar q$ is as follows:

p	q	$p \veebar q$
T	T	F
T	F	T
F	T	T
F	F	F

(ii) We construct the truth table of $(p \vee q) \wedge \sim(p \wedge q)$, by the second method, as follows:

p	q	$(p$	$\vee$	$q)$	$\wedge$	$\sim$	$(p$	$\wedge$	$q)$
T	T	T	T	T	F	F	T	T	T
T	F	T	T	F	T	T	T	F	F
F	T	F	T	T	T	T	F	F	T
F	F	F	F	F	F	T	F	F	F
Step		1	2	1	4	3	1	2	1

Observe that the truth tables of $p \veebar q$ and $(p \vee q) \wedge \sim(p \wedge q)$ are identical; hence $p \veebar q \equiv (p \vee q) \wedge \sim(p \wedge q)$.

2.17. The propositional connective $\downarrow$ is called the *joint denial*; $p \downarrow q$ is read "Neither p nor q".

(i) Construct a truth table for $p \downarrow q$.

(ii) Prove: The three connectives $\vee$, $\wedge$ and $\sim$ may be expressed in terms of the connective $\downarrow$ as follows:

(a) $\sim p \equiv p \downarrow p$, (b) $p \wedge q \equiv (p \downarrow p) \downarrow (q \downarrow q)$, (c) $p \vee q \equiv (p \downarrow q) \downarrow (p \downarrow q)$.

(i) Now $p \downarrow q$ is true only in the case that p is not true and q is not true; hence the truth table of $p \downarrow q$ is the following:

p	q	$p \downarrow q$
T	T	F
T	F	F
F	T	F
F	F	T

(ii) Construct the appropriate truth tables:

(a)

p	$\sim p$	$p \downarrow p$
T	F	F
F	T	T

(b)

p	q	$p \wedge q$	$p \downarrow p$	$q \downarrow q$	$(p \downarrow p) \downarrow (q \downarrow q)$
T	T	T	F	F	T
T	F	F	F	T	F
F	T	F	T	F	F
F	F	F	T	T	F

(c)

p	q	$p \vee q$	$p \downarrow q$	$(p \downarrow q) \downarrow (p \downarrow q)$
T	T	T	F	T
T	F	T	F	T
F	T	T	F	T
F	F	F	T	F

Supplementary Problems

LOGICAL EQUIVALENCE

2.18. Prove the associative law for disjunction: $(p \vee q) \vee r \equiv p \vee (q \vee r)$.

2.19. Prove that conjunction distributes over disjunction: $p \wedge (q \vee r) \equiv (p \wedge q) \vee (p \wedge r)$.

2.20. Prove $(p \vee q) \wedge \sim p \equiv \sim p \wedge q$ by constructing the appropriate truth tables (see Problem 2.13).

2.21. Prove $p \vee (p \wedge q) \equiv p$ by constructing the appropriate truth tables (see Problem 2.14).

2.22. Prove $\sim(p \vee q) \vee (\sim p \wedge q) \equiv \sim p$ by constructing the appropriate truth tables (see Problem 2.15).

2.23. (i) Express $\vee$ in terms of $\wedge$ and $\sim$.
(ii) Express $\wedge$ in terms of $\vee$ and $\sim$.

NEGATION

2.24. Simplify: (i) $\sim(p \wedge \sim q)$, (ii) $\sim(\sim p \vee q)$, (iii) $\sim(\sim p \wedge \sim q)$.

2.25. Write the negation of each of the following statements as simply as possible.
(i) He is tall but handsome.
(ii) He has blond hair or blue eyes.
(iii) He is neither rich nor happy.
(iv) He lost his job or he did not go to work today.
(v) Neither Marc nor Erik is unhappy.
(vi) Audrey speaks Spanish or French, but not German.

ALGEBRA OF PROPOSITIONS

2.26. Prove the following equivalences by using the laws of the algebra of propositions listed on Page 11:
(i) $p \wedge (p \vee q) \equiv p$, (ii) $(p \wedge q) \vee \sim p \equiv \sim p \vee q$, (iii) $p \wedge (\sim p \vee q) \equiv p \wedge q$.

Answers to Supplementary Problems

2.23. (i) $p \vee q \equiv \sim(\sim p \wedge \sim q)$, (ii) $p \wedge q \equiv \sim(\sim p \vee \sim q)$.

2.24. (i) $\sim p \vee q$, (ii) $p \wedge \sim q$, (iii) $p \vee q$.

2.25. (iii) He is rich or happy. (vi) Audrey speaks German but neither Spanish nor French.

2.26. (i) $p \wedge (p \vee q) \equiv (p \vee f) \wedge (p \vee q) \equiv p \vee (f \wedge q) \equiv p \vee f \equiv p$.

Chapter 3

Conditional Statements

CONDITIONAL, $p \rightarrow q$

Many statements, particularly in mathematics, are of the form "If p then q". Such statements are called *conditional* statements and are denoted by

$$p \rightarrow q$$

The conditional $p \rightarrow q$ can also be read:

(i) p implies q (iii) p is sufficient for q

(ii) p only if q (iv) q is necessary for p.

The truth value of $p \rightarrow q$ satisfies:

[T_4] The conditional $p \rightarrow q$ is true except in the case that p is true and q is false.

The truth table of the conditional statement follows:

p	q	$p \rightarrow q$
T	T	T
T	F	F
F	T	T
F	F	T

Example 1.1: Consider the following statements:

(i) If Paris is in France, then $2+2 = 4$.

(ii) If Paris is in France, then $2+2 = 5$.

(iii) If Paris is in England, then $2+2 = 4$.

(iv) If Paris is in England, then $2+2 = 5$.

By the property **[T_4]**, only (ii) is a false statement; the others are true. We emphasize that, by definition, (iv) is a true statement even though its substatements "Paris is in England" and "$2+2 = 5$" are false. It is a statement of the type: If monkeys are human, then the earth is flat.

Now consider the truth table of the proposition $\sim p \vee q$:

p	q	$\sim p$	$\sim p \vee q$
T	T	F	T
T	F	F	F
F	T	T	T
F	F	T	T

Observe that the above truth table is identical to the truth table of $p \rightarrow q$. Hence $p \rightarrow q$ is logically equivalent to the proposition $\sim p \vee q$:

$$p \rightarrow q \equiv \sim p \vee q$$

In other words, the conditional statement "If p then q" is logically equivalent to the statement "Not p or q" which only involves the connectives $\vee$ and $\sim$ and thus was already a part of our language. We may regard $p \rightarrow q$ as an abbreviation for an oft-recurring statement.

BICONDITIONAL, $p \leftrightarrow q$

Another common statement is of the form "p if and only if q" or, simply, "p iff q". Such statements, denoted by

$$p \leftrightarrow q$$

are called *biconditional* statements. The truth value of the biconditional statement $p \leftrightarrow q$ satisfies the following property:

[T_5] If p and q have the same truth value, then $p \leftrightarrow q$ is true; if p and q have opposite truth values, then $p \leftrightarrow q$ is false.

The truth table of the biconditional follows:

p	q	$p \leftrightarrow q$
T	T	T
T	F	F
F	T	F
F	F	T

Example 2.1: Consider the following statements:

(i) Paris is in France if and only if $2+2 = 4$.

(ii) Paris is in France if and only if $2+2 = 5$.

(iii) Paris is in England if and only if $2+2 = 4$.

(iv) Paris is in England if and only if $2+2 = 5$.

By property [T_5], the statements (i) and (iv) are true, and (ii) and (iii) are false.

Recall that propositions $P(p,q,\ldots)$ and $Q(p,q,\ldots)$ are logically equivalent if and only if they have the same truth table; but then, by property [T_5], the composite proposition $P(p,q,\ldots) \leftrightarrow Q(p,q,\ldots)$ is always true, i.e. is a tautology. In other words,

Theorem 3.1: $P(p,q,\ldots) \equiv Q(p,q,\ldots)$ if and only if the proposition

$$P(p,q,\ldots) \leftrightarrow Q(p,q,\ldots)$$

is a tautology.

CONDITIONAL STATEMENTS AND VARIATIONS

Consider the conditional proposition $p \rightarrow q$ and the other simple conditional propositions which contain p and q:

$$q \rightarrow p, \quad \sim p \rightarrow \sim q \quad \text{and} \quad \sim q \rightarrow \sim p$$

called respectively the *converse*, *inverse*, and *contrapositive* propositions. The truth tables of these four propositions follow:

p	q	Conditional $p \rightarrow q$	Converse $q \rightarrow p$	Inverse $\sim p \rightarrow \sim q$	Contrapositive $\sim q \rightarrow \sim p$
T	T	T	T	T	T
T	F	F	T	T	F
F	T	T	F	F	T
F	F	T	T	T	T

Observe first that a conditional statement and its converse or inverse are not logically equivalent. On the other hand, the above truth table establishes

Theorem 3.2: A conditional statement $p \rightarrow q$ and its contrapositive $\sim q \rightarrow \sim p$ are logically equivalent.

Example 3.1: Consider the following statements about a triangle A:

$p \to q$: If A is equilateral, then A is isosceles.

$q \to p$: If A is isosceles, then A is equilateral.

Note that $p \to q$ is true, but $q \to p$ is false.

Example 3.2: Prove: $(p \to q)$ If x^2 is odd then x is odd.

We show that the contrapositive $\sim q \to \sim p$, "If x is even then x^2 is even", is true. Let x be even; then $x = 2n$ where n is an integer. Hence $x^2 = (2n)(2n) = 2(2n^2)$ is also even. Since the contrapositive statement $\sim q \to \sim p$ is true, the original conditional statement $p \to q$ is also true.

Solved Problems

CONDITIONAL

3.1. Let p denote "It is cold" and let q denote "It rains". Write the following statements in symbolic form.

(i) It rains only if it is cold.
(ii) A necessary condition for it to be cold is that it rain.
(iii) A sufficient condition for it to be cold is that it rain.
(iv) Whenever it rains it is cold.
(v) It never rains when it is cold.

Recall that $p \to q$ can be read "p only if q", "p is sufficient for q" or "q is necessary for p".

(i) $q \to p$ (ii) $p \to q$ (iii) $q \to p$

(iv) Now the statement "Whenever it rains it is cold" is equivalent to "If it rains then it is cold". That is, $q \to p$.

(v) The statement "It never rains when it is cold" is equivalent to "If it is cold then it does not rain". That is, $p \to \sim q$.

3.2. Rewrite the following statements without using the conditional.

(i) If it is cold, he wears a hat.
(ii) If productivity increases, then wages rise.

Recall that "If p then q" is equivalent to "Not p or q".

(i) It is not cold or he wears a hat.

(ii) Productivity does not increase or wages rise.

3.3. Determine the truth table of $(p \to q) \to (p \wedge q)$.

p	q	$p \to q$	$p \wedge q$	$(p \to q) \to (p \wedge q)$
T	T	T	T	T
T	F	F	F	T
F	T	T	F	F
F	F	T	F	F

3.4. Determine the truth table of $\sim p \rightarrow (q \rightarrow p)$.

p	q	$\sim p$	$q \rightarrow p$	$\sim p \rightarrow (q \rightarrow p)$
T	T	F	T	T
T	F	F	T	T
F	T	T	F	F
F	F	T	T	T

3.5. Verify that $(p \wedge q) \rightarrow (p \vee q)$ is a tautology.

p	q	$p \wedge q$	$p \vee q$	$(p \wedge q) \rightarrow (p \vee q)$
T	T	T	T	T
T	F	F	T	T
F	T	F	T	T
F	F	F	F	T

3.6. Prove that the conditional operation distributes over conjunction:

$$p \rightarrow (q \wedge r) \equiv (p \rightarrow q) \wedge (p \rightarrow r)$$

p	q	r	$q \wedge r$	$p \rightarrow (q \wedge r)$	$p \rightarrow q$	$p \rightarrow r$	$(p \rightarrow q) \wedge (p \rightarrow r)$
T	T	T	T	T	T	T	T
T	T	F	F	F	T	F	F
T	F	T	F	F	F	T	F
T	F	F	F	F	F	F	F
F	T	T	T	T	T	T	T
F	T	F	F	T	T	T	T
F	F	T	F	T	T	T	T
F	F	F	F	T	T	T	T

BICONDITIONAL

3.7. Show that "p implies q and q implies p" is logically equivalent to the biconditional "p if and only if q"; that is, $(p \rightarrow q) \wedge (q \rightarrow p) \equiv p \leftrightarrow q$.

p	q	$p \leftrightarrow q$	$p \rightarrow q$	$q \rightarrow p$	$(p \rightarrow q) \wedge (q \rightarrow p)$
T	T	T	T	T	T
T	F	F	F	T	F
F	T	F	T	F	F
F	F	T	T	T	T

3.8. Determine the truth value of each statement.

(i) $2+2 = 4$ iff $3+6 = 9$

(ii) $2+2 = 7$ if and only if $5+1 = 2$

(iii) $1+1 = 2$ iff $3+2 = 8$

(iv) $1+2 = 5$ if and only if $3+1 = 4$

Now $p \leftrightarrow q$ is true whenever p and q have the same truth value; hence (i) and (ii) are true statements, but (iii) and (iv) are false. (Observe that (ii) is a true statement by definition of the conditional, even though both substatements $2+2 = 7$ and $5+1 = 2$ are false.)

3.9. Show that the biconditional $p \leftrightarrow q$ can be written in terms of the original three connectives $\vee$, $\wedge$ and $\sim$.

Now $p \rightarrow q \equiv \sim p \vee q$ and $q \rightarrow p \equiv \sim q \vee p$; hence by Problem 3.7,

$$p \leftrightarrow q \;\equiv\; (p \rightarrow q) \wedge (q \rightarrow p) \;\equiv\; (\sim p \vee q) \wedge (\sim q \vee p)$$

3.10. Determine the truth value of $(p \rightarrow q) \vee \sim(p \leftrightarrow \sim q)$.

p	q	$(p$	$\rightarrow$	$q)$	$\vee$	$\sim$	$(p$	$\leftrightarrow$	$\sim$	$q)$
T	T	T	T	T	T	T	T	F	F	T
T	F	T	F	F	F	F	T	T	T	F
F	T	F	T	T	T	F	F	T	F	T
F	F	F	T	F	T	T	F	F	T	F
Step		1	2	1	5	4	1	3	2	1

3.11. Determine the truth value of $(p \leftrightarrow \sim q) \leftrightarrow (q \rightarrow p)$.

p	q	$\sim q$	$p \leftrightarrow \sim q$	$q \rightarrow p$	$(p \leftrightarrow \sim q) \leftrightarrow (q \rightarrow p)$
T	T	F	F	T	F
T	F	T	T	T	T
F	T	F	T	F	F
F	F	T	F	T	F

Method 1

p	q	$(p$	$\leftrightarrow$	$\sim$	$q)$	$\leftrightarrow$	$(q$	$\rightarrow$	$p)$
T	T	T	F	F	T	F	T	T	T
T	F	T	T	T	F	T	F	T	T
F	T	F	T	F	T	F	T	F	F
F	F	F	F	T	F	F	F	T	F
Step		1	3	2	1	4	1	2	1

Method 2

NEGATION

3.12. Verify by truth tables that the negation of the conditional and biconditional are as follows: (i) $\sim(p \rightarrow q) \equiv p \wedge \sim q$, (ii) $\sim(p \leftrightarrow q) \equiv p \leftrightarrow \sim q \equiv \sim p \leftrightarrow q$.

(i)

p	q	$p \rightarrow q$	$\sim(p \rightarrow q)$	$\sim q$	$p \wedge \sim q$
T	T	T	F	F	F
T	F	F	T	T	T
F	T	T	F	F	F
F	F	T	F	T	F

(ii)

p	q	$p \leftrightarrow q$	$\sim(p \leftrightarrow q)$	$\sim p$	$\sim p \leftrightarrow q$	$\sim q$	$p \leftrightarrow \sim q$
T	T	T	F	F	F	F	F
T	F	F	T	F	T	T	T
F	T	F	T	T	T	F	T
F	F	T	F	T	F	T	F

Remark: Since $p \rightarrow q \equiv \sim p \vee q$, we could have used De Morgan's law to verify (i) as follows:

$$\sim(p \rightarrow q) \;\equiv\; \sim(\sim p \vee q) \;\equiv\; \sim\sim p \wedge \sim q \;\equiv\; p \wedge \sim q$$

3.13. Simplify: (i) $\sim(p \leftrightarrow \sim q)$, (ii) $\sim(\sim p \leftrightarrow q)$, (iii) $\sim(\sim p \rightarrow \sim q)$.

(i) $\sim(p \leftrightarrow \sim q) \equiv p \leftrightarrow \sim\sim q \equiv p \leftrightarrow q$

(ii) $\sim(\sim p \leftrightarrow q) \equiv \sim\sim p \leftrightarrow q \equiv p \leftrightarrow q$

(iii) $\sim(\sim p \rightarrow \sim q) \equiv \sim p \wedge \sim\sim q \equiv \sim p \wedge q$

3.14. Write the negation of each statement as simply as possible.

(i) If he studies, he will pass the exam.

(ii) He swims if and only if the water is warm.

(iii) If it snows, then he does not drive the car.

(i) By Problem 3.12, $\sim(p \rightarrow q) \equiv p \wedge \sim q$; hence the negation of (i) is

He studies and he will not pass the exam.

(ii) By Problem 3.12, $\sim(p \leftrightarrow q) \equiv p \leftrightarrow \sim q \equiv \sim p \leftrightarrow q$; hence the negation of (ii) is either of the following:

He swims if and only if the water is not warm.

He does not swim if and only if the water is warm.

(iii) Note that $\sim(p \rightarrow \sim q) \equiv p \wedge \sim\sim q \equiv p \wedge q$. Hence the negation of (iii) is

It snows and he drives the car.

3.15. Write the negation of each statement in as simple a sentence as possible.

(i) If it is cold, then he wears a coat but no sweater.

(ii) If he studies, then he will go to college or to art school.

(i) Let p be "It is cold", q be "He wears a coat" and r be "He wears a sweater". Then the given statement can be written as $p \rightarrow (q \wedge \sim r)$. Now

$$\sim[p \rightarrow (q \wedge \sim r)] \equiv p \wedge \sim(q \wedge \sim r) \equiv p \wedge (\sim q \vee r)$$

Hence the negation of (i) is

It is cold and he wears a sweater or no coat.

(ii) The given statement is of the form $p \rightarrow (q \vee r)$. But

$$\sim[p \rightarrow (q \vee r)] \equiv p \wedge \sim(q \vee r) \equiv p \wedge \sim q \wedge \sim r$$

Thus the negation of (ii) is

He studies and he does not go to college or to art school.

CONDITIONAL STATEMENTS AND VARIATIONS

3.16. Determine the contrapositive of each statement.

(i) If John is a poet, then he is poor.

(ii) Only if Marc studies will he pass the test.

(iii) It is necessary to have snow in order for Eric to ski.

(iv) If x is less than zero, then x is not positive.

(i) The contrapositive of $p \rightarrow q$ is $\sim q \rightarrow \sim p$. Hence the contrapositive of (i) is

If John is not poor, then he is not a poet.

(ii) The given statement is equivalent to "If Marc passes the test, then he studied". Hence the contrapositive of (ii) is

If Marc does not study, then he will not pass the test.

(iii) The given statement is equivalent to "If Eric skis, then it snowed". Hence the contrapositive of (iii) is

If it did not snow, then Eric will not ski.

(iv) The contrapositive of $p \to \sim q$ is $\sim\sim q \to \sim p \equiv q \to \sim p$. Hence the contrapositive of (iv) is

If x is positive, then x is not less than zero.

3.17. Find and simplify: (i) Contrapositive of the contrapositive of $p \to q$. (ii) Contrapositive of the converse of $p \to q$. (iii) Contrapositive of the inverse of $p \to q$.

(i) The contrapositive of $p \to q$ is $\sim q \to \sim p$. The contrapositive of $\sim q \to \sim p$ is $\sim\sim p \to \sim\sim q \equiv p \to q$, which is the original conditional proposition.

(ii) The converse of $p \to q$ is $q \to p$. The contrapositive of $q \to p$ is $\sim p \to \sim q$, which is the inverse of $p \to q$.

(iii) The inverse of $p \to q$ is $\sim p \to \sim q$. The contrapositive of $\sim p \to \sim q$ is $\sim\sim q \to \sim\sim p \equiv q \to p$, which is the converse of $p \to q$.

In other words, the inverse and converse are contrapositives of each other, and the conditional and contrapositive are contrapositives of each other!

Supplementary Problems

STATEMENTS

3.18. Let p denote "He is rich" and let q denote "He is happy". Write each statement in symbolic form using p and q.

(i) If he is rich then he is unhappy.

(ii) He is neither rich nor happy.

(iii) It is necessary to be poor in order to be happy.

(iv) To be poor is to be unhappy.

(v) Being rich is a sufficient condition to being happy.

(vi) Being rich is a necessary condition to being happy.

(vii) One is never happy when one is rich.

(viii) He is poor only if he is happy.

(ix) To be rich means the same as to be happy.

(x) He is poor or else he is both rich and happy.

Note. Assume "He is poor" is equivalent to $\sim p$.

3.19. Determine the truth value of each statement.

(i) If $5 < 3$, then $-3 < -5$.

(ii) It is not true that $1+1 = 2$ iff $3+4 = 5$.

(iii) A necessary condition that $1+2 = 3$ is that $4+4 = 4$.

(iv) It is not true that $1+1 = 5$ iff $3+3 = 1$.

(v) If $3 < 5$, then $-3 < -5$.

(vi) A sufficient condition that $1+2 = 3$ is that $4+4 = 4$.

3.20. Determine the truth value of each statement.

(i) It is not true that if $2+2 = 4$, then $3+3 = 5$ or $1+1 = 2$.

(ii) If $2+2 = 4$, then it is not true that $2+1 = 3$ and $5+5 = 10$.

(iii) If $2+2 = 4$, then $3+3 = 7$ iff $1+1 = 4$.

3.21. Write the negation of each statement in as simple a sentence as possible.
(i) If stock prices fall, then unemployment rises.
(ii) He has blond hair if and only if he has blue eyes.
(iii) If Marc is rich, then both Eric and Audrey are happy.
(iv) Betty smokes *Kent* or *Salem* only if she doesn't smoke *Camels*.
(v) Mary speaks Spanish or French if and only if she speaks Italian.
(vi) If John reads *Newsweek* then he reads neither *Life* nor *Time*.

TRUTH TABLES

3.22. Find the truth table of each proposition: (i) $(\sim p \vee q) \rightarrow p$, (ii) $q \leftrightarrow (\sim q \wedge p)$.

3.23. Find the truth table of each proposition:
(i) $(p \leftrightarrow \sim q) \rightarrow (\sim p \wedge q)$, (ii) $(\sim q \vee p) \leftrightarrow (q \rightarrow \sim p)$.

3.24. Find the truth table of each proposition:
(i) $[p \wedge (\sim q \rightarrow p)] \wedge \sim[(p \leftrightarrow \sim q) \rightarrow (q \vee \sim p)]$, (ii) $[q \leftrightarrow (r \rightarrow \sim p)] \vee [(\sim q \rightarrow p) \leftrightarrow r]$.

3.25. Prove: (i) $(p \wedge q) \rightarrow r \equiv (p \rightarrow r) \vee (q \rightarrow r)$, (ii) $p \rightarrow (q \rightarrow r) \equiv (p \wedge \sim r) \rightarrow \sim q$.

CONDITIONAL AND VARIATIONS

3.26. Determine the contrapositive of each statement.
(i) If he has courage he will win.
(ii) It is necessary to be strong in order to be a sailor.
(iii) Only if he does not tire will he win.
(iv) It is sufficient for it to be a square in order to be a rectangle.

3.27. Find: (i) Contrapositive of $p \rightarrow \sim q$. (iii) Contrapositive of the converse of $p \rightarrow \sim q$.
(ii) Contrapositive of $\sim p \rightarrow q$. (iv) Converse of the contrapositive of $\sim p \rightarrow \sim q$.

Answers to Supplementary Problems

3.18. (i) $p \rightarrow \sim q$ (iii) $q \rightarrow \sim p$ (v) $p \rightarrow q$ (vii) $p \rightarrow \sim q$ (ix) $p \leftrightarrow q$
(ii) $\sim p \wedge \sim q$ (iv) $\sim p \leftrightarrow \sim q$ (vi) $q \rightarrow p$ (viii) $\sim p \rightarrow q$ (x) $\sim p \vee (p \wedge q)$

3.19. (i) T, (ii) T, (iii) F, (iv) F, (v) F, (vi) T

3.20. (i) F, (ii) F, (iii) T

3.21. (i) Stock prices fall and unemployment does not rise.
(ii) He has blond hair but does not have blue eyes.
(iii) Marc is rich and Eric or Audrey is unhappy.
(iv) Betty smokes *Kent* or *Salem*, and *Camels*.
(v) Mary speaks Spanish or French, but not Italian.
(vi) John reads *Newsweek*, and *Life* or *Time*.

3.22. (i) TTFF, (ii) FFFT

3.23. (i) TFTT, (ii) FTFT

3.24. (i) FTFF, (ii) TTTFTTFT

3.25. *Hint.* Construct the appropriate truth tables.

3.26. (i) If he does not win, then he does not have courage.
(ii) If he is not strong, then he is not a sailor.
(iii) If he tires, then he will not win.
(iv) If it is not a rectangle, then it is not a square.

3.27. (i) $q \rightarrow \sim p$, (ii) $\sim q \rightarrow p$, (iii) $\sim p \rightarrow q$, (iv) $p \rightarrow q$

Chapter 4

Arguments, Logical Implication

ARGUMENTS

An *argument* is an assertion that a given set of propositions $P_1, P_2, \ldots, P_n$, called *premises*, yields (has as a consequence) another proposition Q, called the *conclusion*. Such an argument is denoted by

$$P_1, P_2, \ldots, P_n \vdash Q$$

The truth value of an argument is determined as follows:

[T_6] An argument $P_1, P_2, \ldots, P_n \vdash Q$ is true if Q is true whenever all the premises $P_1, P_2, \ldots, P_n$ are true; otherwise the argument is false.

Thus an argument is a statement, i.e. has a truth value. If an argument is true it is called a *valid* argument; if an argument is false it is called a *fallacy*.

Example 1.1: The following argument is valid:

$$p,\ p \to q \vdash q \quad \text{(Law of Detachment)}$$

The proof of this rule follows from the following truth table.

p	q	$p \to q$
T	T	T
T	F	F
F	T	T
F	F	T

For p is true in Cases (lines) 1 and 2, and $p \to q$ is true in Cases 1, 3 and 4; hence p and $p \to q$ are true simultaneously in Case 1. Since in this Case q is true, the argument is valid.

Example 1.2: The following argument is a fallacy:

$$p \to q,\ q \vdash p$$

For $p \to q$ and q are both true in Case (line) 3 in the above truth table, but in this Case p is false.

Now the propositions $P_1, P_2, \ldots, P_n$ are true simultaneously if and only if the proposition $P_1 \wedge P_2 \wedge \cdots \wedge P_n$ is true. Thus the argument $P_1, P_2, \ldots, P_n \vdash Q$ is valid if and only if Q is true whenever $P_1 \wedge P_2 \wedge \cdots \wedge P_n$ is true or, equivalently, if the proposition $(P_1 \wedge P_2 \wedge \cdots \wedge P_n) \to Q$ is a tautology. We state this result formally.

Theorem 4.1: The argument $P_1, P_2, \ldots, P_n \vdash Q$ is valid if and only if the proposition $(P_1 \wedge P_2 \wedge \cdots \wedge P_n) \to Q$ is a tautology.

Example 1.3: A fundamental principle of logical reasoning states:

"If p implies q and q implies r, then p implies r"

that is, the following argument is valid:

$$p \to q,\ q \to r \vdash p \to r \quad \text{(Law of Syllogism)}$$

This fact is verified by the following truth table which shows that the proposition $[(p \to q) \wedge (q \to r)] \to (p \to r)$ is a tautology:

p	q	r	$[(p$	$\rightarrow$	$q)$	$\wedge$	$(q$	$\rightarrow$	$r)]$	$\rightarrow$	$(p$	$\rightarrow$	$r)$
T	T	T	T	T	T	T	T	T	T	T	T	T	T
T	T	F	T	T	T	F	T	F	F	T	T	F	F
T	F	T	T	F	F	F	F	T	T	T	T	T	T
T	F	F	T	F	F	F	F	T	F	T	T	F	F
F	T	T	F	T	T	T	T	T	T	T	F	T	T
F	T	F	F	T	T	F	T	F	F	T	F	T	F
F	F	T	F	T	F	T	F	T	T	T	F	T	T
F	F	F	F	T	F	T	F	T	F	T	F	T	F
Step			1	2	1	3	1	2	1	4	1	2	1

Example 1.4: The following argument is a fallacy:

$$p \rightarrow q,\ \sim p \vdash \sim q$$

For the proposition $[(p \rightarrow q) \wedge \sim p] \rightarrow \sim q$ is not a tautology, as seen in the truth table below.

p	q	$p \rightarrow q$	$\sim p$	$(p \rightarrow q) \wedge \sim p$	$\sim q$	$[(p \rightarrow q) \wedge \sim p] \rightarrow \sim q$
T	T	T	F	F	F	T
T	F	F	F	F	T	T
F	T	T	T	T	F	F
F	F	T	T	T	T	T

Equivalently, the argument is a fallacy since, in Case (line) 3 of the truth table, $p \rightarrow q$ and $\sim p$ are true but $\sim q$ is false.

An argument can also be shown to be valid by using previous results as illustrated in the next example.

Example 1.5: We prove that the argument $p \rightarrow \sim q,\ q \vdash \sim p$ is valid:

Statement	Reason
(1) q is true.	(1) Given
(2) $p \rightarrow \sim q$ is true.	(2) Given
(3) $q \rightarrow \sim p$ is true.	(3) Contrapositive of (2)
(4) $\sim p$ is true.	(4) Law of Detachment (Example 1.1) using (1) and (3)

ARGUMENTS AND STATEMENTS

We now apply the theory of the preceding section to arguments involving specific statements. We emphasize that the validity of an argument does not depend upon the truth values nor the content of the statements appearing in the argument, but upon the particular form of the argument. This is illustrated in the following examples.

Example 2.1: Consider the following argument:

S_1: If a man is a bachelor, he is unhappy.
S_2: If a man is unhappy, he dies young.
....
S: Bachelors die young.

Here the statement S below the line denotes the conclusion of the argument, and the statements S_1 and S_2 above the line denote the premises. We claim that the argument $S_1, S_2 \vdash S$ is valid. For the argument is of the form

$$p \rightarrow q,\ q \rightarrow r \ \vdash\ p \rightarrow r$$

where p is "He is a bachelor", q is "He is unhappy" and r is "He dies young"; and by Example 1.3 this argument (Law of Syllogism) is valid.

Example 2.2: We claim that the following argument is not valid:

S_1: If two sides of a triangle are equal, then the opposite angles are equal.

S_2: Two sides of a triangle are not equal.

....

S: The opposite angles are not equal.

For the argument is of the form $p \to q, \sim p \vdash \sim q$, where p is "Two sides of a triangle are equal" and q is "The opposite angles are equal"; and by Example 1.4 this argument is a fallacy.

Although the conclusion S does follow from S_2 and axioms of Euclidean geometry, the above argument does not constitute such a proof since the argument is a fallacy.

Example 2.3: We claim that the following argument is valid:

S_1: If 5 is a prime number, then 5 does not divide 15.

S_2: 5 divides 15.

....

S: 5 is not a prime number.

For the argument is of the form $p \to \sim q, q \vdash \sim p$ where p is "5 is a prime number" and q is "5 divides 15"; and we proved this argument is valid in Example 1.5.

We remark that although the conclusion here is obviously a false statement, the argument as given is still valid. It is because of the false premise S_1 that we can logically arrive at the false conclusion.

Example 2.4: Determine the validity of the following argument:

S_1: If 7 is less than 4, then 7 is not a prime number.

S_2: 7 is not less than 4.

....

S: 7 is a prime number.

We translate the argument into symbolic form. Let p be "7 is less than 4" and q be "7 is a prime number". Then the argument is of the form

$$p \to \sim q, \sim p \vdash q$$

The argument is a fallacy since in Case (line) 4 of the adjacent truth table, $p \to \sim q$ and $\sim p$ are true but q is false.

p	q	$\sim q$	$p \to \sim q$	$\sim p$
T	T	F	F	F
T	F	T	T	F
F	T	F	T	T
F	F	T	T	T

The fact that the conclusion of the argument happens to be a true statement is irrelevant to the fact that the argument is a fallacy.

LOGICAL IMPLICATION

A proposition $P(p, q, \ldots)$ is said to *logically imply* a proposition $Q(p, q, \ldots)$ if $Q(p, q, \ldots)$ is true whenever $P(p, q, \ldots)$ is true.

Example 3.1: We claim that p logically implies $p \vee q$. For consider the truth tables of p and $p \vee q$ in the adjacent table. Observe that p is true in Cases (lines) 1 and 2, and in these Cases $p \vee q$ is also true. In other words, p logically implies $p \vee q$.

p	q	$p \vee q$
T	T	T
T	F	T
F	T	T
F	F	F

Now if $Q(p, q, \ldots)$ is true whenever $P(p, q, \ldots)$ is true, then the argument

$$P(p, q, \ldots) \vdash Q(p, q, \ldots)$$

is valid; and conversely. Furthermore, the argument $P \vdash Q$ is valid if and only if the conditional statement $P \to Q$ is always true, i.e. a tautology. We state this result formally.

Theorem 4.2: The proposition $P(p, q, \ldots)$ logically implies the proposition $Q(p, q, \ldots)$ if and only if

(i) the argument $P(p, q, \ldots) \vdash Q(p, q, \ldots)$ is valid

or, equivalently,

(ii) the proposition $P(p, q, \ldots) \rightarrow Q(p, q, \ldots)$ is a tautology.

Remark: The reader should be warned that logicians and many texts use the word "implies" in the same sense as we use "logically implies", and so they distinguish between "implies" and "if ... then". These two distinct concepts are, of course, intimately related as seen in the above Theorem 4.2.

Solved Problems

ARGUMENTS

4.1. Show that the following argument is valid: $p \leftrightarrow q, q \vdash p$.

Method 1.

Construct the truth table on the right. Now $p \leftrightarrow q$ is true in Cases (lines) 1 and 4, and q is true in Cases 1 and 3; hence $p \leftrightarrow q$ and q are true simultaneously only in Case 1 where p is also true. Thus the argument $p \leftrightarrow q, q \vdash p$ is valid.

p	q	$p \leftrightarrow q$
T	T	T
T	F	F
F	T	F
F	F	T

Method 2.

Construct the truth table of $[(p \leftrightarrow q) \wedge q] \rightarrow p$:

p	q	$p \leftrightarrow q$	$(p \leftrightarrow q) \wedge q$	$[(p \leftrightarrow q) \wedge q] \rightarrow p$
T	T	T	T	T
T	F	F	F	T
F	T	F	F	T
F	F	T	F	T

Since $[(p \leftrightarrow q) \wedge q] \rightarrow p$ is a tautology, the argument is valid.

4.2. Determine the validity of the argument $p \rightarrow q, \sim q \vdash \sim p$.

Construct the truth table of $[(p \rightarrow q) \wedge \sim q] \rightarrow \sim p$:

p	q	$[(p$	$\rightarrow$	$q)$	$\wedge$	$\sim$	$q]$	$\rightarrow$	$\sim$	p
T	T	T	T	T	F	F	T	T	F	T
T	F	T	F	F	F	T	F	T	F	T
F	T	F	T	T	F	F	T	T	T	F
F	F	F	T	F	T	T	F	T	T	F
Step		1	2	1	3	2	1	4	2	1

Since the proposition $[(p \rightarrow q) \wedge \sim q] \rightarrow \sim p$ is a tautology, the given argument is valid.

4.3. Determine the validity of the argument $\sim p \rightarrow q, p \vdash \sim q$.

Construct the truth table of $[(\sim p \rightarrow q) \wedge p] \rightarrow \sim q$:

p	q	$\sim p$	$\sim p \to q$	$(\sim p \to q) \wedge p$	$\sim q$	$[(\sim p \to q) \wedge p] \to \sim q$
T	T	F	T	T	F	F
T	F	F	T	T	T	T
F	T	T	T	F	F	T
F	F	T	F	F	T	T

Since the proposition $[(\sim p \to q) \wedge p] \to \sim q$ is not a tautology, the argument $\sim p \to q,\ p \vdash \sim q$ is a fallacy.

Observe that $\sim p \to q$ and p are both true in Case (line) 1 but in this Case $\sim q$ is false.

4.4. Prove that the following argument is valid: $p \to \sim q,\ r \to q,\ r \vdash \sim p$.

Method 1. Construct the following truth tables:

	p	q	r	$p \to \sim q$	$r \to q$	$\sim p$
1	T	T	T	F	T	F
2	T	T	F	F	T	F
3	T	F	T	T	F	F
4	T	F	F	T	T	F
5	F	T	T	T	T	T
6	F	T	F	T	T	T
7	F	F	T	T	F	T
8	F	F	F	T	T	T

Now $p \to \sim q$, $r \to q$ and r are true simultaneously only in Case (line) 5, where $\sim p$ is also true; hence the given argument is valid.

Method 2. Construct the truth table for the proposition

$$[(p \to \sim q) \wedge (r \to q) \wedge r] \to \sim p$$

It will be a tautology, and so the given argument is valid.

Method 3.

Statement	Reason
(1) $p \to \sim q$ is true.	(1) Given
(2) $r \to q$ is true.	(2) Given
(3) $\sim q \to \sim r$ is true.	(3) Contrapositive of (2)
(4) $p \to \sim r$ is true.	(4) Law of Syllogism, using (1) and (3)
(5) $r \to \sim p$ is true.	(5) Contrapositive of (4)
(6) r is true.	(6) Given
(7) Hence $\sim p$ is true.	(7) Law of Detachment, using (5) and (6)

ARGUMENTS AND STATEMENTS

4.5. Test the validity of each argument:

(i) If it rains, Erik will be sick.
It did not rain.
....
Erik was not sick.

(ii) If it rains, Erik will be sick.
Erik was not sick.
....
It did not rain.

First translate the arguments into symbolic form:

(i) $p \to q,\ \sim p \vdash \sim q$ (ii) $p \to q,\ \sim q \vdash \sim p$

where p is "It rains" and q is "Erik is sick". By Example 1.4, the argument (i) is a fallacy; by Problem 4.2, the argument (ii) is valid.

4.6. Test the validity of the following argument:

If 6 is not even, then 5 is not prime.

But 6 is even.

....

Therefore 5 is prime.

Translate the argument into symbolic form. Let p be "6 is even" and let q be "5 is prime." Then the argument is of the form

$$\sim p \rightarrow \sim q,\ p \vdash q$$

Now in the adjacent truth table, $\sim p \rightarrow \sim q$ and p are both true in Case (line) 2; but in this Case q is false. Hence the argument is a fallacy.

p	q	$\sim p$	$\sim q$	$\sim p \rightarrow \sim q$
T	T	F	F	T
T	F	F	T	T
F	T	T	F	F
F	F	T	T	T

The argument can also be shown to be a fallacy by constructing the truth table of the proposition $[(\sim p \rightarrow \sim q) \wedge p] \rightarrow q$ and observing that the proposition is not a tautology.

The fact that the conclusion is a true statement does not affect the fact that the argument is a fallacy.

4.7. Test the validity of the following argument:

If I like mathematics, then I will study.

Either I study or I fail.

....

If I fail, then I do not like mathematics.

First translate the argument into symbolic form. Let p be "I like mathematics", q be "I study" and r be "I fail". Then the given argument is of the form

$$p \rightarrow q,\ q \vee r \vdash r \rightarrow \sim p$$

To test the validity of the argument, construct the truth tables of the propositions $p \rightarrow q$, $q \vee r$ and $r \rightarrow \sim p$:

p	q	r	$p \rightarrow q$	$q \vee r$	$\sim p$	$r \rightarrow \sim p$
T	T	T	T	T	F	F
T	T	F	T	T	F	T
T	F	T	F	T	F	F
T	F	F	F	F	F	T
F	T	T	T	T	T	T
F	T	F	T	T	T	T
F	F	T	T	T	T	T
F	F	F	T	F	T	T

Recall that an argument is valid if the conclusion is true whenever the premises are true. Now in Case (line) 1 of the above truth table, the premises $p \rightarrow q$ and $q \vee r$ are both true but the conclusion $r \rightarrow \sim p$ is false; hence the argument is a fallacy.

4.8. Test the validity of the following argument:

If I study, then I will not fail mathematics.

If I do not play basketball, then I will study.

But I failed mathematics.

....

Therefore, I played basketball.

First translate the argument into symbolic form. Let p be "I study", q be "I fail mathematics" and r be "I play basketball". Then the given argument is as follows:

$$p \to \sim q,\ \sim r \to p,\ q \vdash r$$

To test the validity of the argument, construct the truth tables of the given propositions $p \to \sim q$, $\sim r \to p$, q and r:

p	q	r	$\sim q$	$p \to \sim q$	$\sim r$	$\sim r \to p$
T	T	T	F	F	F	T
T	T	F	F	F	T	T
T	F	T	T	T	F	T
T	F	F	T	T	T	T
F	T	T	F	T	F	T
F	T	F	F	T	T	F
F	F	T	T	T	F	T
F	F	F	T	T	T	F

Now the premises $p \to \sim q$, $\sim r \to p$ and q are true simultaneously only in Case (line) 5, and in that case the conclusion r is also true; hence the argument is valid.

LOGICAL IMPLICATION

4.9. Show that $p \wedge q$ logically implies $p \leftrightarrow q$.

Construct the truth table for $(p \wedge q) \to (p \leftrightarrow q)$:

p	q	$p \wedge q$	$p \leftrightarrow q$	$(p \wedge q) \to (p \leftrightarrow q)$
T	T	T	T	T
T	F	F	F	T
F	T	F	F	T
F	F	F	T	T

Since $(p \wedge q) \to (p \leftrightarrow q)$ is a tautology, $p \wedge q$ logically implies $p \leftrightarrow q$.

4.10. Show that $p \leftrightarrow q$ logically implies $p \to q$.

Consider the truth tables of $p \leftrightarrow q$ and $p \to q$:

p	q	$p \leftrightarrow q$	$p \to q$
T	T	T	T
T	F	F	F
F	T	F	T
F	F	T	T

Now $p \leftrightarrow q$ is true in lines 1 and 4, and in these cases $p \to q$ is also true. Hence $p \leftrightarrow q$ logically implies $p \to q$.

4.11. Prove: Let $P(p, q, \ldots)$ logically imply $Q(p, q, \ldots)$. Then for any propositions $P_1, P_2, \ldots$, $P(P_1, P_2, \ldots)$ logically implies $Q(P_1, P_2, \ldots)$.

By Theorem 4.2, if $P(p, q, \ldots)$ logically implies $Q(p, q, \ldots)$ then the proposition $P(p, q, \ldots) \to Q(p, q, \ldots)$ is a tautology. By the Principle of Substitution (Theorem 2.2), the proposition $P(P_1, P_2, \ldots) \to Q(P_1, P_2, \ldots)$ is also a tautology. Accordingly, $P(P_1, P_2, \ldots)$ logically implies $Q(P_1, P_2, \ldots)$.

4.12. Determine the number of nonequivalent propositions $P(p,q)$ which logically imply the proposition $p \leftrightarrow q$.

p	q	$p \leftrightarrow q$
T	T	T
T	F	F
F	T	F
F	F	T

Consider the adjacent truth table of $p \leftrightarrow q$. Now $P(p, q)$ logically implies $p \leftrightarrow q$ if $p \leftrightarrow q$ is true whenever $P(p, q)$ is true. But $p \leftrightarrow q$ is true only in Cases (lines) 1 and 4; hence $P(p, q)$ cannot be true in Cases 2 and 3. There are four such propositions which are listed below:

P_1	P_2	P_3	P_4	$p \leftrightarrow q$
F	T	F	T	T
F	F	F	F	F
F	F	F	F	F
F	F	T	T	T

4.13. Show that $p \leftrightarrow \sim q$ does not logically imply $p \rightarrow q$.

Method 1. Construct the truth tables of $p \leftrightarrow \sim q$ and $p \rightarrow q$:

p	q	$\sim q$	$p \leftrightarrow \sim q$	$p \rightarrow q$
T	T	F	F	T
T	F	T	T	F
F	T	F	T	T
F	F	T	F	T

Recall that $p \leftrightarrow \sim q$ logically implies $p \rightarrow q$ if $p \rightarrow q$ is true whenever $p \leftrightarrow \sim q$ is true. But $p \leftrightarrow \sim q$ is true in Case (line) 2 in the above table, and in that Case $p \rightarrow q$ is false. Hence $p \leftrightarrow \sim q$ does not logically imply $p \rightarrow q$.

Method 2. Construct the truth table of the proposition $(p \leftrightarrow \sim q) \rightarrow (p \rightarrow q)$. It will not be a tautology; hence, by Theorem 4.2, $p \leftrightarrow \sim q$ does not logically imply $p \rightarrow q$.

Supplementary Problems

ARGUMENTS

4.14. Test the validity of each argument: (i) $\sim p \rightarrow q,\ p \vdash \sim q$; (ii) $\sim p \rightarrow \sim q,\ q \vdash p$.

4.15. Test the validity of each argument: (i) $p \rightarrow q,\ r \rightarrow \sim q \vdash r \rightarrow \sim p$; (ii) $p \rightarrow \sim q,\ \sim r \rightarrow \sim q \vdash p \rightarrow \sim r$.

4.16. Test the validity of each argument: (i) $p \rightarrow \sim q,\ r \rightarrow p,\ q \vdash \sim r$; (ii) $p \rightarrow q,\ r \vee \sim q,\ \sim r \vdash \sim p$.

ARGUMENTS AND STATEMENTS

4.17. Test the validity of the argument:

If London is not in Denmark, then Paris is not in France.

But Paris is in France.

....

Therefore, London is in Denmark.

4.18. Test the validity of the argument:

If I study, then I will not fail mathematics.

I did not study.

....

I failed mathematics.

4.19. Translate into symbolic form and test the validity of the argument:

(*a*) If 6 is even, then 2 does not divide 7.
Either 5 is not prime or 2 divides 7.
But 5 is prime.
....
Therefore, 6 is odd (not even).

(*b*) On my wife's birthday, I bring her flowers.
Either it's my wife's birthday or I work late.
I did not bring my wife flowers today.
....
Therefore, today I worked late.

(*c*) If I work, I cannot study.
Either I work, or I pass mathematics.
I passed mathematics.
....
Therefore, I studied.

(*d*) If I work, I cannot study.
Either I study, or I pass mathematics.
I worked.
....
Therefore, I passed mathematics.

LOGICAL IMPLICATION

4.20. Show that (i) $p \wedge q$ logically implies p, (ii) $p \vee q$ does not logically imply p.

4.21. Show that (i) q logically implies $p \rightarrow q$, (ii) $\sim p$ logically implies $p \rightarrow q$.

4.22. Show that $p \wedge (q \vee r)$ logically implies $(p \wedge q) \vee r$.

4.23. Determine those propositions which logically imply (i) a tautology, (ii) a contradiction.

4.24. Determine the number of nonequivalent propositions $P(p, q)$ which logically imply the proposition $p \rightarrow q$, and construct truth tables for such propositions (see Problem 4.12).

Answers to Supplementary Problems

4.14. (i) fallacy, (ii) valid

4.15. (i) valid, (ii) fallacy

4.16. (i) valid, (ii) valid

4.17. valid

4.18. fallacy

4.19. (*a*) $p \rightarrow \sim q,\ \sim r \vee q,\ r \vdash \sim p$; valid.
(*b*) $p \rightarrow q,\ p \vee r,\ \sim q \vdash r$; valid.
(*c*) $p \rightarrow \sim q,\ p \vee r,\ r \vdash q$; fallacy.
(*d*) $p \rightarrow \sim q,\ q \vee r,\ p \vdash r$; valid.

4.23. (i) Every proposition logically implies a tautology. (ii) Only a contradiction logically implies a contradiction.

4.24. There are eight such propositions:

p	q	P_1	P_2	P_3	P_4	P_5	P_6	P_7	P_8	$p \rightarrow q$
T	T	T	T	T	T	F	F	F	F	T
T	F	F	F	F	F	F	F	F	F	F
F	T	T	T	F	F	T	T	F	F	T
F	F	T	F	T	F	T	F	T	F	T

Chapter 5

Set Theory

SETS AND ELEMENTS

The concept of a *set* appears in all branches of mathematics. Intuitively, a set is any well-defined list or collection of objects, and will be denoted by capital letters $A, B, X, Y, \ldots$. The objects comprising the set are called its *elements* or *members* and will be denoted by lower case letters $a, b, x, y, \ldots$. The statement "p is an element of A" or, equivalently, "p belongs to A" is written

$$p \in A$$

The negation of $p \in A$ is written $p \notin A$.

There are essentially two ways to specify a particular set. One way, if it is possible, is to list its members. For example,

$$A = \{a, e, i, o, u\}$$

denotes the set A whose elements are the letters a, e, i, o, u. Note that the elements are separated by commas and enclosed in braces { }. The second way is to state those properties which characterize the elements in the set. For example,

$$B = \{x : x \text{ is an integer, } x > 0\}$$

which reads "B is the set of x such that x is an integer and x is greater than zero," denotes the set B whose elements are the positive integers. A letter, usually x, is used to denote a typical member of the set; the colon is read as "such that" and the comma as "and".

Example 1.1: The set B above can also be written as $B = \{1, 2, 3, \ldots\}$.

Observe that $-6 \notin B$, $3 \in B$ and $\pi \notin B$.

Example 1.2: The set A above can also be written as

$$A = \{x : x \text{ is a letter in the English alphabet, } x \text{ is a vowel}\}$$

Observe that $b \notin A$, $e \in A$ and $p \notin A$.

Example 1.3: Let $E = \{x : x^2 - 3x + 2 = 0\}$. In other words, E consists of those numbers which are solutions of the equation $x^2 - 3x + 2 = 0$, sometimes called the solution set of the given equation. Since the solutions of the equation are 1 and 2, we could also write $E = \{1, 2\}$.

Two sets A and B are *equal*, written $A = B$, if they consist of the same elements, i.e. if each member of A belongs to B and each member of B belongs to A. The negation of $A = B$ is written $A \neq B$.

Example 1.4: Let $E = \{x : x^2 - 3x + 2 = 0\}$, $F = \{2, 1\}$ and $G = \{1, 2, 2, 1, 6/3\}$.

Then $E = F = G$. Observe that a set does not depend on the way in which its elements are displayed. A set remains the same if its elements are repeated or rearranged.

FINITE AND INFINITE SETS

Sets can be finite or infinite. A set is finite if it consists of exactly n different elements, where n is some positive integer; otherwise it is infinite.

Example 2.1: Let M be the set of the days of the week. In other words,
$$M = \{\text{Monday, Tuesday, Wednesday, Thursday, Friday, Saturday, Sunday}\}$$
Then M is finite.

Example 2.2: Let $Y = \{2, 4, 6, 8, \ldots\}$. Then Y is infinite.

Example 2.3: Let $P = \{x : x \text{ is a river on the earth}\}$. Although it may be difficult to count the number of rivers on the earth, P is a finite set.

SUBSETS

A set A is a *subset* of a set B or, equivalently, B is a *superset* of A, written
$$A \subset B \quad \text{or} \quad B \supset A$$
iff each element in A also belongs to B; that is, $x \in A$ implies $x \in B$. We also say that A is *contained* in B or B *contains* A. The negation of $A \subset B$ is written $A \not\subset B$ or $B \not\supset A$ and states that there is an $x \in A$ such that $x \notin B$.

Example 3.1: Consider the sets
$$A = \{1, 3, 5, 7, \ldots\}, \quad B = \{5, 10, 15, 20, \ldots\}$$
$$C = \{x : x \text{ is prime}, \ x > 2\} = \{3, 5, 7, 11, \ldots\}$$
Then $C \subset A$ since every prime number greater than 2 is odd. On the other hand, $B \not\subset A$ since $10 \in B$ but $10 \notin A$.

Example 3.2: Let N denote the set of positive integers, Z denote the set of integers, Q denote the set of rational numbers and R denote the set of real numbers. Then
$$N \subset Z \subset Q \subset R$$

Example 3.3: The set $E = \{2, 4, 6\}$ is a subset of the set $F = \{6, 2, 4\}$, since each number 2, 4 and 6 belonging to E also belongs to F. In fact, $E = F$. In a similar manner it can be shown that every set is a subset of itself.

As noted in the preceding example, $A \subset B$ does not exclude the possibility that $A = B$. In fact, we may restate the definition of equality of sets as follows:

Definition: Two sets A and B are equal if $A \subset B$ and $B \subset A$.

In the case that $A \subset B$ but $A \neq B$, we say that A is a *proper subset* of B or B contains A properly. The reader should be warned that some authors use the symbol $\subseteq$ for a subset and the symbol $\subset$ only for a proper subset.

The following theorem is a consequence of the preceding definitions:

Theorem 5.1: Let A, B and C be sets. Then: (i) $A \subset A$; (ii) if $A \subset B$ and $B \subset A$, then $A = B$; and (iii) if $A \subset B$ and $B \subset C$, then $A \subset C$.

UNIVERSAL AND NULL SETS

In any application of the theory of sets, all sets under investigation are regarded as subsets of a fixed set. We call this set the *universal set* or *universe of discourse* and denote it (in this chapter) by U.

Example 4.1: In plane geometry, the universal set consists of all the points in the plane.

Example 4.2: In human population studies, the universal set consists of all the people in the world.

It is also convenient to introduce the concept of the *empty* or *null* set, that is, a set which contains no elements. This set, denoted by $\emptyset$, is considered finite and a subset of every other set. Thus, for any set A, $\emptyset \subset A \subset U$.

Example 4.3: Let $A = \{x : x^2 = 4,\ x \text{ is odd}\}$. Then A is empty, i.e. $A = \emptyset$.

Example 4.4: Let B be the set of people in the world who are older than 200 years. According to known statistics, B is the null set.

CLASS, COLLECTION, FAMILY

Frequently, the members of a set are sets themselves. For example, each line in a set of lines is a set of points. To help clarify these situations, other words, such as "class", "collection" and "family" are used. Usually we use class or collection for a set of sets, and family for a set of classes. The words subclass, subcollection and subfamily have meanings analogous to subset.

Example 5.1: The members of the class $\{\{2,3\}, \{2\}, \{5,6\}\}$ are the sets $\{2,3\}$, $\{2\}$ and $\{5,6\}$.

Example 5.2: Consider any set A. The power set of A, denoted by $\mathcal{P}(A)$ or 2^A, is the class of all subsets of A. In particular, if $A = \{a, b, c\}$, then

$$\mathcal{P}(A) = \{A, \{a,b\}, \{a,c\}, \{b,c\}, \{a\}, \{b\}, \{c\}, \emptyset\}$$

In general, if A is finite and has n elements, then $\mathcal{P}(A)$ will have 2^n elements.

SET OPERATIONS

The *union* of two sets A and B, denoted by $A \cup B$, is the set of all elements which belong to A or to B:

$$A \cup B = \{x : x \in A \text{ or } x \in B\}$$

Here "or" is used in the sense of and/or.

The *intersection* of two sets A and B, denoted by $A \cap B$, is the set of elements which belong to both A and B:

$$A \cap B = \{x : x \in A \text{ and } x \in B\}$$

If $A \cap B = \emptyset$, that is, if A and B do not have any elements in common, then A and B are said to be *disjoint* or *non-intersecting*.

The *relative complement* of a set B with respect to a set A or, simply, the *difference* of A and B, denoted by $A \setminus B$, is the set of elements which belong to A but which do not belong to B:

$$A \setminus B = \{x : x \in A,\ x \notin B\}$$

Observe that $A \setminus B$ and B are disjoint, i.e. $(A \setminus B) \cap B = \emptyset$.

The *absolute complement* or, simply, *complement* of a set A, denoted by A^c, is the set of elements which do not belong to A:

$$A^c = \{x : x \in U,\ x \notin A\}$$

That is, A^c is the difference of the universal set U and A.

Example 6.1: The following diagrams, called Venn diagrams, illustrate the above set operations. Here sets are represented by simple plane areas and U, the universal set, by the area in the entire rectangle.

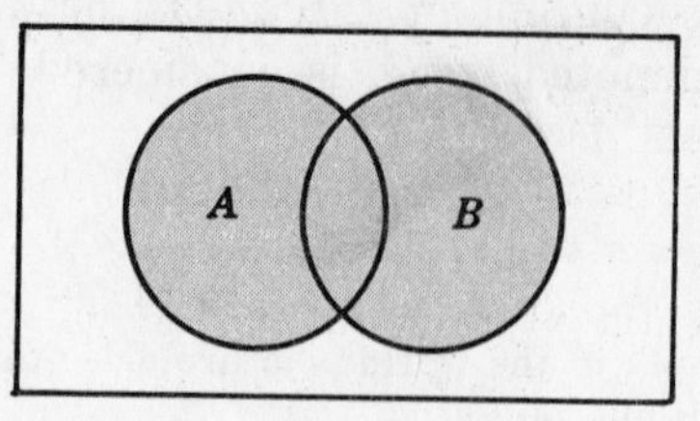

$A \cup B$ is shaded.

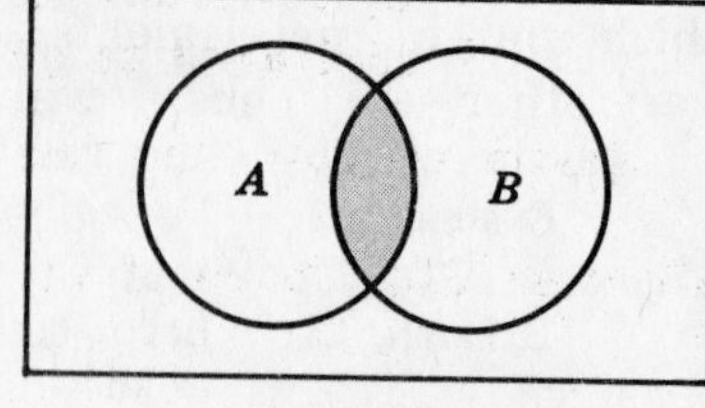

$A \cap B$ is shaded.

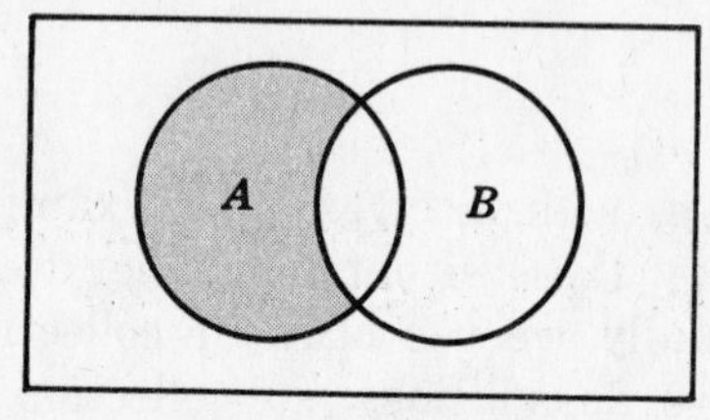

$A \setminus B$ is shaded.

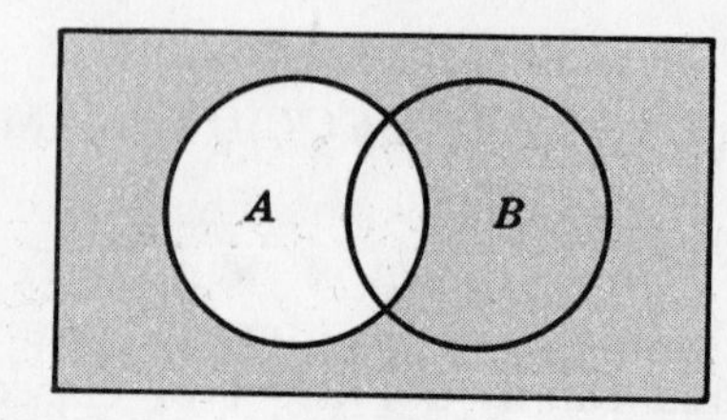

A^c is shaded.

Example 6.2: Let $A = \{1, 2, 3, 4\}$ and $B = \{3, 4, 5, 6\}$ where $U = \{1, 2, 3, \ldots\}$. Then:

$$A \cup B = \{1, 2, 3, 4, 5, 6\} \qquad A \cap B = \{3, 4\}$$
$$A \setminus B = \{1, 2\} \qquad A^c = \{5, 6, 7, \ldots\}$$

Sets under the above operations satisfy various laws or identities which are listed in Table 5.1 below. In fact we state:

Theorem 5.2: Sets satisfy the laws in Table 5.1.

LAWS OF THE ALGEBRA OF SETS	
Idempotent Laws	
1a. $A \cup A = A$	1b. $A \cap A = A$
Associative Laws	
2a. $(A \cup B) \cup C = A \cup (B \cup C)$	2b. $(A \cap B) \cap C = A \cap (B \cap C)$
Commutative Laws	
3a. $A \cup B = B \cup A$	3b. $A \cap B = B \cap A$
Distributive Laws	
4a. $A \cup (B \cap C) = (A \cup B) \cap (A \cup C)$	4b. $A \cap (B \cup C) = (A \cap B) \cup (A \cap C)$
Identity Laws	
5a. $A \cup \emptyset = A$	5b. $A \cap U = A$
6a. $A \cup U = U$	6b. $A \cap \emptyset = \emptyset$
Complement Laws	
7a. $A \cup A^c = U$	7b. $A \cap A^c = \emptyset$
8a. $(A^c)^c = A$	8b. $U^c = \emptyset$, $\emptyset^c = U$
De Morgan's Laws	
9a. $(A \cup B)^c = A^c \cap B^c$	9b. $(A \cap B)^c = A^c \cup B^c$

Table 5.1

Remark: Each of the above laws follows from the analogous logical law in Table 2.1, Page 11. For example,

$$A \cap B = \{x : x \in A \text{ and } x \in B\} = \{x : x \in B \text{ and } x \in A\} = B \cap A$$

Here we use the fact that if p is $x \in A$ and q is $x \in B$, then $p \wedge q$ is logically equivalent to $q \wedge p$: $p \wedge q \equiv q \wedge p$.

Lastly we state the relationship between set inclusion and the above set operations:

Theorem 5.3: Each of the following conditions is equivalent to $A \subset B$:

(i) $A \cap B = A$ (iii) $B^c \subset A^c$ (v) $B \cup A^c = U$

(ii) $A \cup B = B$ (iv) $A \cap B^c = \emptyset$

ARGUMENTS AND VENN DIAGRAMS

Many verbal statements can be translated into equivalent statements about sets which can be described by Venn diagrams. Hence Venn diagrams are very often used to determine the validity of an argument.

Example 7.1: Consider the following argument:

S_1: Babies are illogical.

S_2: Nobody is despised who can manage a crocodile.

S_3: Illogical people are despised.

....

S: Babies cannot manage crocodiles.

(The above argument is adapted from Lewis Carroll, *Symbolic Logic*; he is also the author of *Alice in Wonderland.*) Now by S_1, the set of babies is a subset of the set of illogical people:

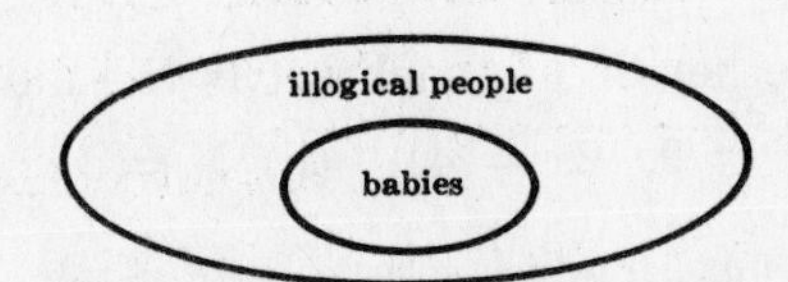

By S_3, the set of illogical people is contained in the set of despised people:

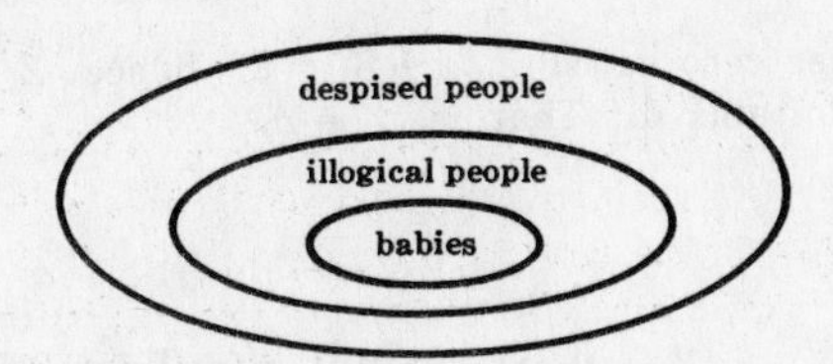

Furthermore, by S_2, the set of despised people and the set of people who can manage a crocodile are disjoint:

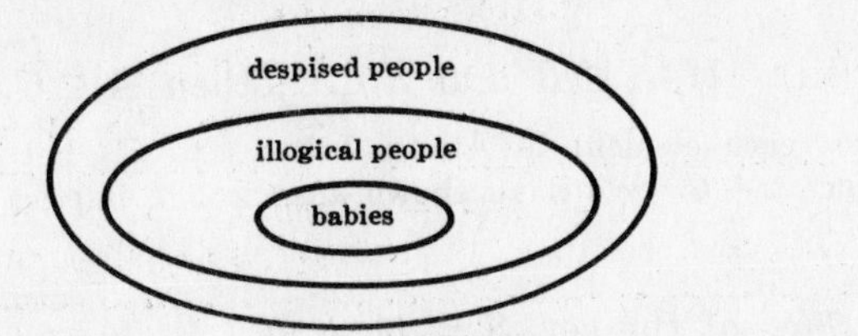

But by the above Venn diagram, the set of babies is disjoint from the set of people who can manage crocodiles, or "Babies cannot manage crocodiles" is a consequence of S_1, S_2 and S_3. Thus the above argument,

$$S_1, S_2, S_3 \vdash S$$

is valid.

Solved Problems

SETS, ELEMENTS

5.1. Let $A = \{x : 3x = 6\}$. Does $A = 2$?

A is the set which consists of the single element 2, that is, $A = \{2\}$. The number 2 belongs to A; it does not equal A. There is a basic difference between an element p and the singleton set $\{p\}$.

5.2. Which of these sets are equal: $\{r, t, s\}$, $\{s, t, r, s\}$, $\{t, s, t, r\}$, $\{s, r, s, t\}$?

They are all equal. Order and repetition do not change a set.

5.3. Which of the following sets are finite?

(i) The months of the year.
(ii) $\{1, 2, 3, \ldots, 99, 100\}$
(iii) The number of people living on the earth.
(iv) $\{x : x \text{ is an even number}\}$
(v) $\{1, 2, 3, \ldots\}$

The first three sets are finite; the last two sets are infinite.

5.4. Determine which of the following sets are equal: $\emptyset$, $\{0\}$, $\{\emptyset\}$.

Each is different from the other. The set $\{0\}$ contains one element, the number zero. The set $\emptyset$ contains no elements; it is the empty set. The set $\{\emptyset\}$ also contains one element, the null set.

5.5. Determine whether or not each set is the null set:

(i) $X = \{x : x^2 = 9,\ 2x = 4\}$, (ii) $Y = \{x : x \neq x\}$, (iii) $Z = \{x : x + 8 = 8\}$.

(i) There is no number which satisfies both $x^2 = 9$ and $2x = 4$; hence X is empty, i.e. $X = \emptyset$.

(ii) We assume that any object is itself, so Y is also empty. In fact, some texts define the null set as follows:

$$\emptyset \equiv \{x : x \neq x\}$$

(iii) The number zero satisfies $x + 8 = 8$; hence $Z = \{0\}$. Accordingly, Z is not the empty set since it contains 0. That is, $Z \neq \emptyset$.

SUBSETS

5.6. Prove that $A = \{2, 3, 4, 5\}$ is not a subset of $B = \{x : x \text{ is even}\}$.

It is necessary to show that at least one element in A does not belong to B. Now $3 \in A$ and, since B consists of even numbers, $3 \notin B$; hence A is not a subset of B.

5.7. Prove Theorem 5.1(iii): If $A \subset B$ and $B \subset C$, then $A \subset C$.

We must show that each element in A also belongs to C. Let $x \in A$. Now $A \subset B$ implies $x \in B$. But $B \subset C$; hence $x \in C$. We have shown that $x \in A$ implies $x \in C$, that is, that $A \subset C$.

5.8. Find the power set $\mathcal{P}(S)$ of the set $S = \{1, 2, 3\}$.

The power set $\mathcal{P}(S)$ of S is the class of all subsets of S; these are $\{1, 2, 3\}$, $\{1, 2\}$, $\{1, 3\}$, $\{2, 3\}$, $\{1\}$, $\{2\}$, $\{3\}$ and the empty set $\emptyset$. Hence

$$\mathcal{P}(S) = \{S, \{1, 3\}, \{2, 3\}, \{1, 2\}, \{1\}, \{2\}, \{3\}, \emptyset\}$$

Note that there are $2^3 = 8$ subsets of S.

5.9. Let $V = \{d\}$, $W = \{c, d\}$, $X = \{a, b, c\}$, $Y = \{a, b\}$ and $Z = \{a, b, d\}$. Determine whether each statement is true or false:
(i) $Y \subset X$, (ii) $W \neq Z$, (iii) $Z \supset V$, (iv) $V \subset X$, (v) $X = W$, (vi) $W \subset Y$.

(i) Since each element in Y is a member of X, $Y \subset X$ is true.

(ii) Now $a \in Z$ but $a \notin W$; hence $W \neq Z$ is true.

(iii) The only element in V is d and it also belongs to Z; hence $Z \supset V$ is true.

(iv) V is not a subset of X since $d \in V$ but $d \notin X$; hence $V \subset X$ is false.

(v) Now $a \in X$ but $a \notin W$; hence $X = W$ is false.

(vi) W is not a subset of Y since $c \in W$ but $c \notin Y$; hence $W \subset Y$ is false.

5.10. Prove: If A is a subset of the empty set $\emptyset$, then $A = \emptyset$.

The null set $\emptyset$ is a subset of every set; in particular, $\emptyset \subset A$. But, by hypothesis, $A \subset \emptyset$; hence $A = \emptyset$.

SET OPERATIONS

5.11. Let $U = \{1, 2, \ldots, 8, 9\}$, $A = \{1, 2, 3, 4\}$, $B = \{2, 4, 6, 8\}$ and $C = \{3, 4, 5, 6\}$. Find:
(i) A^c, (ii) $A \cap C$, (iii) $(A \cap C)^c$, (iv) $A \cup B$, (v) $B \setminus C$.

(i) A^c consists of the elements in U that are not in A; hence $A^c = \{5, 6, 7, 8, 9\}$.

(ii) $A \cap C$ consists of the elements in both A and C; hence $A \cap C = \{3, 4\}$.

(iii) $(A \cap C)^c$ consists of the elements in U that are not in $A \cap C$. Now by (ii), $A \cap C = \{3, 4\}$ and so $(A \cap C)^c = \{1, 2, 5, 6, 7, 8, 9\}$.

(iv) $A \cup B$ consists of the elements in A or B (or both): hence $A \cup B = \{1, 2, 3, 4, 6, 8\}$.

(v) $B \setminus C$ consists of the elements in B which are not in C; hence $B \setminus C = \{2, 8\}$.

5.12. In each Venn diagram below, shade: (i) $A \cup B$, (ii) $A \cap B$.

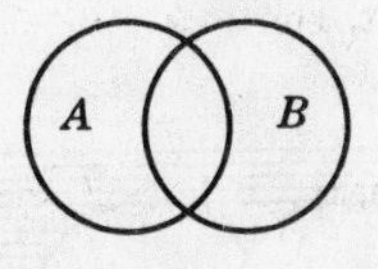

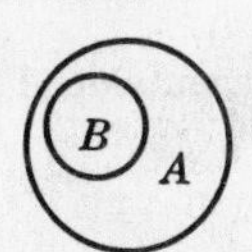

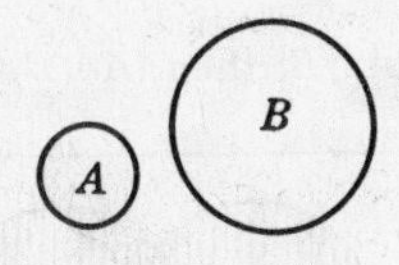

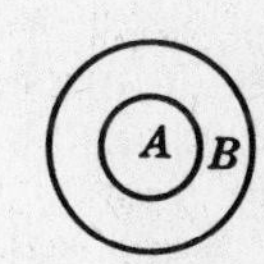

(i) $A \cup B$ consists of those elements which belong to A or B (or both); hence shade the area in A and in B as follows:

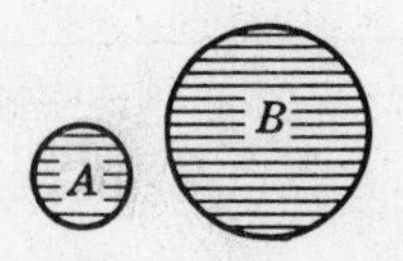

$A \cup B$ is shaded.

(ii) $A \cap B$ consists of the area that is common to both A and B. To compute $A \cap B$, first shade A with strokes slanting upward to the right (////) and then shade B with strokes slanting downward to the right (\\\\), as follows:

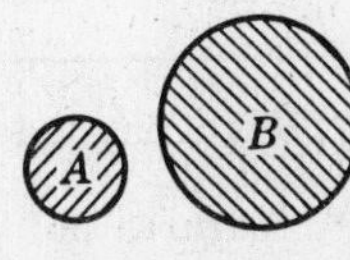

Then $A \cap B$ consists of the cross-hatched area which is shaded below:

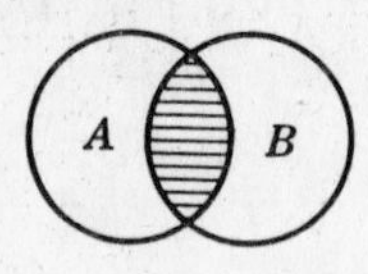

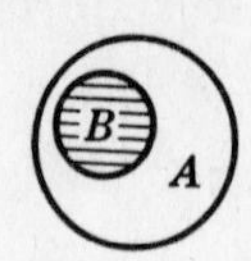

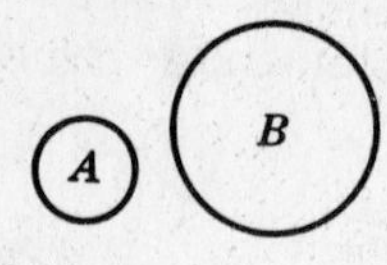

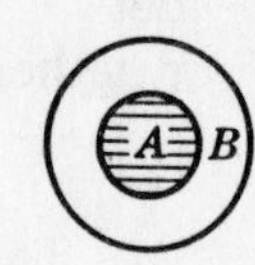

$A \cap B$ is shaded.

Observe the following:

(a) $A \cap B$ is empty if A and B are disjoint.

(b) $A \cap B = B$ if $B \subset A$.

(c) $A \cap B = A$ if $A \subset B$.

5.13. In the Venn diagram below, shade: (i) B^c, (ii) $(A \cup B)^c$, (iii) $(B \setminus A)^c$, (iv) $A^c \cap B^c$.

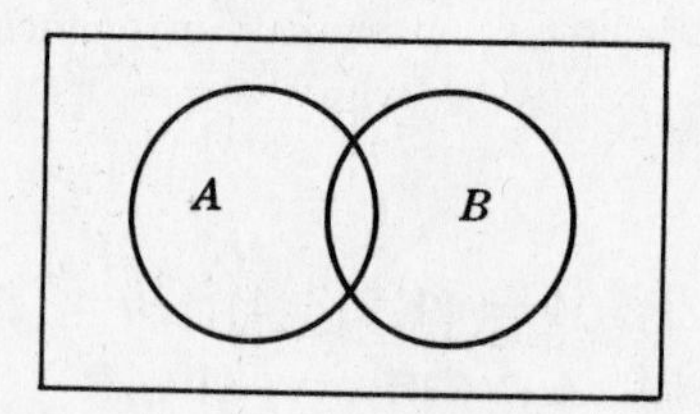

(i) B^c consists of the elements which do not belong to B; hence shade the area outside B as follows:

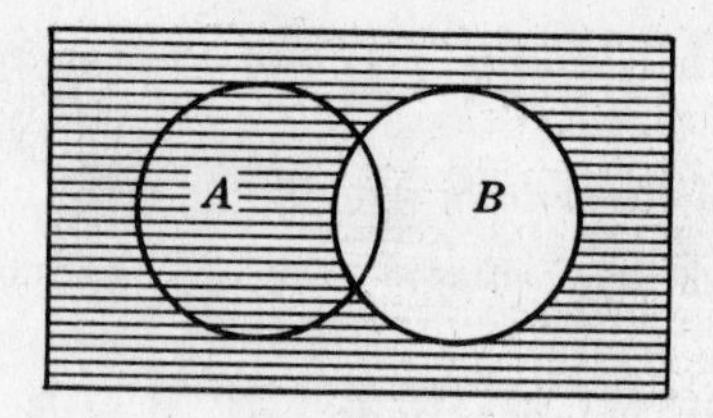

B^c is shaded.

(ii) First shade $A \cup B$; then $(A \cup B)^c$ is the area outside $A \cup B$:

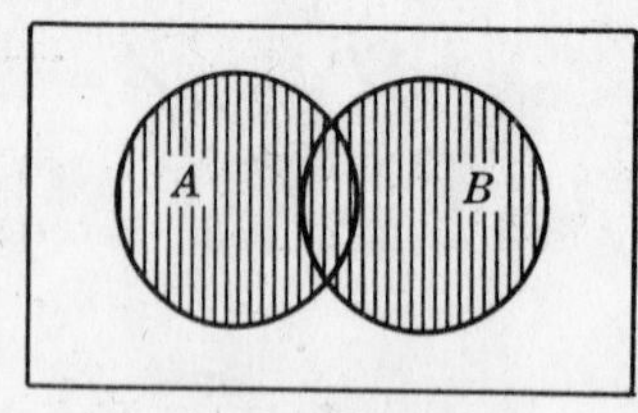

$A \cup B$ is shaded.

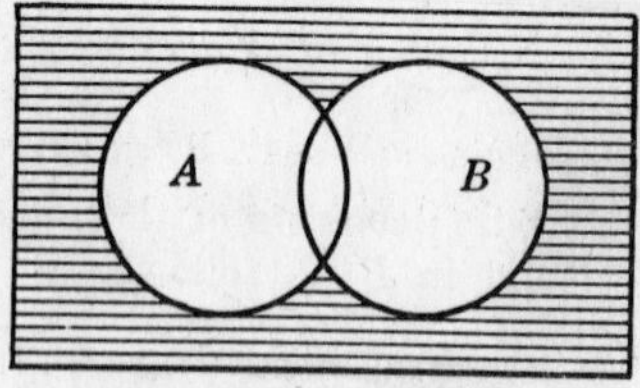

$(A \cup B)^c$ is shaded.

(iii) First shade $B \setminus A$, the area in B which does not lie in A; then $(B \setminus A)^c$ is the area outside $B \setminus A$:

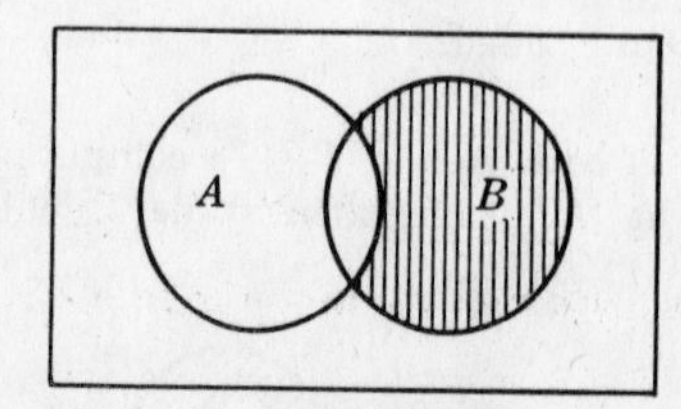

$B \setminus A$ is shaded.

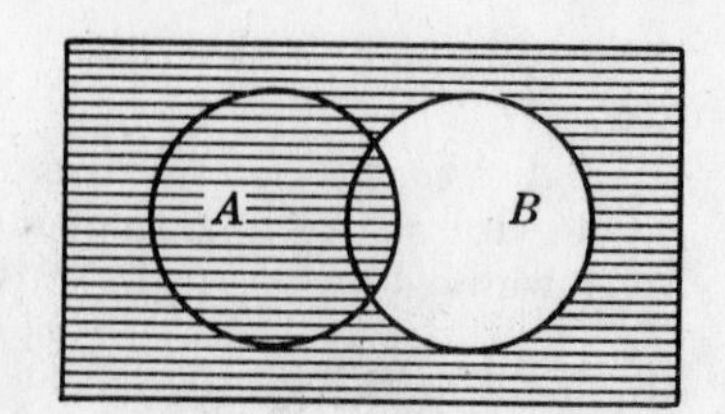

$(B \setminus A)^c$ is shaded.

(iv) First shade A^c, the area outside of A, with strokes slanting upward to the right (////), and then shade B^c with strokes slanting downward to the right (\\\\); then $A^c \cap B^c$ is the cross-hatched area:

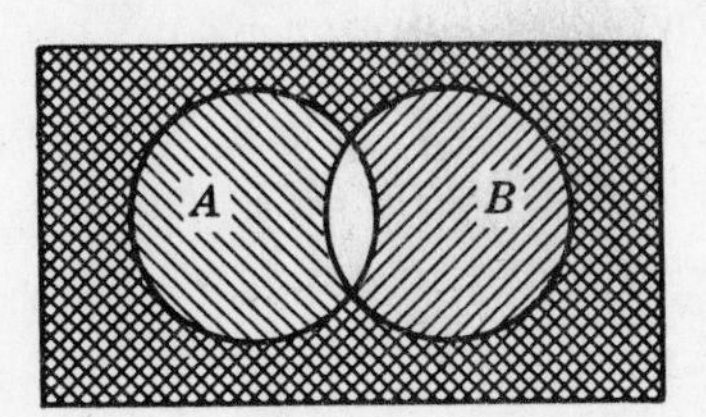

A^c and B^c are shaded.

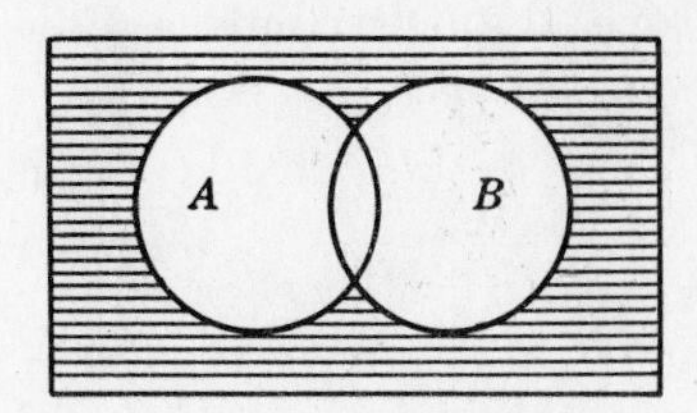

$A^c \cap B^c$ is shaded.

Observe that $(A \cup B)^c = A^c \cap B^c$, as expected by De Morgan's law.

5.14. In the Venn diagram below, shade (i) $A \cap (B \cup C)$, (ii) $(A \cap B) \cup (A \cap C)$.

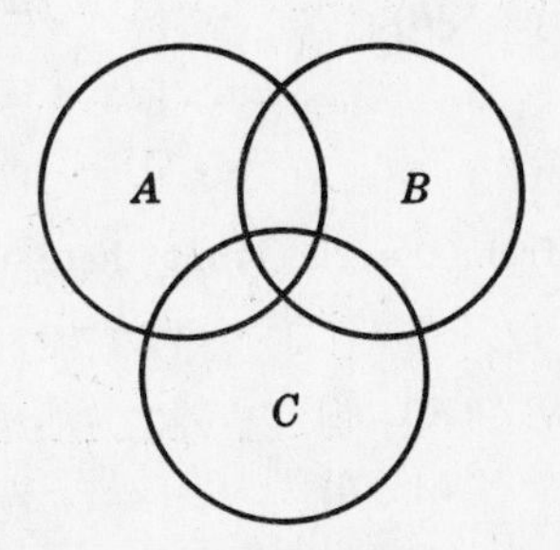

(i) First shade A with upward slanted strokes, and then shade $B \cup C$ with downward slanted strokes; now $A \cap (B \cup C)$ is the cross-hatched area:

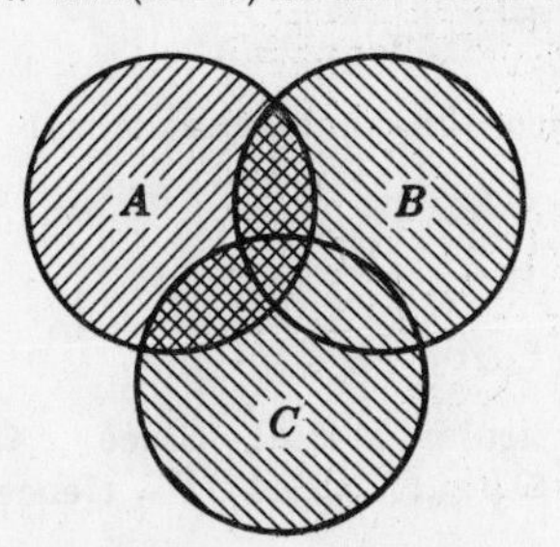

A and $B \cup C$ are shaded.

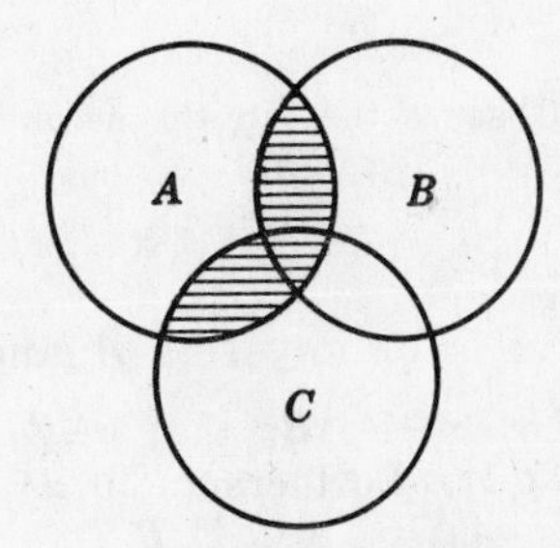

$A \cap (B \cup C)$ is shaded.

(ii) First shade $A \cap B$ with upward slanted strokes, and then shade $A \cap C$ with downward slanted strokes; now $(A \cap B) \cup (A \cap C)$ is the total area shaded:

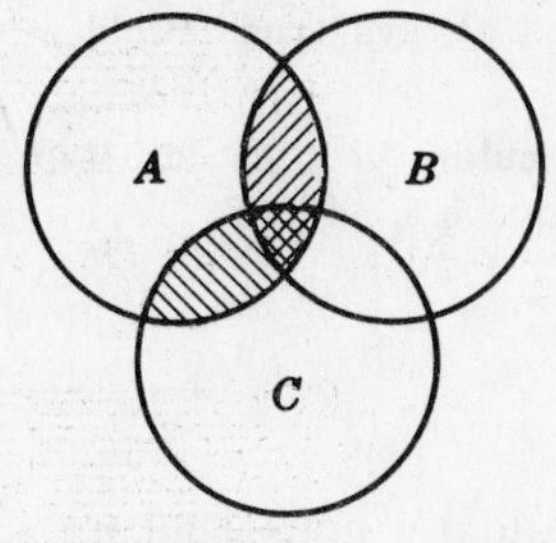

$A \cap B$ and $A \cap C$ are shaded.

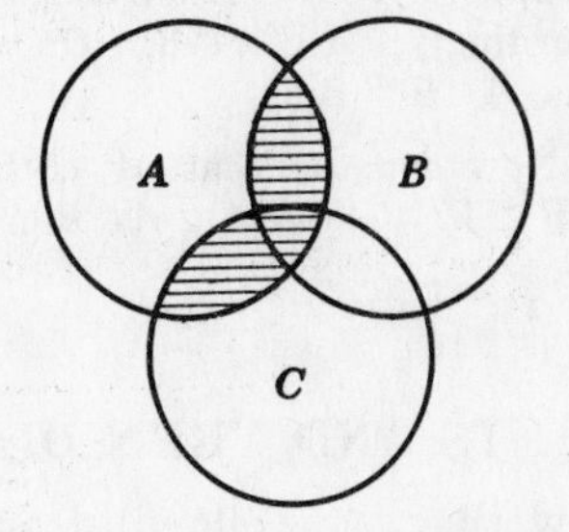

$(A \cap B) \cup (A \cap C)$ is shaded.

Notice that $A \cap (B \cup C) = (A \cap B) \cup (A \cap C)$, as expected by the distributive law.

5.15. Prove: $B \setminus A = B \cap A^c$. Thus the set operation of difference can be written in terms of the operations of intersection and complementation.

$$B \setminus A \;=\; \{x : x \in B, x \notin A\} \;=\; \{x : x \in B, x \in A^c\} \;=\; B \cap A^c$$

5.16. Prove the Distributive Law: $A \cap (B \cup C) = (A \cap B) \cup (A \cap C)$.

$$\begin{aligned} A \cap (B \cup C) &= \{x : x \in A;\ x \in B \cup C\} \\ &= \{x : x \in A;\ x \in B \text{ or } x \in C\} \\ &= \{x : x \in A,\ x \in B;\ \text{or } x \in A,\ x \in C\} \\ &= \{x : x \in A \cap B \text{ or } x \in A \cap C\} \\ &= (A \cap B) \cup (A \cap C) \end{aligned}$$

Observe that in the third step above we used the analogous logical law

$$p \wedge (q \vee r) \equiv (p \wedge q) \vee (p \wedge r)$$

5.17. Prove: $(A \setminus B) \cap B = \emptyset$.

$$\begin{aligned} (A \setminus B) \cap B &= \{x : x \in A \setminus B,\ x \in B\} \\ &= \{x : x \in A,\ x \notin B;\ x \in B\} \\ &= \emptyset \end{aligned}$$

The last step follows from the fact that there is no element x satisfying $x \in B$ and $x \notin B$.

5.18. Prove De Morgan's Law: $(A \cup B)^c = A^c \cap B^c$.

$$\begin{aligned} (A \cup B)^c &= \{x : x \notin A \cup B\} \\ &= \{x : x \notin A,\ x \notin B\} \\ &= \{x : x \in A^c,\ x \in B^c\} \\ &= A^c \cap B^c \end{aligned}$$

Observe that in the second step above we used the analogous logical law

$$\sim(p \vee q) \equiv \sim p \wedge \sim q$$

5.19. Prove: For any sets A and B, $A \cap B \subset A \subset A \cup B$.

Let $x \in A \cap B$; then $x \in A$ and $x \in B$. In particular, $x \in A$. Since $x \in A \cap B$ implies $x \in A$, $A \cap B \subset A$. Furthermore, if $x \in A$, then $x \in A$ or $x \in B$, i.e. $x \in A \cup B$. Hence $A \subset A \cup B$. In other words, $A \cap B \subset A \subset A \cup B$.

5.20. Prove Theorem 5.3(i): $A \subset B$ if and only if $A \cap B = A$.

Suppose $A \subset B$. Let $x \in A$; then by hypothesis, $x \in B$. Hence $x \in A$ and $x \in B$, i.e. $x \in A \cap B$. Accordingly, $A \subset A \cap B$. On the other hand, it is always true (Problem 5.19) that $A \cap B \subset A$. Thus $A \cap B = A$.

Now suppose that $A \cap B = A$. Then in particular, $A \subset A \cap B$. But it is always true that $A \cap B \subset B$. Thus $A \subset A \cap B \subset B$ and so, by Theorem 5.1, $A \subset B$.

ARGUMENTS AND VENN DIAGRAMS

5.21. Show that the following argument is not valid by constructing a Venn diagram in which the premises hold but the conclusion does not hold:

Some students are lazy.

All males are lazy.

....

Some students are males.

Consider the following Venn diagram:

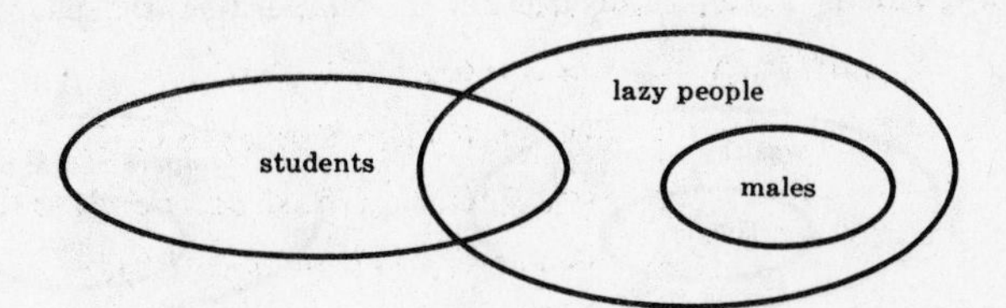

Notice that both premises hold, but the conclusion does not hold. For an argument to be valid, the conclusion must always be true whenever the premises are true. Since the diagram represents a case in which the conclusion is false, even though the premises are true, the argument is false.

It is possible to construct a Venn diagram in which the premises and conclusion hold, such as

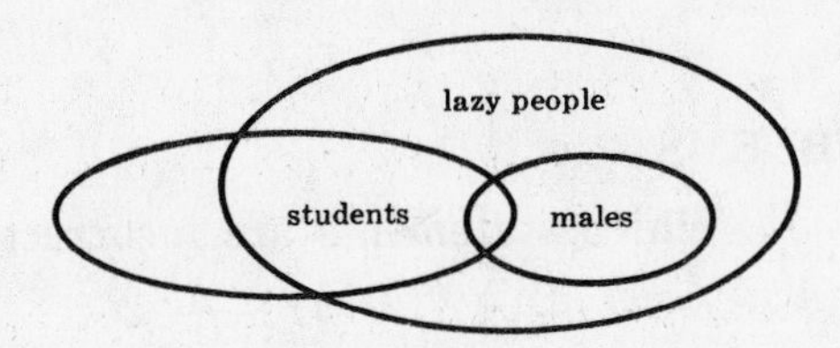

5.22. Show that the following argument is not valid:

All students are lazy.
Nobody who is wealthy is a student.
....
Lazy people are not wealthy.

Consider the following Venn diagram:

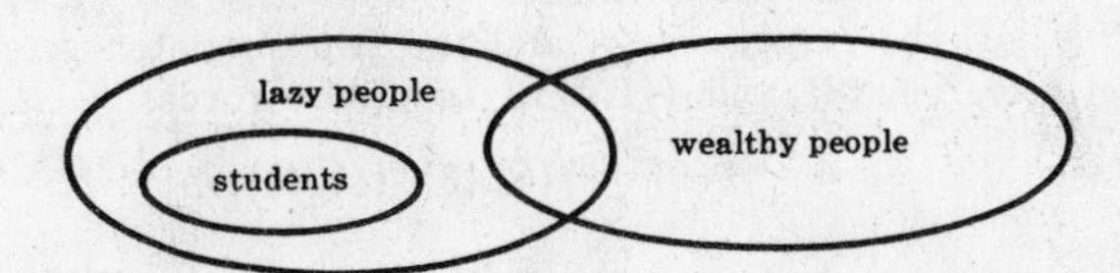

Now the premises hold in the above diagram, but the conclusion does not hold; hence the argument is not valid.

5.23. For the following set of premises, find a conclusion such that the argument is valid:

S_1: All lawyers are wealthy.
S_2: Poets are temperamental.
S_3: No temperamental person is wealthy.
....
S: ____________________________

By S_1, the set of lawyers is a subset of the set of wealthy people; and by S_3, the set of wealthy people and the set of temperamental people are disjoint. Thus

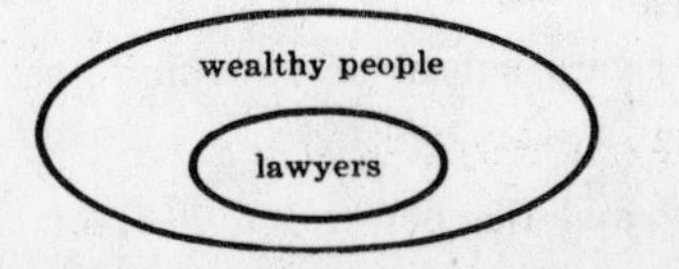

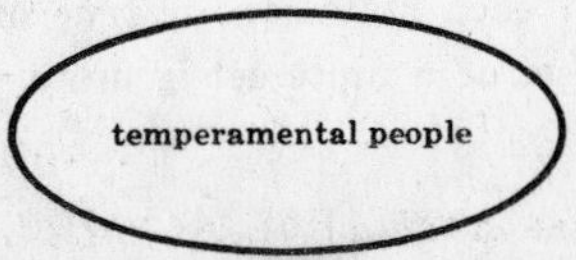

By S_2, the set of poets is a subset of the set of temperamental people; hence

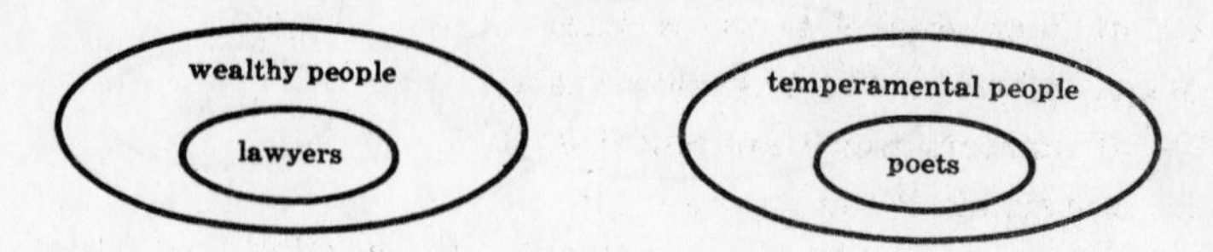

Thus the statement "No poet is a lawyer" or equivalently "No lawyer is a poet" is a valid conclusion.

The statements "No poet is wealthy" and "No lawyer is temperamental" are also valid conclusions which do not make use of all the premises.

MISCELLANEOUS PROBLEMS

5.24. Let $A = \{2, \{4,5\}, 4\}$. Which statements are incorrect and why?
(i) $\{4,5\} \subset A$, (ii) $\{4,5\} \in A$, (iii) $\{\{4,5\}\} \subset A$.

The elements of A are 2, 4 and the set $\{4,5\}$. Therefore (ii) is correct, but (i) is an incorrect statement. Furthermore, (iii) is also a correct statement since the set consisting of the single element $\{4,5\}$ which belongs to A is a subset of A.

5.25. Let $A = \{2, \{4,5\}, 4\}$. Which statements are incorrect and why?
(i) $5 \in A$, (ii) $\{5\} \in A$, (iii) $\{5\} \subset A$.

Each statement is incorrect. The elements of A are 2, 4 and the set $\{4,5\}$; hence (i) and (ii) are incorrect. There are eight subsets of A and $\{5\}$ is not one of them; so (iii) is also incorrect.

5.26. Find the power set $\mathcal{P}(S)$ of the set $S = \{3, \{1,4\}\}$.

Note first that S contains two elements, 3 and the set $\{1,4\}$. Therefore $\mathcal{P}(S)$ contains $2^2 = 4$ elements: S itself, the empty set $\emptyset$, and the two singleton sets which contain the elements 3 and $\{1,4\}$ respectively, i.e. $\{3\}$ and $\{\{1,4\}\}$. In other words,

$$\mathcal{P}(S) = \{S, \{3\}, \{\{1,4\}\}, \emptyset\}$$

Supplementary Problems

SETS, SUBSETS

5.27. Let $A = \{1,2,\ldots,8,9\}$, $B = \{2,4,6,8\}$, $C = \{1,3,5,7,9\}$, $D = \{3,4,5\}$ and $E = \{3,5\}$. Which sets can equal X if we are given the following information?
(i) X and B are disjoint. (ii) $X \subset D$ but $X \not\subset B$. (iii) $X \subset A$ but $X \not\subset C$. (iv) $X \subset C$ but $X \not\subset A$.

5.28. State whether each statement is true or false:
(i) Every subset of a finite set is finite. (ii) Every subset of an infinite set is infinite.

5.29. Find the power set $\mathcal{P}(A)$ of $A = \{1,2,3,4\}$ and the power set $\mathcal{P}(B)$ of $B = \{1, \{2,3\}, 4\}$.

5.30. State whether each set is finite or infinite:

(i) The set of lines parallel to the x axis.

(ii) The set of letters in the English alphabet.

(iii) The set of numbers which are multiples of 5.

(iv) The set of animals living on the earth.

(v) The set of numbers which are solutions of the equation $x^{27} + 26x^{18} - 17x^{11} + 7x^3 - 10 = 0$.

(vi) The set of circles through the origin $(0, 0)$.

5.31. State whether each statement is true or false:

(i) $\{1, 4, 3\} = \{3, 4, 1\}$ (iv) $\{4\} \in \{\{4\}\}$

(ii) $\{1, 3, 1, 2, 3, 2\} \subset \{1, 2, 3\}$ (v) $\{4\} \subset \{\{4\}\}$

(iii) $\{1, 2\} = \{2, 1, 1, 2, 1\}$ (vi) $\emptyset \subset \{\{4\}\}$

SET OPERATIONS

5.32. Let $U = \{a, b, c, d, e, f, g\}$, $A = \{a, b, c, d, e\}$, $B = \{a, c, e, g\}$ and $C = \{b, e, f, g\}$. Find:

(i) $A \cup C$ (iii) $C \setminus B$ (v) $C^c \cap A$ (vii) $(A \setminus B^c)^c$

(ii) $B \cap A$ (iv) $B^c \cup C$ (vi) $(A \setminus C)^c$ (viii) $(A \cap A^c)^c$

5.33. In the Venn diagrams below, shade (i) $W \setminus V$ (ii) $V^c \cup W$ (iii) $V \cap W^c$ (iv) $V^c \setminus W^c$.

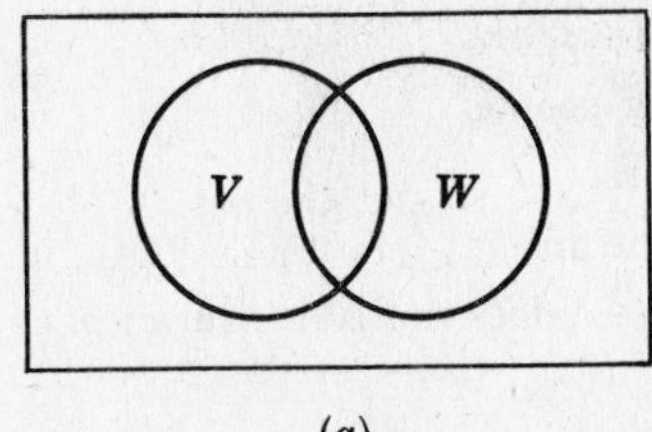

(a)

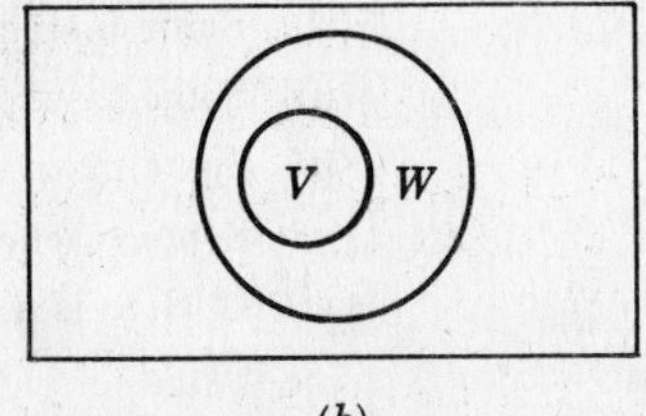

(b)

5.34. Draw a Venn diagram of three non-empty sets A, B and C so that A, B and C have the following properties:

(i) $A \subset B$, $C \subset B$, $A \cap C = \emptyset$ (iii) $A \subset C$, $A \neq C$, $B \cap C = \emptyset$

(ii) $A \subset B$, $C \not\subset B$, $A \cap C \neq \emptyset$ (iv) $A \subset (B \cap C)$, $B \subset C$, $C \neq B$, $A \neq C$

5.35. The formula $A \setminus B = A \cap B^c$ defines the difference operation in terms of the operations of intersection and complement. Find a formula that defines the union of two sets, $A \cup B$, in terms of the operations of intersection and complement.

5.36. Prove Theorem 5.3(ii): $A \subset B$ if and only if $A \cup B = B$.

5.37. Prove: If $A \cap B = \emptyset$, then $A \subset B^c$.

5.38. Prove: $A^c \setminus B^c = B \setminus A$.

5.39. Prove: $A \subset B$ implies $A \cup (B \setminus A) = B$.

5.40. (i) Prove: $A \cap (B \setminus C) = (A \cap B) \setminus (A \cap C)$.

(ii) Give an example to show that $A \cup (B \setminus C) \neq (A \cup B) \setminus (A \cup C)$.

ARGUMENTS AND VENN DIAGRAMS

5.41. Determine the validity of each argument for each proposed conclusion.

No college professor is wealthy.
Some poets are wealthy.

(i) Some poets are college professors.

(ii) Some poets are not college professors.

5.42. Determine the validity of each argument for each proposed conclusion.

All poets are interesting people.
Audrey is an interesting person.

(i) Audrey is a poet.
(ii) Audrey is not a poet.

5.43. Determine the validity of the argument for each proposed conclusion.

All poets are poor.
In order to be a teacher, one must graduate from college.
No college graduate is poor.

(i) Teachers are not poor.
(ii) Poets are not teachers.
(iii) If Marc is a college graduate, then he is not a poet.

5.44. Determine the validity of the argument for each proposed conclusion.

All mathematicians are interesting people.
Some teachers sell insurance.
Only uninteresting people become insurance salesmen.

(i) Insurance salesmen are not mathematicians.
(ii) Some interesting people are not teachers.
(iii) Some teachers are not interesting people.
(iv) Some mathematicians are teachers.
(v) Some teachers are not mathematicians.
(vi) If Eric is a mathematician, then he does not sell insurance.

Answers to Supplementary Problems

5.27. (i) C and E, (ii) D and E, (iii) A, B and D, (iv) None

5.28. (i) T, (ii) F

5.29. $\mathcal{P}(B) = \{B, \{1, \{2,3\}\}, \{1,4\}, \{\{2,3\}, 4\}, \{1\}, \{\{2,3\}\}, \{4\}, \emptyset\}$

5.30. (i) infinite, (ii) finite, (iii) infinite, (iv) finite, (v) finite, (vi) infinite

5.31. (i) T, (ii) T, (iii) T, (iv) T, (v) F, (vi) T

5.32. (i) $A \cup C = U$ (v) $C^c \cap A = \{a, c, d\} = C^c$
(ii) $B \cap A = \{a, c, e\}$ (vi) $(A \setminus C)^c = \{b, e, f, g\}$
(iii) $C \setminus B = \{b, f\}$ (vii) $(A \setminus B^c)^c = \{b, d, f, g\}$
(iv) $B^c \cup C = \{b, d, e, f, g\}$ (viii) $(A \cap A^c)^c = U$

5.33. (*a*)

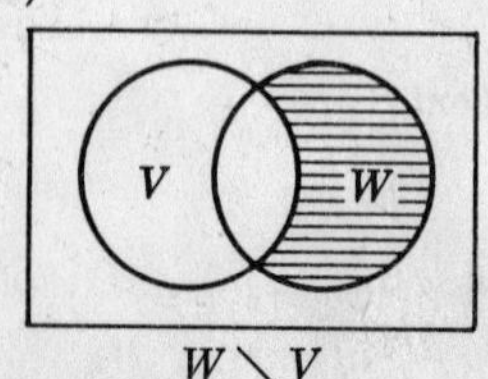

$W \setminus V$

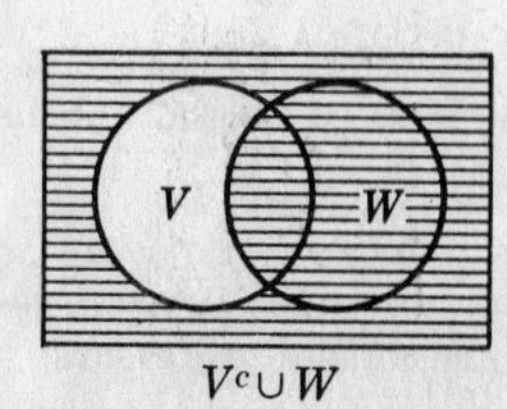

$V^c \cup W$

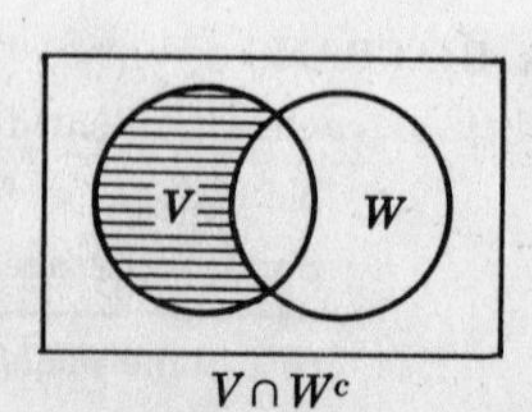

$V \cap W^c$

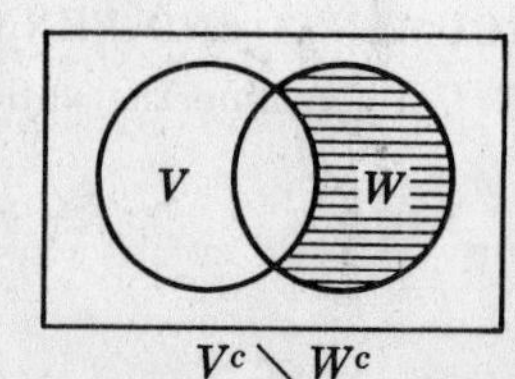

$V^c \setminus W^c$

(*b*)

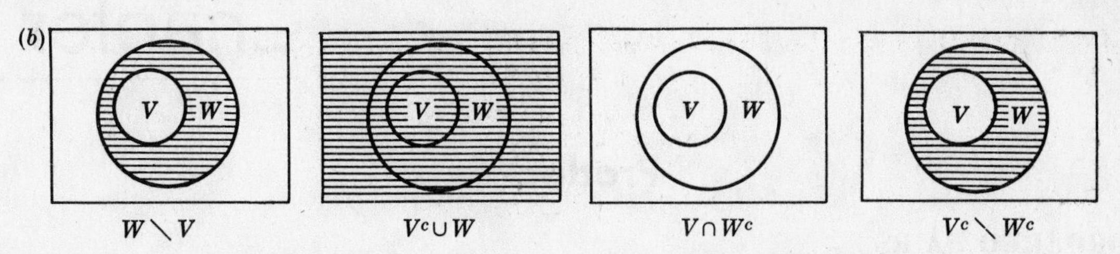

$W \setminus V$ $\quad$ $V^c \cup W$ $\quad$ $V \cap W^c$ $\quad$ $V^c \setminus W^c$

Observe that $V^c \cup W = U$ and $V \cap W^c = \emptyset$ in case (*b*) where $V \subset W$.

5.34. (i) (iii) (ii) (iv)

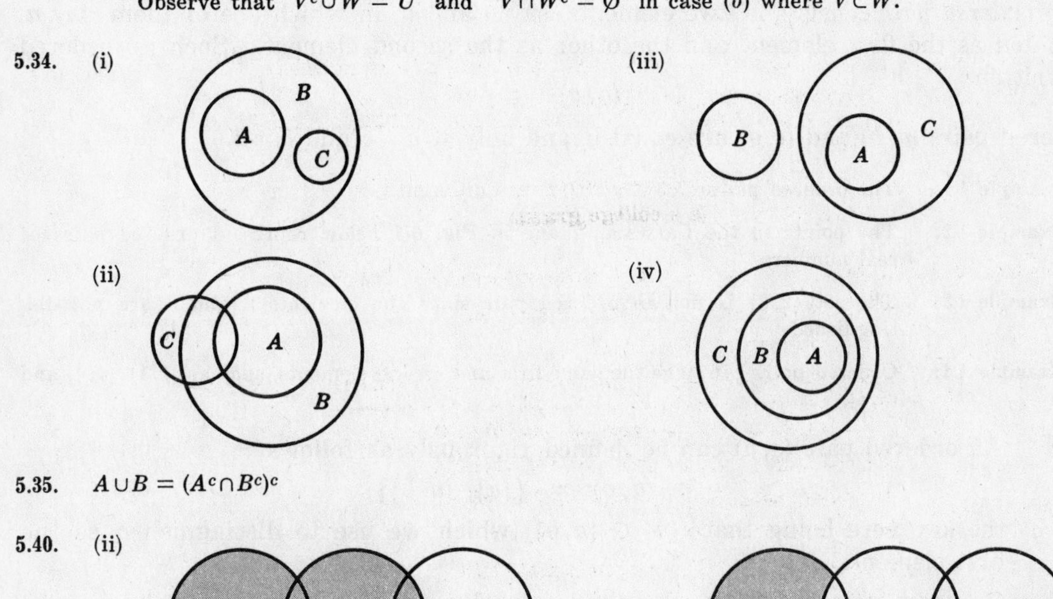

5.35. $A \cup B = (A^c \cap B^c)^c$

5.40. (ii)

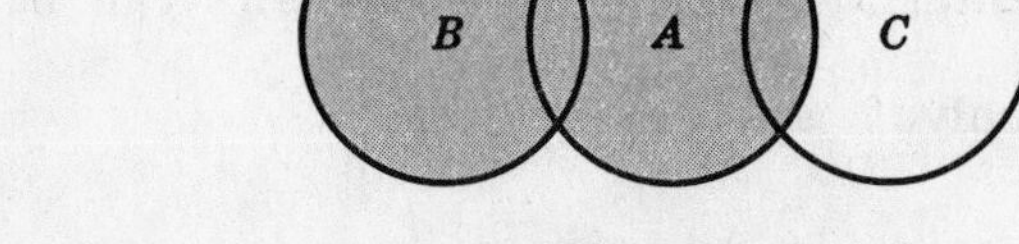

$A \cup (B \setminus C)$ is shaded.

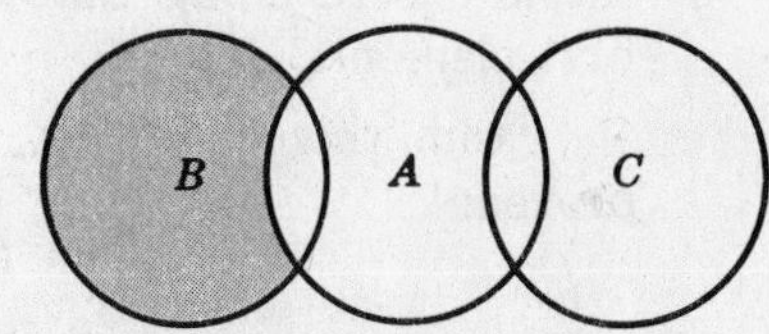

$(A \cup B) \setminus (A \cup C)$ is shaded.

5.41. (i) fallacy, (ii) valid

5.42. (i) fallacy, (ii) fallacy

5.43. (i) valid, (ii) valid, (iii) valid

5.44. (i) valid, (ii) fallacy, (iii) valid, (iv) fallacy, (v) valid, (vi) valid

The following Venn diagrams show why (ii) and (iv) are fallacies:

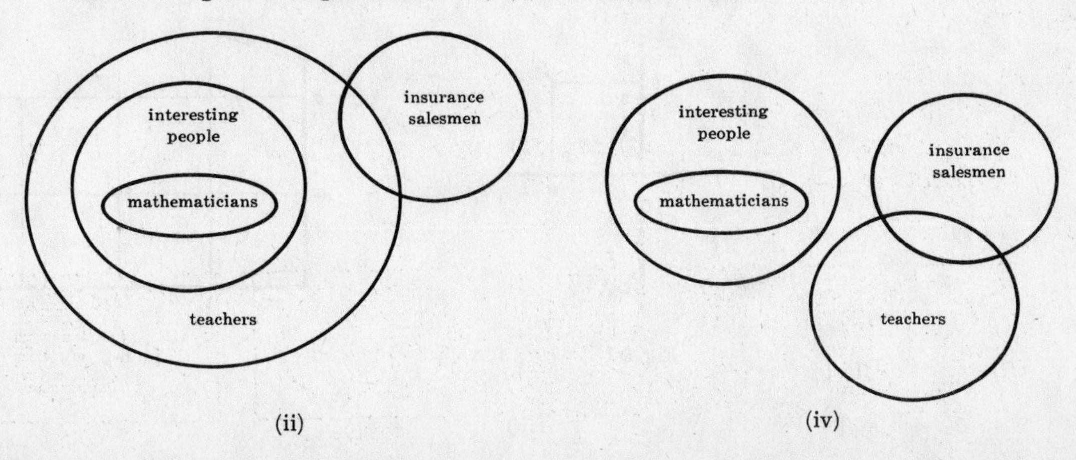

(ii) $\quad$ (iv)

Chapter 6

Product Sets

ORDERED PAIRS

An *ordered pair* consists of two elements, say a and b, in which one of them, say a, is designated as the first element and the other as the second element. Such an ordered pair is written

$$(a, b)$$

Two ordered pairs (a, b) and (c, d) are equal if and only if $a = c$ and $b = d$.

Example 1.1: The ordered pairs $(2, 3)$ and $(3, 2)$ are different.

Example 1.2: The points in the Cartesian plane in Fig. 6-1 below represent ordered pairs of real numbers.

Example 1.3: The set $\{2, 3\}$ is not an ordered pair since the elements 2 and 3 are not distinguished.

Example 1.4: Ordered pairs can have the same first and second elements, such as $(1, 1)$, $(4, 4)$ and $(5, 5)$.

Remark: An ordered pair (a, b) can be defined rigorously as follows:

$$(a, b) \;=\; \{\{a\}, \{a, b\}\}$$

the key here being that $\{a\} \subset \{a, b\}$ which we use to distinguish a as the first element.

From this definition, the fundamental property of ordered pairs can be proven:

$$(a, b) = (c, d) \text{ if and only if } a = c \text{ and } b = d$$

PRODUCT SETS

Let A and B be two sets. The *product set* (or *Cartesian product*) of A and B, written $A \times B$, consists of all ordered pairs (a, b) where $a \in A$ and $b \in B$:

$$A \times B \;=\; \{(a, b) : a \in A,\ b \in B\}$$

The product of a set with itself, say $A \times A$, is sometimes denoted by A^2.

Example 2.1: The reader is familiar with the Cartesian plane $R^2 = R \times R$ (Fig. 6-1 below). Here each point P represents an ordered pair (a, b) of real numbers and vice versa; the vertical line through P meets the x axis at a, and the horizontal line through P meets the y axis at b.

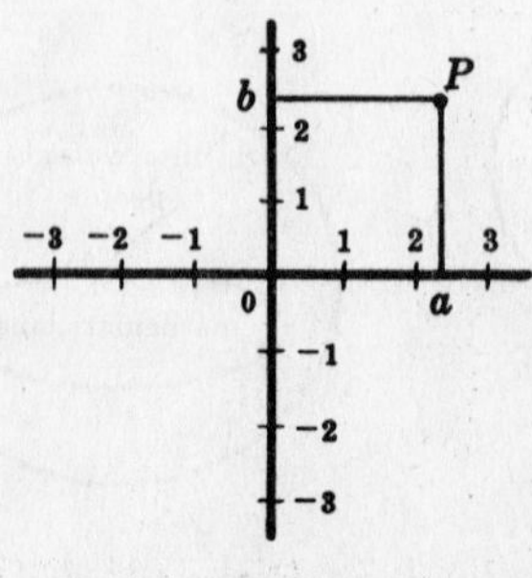

Fig. 6-1

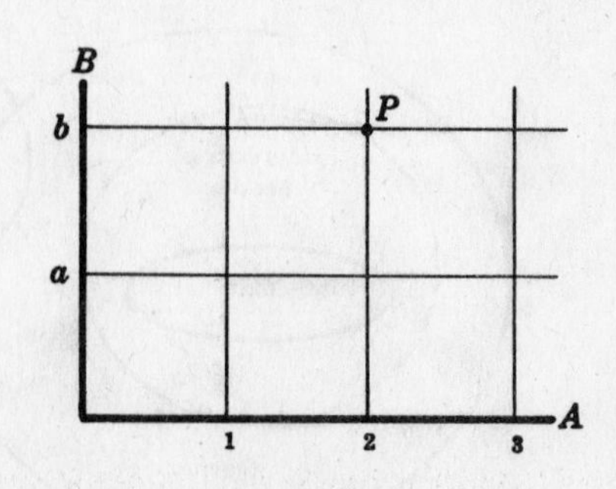

Fig. 6-2

Example 2.2: Let $A = \{1, 2, 3\}$ and $B = \{a, b\}$. Then

$$A \times B = \{(1, a), (1, b), (2, a), (2, b), (3, a), (3, b)\}$$

Since A and B do not contain many elements, it is possible to represent $A \times B$ by a coordinate diagram as shown in Fig. 6-2 above. Here the vertical lines through the points of A and the horizontal lines through the points of B meet in 6 points which represent $A \times B$ in the obvious way. The point P is the ordered pair $(2, b)$.

In general, if a finite set A has s elements and a finite set B has t elements, then $A \times B$ has s times t elements. If either A or B is empty, then $A \times B$ is empty. Lastly, if either A or B is infinite, and the other is not empty, then $A \times B$ is also infinite.

PRODUCT SETS IN GENERAL

The concept of a product set can be extended to more than two sets in a natural way. The Cartesian product of sets A, B, C, denoted by $A \times B \times C$, consists of all ordered triplets (a, b, c) where $a \in A$, $b \in B$ and $c \in C$:

$$A \times B \times C = \{(a, b, c) : a \in A,\ b \in B,\ c \in C\}$$

Analogously, the Cartesian product of n sets $A_1, A_2, \ldots, A_n$, denoted by $A_1 \times A_2 \times \cdots \times A_n$, consists of all ordered n-tuples $(a_1, a_2, \ldots, a_n)$ where $a_1 \in A_1, \ldots, a_n \in A_n$:

$$A_1 \times A_2 \times \cdots \times A_n = \{(a_1, \ldots, a_n) : a_1 \in A_1,\ \ldots,\ a_n \in A_n\}$$

Here an ordered n-tuple has the obvious intuitive meaning, that is, it consists of n elements, not necessarily distinct, in which one of them is designated as the first element, another as the second element, etc. Furthermore,

$$(a_1, \ldots, a_n) = (b_1, \ldots, b_n) \quad \text{iff} \quad a_1 = b_1,\ \ldots,\ a_n = b_n$$

Example 3.1: In three dimensional Euclidean geometry each point represents an ordered triplet: its x-component, its y-component and its z-component.

Example 3.2: Let $A = \{a, b\}$, $B = \{1, 2, 3\}$ and $C = \{x, y\}$. Then:

$$A \times B \times C = \{(a, 1, x), (a, 1, y), (a, 2, x), (a, 2, y), (a, 3, x), (a, 3, y), (b, 1, x), (b, 1, y), (b, 2, x), (b, 2, y), (b, 3, x), (b, 3, y)\}$$

TRUTH SETS OF PROPOSITIONS

Recall that any proposition P containing, say, three variables p, q and r, assigns a truth value to each of the eight cases below:

p	q	r
T	T	T
T	T	F
T	F	T
T	F	F
F	T	T
F	T	F
F	F	T
F	F	F

Let U denote the set consisting of the eight 3-tuples appearing in the table above:

$$U = \{\text{TTT, TTF, TFT, TFF, FTT, FTF, FFT, FFF}\}$$

(For notational convenience we have written, say, TTT for (T, T, T).)

Definition: The *truth set* of a proposition P, written $\mathcal{T}(P)$, consists of those elements of U for which the proposition P is true.

Example 4.1: Following is the truth table of $(p \to q) \wedge (q \to r)$:

p	q	r	$p \to q$	$q \to r$	$(p \to q) \wedge (q \to r)$
T	T	T	T	T	T
T	T	F	T	F	F
T	F	T	F	T	F
T	F	F	F	T	F
F	T	T	T	T	T
F	T	F	T	F	F
F	F	T	T	T	T
F	F	F	T	T	T

Accordingly, the truth set of $(p \to q) \wedge (q \to r)$ is

$$\mathcal{T}((p \to q) \wedge (q \to r)) = \{\text{TTT, FTT, FFT, FFF}\}$$

The next theorem shows the intimate relationship between the set operations and the logical connectives.

Theorem 6.1: Let P and Q be propositions. Then:

(i) $\mathcal{T}(P \wedge Q) = \mathcal{T}(P) \cap \mathcal{T}(Q)$

(ii) $\mathcal{T}(P \vee Q) = \mathcal{T}(P) \cup \mathcal{T}(Q)$

(iii) $\mathcal{T}(\sim P) = (\mathcal{T}(P))^c$

(iv) P logically implies Q if and only if $\mathcal{T}(P) \subset \mathcal{T}(Q)$

The proof of this theorem follows directly from the definitions of the logical connectives and the set operations.

Solved Problems

ORDERED PAIRS

6.1. Let $W = \{\text{John, Jim, Tom}\}$ and let $V = \{\text{Betty, Mary}\}$. Find $W \times V$.

$W \times V$ consists of all ordered pairs (a, b) where $a \in W$ and $b \in V$. Hence,

$$W \times V = \{(\text{John, Betty}), (\text{John, Mary}), (\text{Jim, Betty}), (\text{Jim, Mary}), (\text{Tom, Betty}), (\text{Tom, Mary})\}$$

6.2. Suppose $(x + y, 1) = (3, x - y)$. Find x and y.

If $(x + y, 1) = (3, x - y)$ then, by the fundamental property of ordered pairs,

$$x + y = 3 \quad \text{and} \quad 1 = x - y$$

The solution of these simultaneous equations is given by $x = 2$, $y = 1$.

6.3. Let $A = \{a, b, c, d, e, f\}$ and $B = \{a, e, i, o, u\}$. Determine the ordered pairs corresponding to the points P_1, P_2, P_3 and P_4 which appear in the coordinate diagram of $A \times B$ on the right.

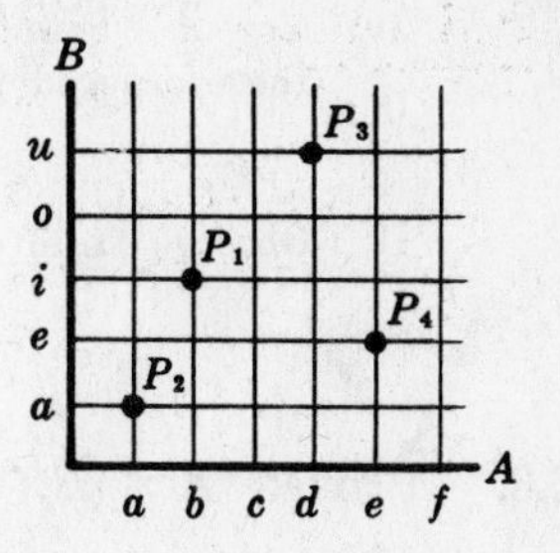

The vertical line through P_1 crosses the A axis at b and the horizontal line through P_1 crosses the B axis at i; hence P_1 corresponds to the ordered pair (b, i). Similarly, $P_2 = (a, a)$, $P_3 = (d, u)$ and $P_4 = (e, e)$.

PRODUCT SETS

6.4. Let $A = \{1,2,3\}$, $B = \{2,4\}$ and $C = \{3,4,5\}$. Find $A \times B \times C$.

A convenient method of finding $A \times B \times C$ is through the so-called "tree diagram" shown below:

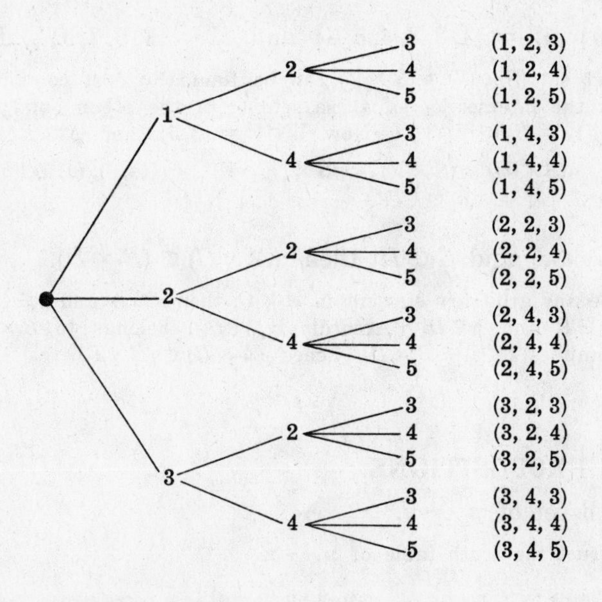

The "tree" is constructed from the left to the right. $A \times B \times C$ consists of the ordered triples listed to the right of the "tree".

6.5. Let $A = \{a, b\}$, $B = \{2,3\}$ and $C = \{3,4\}$. Find:
(i) $A \times (B \cup C)$, (ii) $(A \times B) \cup (A \times C)$, (iii) $A \times (B \cap C)$, (iv) $(A \times B) \cap (A \times C)$.

(i) First compute $B \cup C = \{2, 3, 4\}$. Then

$$A \times (B \cup C) = \{(a, 2), (a, 3), (a, 4), (b, 2), (b, 3), (b, 4)\}$$

(ii) First find $A \times B$ and $A \times C$:

$$A \times B = \{(a, 2), (a, 3), (b, 2), (b, 3)\}$$
$$A \times C = \{(a, 3), (a, 4), (b, 3), (b, 4)\}$$

Then compute the union of the two sets:

$$(A \times B) \cup (A \times C) = \{(a, 2), (a, 3), (b, 2), (b, 3), (a, 4), (b, 4)\}$$

Observe, from (i) and (ii), that

$$A \times (B \cup C) = (A \times B) \cup (A \times C)$$

(iii) First compute $B \cap C = \{3\}$. Then

$$A \times (B \cap C) = \{(a, 3), (b, 3)\}$$

(iv) Now $A \times B$ and $A \times C$ were computed above. The intersection of $A \times B$ and $A \times C$ consists of those ordered pairs which belong to both sets:

$$(A \times B) \cap (A \times C) = \{(a, 3), (b, 3)\}$$

Observe from (iii) and (iv) that

$$A \times (B \cap C) = (A \times B) \cap (A \times C)$$

6.6. Prove: $A \times (B \cap C) = (A \times B) \cap (A \times C)$.

$$\begin{aligned} A \times (B \cap C) &= \{(x, y) : x \in A,\ y \in B \cap C\} \\ &= \{(x, y) : x \in A,\ y \in B,\ y \in C\} \\ &= \{(x, y) : (x, y) \in A \times B,\ (x, y) \in A \times C\} \\ &= (A \times B) \cap (A \times C) \end{aligned}$$

6.7. Let $S = \{a, b\}$, $W = \{1, 2, 3, 4, 5, 6\}$ and $V = \{3, 5, 7, 9\}$. Find $(S \times W) \cap (S \times V)$.

The product set $(S \times W) \cap (S \times V)$ can be found by first computing $S \times W$ and $S \times V$, and then computing the intersection of these sets. On the other hand, by the preceding problem, $(S \times W) \cap (S \times V) = S \times (W \cap V)$. Now $W \cap V = \{3, 5\}$, and so

$$(S \times W) \cap (S \times V) = S \times (W \cap V) = \{(a, 3), (a, 5), (b, 3), (b, 5)\}$$

6.8. Prove: Let $A \subset B$ and $C \subset D$; then $(A \times C) \subset (B \times D)$.

Let (x, y) be any arbitrary element in $A \times C$; then $x \in A$ and $y \in C$. By hypothesis, $A \subset B$ and $C \subset D$; hence $x \in B$ and $y \in D$. Accordingly, (x, y) belongs to $B \times D$. We have shown that $(x, y) \in A \times C$ implies $(x, y) \in B \times D$; hence $(A \times C) \subset (B \times D)$.

TRUTH SETS OF PROPOSITIONS

6.9. Find the truth set of $p \wedge \sim q$.

First construct the truth table of $p \wedge \sim q$:

p	q	$\sim q$	$p \wedge \sim q$
T	T	F	F
T	F	T	T
F	T	F	F
F	F	T	F

Note that $p \wedge \sim q$ is true only in the case that p is true and q is false; hence

$$\mathcal{T}(p \wedge \sim q) = \{\text{TF}\}$$

6.10. Find the truth set of $\sim p \rightarrow q$.

First construct the truth table of $\sim p \rightarrow q$:

p	q	$\sim p$	$\sim p \rightarrow q$
T	T	F	T
T	F	F	T
F	T	T	T
F	F	T	F

Now $\sim p \rightarrow q$ is true in the first three cases; hence

$$\mathcal{T}(\sim p \rightarrow q) = \{\text{TT, TF, FT}\}$$

6.11. Find the truth set of the proposition $(p \vee q) \wedge r$.

First construct the truth table of $(p \vee q) \wedge r$:

p	q	r	$p \vee q$	$(p \vee q) \wedge r$
T	T	T	T	T
T	T	F	T	F
T	F	T	T	T
T	F	F	T	F
F	T	T	T	T
F	T	F	T	F
F	F	T	F	F
F	F	F	F	F

Now the proposition $(p \vee q) \wedge r$ is true only in the first, third and fifth cases. Hence

$$\mathcal{T}((p \vee q) \wedge r) = \{\text{TTT, TFT, FTT}\}$$

6.12. Suppose the proposition $P = P(p, q, \ldots)$ is a tautology. Determine the truth set $\mathcal{T}(P)$ of the proposition P.

If P is a tautology, then P is true for any truth values of its variables. Hence the truth set of P is the universal set: $\mathcal{T}(P) = U$.

6.13. Suppose the proposition $P = P(p, q, \ldots)$ is a contradiction. Determine the truth set $\mathcal{T}(P)$ of the proposition P.

If P is a contradiction, then there is no case in which P is true, i.e. P is false for any truth values of its variables. Hence the truth set of P is empty: $\mathcal{T}(P) = \emptyset$.

6.14. Let $P = P(p, q, \ldots)$ and $Q = Q(p, q, \ldots)$ be propositions such that $P \wedge Q$ is a contradiction. Show that the truth sets $\mathcal{T}(P)$ and $\mathcal{T}(Q)$ are disjoint.

If $P \wedge Q$ is a contradiction, then its truth set is empty: $\mathcal{T}(P \wedge Q) = \emptyset$. Hence, by Theorem 6.1(i),

$$\mathcal{T}(P) \cap \mathcal{T}(Q) = \mathcal{T}(P \wedge Q) = \emptyset$$

6.15. Suppose that the proposition $P = P(p, q, \ldots)$ logically implies the proposition $Q = Q(p, q, \ldots)$. Show that the truth sets $\mathcal{T}(P)$ and $\mathcal{T}(\sim Q)$ are disjoint.

Since P logically implies Q, $\mathcal{T}(P)$ is a subset of $\mathcal{T}(Q)$. But by Theorem 5.3,

$$\mathcal{T}(P) \subset \mathcal{T}(Q) \quad \text{is equivalent to} \quad \mathcal{T}(P) \cap (\mathcal{T}(Q))^c = \emptyset$$

Furthermore, by Theorem 6.1(iii), $(\mathcal{T}(Q))^c = \mathcal{T}(\sim Q)$. Hence

$$\mathcal{T}(P) \cap \mathcal{T}(\sim Q) = \emptyset$$

Supplementary Problems

ORDERED PAIRS, PRODUCT SETS

6.16. Suppose $(y-2, 2x+1) = (x-1, y+2)$. Find x and y.

6.17. Find the ordered pairs corresponding to the points P_1, P_2, P_3 and P_4 which appear below in the coordinate diagram of $\{1,2,3,4\} \times \{2,4,6,8\}$.

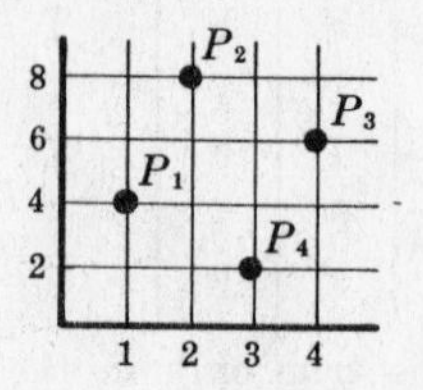

6.18. Let $W = \{\text{Mark, Eric, Paul}\}$ and let $V = \{\text{Eric, David}\}$. Find: (i) $W \times V$, (ii) $V \times W$, (iii) $V^2 = V \times V$.

6.19. Let $A = \{2,3\}$, $B = \{1,3,5\}$ and $C = \{3,4\}$. Construct the "tree diagram" of $A \times B \times C$, as in Problem 6.4, and then find $A \times B \times C$.

6.20. Let $S = \{a,b,c\}$, $T = \{b,c,d\}$ and $W = \{a,d\}$. Construct the tree diagram of $S \times T \times W$ and then find $S \times T \times W$.

6.21. Suppose that the sets V, W and Z have 3, 4 and 5 elements respectively. Determine the number of elements in (i) $V \times W \times Z$, (ii) $Z \times V \times W$, (iii) $W \times Z \times V$.

6.22. Let $A = B \cap C$. Determine if either statement is true:
(i) $A \times A = (B \times B) \cap (C \times C)$, (ii) $A \times A = (B \times C) \cap (C \times B)$.

6.23. Prove: $A \times (B \cup C) = (A \times B) \cup (A \times C)$.

TRUTH SETS OF PROPOSITIONS

6.24. Find the truth set of $p \leftrightarrow {\sim}q$.

6.25. Find the truth set of ${\sim}p \vee {\sim}q$.

6.26. Find the truth set of $(p \vee q) \rightarrow {\sim}r$.

6.27. Find the truth set of $(p \rightarrow q) \wedge (p \leftrightarrow r)$.

6.28. Let $P = P(p,q,\ldots)$ and $Q = Q(p,q,\ldots)$ be propositions such that $P \vee Q$ is a tautology. Show that the union of the truth sets $T(P)$ and $T(Q)$ is the universal set: $T(P) \cup T(Q) = U$.

6.29. Let the proposition $P = P(p,q,\ldots)$ logically imply the proposition $Q = Q(p,q,\ldots)$. Show that the union of the truth sets $T({\sim}P)$ and $T(Q)$ is the universal set: $T({\sim}P) \cup T(Q) = U$.

Answers to Supplementary Problems

6.16. $x = 2$, $y = 3$.

6.17. $P_1 = (1,4)$, $P_2 = (2,8)$, $P_3 = (4,6)$ and $P_4 = (3,2)$.

6.18. (i) $W \times V = \{$(Mark, Eric), (Mark, David), (Eric, Eric), (Eric, David), (Paul, Eric), (Paul, David)$\}$
(ii) $V \times W = \{$(Eric, Mark), (David, Mark), (Eric, Eric), (David, Eric), (Eric, Paul), (David, Paul)$\}$
(iii) $V \times V = \{$(Eric, Eric), (Eric, David), (David, Eric), (David, David)$\}$

6.19.

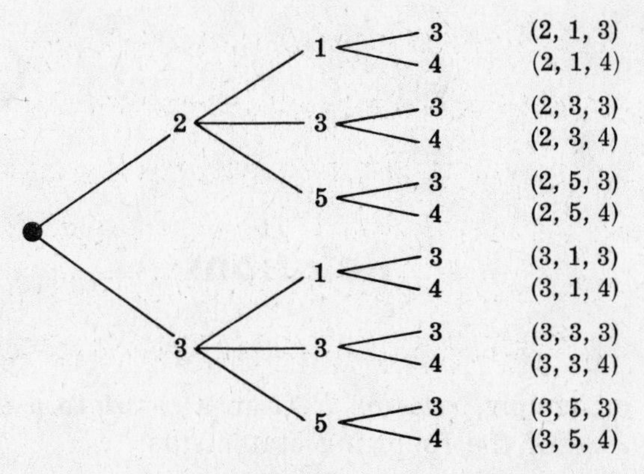

The elements of $A \times B \times C$ are the ordered triplets to the right of the tree diagram above.

6.20.

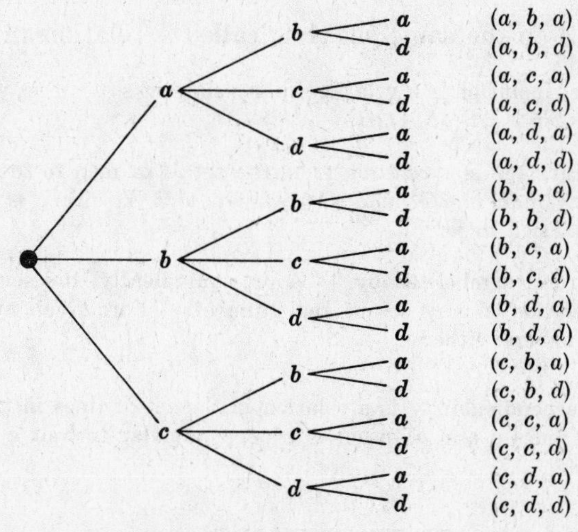

The elements of $S \times T \times W$ are the ordered triplets listed to the right of the tree diagram.

6.21. Each has 60 elements.

6.22. Both are true: $A \times A = (B \times B) \cap (C \times C) = (B \times C) \cap (C \times B)$.

6.23.

$$\begin{aligned} A \times (B \cup C) &= \{(x, y) : x \in A,\ y \in B \cup C\} \\ &= \{(x, y) : x \in A;\ y \in B \text{ or } y \in C\} \\ &= \{(x, y) : x \in A,\ y \in B; \text{ or } x \in A,\ y \in C\} \\ &= \{(x, y) : (x, y) \in A \times B; \text{ or } (x, y) \in A \times C\} \\ &= (A \times B) \cup (A \times C) \end{aligned}$$

Observe that the logical law $p \wedge (q \vee r) \equiv (p \wedge q) \vee (p \wedge r)$ was used in the third step above.

6.24. $\mathcal{T}(p \leftrightarrow \sim q) = \{\text{TF, FT}\}$

6.25. $\mathcal{T}(\sim p \vee \sim q) = \{\text{TF, FT, FF}\}$

6.26. $\mathcal{T}((p \vee q) \rightarrow \sim r) = \{\text{TTF, TFF, FTF, FFT, FFF}\}$

6.27. $\mathcal{T}((p \rightarrow q) \wedge (p \leftrightarrow r)) = \{\text{TTT, FTF, FFF}\}$

Chapter 7

Relations

RELATIONS

A *binary relation* or, simply, *relation* R from a set A to a set B assigns to each pair (a, b) in $A \times B$ exactly one of the following statements:

(i) "a is related to b", written $a R b$

(ii) "a is not related to b", written $a \not R b$

A relation from a set A to the same set A is called a relation in A.

Example 1.1: Set inclusion is a relation in any class of sets. For, given any pair of sets A and B, either $A \subset B$ or $A \not\subset B$.

Example 1.2: Marriage is a relation from the set M of men to the set W of women. For, given any man $m \in M$ and any woman $w \in W$, either m is married to w or m is not married to w.

Example 1.3: Order, symbolized by "$<$", or, equivalently, the sentence "x is less than y", is a relation in any set of real numbers. For, given any ordered pair (a, b) of real numbers, either

$$a < b \quad \text{or} \quad a \not< b$$

Example 1.4: Perpendicularity is a relation in the set of lines in the plane. For, given any pair of lines a and b, either a is perpendicular to b or a is not perpendicular to b.

RELATIONS AS SETS OF ORDERED PAIRS

Now any relation R from a set A to a set B uniquely defines a subset R^* of $A \times B$ as follows:

$$R^* = \{(a, b) : a \text{ is related to } b\} = \{(a, b) : a R b\}$$

On the other hand, any subset R^* of $A \times B$ defines a relation R from A to B as follows:

$$a R b \quad \text{iff} \quad (a, b) \in R^*$$

In view of the correspondence between relations R from A to B and subsets of $A \times B$, we redefine a relation by

Definition: A relation R from A to B is a subset of $A \times B$.

Example 2.1: Let R be the following relation from $A = \{1, 2, 3\}$ to $B = \{a, b\}$:

$$R = \{(1, a), (1, b), (3, a)\}$$

Then $1 R a$, $2 \not R b$, $3 R a$, and $3 \not R b$. The relation R is displayed on the coordinate diagram of $A \times B$ on the right.

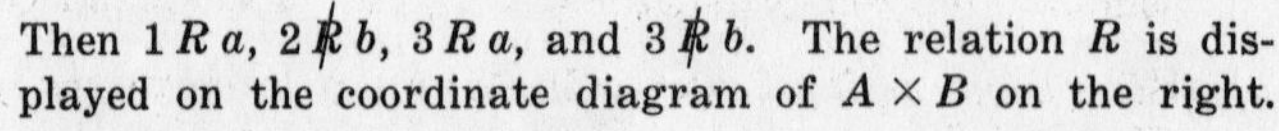

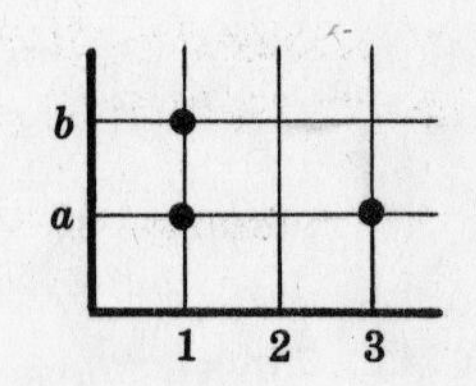

Example 2.2: Let R be the following relation in $W = \{a, b, c\}$:

$$R = \{(a, b), (a, c), (c, c), (c, b)\}$$

Then $a \not R a$, $a R b$, $c R c$ and $c \not R a$.

Example 2.3: Let A be any set. The *identity relation* in A, denoted by Δ or Δ_A, is the set of all pairs in $A \times A$ with equal coordinates:

$$\Delta_A = \{(a, a) : a \in A\}$$

The identity relation is also called the *diagonal* by virtue of its position in a coordinate diagram of $A \times A$.

INVERSE RELATION

Let R be a relation from A to B. The *inverse* of R, denoted by R^{-1}, is the relation from B to A which consists of those ordered pairs which when reversed belong to R:

$$R^{-1} = \{(b, a) : (a, b) \in R\}$$

Example 3.1: Consider the relation

$$R = \{(1,2), (1,3), (2,3)\}$$

in $A = \{1, 2, 3\}$. Then

$$R^{-1} = \{(2,1), (3,1), (3,2)\}$$

Observe that R and R^{-1} are identical, respectively, to the relations $<$ and $>$ in A, i.e.,

$$(a, b) \in R \text{ iff } a < b \quad \text{and} \quad (a, b) \in R^{-1} \text{ iff } a > b$$

Example 3.2: The inverses of the relations defined by

"x is the husband of y" and "x is taller than y"

are respectively

"x is the wife of y" and "x is shorter than y"

EQUIVALENCE RELATIONS

A relation R in a set A is called *reflexive* if $a R a$, i.e. $(a, a) \in R$, for every $a \in A$.

Example 4.1: (i) Let R be the relation of similarity in the set of triangles in the plane. Then R is reflexive since every triangle is similar to itself.

(ii) Let R be the relation $<$ in any set of real numbers, i.e. $(a, b) \in R$ iff $a < b$. Then R is not reflexive since $a \not< a$ for any real number a.

A relation R in a set A is called *symmetric* if whenever $a R b$ then $b R a$, i.e. if $(a, b) \in R$ implies $(b, a) \in R$.

Example 4.2: (i) The relation R of similarity of triangles is symmetric. For if triangle α is similar to triangle β, then β is similar to α.

(ii) The relation $R = \{(1,3), (2,3), (2,2), (3,1)\}$ in $A = \{1, 2, 3\}$ is not symmetric since $(2,3) \in R$ but $(3,2) \notin R$.

A relation R in a set A is called *transitive* if whenever $a R b$ and $b R c$ then $a R c$, i.e. if $(a, b) \in R$ and $(b, c) \in R$ implies $(a, c) \in R$.

Example 4.3: (i) The relation R of similarity of triangles is transitive since if triangle α is similar to β and β is similar to γ, then α is similar to γ.

(ii) The relation R of perpendicularity of lines in the plane is not transitive. Since if line a is perpendicular to line b and line b is perpendicular to line c, then a is parallel and not perpendicular to c.

A relation R is an *equivalence relation* if R is (i) reflexive, (ii) symmetric, and (iii) transitive.

Example 4.4: By the three preceding examples, the relation R of similarity of triangles is an equivalence relation since it is reflexive, symmetric and transitive.

PARTITIONS

A *partition* of a set X is a subdivision of X into subsets which are disjoint and whose union is X, i.e. such that each $a \in X$ belongs to one and only one of the subsets. The subsets in a partition are called *cells*.

Thus the collection $\{A_1, A_2, \ldots, A_m\}$ of subsets of X is a partition of X iff:

(i) $X = A_1 \cup A_2 \cup \cdots \cup A_m$; (ii) for any A_i, A_j, either $A_i = A_j$ or $A_i \cap A_j = \emptyset$

Example 5.1: Consider the following classes of subsets of $X = \{1, 2, \ldots, 8, 9\}$:

(i) $[\{1,3,5\}, \{2,6\}, \{4,8,9\}]$

(ii) $[\{1,3,5\}, \{2,4,6,8\}, \{5,7,9\}]$

(iii) $[\{1,3,5\}, \{2,4,6,8\}, \{7,9\}]$

Then (i) is not a partition of X since $7 \in X$ but 7 does not belong to any of the cells. Furthermore, (ii) is not a partition of X since $5 \in X$ and 5 belongs to both $\{1,3,5\}$ and $\{5,7,9\}$. On the other hand, (iii) is a partition of X since each element of X belongs to exactly one cell.

EQUIVALENCE RELATIONS AND PARTITIONS

Let R be an equivalence relation in a set A and, for each $a \in A$, let $[a]$, called the *equivalence class* of A, be the set of elements to which a is related:

$$[a] = \{x : (a, x) \in R\}$$

The collection of equivalence classes of A, denoted by A / R, is called the *quotient* of A by R:

$$A / R = \{[a] : a \in A\}$$

The fundamental property of a quotient set is contained in the following theorem.

Theorem 7.1: Let R be an equivalence relation in a set A. Then the quotient set A / R is a partition of A. Specifically,

(i) $a \in [a]$, for every $a \in A$;

(ii) $[a] = [b]$ if and only if $(a, b) \in R$;

(iii) if $[a] \neq [b]$, then $[a]$ and $[b]$ are disjoint.

Example 6.1: Let R_5 be the relation in Z, the set of integers, defined by

$$x \equiv y \pmod 5$$

which reads "x is congruent to y modulo 5" and which means that the difference $x - y$ is divisible by 5. Then R_5 is an equivalence relation in Z. There are exactly five distinct equivalence classes in Z / R_5:

$$A_0 = \{\ldots, -10, -5, 0, 5, 10, \ldots\}$$

$$A_1 = \{\ldots, -9, -4, 1, 6, 11, \ldots\}$$

$$A_2 = \{\ldots, -8, -3, 2, 7, 12, \ldots\}$$

$$A_3 = \{\ldots, -7, -2, 3, 8, 13, \ldots\}$$

$$A_4 = \{\ldots, -6, -1, 4, 9, 14, \ldots\}$$

Observe that each integer x which is uniquely expressible in the form $x = 5q + r$ where $0 \leq r < 5$ is a member of the equivalence A_r where r is the remainder. Note that the equivalence classes are pairwise disjoint and that

$$Z = A_0 \cup A_1 \cup A_2 \cup A_3 \cup A_4$$

Solved Problems

RELATIONS

7.1. Let R be the relation $<$ from $A = \{1,2,3,4\}$ to $B = \{1,3,5\}$, i.e. defined by "x is less than y".

(i) Write R as a set of ordered pairs.

(ii) Plot R on a coordinate diagram of $A \times B$.

(iii) Find the inverse relation R^{-1}.

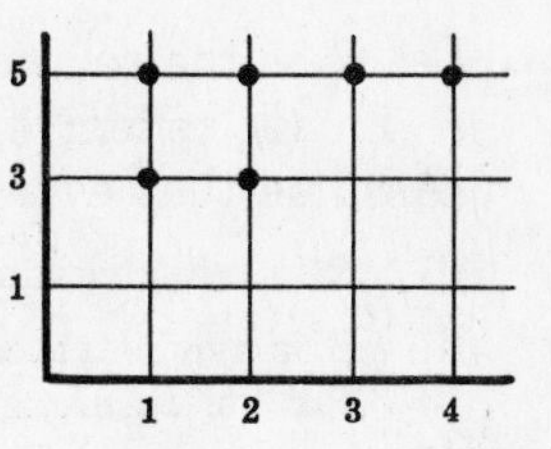

(i) R consists of those ordered pairs $(a,b) \in A \times B$ for which $a < b$; hence

$$R = \{(1,3), (1,5), (2,3), (2,5), (3,5), (4,5)\}$$

(ii) R is sketched on the coordinate diagram of $A \times B$ as shown in the figure.

(iii) The inverse of R consists of the same pairs as are in R but in the reverse order; hence

$$R^{-1} = \{(3,1), (5,1), (3,2), (5,2), (5,3), (5,4)\}$$

Observe that R^{-1} is the relation $>$, i.e. defined by "x is greater than y".

7.2. Let R be the relation from $E = \{2,3,4,5\}$ to $F = \{3,6,7,10\}$ defined by "x divides y".

(i) Write R as a set of ordered pairs.

(ii) Plot R on a coordinate diagram of $E \times F$.

(iii) Find the inverse relation R^{-1}.

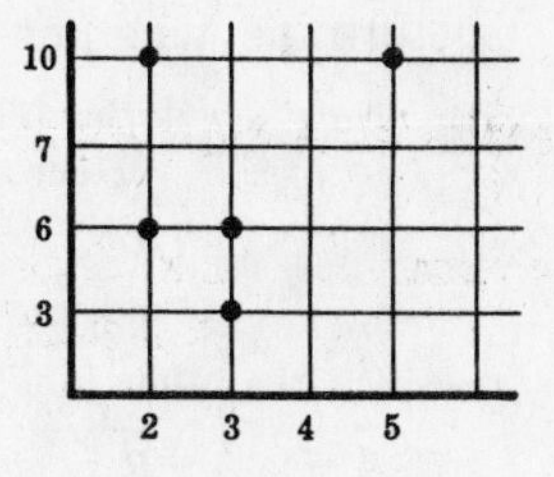

(i) Choose from the sixteen ordered pairs in $E \times F$ those in which the first element divides the second; then

$$R = \{(2,6), (2,10), (3,3), (3,6), (5,10)\}$$

(ii) R is sketched on the coordinate diagram of $E \times F$ as shown in the figure.

(iii) To find the inverse of R, write the elements of R but in reverse order:

$$R^{-1} = \{(6,2), (10,2), (3,3), (6,3), (10,5)\}$$

7.3. Let $M = \{a,b,c,d\}$ and let R be the relation in M consisting of those points which are displayed on the coordinate diagram of $M \times M$ on the right.

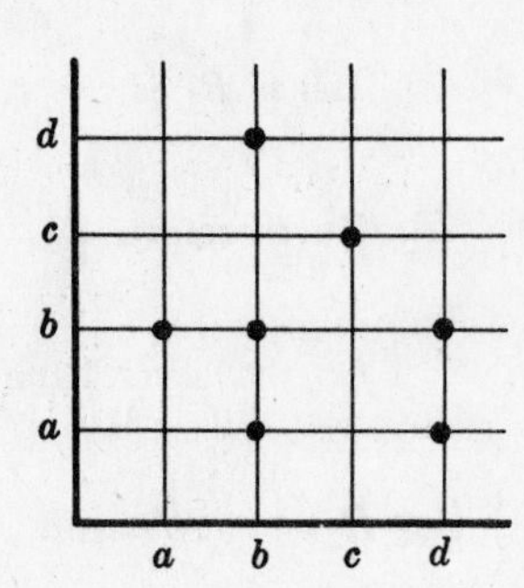

(i) Find all the elements in M which are related to b, that is, $\{x : (x,b) \in R\}$.

(ii) Find all those elements in M to which d is related, that is, $\{x : (d,x) \in R\}$.

(iii) Find the inverse relation R^{-1}.

(i) The horizontal line through b contains all points of R in which b appears as the second element: (a,b), (b,b) and (d,b). Hence the desired set is $\{a,b,d\}$.

(ii) The vertical line through d contains all the points of R in which d appears as the first element: (d,a) and (d,b). Hence $\{a,b\}$ is the desired set.

(iii) First write R as a set of ordered pairs, and then write the pairs in reverse order:

$$R = \{(a,b), (b,a), (b,b), (b,d), (c,c), (d,a), (d,b)\}$$
$$R^{-1} = \{(b,a), (a,b), (b,b), (d,b), (c,c), (a,d), (b,d)\}$$

EQUIVALENCE RELATIONS

7.4. Let R be the relation $\leq$ in $N = \{1, 2, 3, \ldots\}$, i.e. $(a, b) \in R$ iff $a \leq b$. Determine whether R is (i) reflexive, (ii) symmetric, (iii) transitive, (iv) an equivalence relation.

(i) R is reflexive since $a \leq a$ for every $a \in N$.

(ii) R is not symmetric since, for example, $3 \leq 5$ but $5 \not\leq 3$, i.e. $(3, 5) \in R$ but $(5, 3) \notin R$.

(iii) R is transitive since $a \leq b$ and $b \leq c$ implies $a \leq c$.

(iv) R is not an equivalence relation since it is not symmetric.

7.5. Let R be the relation $\|$ (parallel) in the set of lines in the plane. Determine whether R is (i) reflexive, (ii) symmetric, (iii) transitive, (iv) an equivalence relation. (Assume that every line *is* parallel to itself.)

(i) R is reflexive since, by assumption, $\alpha \| \alpha$ for every line α.

(ii) R is symmetric since if $\alpha \| \beta$ then $\beta \| \alpha$, i.e. if the line α is parallel to the line β then β is parallel to α.

(iii) R is transitive since if $\alpha \| \beta$ and $\beta \| \gamma$ then $\alpha \| \gamma$.

(iv) R is an equivalence relation since it is reflexive, symmetric and transitive.

7.6. Let $W = \{1, 2, 3, 4\}$. Consider the following relations in W:

$$R_1 = \{(1,2), (4,3), (2,2), (2,1), (3,1)\}$$
$$R_2 = \{(2,2), (2,3), (3,2)\}$$
$$R_3 = \{(1,3)\}$$

Determine whether each relation is (i) symmetric, (ii) transitive.

(i) Now a relation R is not symmetric if there exists an ordered pair $(a, b) \in R$ such that $(b, a) \notin R$. Hence:

$$R_1 \text{ is not symmetric since } (4,3) \in R_1 \text{ but } (3,4) \notin R_1,$$
$$R_3 \text{ is not symmetric since } (1,3) \in R_3 \text{ but } (3,1) \notin R_3.$$

On the other hand, R_2 is symmetric.

(ii) A relation R is *not* transitive if there exist elements a, b and c, not necessarily distinct, such that

$$(a, b) \in R \text{ and } (b, c) \in R \text{ but } (a, c) \notin R$$

Hence R_1 is not transitive since

$$(4,3) \in R_1 \text{ and } (3,1) \in R_1 \text{ but } (4,1) \notin R_1$$

Furthermore, R_2 is not transitive since

$$(3,2) \in R_2 \text{ and } (2,3) \in R_2 \text{ but } (3,3) \notin R_2$$

On the other hand, R_3 is transitive.

7.7. Let R be a relation in A. Show that:
(i) R is reflexive iff $\Delta \subset R$; (ii) R is symmetric iff $R = R^{-1}$.

(i) Recall that $\Delta = \{(a, a) : a \in A\}$. Thus R is reflexive iff $(a, a) \in R$ for every $a \in A$ iff $\Delta \subset R$.

(ii) Suppose R is symmetric. Let $(a, b) \in R$, then $(b, a) \in R$ by symmetry. Hence $(a, b) \in R^{-1}$, and so $R \subset R^{-1}$. On the other hand, let $(a, b) \in R^{-1}$; then $(b, a) \in R$ and, by symmetry, $(a, b) \in R$. Thus $R^{-1} \subset R$, and so $R = R^{-1}$.

Now suppose $R = R^{-1}$. Let $(a, b) \in R$; then $(b, a) \in R^{-1} = R$. Accordingly R is symmetric.

7.8. Consider the relation $R = \{(1,1),(2,3),(3,2)\}$ in $X = \{1,2,3\}$. Determine whether or not R is (i) reflexive, (ii) symmetric, (iii) transitive.

(i) R is not reflexive since $2 \in X$ but $(2,2) \notin R$.

(ii) R is symmetric since $R^{-1} = \{(1,1),(3,2),(2,3)\} = R$.

(iii) R is not transitive since $(3,2) \in R$ and $(2,3) \in R$ but $(3,3) \notin R$.

7.9. Let $N = \{1,2,3,\ldots\}$, and let R be the relation $\cong$ in $N \times N$ defined by

$$(a,b) \cong (c,d) \quad \text{iff} \quad ad = bc$$

Prove that R is an equivalence relation.

For every $(a,b) \in N \times N$, $(a,b) \cong (a,b)$ since $ab = ba$; hence R is reflexive.

Suppose $(a,b) \cong (c,d)$. Then $ad = bc$, which implies that $cb = da$. Hence $(c,d) \cong (a,b)$ and so R is symmetric.

Now suppose $(a,b) \cong (c,d)$ and $(c,d) \cong (e,f)$. Then $ad = bc$ and $cf = de$. Thus

$$(ad)(cf) = (bc)(de)$$

and, by cancelling from both sides, $af = be$. Accordingly, $(a,b) \cong (e,f)$ and so R is transitive.

Since R is reflexive, symmetric and transitive, R is an equivalence relation.

Observe that if the ordered pair (a,b) is written as a fraction $\frac{a}{b}$, then the above relation R is, in fact, the usual definition of equality between fractions, i.e. $\frac{a}{b} = \frac{c}{d}$ iff $ad = bc$.

7.10. Prove Theorem 7.1: Let R be an equivalence relation in a set A. Then the quotient set A/R is a partition of A. Specifically,

(i) $a \in [a]$, for every $a \in A$;

(ii) $[a] = [b]$ if and only if $(a,b) \in R$;

(iii) if $[a] \neq [b]$, then $[a]$ and $[b]$ are disjoint.

Proof of (i). Since R is reflexive, $(a,a) \in R$ for every $a \in A$ and therefore $a \in [a]$.

Proof of (ii). Suppose $(a,b) \in R$. We want to show that $[a] = [b]$. Let $x \in [b]$; then $(b,x) \in R$. But by hypothesis $(a,b) \in R$ and so, by transitivity, $(a,x) \in R$. Accordingly $x \in [a]$. Thus $[b] \subset [a]$. To prove that $[a] \subset [b]$, we observe that $(a,b) \in R$ implies, by symmetry, that $(b,a) \in R$. Then by a similar argument, we obtain $[a] \subset [b]$. Consequently, $[a] = [b]$.

On the other hand, if $[a] = [b]$, then, by (i), $b \in [b] = [a]$; hence $(a,b) \in R$.

Proof of (iii). We prove the equivalent contrapositive statement:

$$\text{if} \quad [a] \cap [b] \neq \emptyset \quad \text{then} \quad [a] = [b]$$

If $[a] \cap [b] \neq \emptyset$, then there exists an element $x \in A$ with $x \in [a] \cap [b]$. Hence $(a,x) \in R$ and $(b,x) \in R$. By symmetry, $(x,b) \in R$ and by transitivity, $(a,b) \in R$. Consequently by (ii), $[a] = [b]$.

PARTITIONS

7.11. Let $X = \{a,b,c,d,e,f,g\}$, and let:

(i) $A_1 = \{a,c,e\}$, $A_2 = \{b\}$, $A_3 = \{d,g\}$;

(ii) $B_1 = \{a,e,g\}$, $B_2 = \{c,d\}$, $B_3 = \{b,e,f\}$;

(iii) $C_1 = \{a,b,e,g\}$, $C_2 = \{c\}$, $C_3 = \{d,f\}$;

(iv) $D_1 = \{a,b,c,d,e,f,g\}$.

Which of $\{A_1,A_2,A_3\}$, $\{B_1,B_2,B_3\}$, $\{C_1,C_2,C_3\}$, $\{D_1\}$ are partitions of X?

(i) $\{A_1,A_2,A_3\}$ is not a partition of X since $f \in X$ but f does not belong to either A_1, A_2, or A_3.

(ii) $\{B_1,B_2,B_3\}$ is not a partition of X since $e \in X$ belongs to both B_1 and B_3.

(iii) $\{C_1,C_2,C_3\}$ is a partition of X since each element in X belongs to exactly one cell, i.e. $X = C_1 \cup C_2 \cup C_3$ and the sets are pairwise disjoint.

(iv) $\{D_1\}$ is a partition of X.

7.12. Find all the partitions of $X = \{a, b, c, d\}$.

Note first that each partition of X contains either 1, 2, 3, or 4 distinct sets. The partitions are as follows:

(1) $[\{a, b, c, d\}]$

(2) $[\{a\}, \{b, c, d\}]$, $[\{b\}, \{a, c, d\}]$, $[\{c\}, \{a, b, d\}]$, $[\{d\}, \{a, b, c\}]$,
$[\{a, b\}, \{c, d\}]$, $[\{a, c\}, \{b, d\}]$, $[\{a, d\}, \{b, c\}]$

(3) $[\{a\}, \{b\}, \{c, d\}]$, $[\{a\}, \{c\}, \{b, d\}]$, $[\{a\}, \{d\}, \{b, c\}]$,
$[\{b\}, \{c\}, \{a, d\}]$, $[\{b\}, \{d\}, \{a, c\}]$, $[\{c\}, \{d\}, \{a, b\}]$

(4) $[\{a\}, \{b\}, \{c\}, \{d\}]$

There are fifteen different partitions of X.

Supplementary Problems

RELATIONS

7.13. Let R be the relation in $A = \{2, 3, 4, 5\}$ defined by "x and y are relatively prime", i.e. "the only common divisor of x and y is 1".

(i) Write R as a set of ordered pairs. (ii) Plot R on a coordinate diagram of $A \times A$. (iii) Find R^{-1}.

7.14. Let $N = \{1, 2, 3, \ldots\}$ and let R be the relation in N defined by $x + 2y = 8$, i.e.,

$$R = \{(x, y) : x, y \in N,\ x + 2y = 8\}$$

(i) Write R as a set of ordered pairs. (ii) Find R^{-1}.

7.15. Let $C = \{1, 2, 3, 4, 5\}$ and let R be the relation in C consisting of the points displayed in the following coordinate diagram of $C \times C$:

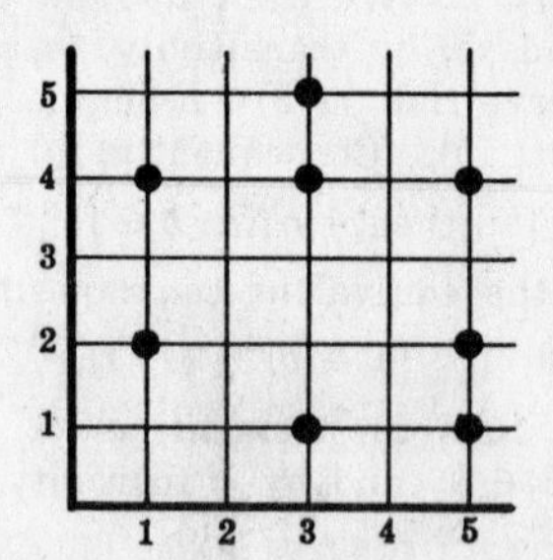

(i) State whether each is true or false: (a) $1\,R\,4$, (b) $2\,R\,5$, (c) $3\,\not R\,1$, (d) $5\,\not R\,3$.

(ii) Find the elements of each of the following subsets of C:
(a) $\{x : 3\,R\,x\}$, (b) $\{x : (4, x) \in R\}$, (c) $\{x : (x, 2) \notin R\}$, (d) $\{x : x\,R\,5\}$.

7.16. Consider the relations $<$ and $\leq$ in $N = \{1, 2, 3, \ldots\}$. Show that $< \cup \Delta = \leq$ where Δ is the diagonal relation.

EQUIVALENCE RELATIONS

7.17. Let $W = \{1, 2, 3, 4\}$. Consider the following relations in W:

$R_1 = \{(1, 1), (1, 2)\}$ $\quad$ $R_4 = \{(1, 1), (2, 2), (3, 3)\}$

$R_2 = \{(1, 1), (2, 3), (4, 1)\}$ $\quad$ $R_5 = W \times W$

$R_3 = \{(1, 3), (2, 4)\}$ $\quad$ $R_6 = \emptyset$

Determine whether or not each relation is reflexive.

7.18. Determine whether or not each relation in Problem 7.17 is symmetric.

7.19. Determine whether or not each relation in Problem 7.17 is transitive.

7.20. Let R be the relation $\perp$ of perpendicularity in the set of lines in the plane. Determine whether R is (i) reflexive, (ii) symmetric, (iii) transitive, (iv) an equivalence relation.

7.21. Let $N = \{1, 2, 3, \ldots\}$ and let $\cong$ be the relation in $N \times N$ defined by

$$(a, b) \cong (c, d) \quad \text{iff} \quad a + d = b + c$$

(i) Prove $\cong$ is an equivalence relation. (ii) Find the equivalence class of $(2, 5)$, i.e. $[(2, 5)]$.

7.22. Prove: If R and S are equivalence relations in a set X, then $R \cap S$ is also an equivalence relation in X.

7.23. (i) Show that if relations R and S are each reflexive and symmetric, then $R \cup S$ is also reflexive and symmetric.

(ii) Give an example of transitive relations R and S for which $R \cup S$ is not transitive.

PARTITIONS

7.24. Let $W = \{1, 2, 3, 4, 5, 6\}$. Determine whether each of the following is a partition of W:

(i) $[\{1, 3, 5\}, \{2, 4\}, \{3, 6\}]$ (iii) $[\{1, 5\}, \{2\}, \{4\}, \{1, 5\}, \{3, 6\}]$

(ii) $[\{1, 5\}, \{2\}, \{3, 6\}]$ (iv) $[\{1, 2, 3, 4, 5, 6\}]$

7.25. Find all partitions of $V = \{1, 2, 3\}$.

7.26. Let $[A_1, A_2, \ldots, A_m]$ and $[B_1, B_2, \ldots, B_n]$ be partitions of a set X. Show that the collection of sets

$$[A_i \cap B_j : i = 1, \ldots, m, j = 1, \ldots, n]$$

is also a partition (called the *cross partition*) of X.

Answers to Supplementary Problems

7.13. $R = R^{-1} = \{(2, 3), (3, 2), (2, 5), (5, 2), (3, 4), (4, 3), (3, 5), (5, 3), (4, 5), (5, 4)\}$

7.14. $R = \{(2, 3), (4, 2), (6, 1)\}$; $R^{-1} = \{(3, 2), (2, 4), (1, 6)\}$

7.15. (i) T, F, F, T. (ii) (*a*) $\{1, 4, 5\}$, (*b*) $\emptyset$, (*c*) $\{2, 3, 4\}$, (*d*) $\{3\}$

7.17. R_5 is the only reflexive relation.

7.18. R_4, R_5 and R_6 are the only symmetric relations.

7.19. All the relations are transitive.

7.20. (i) no, (ii) yes, (iii) no, (iv) no

7.21. (ii)
$$[(2, 5)] = \{(a, b) : a + 5 = b + 2,\ a, b \in N\}$$
$$= \{(a, a + 3) : a \in N\} = \{(1, 4), (2, 5), (3, 6), (4, 7), \ldots\}$$

7.23. (ii) $R = \{(1, 2)\}$ and $S = \{(2, 3)\}$

7.24. (i) no, (ii) no, (iii) yes, (iv) yes

7.25. $[\{1, 2, 3\}]$, $[\{1\}, \{2, 3\}]$, $[\{2\}, \{1, 3\}]$, $[\{3\}, \{1, 2\}]$ and $[\{1\}, \{2\}, \{3\}]$

Chapter 8

Functions

DEFINITION OF A FUNCTION

Suppose that to each element of a set A there is assigned a unique element of a set B; the collection, f, of such assignments is called a *function* (or mapping) from (or on) A into B and is written

$$f: A \to B \quad \text{or} \quad A \xrightarrow{f} B$$

The unique element in B assigned to $a \in A$ by f is denoted by $f(a)$, and called the *image* of a under f or the value of f at a. The *domain* of f is A, the *co-domain* B. The *range* of f, denoted by $f[A]$ is the set of images, i.e.,

$$f[A] = \{f(a) : a \in A\}$$

Example 1.1: Let f assign to each real number its square, that is, for every real number x let $f(x) = x^2$. Then the image of -3 is 9 and so we may write $f(-3) = 9$ or $f: -3 \to 9$.

Example 1.2: Let f assign to each country in the world its capital city. Here the domain of f is the set of countries in the world; the co-domain is the list of cities of the world. The image of France is Paris, that is, $f(\text{France}) = \text{Paris}$.

Example 1.3: Let $A = \{a, b, c, d\}$ and $B = \{a, b, c\}$. The assignments

$$a \to b, \quad b \to c, \quad c \to c \quad \text{and} \quad d \to b$$

define a function f from A into B. Here $f(a) = b$, $f(b) = c$, $f(c) = c$ and $f(d) = b$. The range of f is $\{b, c\}$, that is, $f[A] = \{b, c\}$.

Example 1.4: Let R be the set of real numbers, and let $f: R \to R$ assign to each rational number the number 1, and to each irrational number the number -1. Thus

$$f(x) = \begin{cases} 1 & \text{if } x \text{ is rational} \\ -1 & \text{if } x \text{ is irrational} \end{cases}$$

The range of f consists of 1 and -1: $f[R] = \{1, -1\}$.

Example 1.5: Let $A = \{a, b, c, d\}$ and $B = \{x, y, z\}$. The following diagram defines a function $f: A \to B$.

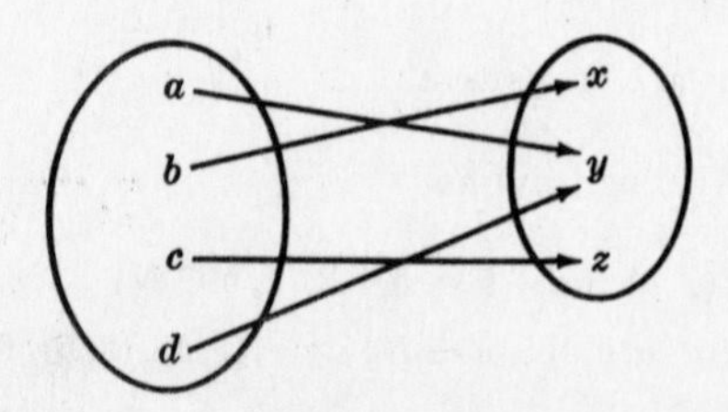

Here $f(a) = y$, $f(b) = x$, $f(c) = z$ and $f(d) = y$. Also $f[A] = B$, that is, the range and the co-domain are identical.

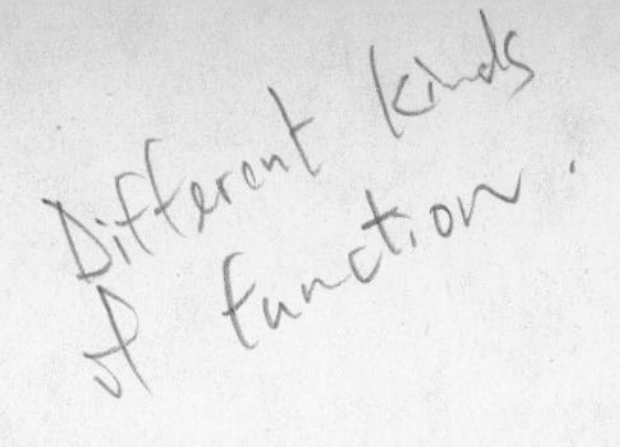

GRAPH OF A FUNCTION

To each function $f: A \to B$ there corresponds the relation in $A \times B$ given by

$$\{(a, f(a)) : a \in A\}$$

We call this set the *graph* of f. Two functions $f: A \to B$ and $g: A \to B$ are defined to be equal, written $f = g$, if $f(a) = g(a)$ for every $a \in A$, that is, if they have the same graph. Accordingly, we do not distinguish between a function and its graph. The negation of $f = g$ is written $f \neq g$ and is the statement:

$$\text{there exists an } a \in A \text{ for which } f(a) \neq g(a)$$

A subset f of $A \times B$, i.e. a relation from A to B, is a function if it possesses the following property:

[F] Each $a \in A$ appears as the first coordinate in exactly one ordered pair (a, b) in f. Accordingly, if $(a, b) \in f$, then $f(a) = b$.

Example 2.1: Let $f: A \to B$ be the function defined by the diagram in Example 1.5. Then the graph of f is the relation

$$\{(a, y), (b, x), (c, z), (d, y)\}$$

Example 2.2: Consider the following relations in $A = \{1, 2, 3\}$:

$$f = \{(1,3), (2,3), (3,1)\}$$
$$g = \{(1,2), (3,1)\}$$
$$h = \{(1,3), (2,1), (1,2), (3,1)\}$$

f is a function from A into A since each member of A appears as the first coordinate in exactly one ordered pair in f; here $f(1) = 3$, $f(2) = 3$ and $f(3) = 1$. g is not a function from A into A since $2 \in A$ is not the first coordinate of any pair in g and so g does not assign any image to 2. Also h is not a function from A into A since $1 \in A$ appears as the first coordinate of two distinct ordered pairs in h, $(1, 3)$ and $(1, 2)$. If h is to be a function it cannot assign both 3 and 2 to the element $1 \in A$.

Example 2.3: A real-valued function $f: R \to R$ of the form

$$f(x) = ax + b \qquad \text{(or: defined by } y = ax + b\text{)}$$

is called a *linear function*; its graph is a line in the Cartesian plane R^2. The graph of a linear function can be obtained by plotting (at least) two of its points. For example, to obtain the graph of $f(x) = 2x - 1$, set up a table with at least two values of x and the corresponding values of $f(x)$ as in the adjoining table. The line through these points, $(-2, -5)$, $(0, -1)$ and $(2, 3)$, is the graph of f as shown in the diagram.

x	$f(x)$
-2	-5
0	-1
2	3

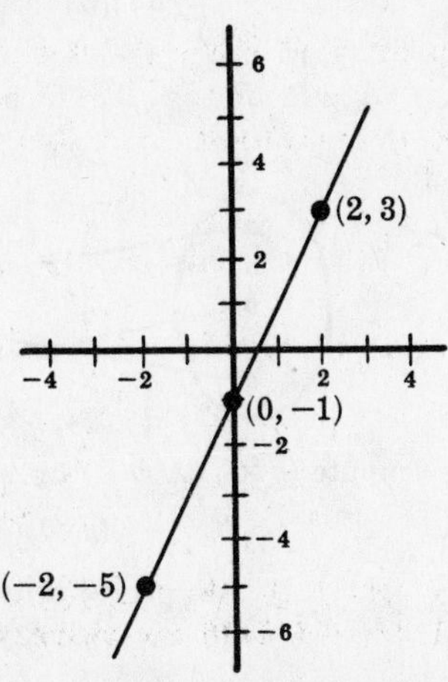

Graph of $f(x) = 2x - 1$

Example 2.4: A real-valued function $f:R \to R$ of the form

$$f(x) = a_0x^n + a_1x^{n-1} + \cdots + a_{n-1}x + a_n$$

is called a *polynomial function*. The graph of such a function is sketched by plotting various points and drawing a smooth continuous curve through them.

Consider, for example, the function $f(x) = x^2 - 2x - 3$. Set up a table of values for x and then find the corresponding values for $f(x)$ as in the adjoining table. The diagram shows these points and the sketch of the graph.

x	$f(x)$
-2	5
-1	0
0	-3
1	-4
2	-3
3	0
4	5

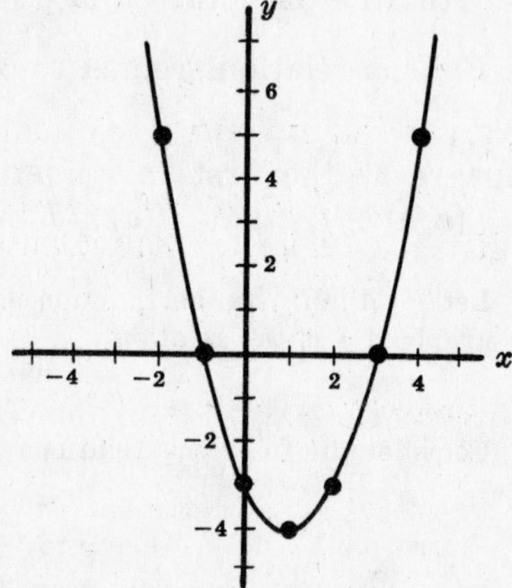

Graph of $f(x) = x^2 - 2x - 3$

COMPOSITION FUNCTION

Consider now functions $f:A \to B$ and $g:B \to C$ illustrated below:

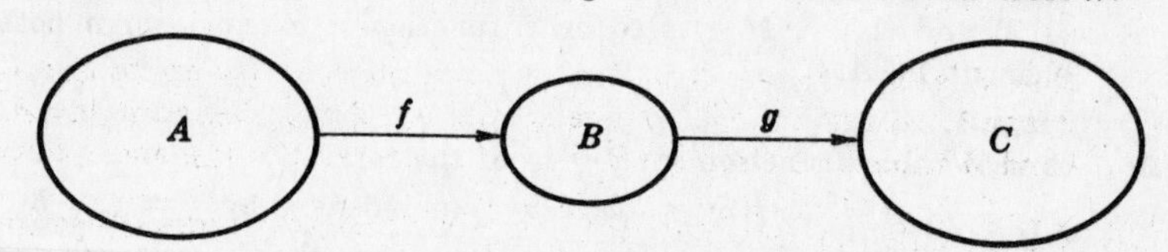

Let $a \in A$; then its image $f(a)$ is in B, the domain of g. Hence we can find the image of $f(a)$ under the function g, i.e. $g(f(a))$. The function from A into C which assigns to each $a \in A$ the element $g(f(a)) \in C$ is called the composition or product of f and g and is denoted by $g \circ f$. Hence, by definition,

$$(g \circ f)(a) = g(f(a))$$

Example 3.1: Let $f:A \to B$ and $g:B \to C$ be defined by the following diagrams:

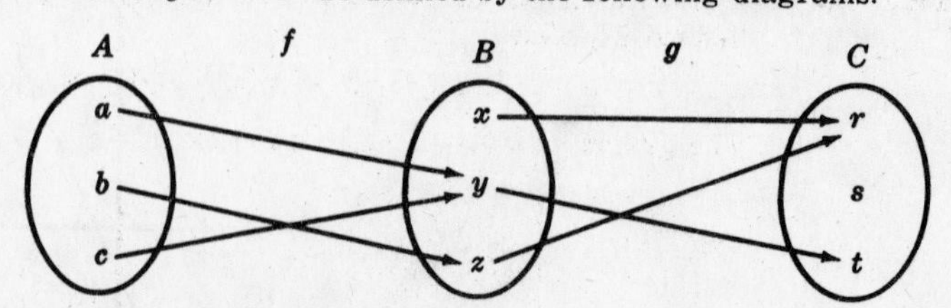

We compute $(g \circ f):A \to C$ by its definition:

$$(g \circ f)(a) = g(f(a)) = g(y) = t$$
$$(g \circ f)(b) = g(f(b)) = g(z) = r$$
$$(g \circ f)(c) = g(f(c)) = g(y) = t$$

Notice that the composition function $g \circ f$ is equivalent to "following the arrows" from A to C in the diagrams of the functions f and g.

Example 3.2: Let R be the set of real numbers, and let $f: R \to R$ and $g: R \to R$ be defined as follows:

$$f(x) = x^2 \quad \text{and} \quad g(x) = x + 3$$

Then
$$(f \circ g)(2) = f(g(2)) = f(5) = 25$$
$$(g \circ f)(2) = g(f(2)) = g(4) = 7$$

Observe that the product functions $f \circ g$ and $g \circ f$ are not the same function. We compute a general formula for these functions:

$$(f \circ g)(x) = f(g(x)) = f(x+3) = (x+3)^2 = x^2 + 6x + 9$$
$$(g \circ f)(x) = g(f(x)) = g(x^2) = x^2 + 3$$

ONE-ONE AND ONTO FUNCTIONS

A function $f: A \to B$ is said to be *one-to-one* (or: *one-one* or 1-1) if different elements in the domain have distinct images. Equivalently, $f: A \to B$ is one-one if

$$f(a) = f(a') \quad \text{implies} \quad a = a'$$

Example 4.1: Consider the functions $f: A \to B$, $g: B \to C$ and $h: C \to D$ defined by the following diagram:

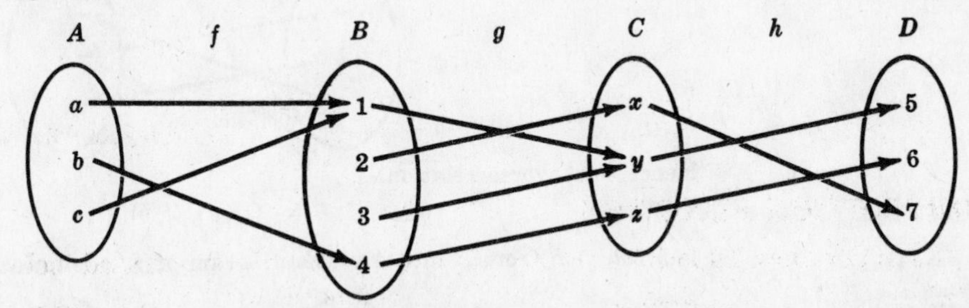

Now f is not one-one since the two elements a and c in its domain have the same image 1. Also, g is not one-one since 1 and 3 have the same image y. On the other hand, h is one-one since the elements in the domain, x, y and z, have distinct images.

A function $f: A \to B$ is said to be *onto* (or: f is a function from A *onto* B or f maps A *onto* B) if every $b \in B$ is the image of some $a \in A$. Hence $f: A \to B$ is onto iff the range of f is the entire co-domain, i.e. $f[A] = B$.

Example 4.2: Consider the functions f, g and h in the preceding example. Then f is not onto since $2 \in B$ is not the image of any element in the domain A. On the other hand, both g and h are onto functions.

Example 4.3: Let R be the set of real numbers and let $f: R \to R$, $g: R \to R$ and $h: R \to R$ be defined as follows:

$$f(x) = 2^x, \quad g(x) = x^3 - x \quad \text{and} \quad h(x) = x^2$$

The graphs of these functions follow:

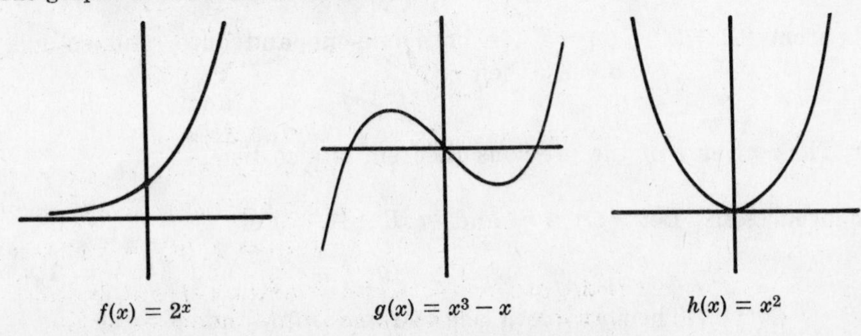

$f(x) = 2^x$ $\qquad$ $g(x) = x^3 - x$ $\qquad$ $h(x) = x^2$

The function f is one-one; geometrically, this means that each horizontal line does not contain more than one point of f. The function g is onto; geometrically, this means that each horizontal line contains at least one point of g. The function h is neither one-one or onto; for $h(2) = h(-2) = 4$, i.e. the two elements 2 and -2 have the same image 4, and $h[R]$ is a proper subset of R; for example, $-16 \notin h[R]$.

INVERSE AND IDENTITY FUNCTIONS

In general, the inverse relation f^{-1} of a function $f \subset A \times B$ need not be a function. However, if f is both one-one and onto, then f^{-1} is a function from B onto A and is called the inverse function.

Example 5.1: Let $A = \{a, b, c\}$ and $B = \{r, s, t\}$. Then

$$f = \{(a, s), (b, t), (c, r)\}$$

is a function from A into B which is both one-one and onto. This can easily be seen by the following diagram of f:

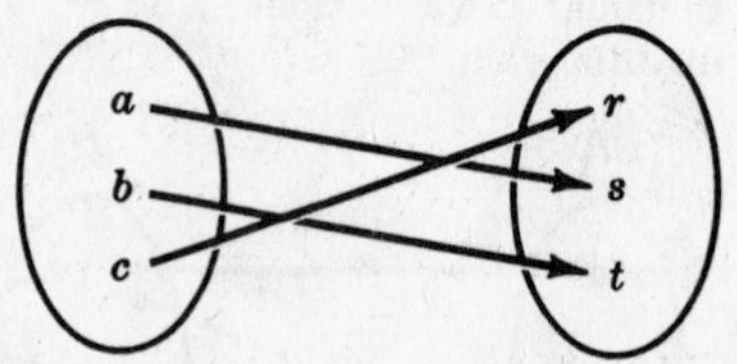

Hence the inverse relation

$$f^{-1} = \{(s, a), (t, b), (r, c)\}$$

is a function from B into A. The diagram of f^{-1} follows:

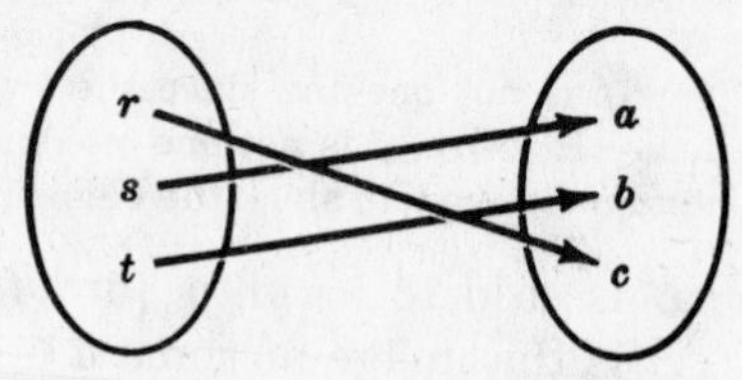

Observe that the diagram of f^{-1} can be obtained from the diagram of f by reversing the arrows.

For any set A, the function $f: A \to A$ defined by $f(x) = x$, i.e. which assigns to each element in A itself, is called the *identity function* on A and is usually denoted by 1_A or simply 1. Note that the identity function 1_A is the same as the diagonal relation: $1_A = \Delta_A$. The identity function satisfies the following properties:

Theorem 8.1: For any function $f: A \to B$,

$$1_B \circ f = f = f \circ 1_A$$

Theorem 8.2: If $f: A \to B$ is both one-one and onto, and so has an inverse function $f^{-1}: B \to A$, then

$$f^{-1} \circ f = 1_A \quad \text{and} \quad f \circ f^{-1} = 1_B$$

The converse of the previous theorem is also true:

Theorem 8.3: Let $f: A \to B$ and $g: B \to A$ satisfy

$$g \circ f = 1_A \quad \text{and} \quad f \circ g = 1_B$$

Then f is both one-one and onto, and $g = f^{-1}$.

Solved Problems

FUNCTIONS

8.1. State whether or not each diagram defines a function from $A = \{a, b, c\}$ into $B = \{x, y, z\}$.

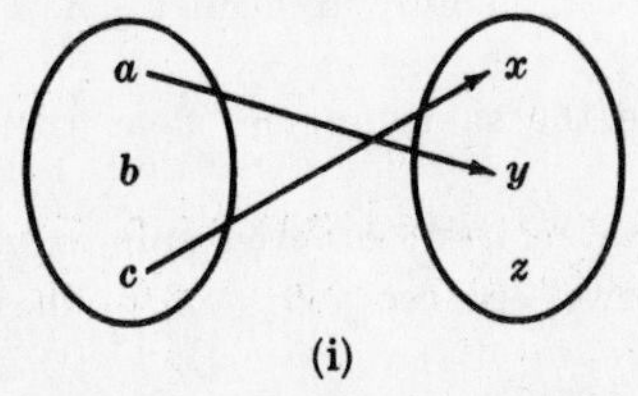

(i)

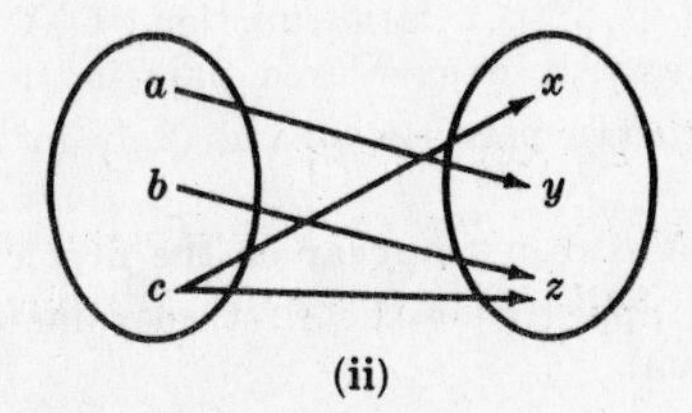

(ii)

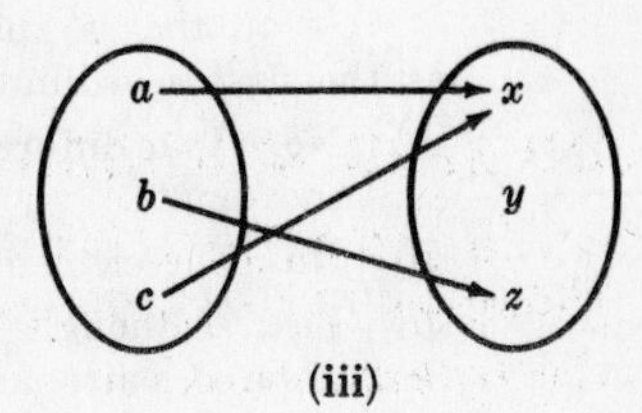

(iii)

(i) No. There is nothing assigned to the element $b \in A$.
(ii) No. Two elements, x and z, are assigned to $c \in A$.
(iii) Yes.

8.2. Rewrite each of the following functions using a formula:
(i) To each number let f assign its cube.
(ii) To each number let g assign the number 5.
(iii) To each positive number let h assign its square, and to each non-positive number let h assign the number 4.

(i) Since f assigns to any number x its cube x^3, f can be defined by $f(x) = x^3$.
(ii) Since g assigns 5 to any number x, we can define g by $g(x) = 5$.
(iii) Two different rules are used to define h as follows:

$$h(x) = \begin{cases} x^2 & \text{if } x > 0 \\ 4 & \text{if } x \leqq 0 \end{cases}$$

8.3. Let f, g and h be the functions of the preceding problem. Find:
(i) $f(4), f(-2), f(0)$; (ii) $g(4), g(-2), g(0)$; (iii) $h(4), h(-2), h(0)$.

(i) Now $f(x) = x^3$ for every number x; hence $f(4) = 4^3 = 64$, $f(-2) = (-2)^3 = -8$, $f(0) = 0^3 = 0$.
(ii) Since $g(x) = 5$ for every number x, $g(4) = 5$, $g(-2) = 5$ and $g(0) = 5$.
(iii) If $x > 0$, then $h(x) = x^2$; hence $h(4) = 4^2 = 16$. On the other hand, if $x \leqq 0$, then $h(x) = 4$; thus $h(-2) = 4$ and $h(0) = 4$.

8.4. Let $A = \{1, 2, 3, 4, 5\}$ and let $f: A \to A$ be the function defined in the diagram:

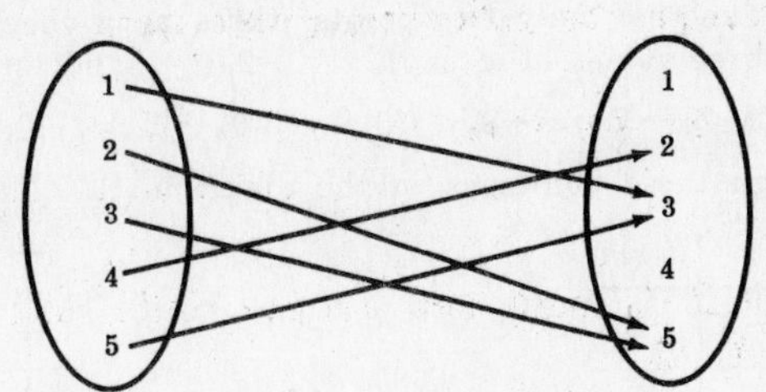

(i) Find the range of f.
(ii) Find the graph of f, i.e. write f as a set of ordered pairs.

(i) The range $f[A]$ of the function f consists of all the image points. Now only 2, 3 and 5 appear as the image of any elements of A; hence $f[A] = \{2, 3, 5\}$.
(ii) The ordered pairs $(a, f(a))$, where $a \in A$, form the graph of f. Now $f(1) = 3$, $f(2) = 5$, $f(3) = 5$, $f(4) = 2$ and $f(5) = 3$; hence $f = \{(1,3), (2,5), (3,5), (4,2), (5,3)\}$.

8.5. Let $X = \{1, 2, 3, 4\}$. Determine whether or not each relation is a function from X into X.

(i) $f = \{(2,3), (1,4), (2,1), (3,2), (4,4)\}$

(ii) $g = \{(3,1), (4,2), (1,1)\}$

(iii) $h = \{(2,1), (3,4), (1,4), (2,1), (4,4)\}$

Recall that a subset f of $X \times X$ is a function $f: X \to X$ if and only if each $a \in X$ appears as the first coordinate in exactly one ordered pair in f.

(i) No. Two different ordered pairs $(2,3)$ and $(2,1)$ in f have the same number 2 as their first coordinate.

(ii) No. The element $2 \in X$ does not appear as the first coordinate in any ordered pair in g.

(iii) Yes. Although $2 \in X$ appears as the first coordinate in two ordered pairs in h, these two ordered pairs are equal.

8.6. Find the geometric conditions under which a set f of points on the coordinate diagram of $A \times B$ defines a function $f: A \to B$.

The requirement that each $a \in A$ appear as the first coordinate in exactly one ordered pair in f is equivalent to the geometric condition that each vertical line contains exactly one point in f.

8.7. Let $W = \{a, b, c, d\}$. Determine whether the set of points in each coordinate diagram of $W \times W$ is a function from W into W.

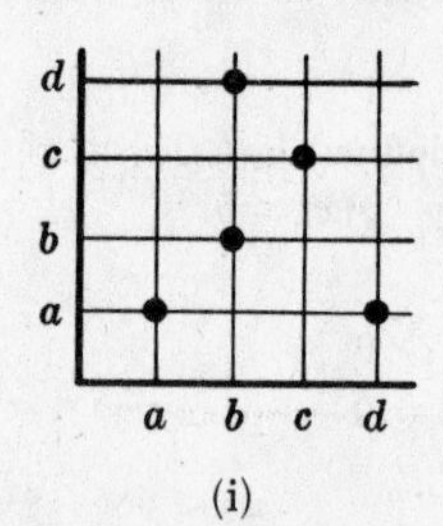

(i)

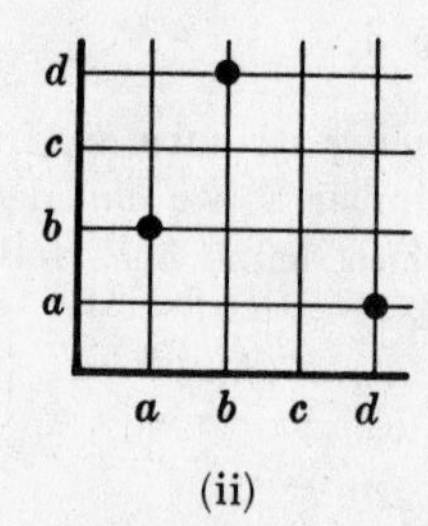

(ii)

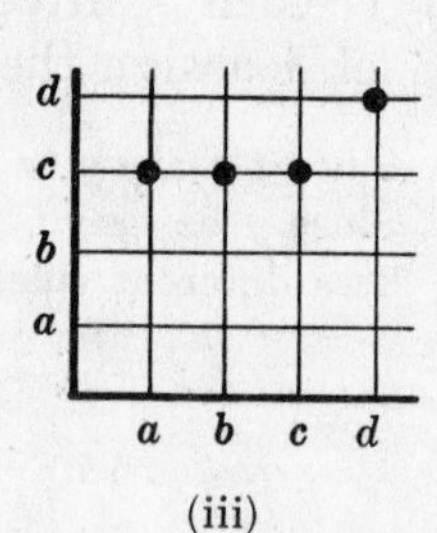

(iii)

(i) No. The vertical line through b contains two points of the set, i.e. two different ordered pairs (b, b) and (b, d) contain the same first element b.

(ii) No. The vertical line through c contains no point of the set, i.e. $c \in W$ does not appear as the first element in any ordered pair.

(iii) Yes. Each vertical line contains exactly one point of the set.

GRAPHS OF REAL-VALUED FUNCTIONS

8.8. Sketch the graph of $f(x) = 3x - 2$.

This is a linear function; only two points (three as a check) are needed to sketch its graph. Set up a table with three values of x, say, $x = -2, 0, 2$ and find the corresponding values of $f(x)$:

$$f(-2) = 3(-2) - 2 = -8, \quad f(0) = 3(0) - 2 = -2, \quad f(2) = 3(2) - 2 = 4$$

Draw the line through these points as in the diagram.

x	$f(x)$
-2	-8
0	-2
2	4

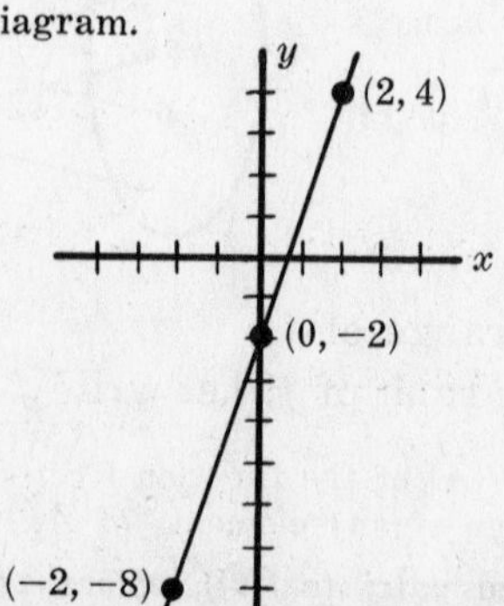

8.9. Sketch the graph of (i) $f(x) = x^2 + x - 6$, (ii) $g(x) = x^3 - 3x^2 - x + 3$.

In each case, set up a table of values for x and then find the corresponding values of $f(x)$. Plot the points in a coordinate diagram, and then draw a smooth continuous curve through the points:

(i)

x	$f(x)$
-4	6
-3	0
-2	-4
-1	-6
0	-6
1	-4
2	0
3	6

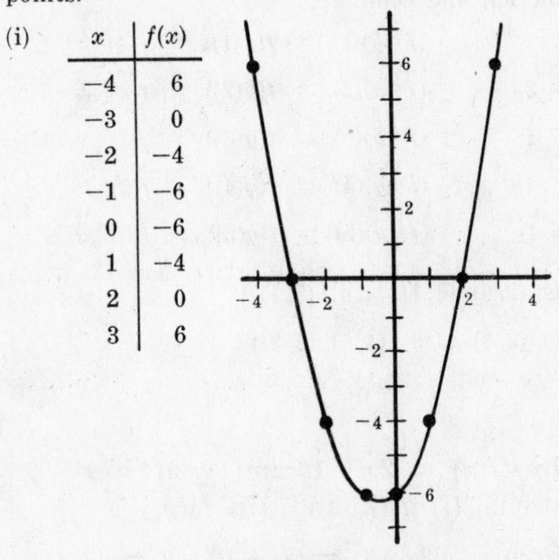

Graph of f.

(ii)

x	$g(x)$
-2	-15
-1	0
0	3
1	0
2	-3
3	0
4	15

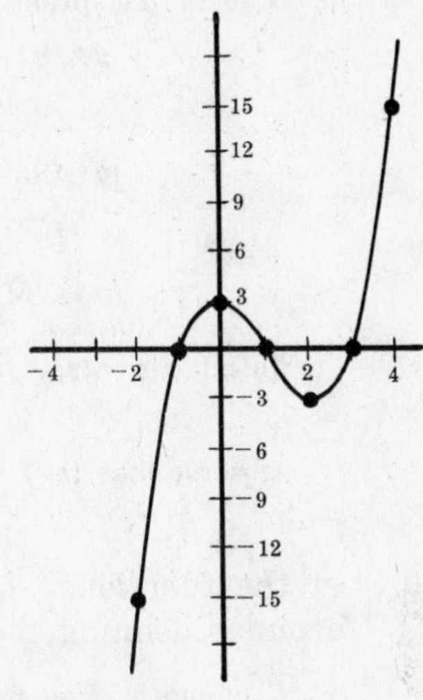

Graph of g.

COMPOSITION OF FUNCTIONS

8.10. Let the functions $f: A \to B$ and $g: B \to C$ be defined by the diagram

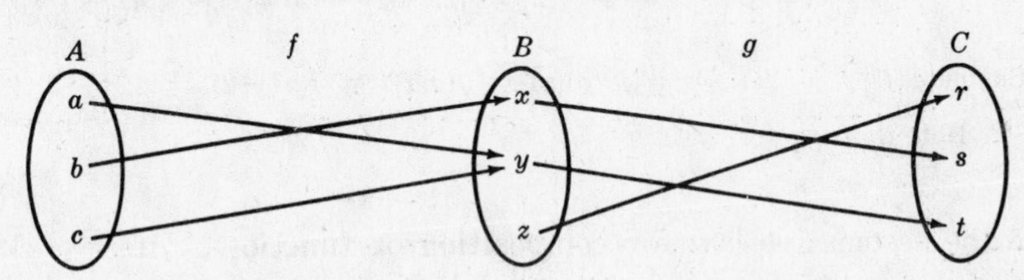

(i) Find the composition function $g \circ f: A \to C$. (ii) Find the ranges of f, g and $g \circ f$.

(i) We use the definition of the composition function to compute:

$$(g \circ f)(a) = g(f(a)) = g(y) = t$$
$$(g \circ f)(b) = g(f(b)) = g(x) = s$$
$$(g \circ f)(c) = g(f(c)) = g(y) = t$$

Note that we arrive at the same answer if we "follow the arrows" in the diagram:

$$a \to y \to t, \quad b \to x \to s, \quad c \to y \to t$$

(ii) By the diagram, the images under the function f are x and y, and the images under g are r, s and t; hence

$$\text{range of } f = \{x, y\} \quad \text{and} \quad \text{range of } g = \{r, s, t\}$$

By (i), the images under the composition function are t and s; hence

$$\text{range of } g \circ f = \{s, t\}$$

Note that the ranges of g and $g \circ f$ are different.

8.11. Consider the functions

$$f = \{(1,3), (2,5), (3,3), (4,1), (5,2)\}$$
$$g = \{(1,4), (2,1), (3,1), (4,2), (5,3)\}$$

from $X = \{1,2,3,4,5\}$ into X. (i) Determine the range of f and of g. (ii) Find the composition functions $g \circ f$ and $f \circ g$.

(i) The range of a function is the set of image values, i.e. the set of second coordinates; hence

$$\text{range of } f = \{3,5,1,2\} \quad \text{and} \quad \text{range of } g = \{4,1,2,3\}$$

(ii) Use the definition of the composition function and compute:

$$(g \circ f)(1) = g(f(1)) = g(3) = 1 \qquad (f \circ g)(1) = f(g(1)) = f(4) = 1$$
$$(g \circ f)(2) = g(f(2)) = g(5) = 3 \qquad (f \circ g)(2) = f(g(2)) = f(1) = 3$$
$$(g \circ f)(3) = g(f(3)) = g(3) = 1 \qquad (f \circ g)(3) = f(g(3)) = f(1) = 3$$
$$(g \circ f)(4) = g(f(4)) = g(1) = 4 \qquad (f \circ g)(4) = f(g(4)) = f(2) = 5$$
$$(g \circ f)(5) = g(f(5)) = g(2) = 1 \qquad (f \circ g)(5) = f(g(5)) = f(3) = 3$$

In other words,

$$g \circ f = \{(1,1), (2,3), (3,1), (4,4), (5,1)\}$$
$$f \circ g = \{(1,1), (2,3), (3,3), (4,5), (5,3)\}$$

Observe that $g \circ f \neq f \circ g$.

8.12. Let the functions f and g be defined by $f(x) = 2x+1$ and $g(x) = x^2-2$. Find formulas defining the composition functions (i) $g \circ f$ and (ii) $f \circ g$.

(i) Compute $g \circ f$ as follows: $(g \circ f)(x) = g(f(x)) = g(2x+1) = (2x+1)^2 - 2 = 4x^2+4x-1$.

Observe that the same answer can be found by writing

$$y = f(x) = 2x+1 \quad \text{and} \quad z = g(y) = y^2 - 2$$

and then eliminating y from both equations:

$$z = y^2 - 2 = (2x+1)^2 - 2 = 4x^2 + 4x - 1$$

(ii) Compute $f \circ g$ as follows: $(f \circ g)(x) = f(g(x)) = f(x^2-2) = 2(x^2-2)+1 = 2x^2-3$.

Note that $f \circ g \neq g \circ f$.

8.13. Prove the associative law for composition of functions: if

$$A \xrightarrow{f} B \xrightarrow{g} C \xrightarrow{h} D$$

then $(h \circ g) \circ f = h \circ (g \circ f)$.

For every $a \in A$,

$$((h \circ g) \circ f)(a) = (h \circ g)(f(a)) = h(g(f(a)))$$
$$(h \circ (g \circ f))(a) = h((g \circ f)(a)) = h(g(f(a)))$$

Hence $(h \circ g) \circ f = h \circ (g \circ f)$, since they each assign the same image to every $a \in A$. Accordingly, we may simply write the composition function without parentheses: $h \circ g \circ f$.

ONE-ONE AND ONTO FUNCTIONS

8.14. Let $A = \{a,b,c,d,e\}$, and let B be the set of letters in the alphabet. Let the functions f, g and h from A into B be defined as follows:

f	g	h
$a \to r$	$a \to z$	$a \to a$
$b \to a$	$b \to y$	$b \to c$
$c \to s$	$c \to x$	$c \to e$
$d \to r$	$d \to y$	$d \to r$
$e \to e$	$e \to z$	$e \to s$
(i)	(ii)	(iii)

Are any of these functions one-one?

Recall that a function is one-one if it assigns distinct image values to distinct elements in the domain.

(i) No. For f assigns r to both a and d.

(ii) No. For g assigns z to both a and e.

(iii) Yes. For h assigns distinct images to different elements in the domain.

8.15. Determine if each function is one-one.

(i) To each person on the earth assign the number which corresponds to his age.

(ii) To each country in the world assign the latitude and longitude of its capital.

(iii) To each book written by only one author assign the author.

(iv) To each country in the world which has a prime minister assign its prime minister.

(i) No. Many people in the world have the same age.

(ii) Yes.

(iii) No. There are different books with the same author.

(iv) Yes. Different countries in the world have different prime ministers.

8.16. Prove: If $f: A \to B$ and $g: B \to C$ are one-one functions, then the composition function $g \circ f: A \to C$ is also one-one.

Let $(g \circ f)(a) = (g \circ f)(a')$; i.e. $g(f(a)) = g(f(a'))$. Then $f(a) = f(a')$ since g is one-one. Furthermore, $a = a'$ since f is one-one. Accordingly, $g \circ f$ is also one-one.

8.17. Let the functions $f: A \to B$, $g: B \to C$ and $h: C \to D$ be defined by the diagram.

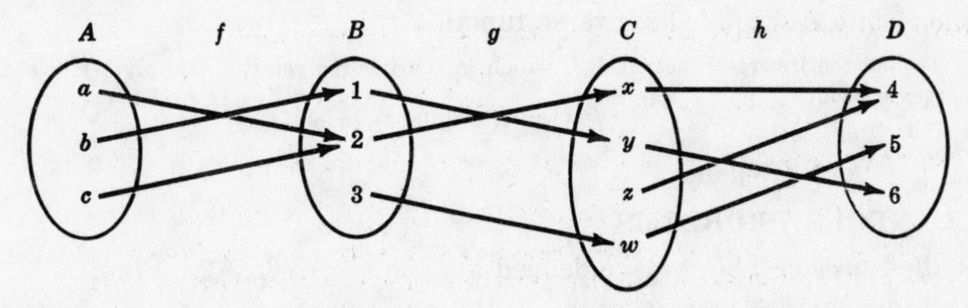

(i) Determine if each function is onto. (ii) Find the composition function $h \circ g \circ f$.

(i) The function $f: A \to B$ is not onto since $3 \in B$ is not the image of any element in A.

The function $g: B \to C$ is not onto since $z \in C$ is not the image of any element in B.

The function $h: C \to D$ is onto since each element in D is the image of some element of C.

(ii) Now $a \to 2 \to x \to 4$, $b \to 1 \to y \to 6$, $c \to 2 \to x \to 4$. Hence $h \circ g \circ f = \{(a, 4), (b, 6), (c, 4)\}$.

8.18. Prove: If $f: A \to B$ and $g: B \to C$ are onto functions, then the composition function $g \circ f: A \to C$ is also onto.

Let c be any arbitrary element of C. Since g is onto, there exists a $b \in B$ such that $g(b) = c$. Since f is onto, there exists an $a \in A$ such that $f(a) = b$. But then

$$(g \circ f)(a) = g(f(a)) = g(b) = c$$

Hence each $c \in C$ is the image of some element $a \in A$. Accordingly, $g \circ f$ is an onto function.

INVERSE FUNCTIONS

8.19. Let $W = \{1,2,3,4,5\}$ and let $f: W \to W$, $g: W \to W$ and $h: W \to W$ be defined by the following diagrams:

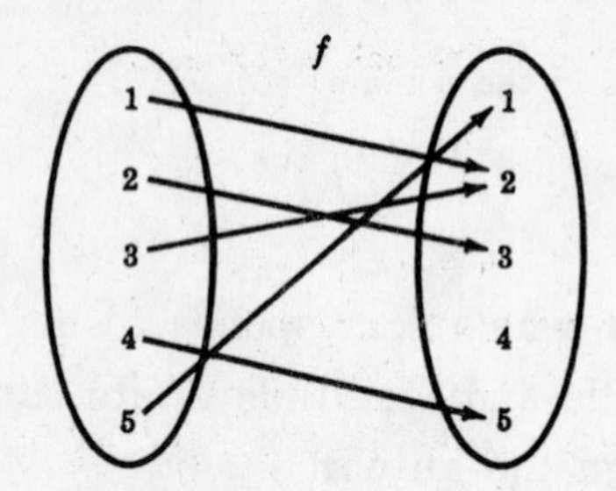

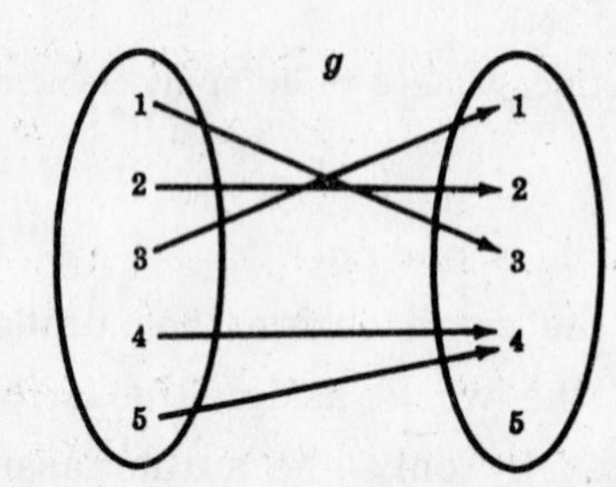

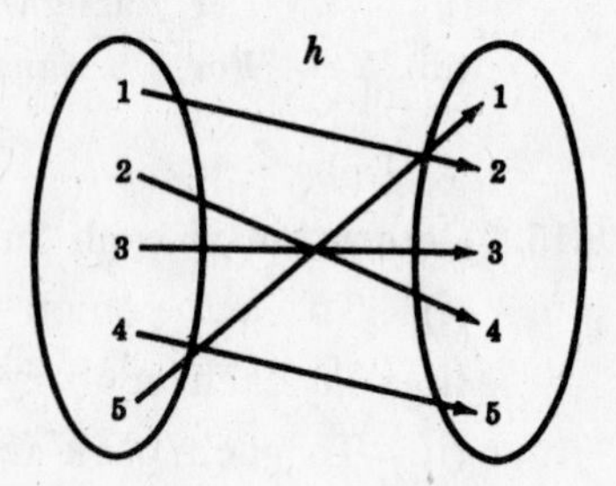

Determine whether each function has an inverse function.

In order for a function to have an inverse, the function must be both one-one and onto. Only h is one-one and onto; hence h, and only h, has an inverse function.

8.20. Let $f: R \to R$ be defined by $f(x) = 2x - 3$. Now f is one-one and onto; hence f has an inverse function f^{-1}. Find a formula for f^{-1}.

Let y be the image of x under the function f:

$$y = f(x) = 2x - 3$$

Consequently, x will be the image of y under the inverse function f^{-1}. Solve for x in terms of y in the above equation:

$$x = (y+3)/2$$

Then $$f^{-1}(y) = (y+3)/2$$

is a formula defining the inverse function.

8.21. Let $A = \{a, b, c, d\}$. Then $f = \{(a,b), (b,d), (c,a), (d,c)\}$ is a one-one, onto function from A into A. Find the inverse function f^{-1}.

To find the inverse function f^{-1}, which is the inverse relation, simply write each ordered pair in reverse order:

$$f^{-1} = \{(b,a), (d,b), (a,c), (c,d)\}$$

MISCELLANEOUS PROBLEMS

8.22. Let the function $f: R \to R$ be defined by $f(x) = x^2 - 3x + 2$. Find:

(a) $f(-3)$ (e) $f(x^2)$ (i) $f(2x-3)$ (m) $f(f(x+1))$

(b) $f(2) - f(-4)$ (f) $f(y-z)$ (j) $f(2x-3) + f(x+3)$ (n) $f(x+h) - f(x)$

(c) $f(y)$ (g) $f(x+h)$ (k) $f(x^2 - 3x + 2)$ (o) $[f(x+h) - f(x)]/h$

(d) $f(a^2)$ (h) $f(x+3)$ (l) $f(f(x))$

The function assigns to any element the square of the element minus 3 times the element plus 2.

(a) $f(-3) = (-3)^2 - 3(-3) + 2 = 9 + 9 + 2 = 20$

(b) $f(2) = (2)^2 - 3(2) + 2 = 0$, $f(-4) = (-4)^2 - 3(-4) + 2 = 30$. Then

$$f(2) - f(-4) = 0 - 30 = -30$$

(c) $f(y) = (y)^2 - 3(y) + 2 = y^2 - 3y + 2$

(d) $f(a^2) = (a^2)^2 - 3(a^2) + 2 = a^4 - 3a^2 + 2$

(e) $f(x^2) = (x^2)^2 - 3(x^2) + 2 = x^4 - 3x^2 + 2$

(f) $f(y-z) = (y-z)^2 - 3(y-z) + 2 = y^2 - 2yz + z^2 - 3y + 3z + 2$

(g) $f(x+h) = (x+h)^2 - 3(x+h) + 2 = x^2 + 2xh + h^2 - 3x - 3h + 2$

(*h*) $f(x+3) = (x+3)^2 - 3(x+3) + 2 = (x^2+6x+9) - 3x - 9 + 2 = x^2 + 3x + 2$

(*i*) $f(2x-3) = (2x-3)^2 - 3(2x-3) + 2 = 4x^2 - 12x + 9 - 6x + 9 + 2 = 4x^2 - 18x + 20$

(*j*) Using (*h*) and (*i*), we have

$$f(2x-3) + f(x+3) = (4x^2 - 18x + 20) + (x^2 + 3x + 2) = 5x^2 - 15x + 22$$

(*k*) $f(x^2-3x+2) = (x^2-3x+2)^2 - 3(x^2-3x+2) + 2 = x^4 - 6x^3 + 10x^2 - 3x$

(*l*) $f(f(x)) = f(x^2-3x+2) = x^4 - 6x^3 + 10x^2 - 3x$

(*m*) $f(f(x+1)) = f([(x+1)^2 - 3(x+1) + 2]) = f([x^2 + 2x + 1 - 3x - 3 + 2])$
$= f(x^2 - x) = (x^2-x)^2 - 3(x^2-x) + 2 = x^4 - 2x^3 - 2x^2 + 3x + 2$

(*n*) By (*g*), $f(x+h) = x^2 + 2xh + h^2 - 3x - 3h + 2$. Hence

$$f(x+h) - f(x) = (x^2 + 2xh + h^2 - 3x - 3h + 2) - (x^2 - 3x + 2) = 2xh + h^2 - 3h$$

(*o*) Using (*n*), we have

$$[f(x+h) - f(x)]/h = (2xh + h^2 - 3h)/h = 2x + h - 3$$

8.23. Prove Theorem 8.1: For any function $f: A \to B$, $1_B \circ f = f = f \circ 1_A$.

Let a be any arbitrary element in A; then

$$(1_B \circ f)(a) = 1_B(f(a)) = f(a) \quad \text{and} \quad (f \circ 1_A)(a) = f(1_A(a)) = f(a)$$

Hence $1_B \circ f = f = f \circ 1_A$ since they each assign $f(a)$ to every element $a \in A$.

Supplementary Problems

FUNCTIONS

8.24. State whether each diagram defines a function from $\{1, 2, 3\}$ into $\{4, 5, 6\}$.

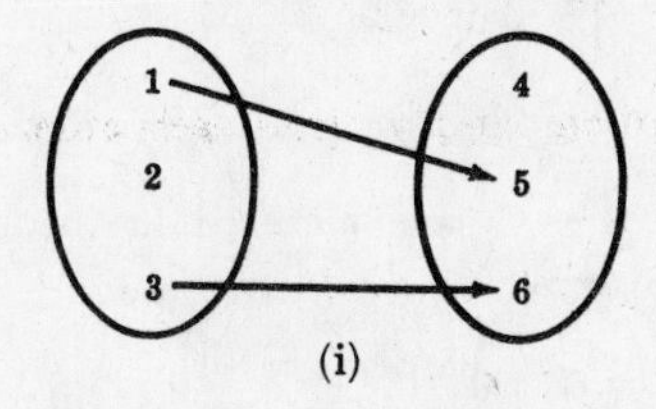

(i)

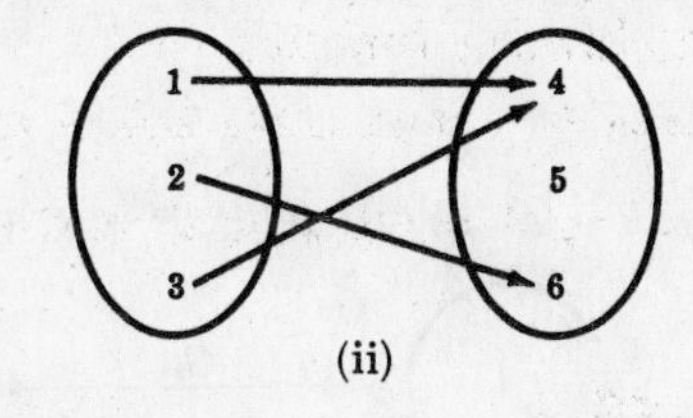

(ii)

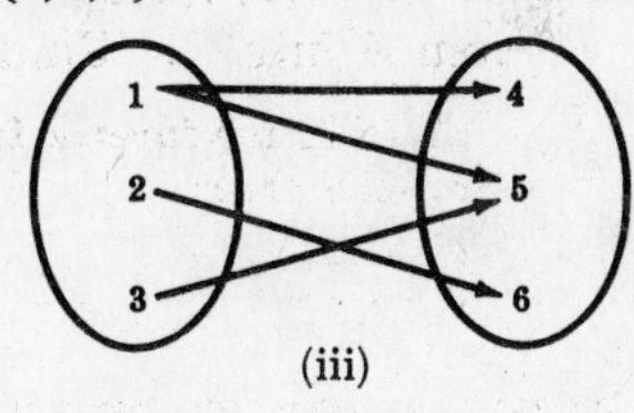

(iii)

8.25. Define each function by a formula:

(i) To each number let f assign its square plus 3.

(ii) To each number let g assign its cube plus twice the number.

(iii) To each number greater than or equal to 3 let h assign the number squared, and to each number less than 3 let h assign the number -2.

8.26. Determine the number of different functions from $\{a, b\}$ into $\{1, 2, 3\}$.

8.27. Let $f: R \to R$ be defined by $f(x) = x^2 - 4x + 3$. Find (i) $f(4)$, (ii) $f(-3)$, (iii) $f(y - 2x)$, (iv) $f(x-2)$.

8.28. Let $g: R \to R$ be defined by $g(x) = \begin{cases} x^2 - 3x & \text{if } x \geq 2 \\ x + 2 & \text{if } x < 2 \end{cases}$. Find (i) $g(5)$, (ii) $g(0)$, (iii) $g(-2)$.

8.29. Let $W = \{a, b, c, d\}$. Determine whether each set of ordered pairs is a function from W into W.

(i) $\{(b, a), (c, d), (d, a), (c, d), (a, d)\}$

(ii) $\{(d, d), (c, a), (a, b), (d, b)\}$

(iii) $\{(a, b), (b, b), (c, b), (d, b)\}$

(iv) $\{(a, a), (b, a), (a, b), (c, d)\}$

8.30. Let the function g assign to each name in the set {Betty, Martin, David, Alan, Rebecca} the number of different letters needed to spell the name. Find the graph of g, i.e. write g as a set of ordered pairs.

8.31. Let $W = \{1, 2, 3, 4\}$ and let $g: W \rightarrow W$ be defined by the diagram

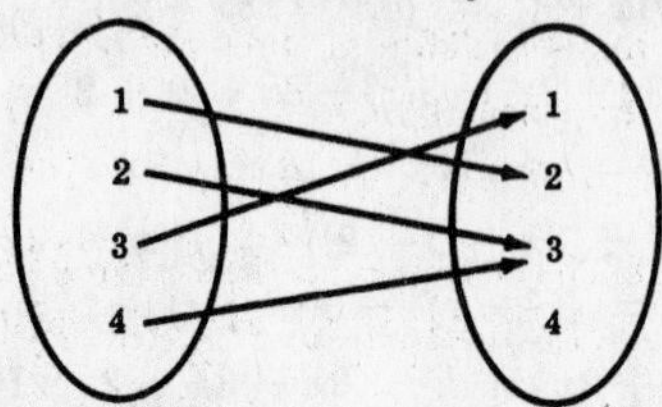

(i) Write g as a set of ordered pairs. (ii) Plot g on the coordinate diagram of $W \times W$. (iii) Find the range of g.

8.32. Let $V = \{1, 2, 3, 4\}$. Determine whether the set of points in each coordinate diagram of $V \times V$ is a function from V into V.

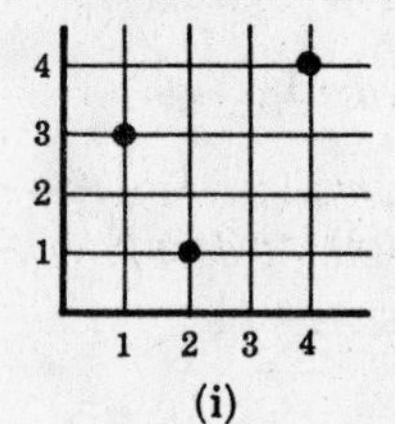
(i)

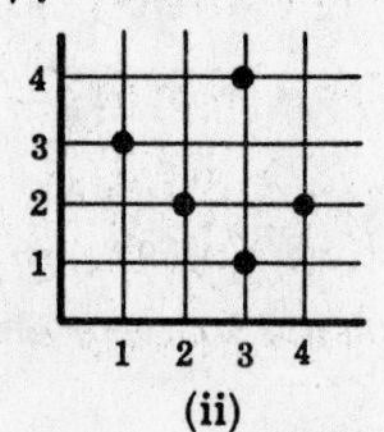
(ii)

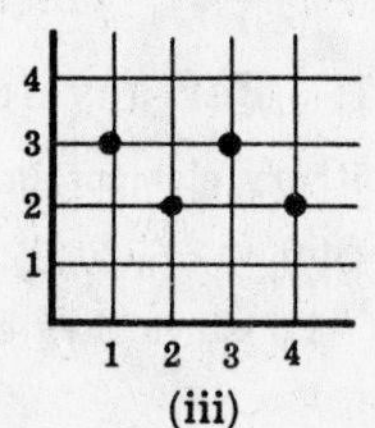
(iii)

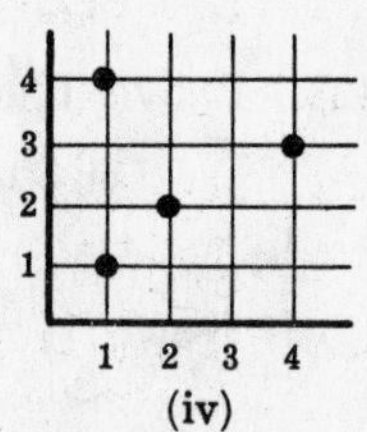
(iv)

GRAPHS OF REAL-VALUED FUNCTIONS

8.33. Sketch the graph of each function:

(i) $f(x) = 2$, (ii) $g(x) = \frac{1}{2}x - 1$, (iii) $h(x) = 2x^2 - 4x - 3$.

8.34. Sketch the graph of each function:

(i) $f(x) = x^3 - 3x + 2$, (ii) $g(x) = x^4 - 10x^2 + 9$, (iii) $h(x) = \begin{cases} 0 & \text{if } x = 0 \\ \dfrac{1}{x} & \text{if } x \neq 0 \end{cases}$.

COMPOSITION OF FUNCTIONS

8.35. The functions $f: A \rightarrow B$, $g: B \rightarrow A$, $h: C \rightarrow B$, $F: B \rightarrow C$, $G: A \rightarrow C$ are pictured in the diagram below.

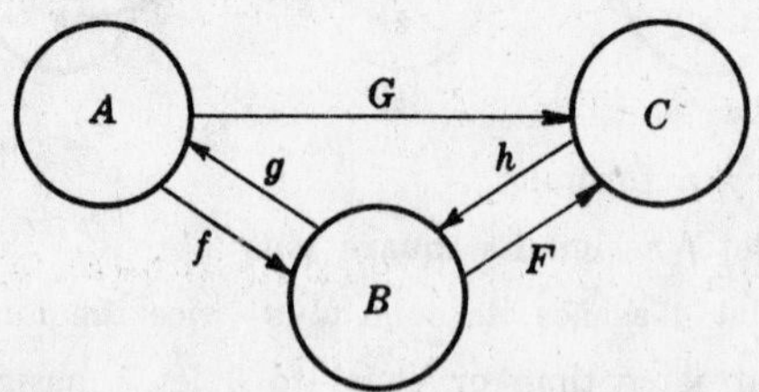

Determine whether each of the following defines a product function and if it does, find its domain and co-domain: (i) $g \circ f$, (ii) $h \circ f$, (iii) $F \circ f$, (iv) $G \circ f$, (v) $g \circ h$, (vi) $F \circ h$, (vii) $h \circ G \circ g$, (viii) $h \circ G$.

8.36. The following diagrams define functions f, g and h which map the set $\{1, 2, 3, 4\}$ into itself.

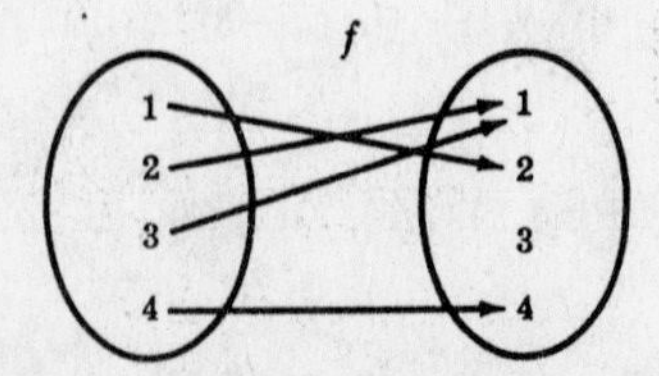

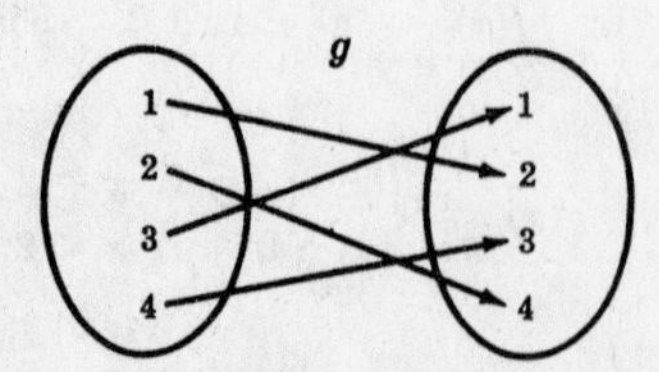

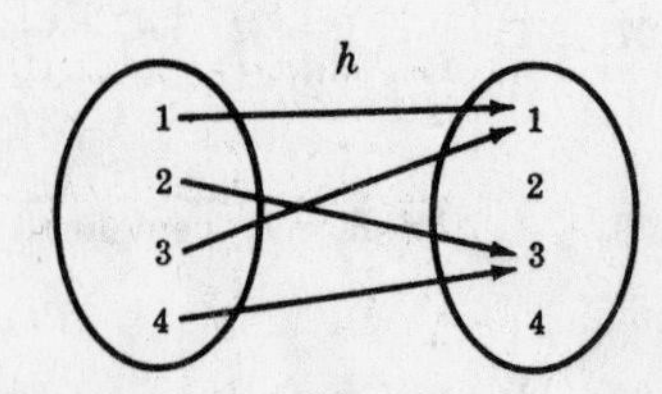

(i) Find the ranges of f, g and h.

(ii) Find the composition functions (1) $f \circ g$, (2) $h \circ f$, (3) g^2, i.e. $g \circ g$.

8.37. Let $f: R \to R$ and $g: R \to R$ be defined by $f(x) = x^2 + 3x + 1$ and $g(x) = 2x - 3$. Find formulas defining the product functions (i) $f \circ g$, (ii) $g \circ f$, (iii) $g \circ g$, (iv) $f \circ f$.

ONE-ONE, ONTO AND INVERSE FUNCTIONS

8.38 Let $f: X \to Y$. Which conditions define a one-one function:

(i) $f(a) = f(b)$ implies $a = b$ (iii) $f(a) \neq f(b)$ implies $a \neq b$

(ii) $a = b$ implies $f(a) = f(b)$ (iv) $a \neq b$ implies $f(a) \neq f(b)$

8.39. (i) State whether or not each function in Problem 8.36 is one-one.

(ii) State whether or not each function in Problem 8.36 is onto.

8.40. Prove Theorem 8.2: If $f: A \to B$ is one-one and onto, and so has an inverse function f^{-1}, then (i) $f^{-1} \circ f = 1_A$ and (ii) $f \circ f^{-1} = 1_B$.

8.41. Prove: If $f: A \to B$ and $g: B \to A$ satisfy $g \circ f = 1_A$, then f is one-one and g is onto.

8.42. Let $f: R \to R$ be defined by $f(x) = 3x - 7$. Find a formula for the inverse function $f^{-1}: R \to R$.

8.43. Let $g: R \to R$ be defined by $g(x) = x^3 + 2$. Find a formula for the inverse function $g^{-1}: R \to R$.

8.44. Let R be an equivalence relation in a non-empty set A. The function η from A into the quotient set A/R is defined by $\eta(a) = [a]$, the equivalence class of a. Show that η is an onto function.

8.45. Prove Theorem 8.3: Let $f: A \to B$ and $g: B \to A$ satisfy $g \circ f = 1_A$ and $f \circ g = 1_B$. Then f is one-one and onto, and $g = f^{-1}$.

Answers to Supplementary Problems

8.24. (i) No, (ii) Yes, (iii) No

8.25. (i) $f(x) = x^2 + 3$, (ii) $g(x) = x^3 + 2x$, (iii) $h(x) = \begin{cases} x^2 & \text{if } x \geq 3 \\ -2 & \text{if } x < 3 \end{cases}$

8.26. Nine.

8.27. (i) 3, (ii) 24, (iii) $y^2 - 4xy + 4x^2 - 4y + 8x + 3$, (iv) $x^2 - 8x + 15$

8.28. (i) $g(5) = 10$, (ii) $g(0) = 2$, (iii) $g(-2) = 0$

8.29. (i) Yes, (ii) No, (iii) Yes, (iv) No

8.30. g = {(Betty, 4), (Martin, 6), (David, 4), (Alan, 3), (Rebecca, 5)}

8.31. (i) $g = \{(1,2), (2,3), (3,1), (4,3)\}$, (iii) $\{2, 3, 1\}$

8.32. (i) No, (ii) No, (iii) Yes, (iv) No

8.33. (i)

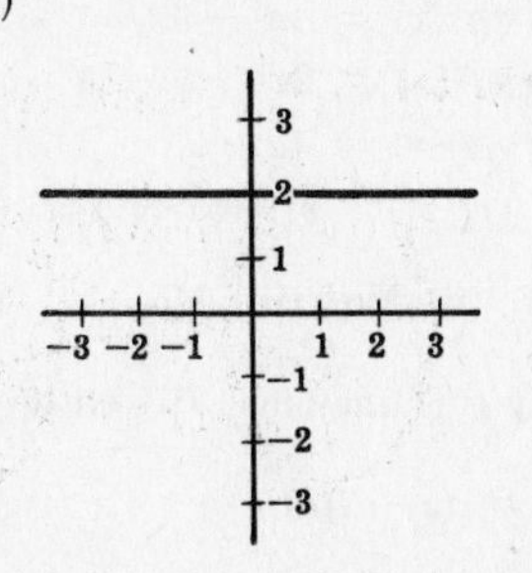

Graph of f

(ii)

x	$g(x)$
−2	−2
0	−1
2	0

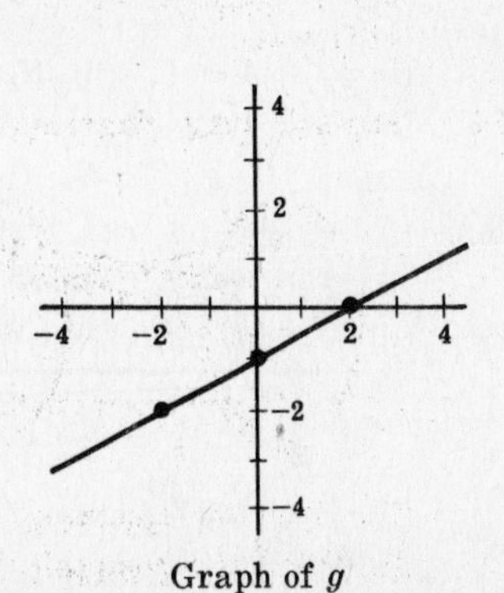

Graph of g

(iii)

x	$h(x)$
-2	13
-1	3
0	-3
1	-5
2	-3
3	3

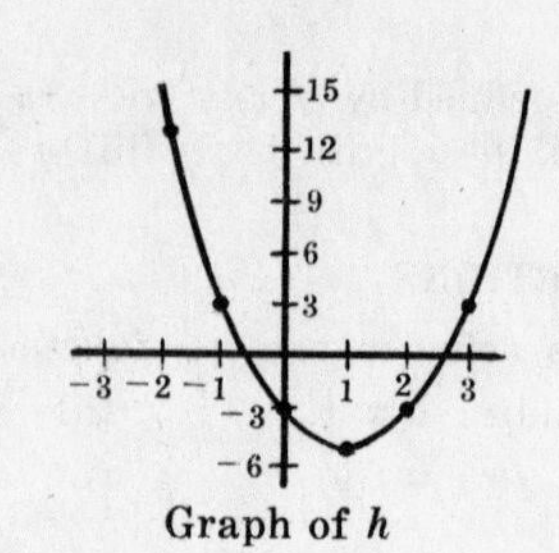

Graph of h

8.34. (i)

x	$f(x)$
-3	-16
-2	0
-1	4
0	2
1	0
2	4
3	20

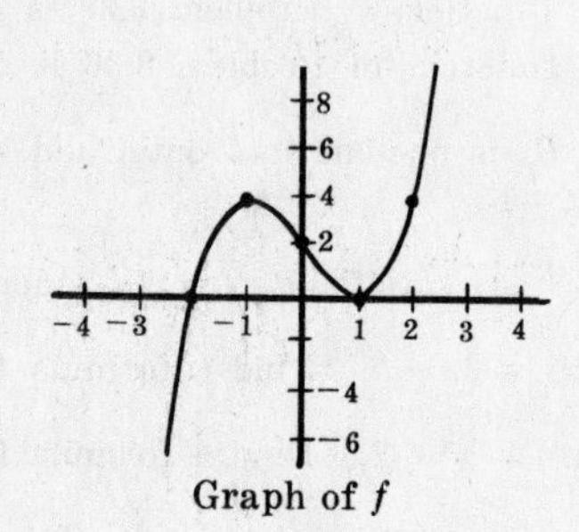

Graph of f

(ii)

x	$g(x)$
-4	105
-3	0
-2	-15
-1	0
0	9
1	0
2	-15
3	0
4	105

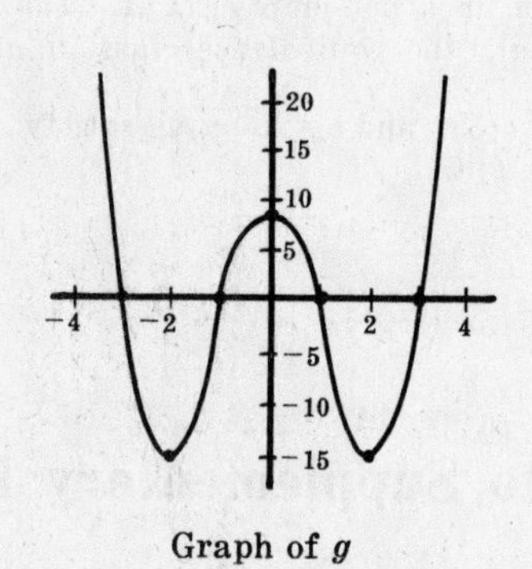

Graph of g

(iii)

x	$h(x)$
4	$\frac{1}{4}$
2	$\frac{1}{2}$
1	1
$\frac{1}{2}$	2
$\frac{1}{4}$	4
0	0
$-\frac{1}{4}$	-4
$-\frac{1}{2}$	-2
-1	-1
-2	$-\frac{1}{2}$
-4	$-\frac{1}{4}$

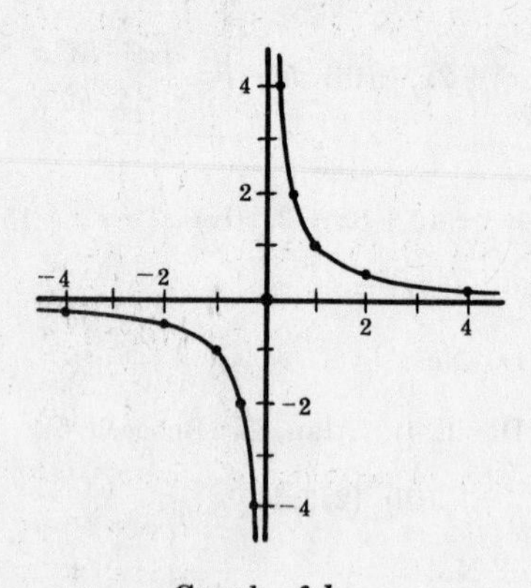

Graph of h

8.35. (i) $g \circ f : A \to A$, (ii) No, (iii) $F \circ f : A \to C$, (iv) No, (v) $g \circ h : C \to A$, (vi) $F \circ h : C \to C$, (vii) $h \circ G \circ g : B \to B$, (viii) $h \circ G : A \to B$

8.36. (i) range of $f = \{1, 2, 4\}$, range of $g = \{1, 2, 3, 4\}$, range of $h = \{1, 3\}$

(ii)

x	$(f \circ g)(x)$	$(h \circ f)(x)$	$g^2(x)$
1	1	3	4
2	4	1	3
3	2	1	2
4	1	3	1

8.37. (i) $(f \circ g)(x) = 4x^2 - 6x + 1$
(ii) $(g \circ f)(x) = 2x^2 + 6x - 1$
(iii) $(g \circ g)(x) = 4x - 9$
(iv) $(f \circ f)(x) = x^4 + 6x^3 + 14x^2 + 15x + 5$

8.38. (i) Yes, (ii) No, (iii) No, (iv) Yes

8.39. (i) Only g is one-one. (ii) Only g is onto.

8.42. $f^{-1}(x) = (x + 7)/3$

8.43. $g^{-1}(x) = \sqrt[3]{x - 2}$

Chapter 9

Vectors

COLUMN VECTORS

A *column vector* u is a set of numbers $u_1, u_2, \ldots, u_n$ written in a column:

$$u = \begin{pmatrix} u_1 \\ u_2 \\ \cdot \\ \cdot \\ \cdot \\ u_n \end{pmatrix}$$

The numbers u_i are called the *components* of the vector u.

Example 1.1: The following are column vectors:

$$\begin{pmatrix} 0 \\ 1 \end{pmatrix}, \quad \begin{pmatrix} 1 \\ -3 \end{pmatrix}, \quad \begin{pmatrix} 1 \\ 7 \\ -8 \end{pmatrix} \quad \text{and} \quad \begin{pmatrix} 1.2 \\ \frac{1}{2} \\ 28 \end{pmatrix}$$

The first two vectors have two components; whereas the last two vectors have three components.

Two column vectors u and v are equal, written $u = v$, if they have the same number of components and if corresponding components are equal. The vectors

$$\begin{pmatrix} 1 \\ 2 \\ 3 \end{pmatrix} \quad \text{and} \quad \begin{pmatrix} 2 \\ 3 \\ 1 \end{pmatrix}$$

are *not* equal since corresponding elements are not equal.

Example 1.2: Let

$$\begin{pmatrix} x - y \\ x + y \\ z - 1 \end{pmatrix} = \begin{pmatrix} 4 \\ 2 \\ 3 \end{pmatrix}$$

Then, by definition of equality of column vectors,

$$\begin{aligned} x - y &= 4 \\ x + y &= 2 \\ z - 1 &= 3 \end{aligned}$$

Solving the above system of equations gives $x = 3$, $y = -1$, and $z = 4$.

Remark: In this chapter we shall frequently refer to numbers as *scalars*.

VECTOR ADDITION

Let u and v be column vectors with the same number of components. The sum of u and v, denoted by $u + v$, is the column vector obtained by adding corresponding components:

$$u + v = \begin{pmatrix} u_1 \\ u_2 \\ \cdot \\ \cdot \\ \cdot \\ u_n \end{pmatrix} + \begin{pmatrix} v_1 \\ v_2 \\ \cdot \\ \cdot \\ \cdot \\ v_n \end{pmatrix} = \begin{pmatrix} u_1 + v_1 \\ u_2 + v_2 \\ \cdot \\ \cdot \\ \cdot \\ u_n + v_n \end{pmatrix}$$

Note that $u + v$ has the same number of components as u and v. The sum of two vectors with different numbers of components is not defined.

Example 2.1:

$$\begin{pmatrix} 1 \\ -2 \\ 3 \end{pmatrix} + \begin{pmatrix} 4 \\ 5 \\ -6 \end{pmatrix} = \begin{pmatrix} 1+4 \\ -2+5 \\ 3-6 \end{pmatrix} = \begin{pmatrix} 5 \\ 3 \\ -3 \end{pmatrix}$$

Example 2.2: The sum

$$\begin{pmatrix} 1 \\ -2 \\ 3 \end{pmatrix} + \begin{pmatrix} 4 \\ -5 \end{pmatrix}$$

is not defined since the vectors have different numbers of components.

A column vector whose components are all zero is called a *zero vector* and is also denoted by 0. The next Example shows that the zero vector is similar to the number zero in that, for any vector u, $u + 0 = u$.

Example 2.3:

$$u + 0 = \begin{pmatrix} u_1 \\ u_2 \\ \cdot \\ \cdot \\ \cdot \\ u_n \end{pmatrix} + \begin{pmatrix} 0 \\ 0 \\ \cdot \\ \cdot \\ \cdot \\ 0 \end{pmatrix} = \begin{pmatrix} u_1 + 0 \\ u_2 + 0 \\ \cdot \\ \cdot \\ \cdot \\ u_n + 0 \end{pmatrix} = \begin{pmatrix} u_1 \\ u_2 \\ \cdot \\ \cdot \\ \cdot \\ u_n \end{pmatrix} = u$$

SCALAR MULTIPLICATION

The product of a scalar k and a column vector u, denoted by $k \cdot u$ or simply ku, is the column vector obtained by multiplying each component of u by k:

$$k \cdot u = k \begin{pmatrix} u_1 \\ u_2 \\ \cdot \\ \cdot \\ \cdot \\ u_n \end{pmatrix} = \begin{pmatrix} ku_1 \\ ku_2 \\ \cdot \\ \cdot \\ \cdot \\ ku_n \end{pmatrix}$$

Observe that u and $k \cdot u$ have the same number of components. We also define:

$$-u = -1 \cdot u \qquad \text{and} \qquad u - v = u + (-v)$$

Example 3.1:

$$4 \begin{pmatrix} 1 \\ -2 \\ 3 \end{pmatrix} = \begin{pmatrix} 4 \\ -8 \\ 12 \end{pmatrix}; \qquad - \begin{pmatrix} 1 \\ -2 \\ 3 \end{pmatrix} = \begin{pmatrix} -1 \\ 2 \\ -3 \end{pmatrix}$$

$$0 \begin{pmatrix} 1 \\ -2 \\ 3 \end{pmatrix} = \begin{pmatrix} 0 \\ 0 \\ 0 \end{pmatrix}; \qquad 1 \begin{pmatrix} 1 \\ -2 \\ 3 \end{pmatrix} = \begin{pmatrix} 1 \\ -2 \\ 3 \end{pmatrix}$$

The main properties of the column vectors under the operations of vector addition and scalar multiplication are contained in the following theorem:

Theorem 9.1: Let V_n be the set of all n-component column vectors, and let $u, v, w \in V_n$ and let k, k' be scalars. Then:

(i) $(u+v)+w = u+(v+w)$, i.e. addition is associative;

(ii) $u+v = v+u$, i.e. addition is commutative;

(iii) $u+0 = 0+u = u$;

(iv) $u+(-u) = (-u)+u = 0$;

(v) $k(u+v) = ku+kv$;

(vi) $(k+k')u = ku+k'u$;

(vii) $(kk')u = k(k'u)$;

(viii) $1u = u$.

The properties listed in the above theorem are those which are used to define an abstract mathematical system called a *linear space* or *vector space*. Accordingly, the theorem can be restated as follows:

Theorem 9.1: The set of all n-component column vectors under the operations of vector addition and scalar multiplication is a vector space.

ROW VECTORS

Analogously, a *row vector* u is a set of numbers $u_1, u_2, \ldots, u_n$ written in a row:

$$u = (u_1, u_2, \ldots, u_n)$$

The numbers u_i are called the *components* of the vector u. Two row vectors are equal if they have the same number of components and if corresponding components are equal. Observe that a row vector is simply an ordered n-tuple of numbers.

The sum of two row vectors u and v with the same number of components, denoted by $u+v$, is the row vector obtained by adding corresponding components from u and v:

$$u+v = (u_1, u_2, \ldots, u_n) + (v_1, v_2, \ldots, v_n) = (u_1+v_1, u_2+v_2, \ldots, u_n+v_n)$$

The product of a scalar k and a row vector u, denoted by $k \cdot u$ or simply ku, is the row vector obtained by multiplying each component of u by k:

$$k \cdot u = k(u_1, u_2, \ldots, u_n) = (ku_1, ku_2, \ldots, ku_n)$$

We also define $\quad -u = -1 \cdot u \quad$ and $\quad u - v = u + (-v)$

as we did for column vectors.

Example 4.1:

$$(1, -2, 3, -4) + (0, 5, -7, 11) = (1, 3, -4, 7)$$
$$5 \cdot (1, -2, 3, -4) = (5, -10, 15, -20)$$

We also have a theorem for row vectors which corresponds to Theorem 9.1.

Theorem 9.2: The set of all n-component row vectors under the operations of vector addition and scalar multiplication satisfies the properties listed in Theorem 9.1, that is, is a vector space.

MULTIPLICATION OF A ROW VECTOR AND A COLUMN VECTOR

If a row vector u and a column vector v have the same number of components, then their product, denoted by $u \cdot v$ or simply uv, is the scalar obtained by multiplying corresponding elements and adding the resulting products:

$$u \cdot v = (u_1, u_2, \ldots, u_n)\begin{pmatrix} v_1 \\ v_2 \\ \cdot \\ \cdot \\ \cdot \\ v_n \end{pmatrix} = u_1v_1 + u_2v_2 + \cdots + u_nv_n = \sum_{i=1}^{n} u_iv_i$$

If a row vector and a column vector have different numbers of components, then their product is not defined.

Example 5.1:

$$(1, -2, 3)\begin{pmatrix} 4 \\ 5 \\ -6 \end{pmatrix} = 1 \cdot 4 + (-2) \cdot 5 + 3 \cdot (-6) = -24$$

Example 5.2: The product

$$(1, -2, 3)\begin{pmatrix} 5 \\ 7 \end{pmatrix}$$

is not defined since the vectors have different numbers of components.

Remark: The product of two (row) vectors $u = (u_1, u_2, \ldots, u_n)$ and $v = (v_1, v_2, \ldots, v_n)$ is also defined as above:

$$\begin{aligned} u \cdot v &= (u_1, u_2, \ldots, u_n) \cdot (v_1, v_2, \ldots, v_n) \\ &= u_1v_1 + u_2v_2 + \cdots + u_nv_n = \sum_{i=1}^{n} u_iv_i \end{aligned}$$

This product is called the *dot product* (or: *scalar product* or *inner product*).

Solved Problems

COLUMN VECTORS

9.1. Compute: (i) $\begin{pmatrix} 3 \\ -4 \\ 5 \end{pmatrix} + \begin{pmatrix} 1 \\ 1 \\ -2 \end{pmatrix}$ (ii) $\begin{pmatrix} 1 \\ 2 \\ -3 \end{pmatrix} + \begin{pmatrix} 4 \\ -5 \end{pmatrix}$ (iii) $-3\begin{pmatrix} 4 \\ -5 \\ -6 \end{pmatrix}$ (iv) $-\begin{pmatrix} -6 \\ 7 \\ -8 \end{pmatrix}$

(i) Add corresponding components:

$$\begin{pmatrix} 3 \\ -4 \\ 5 \end{pmatrix} + \begin{pmatrix} 1 \\ 1 \\ -2 \end{pmatrix} = \begin{pmatrix} 3+1 \\ -4+1 \\ 5-2 \end{pmatrix} = \begin{pmatrix} 4 \\ -3 \\ 3 \end{pmatrix}$$

(ii) The sum is not defined since the column vectors have different numbers of components.

(iii) Multiply each component by the scalar:

$$-3\begin{pmatrix} 4 \\ -5 \\ -6 \end{pmatrix} = \begin{pmatrix} -3 \cdot 4 \\ -3 \cdot -5 \\ -3 \cdot -6 \end{pmatrix} = \begin{pmatrix} -12 \\ 15 \\ 18 \end{pmatrix}$$

(iv) Multiply each component by -1; i.e. change the sign of each component:

$$-\begin{pmatrix} -6 \\ 7 \\ -8 \end{pmatrix} = \begin{pmatrix} 6 \\ -7 \\ 8 \end{pmatrix}$$

9.2. Compute: (i) $2\begin{pmatrix}1\\-1\\3\end{pmatrix} - 3\begin{pmatrix}2\\3\\-4\end{pmatrix}$ (ii) $-2\begin{pmatrix}5\\3\\-4\end{pmatrix} + 4\begin{pmatrix}-1\\5\\2\end{pmatrix} - 3\begin{pmatrix}3\\-1\\-1\end{pmatrix}$

First perform the scalar multiplication and then the vector addition:

(i) $$2\begin{pmatrix}1\\-1\\3\end{pmatrix} - 3\begin{pmatrix}2\\3\\-4\end{pmatrix} = \begin{pmatrix}2\\-2\\6\end{pmatrix} + \begin{pmatrix}-6\\-9\\12\end{pmatrix} = \begin{pmatrix}-4\\-11\\18\end{pmatrix}$$

(ii) $$-2\begin{pmatrix}5\\3\\-4\end{pmatrix} + 4\begin{pmatrix}-1\\5\\2\end{pmatrix} - 3\begin{pmatrix}3\\-1\\-1\end{pmatrix} = \begin{pmatrix}-10\\-6\\8\end{pmatrix} + \begin{pmatrix}-4\\20\\8\end{pmatrix} + \begin{pmatrix}-9\\3\\3\end{pmatrix} = \begin{pmatrix}-23\\17\\19\end{pmatrix}$$

9.3. Find x and y if $\begin{pmatrix}x\\3\end{pmatrix} = \begin{pmatrix}2\\x+y\end{pmatrix}$.

Since the two vectors are equal, the corresponding components are equal to each other:

$$\begin{cases} x = 2 \\ 3 = x + y \end{cases}$$

Substituting $x = 2$ into the second equation and solving for y gives $y = 1$.

9.4. Find x and y if $\begin{pmatrix}4\\y\end{pmatrix} = x\begin{pmatrix}2\\3\end{pmatrix}$.

Multiplying by the scalar x gives

$$\begin{pmatrix}4\\y\end{pmatrix} = x\begin{pmatrix}2\\3\end{pmatrix} = \begin{pmatrix}2x\\3x\end{pmatrix}$$

Set the corresponding components equal to each other:

$$4 = 2x$$
$$y = 3x$$

Solve the linear equations for x and y: $x = 2$ and $y = 6$.

9.5. Find x, y and z if: $x\begin{pmatrix}1\\1\\1\end{pmatrix} + y\begin{pmatrix}1\\1\\0\end{pmatrix} + z\begin{pmatrix}1\\0\\0\end{pmatrix} = \begin{pmatrix}2\\-3\\4\end{pmatrix}$

First multiply by the scalars x, y and z and then add:

$$x\begin{pmatrix}1\\1\\1\end{pmatrix} + y\begin{pmatrix}1\\1\\0\end{pmatrix} + z\begin{pmatrix}1\\0\\0\end{pmatrix} = \begin{pmatrix}x\\x\\x\end{pmatrix} + \begin{pmatrix}y\\y\\0\end{pmatrix} + \begin{pmatrix}z\\0\\0\end{pmatrix} = \begin{pmatrix}x+y+z\\x+y\\x\end{pmatrix} = \begin{pmatrix}2\\-3\\4\end{pmatrix}$$

Set the corresponding components equal to each other:

$$x + y + z = 2$$
$$x + y = -3$$
$$x = 4$$

Solve the system of linear equations. Substitute $x = 4$ into the second equation to obtain $4 + y = -3$ or $y = -7$. Then substitute into the first equation to obtain $z = 5$. Thus $x = 4$, $y = -7$ and $z = 5$.

ROW VECTORS

9.6. Let $u = (2, -7, 1)$, $v = (-3, 0, 4)$ and $w = (0, 5, -8)$. Find: (i) $u + v$, (ii) $v + w$, (iii) $-3u$, (iv) $-w$.

(i) Add corresponding components:

$$u + v = (2, -7, 1) + (-3, 0, 4) = (2 - 3, -7 + 0, 1 + 4) = (-1, -7, 5)$$

(ii) Add corresponding components:

$$v + w = (-3, 0, 4) + (0, 5, -8) = (-3 + 0, 0 + 5, 4 - 8) = (-3, 5, -4)$$

(iii) Multiply each component of u by the scalar -3: $-3u = -3(2, -7, 1) = (-6, 21, -3)$.

(iv) Multiply each component of w by -1, i.e. change the sign of each component:

$$-w = -(0, 5, -8) = (0, -5, 8)$$

9.7. Let u, v and w be the row vectors of the preceding problem. Find (i) $3u - 4v$, (ii) $2u + 3v - 5w$.

First perform the scalar multiplication and then the vector addition.

(i) $3u - 4v = 3(2, -7, 1) - 4(-3, 0, 4) = (6, -21, 3) + (12, 0, -16) = (18, -21, -13)$

(ii)
$$\begin{aligned} 2u + 3v - 5w &= 2(2, -7, 1) + 3(-3, 0, 4) - 5(0, 5, -8) \\ &= (4, -14, 2) + (-9, 0, 12) + (0, -25, 40) \\ &= (4 - 9 + 0, -14 + 0 - 25, 2 + 12 + 40) = (-5, -39, 54) \end{aligned}$$

9.8. Find x, y and z if $(2x, 3, y) = (4, x + z, 2z)$.

Set corresponding components equal to each other to obtain the system of equations

$$\begin{aligned} 2x &= 4 \\ 3 &= x + z \\ y &= 2z \end{aligned}$$

Then $x = 2$, $z = 1$ and $y = 2$.

9.9. Find x and y if $x(1, 1) + y(2, -1) = (1, 4)$.

First multiply by the scalars x and y and then add:

$$x(1, 1) + y(2, -1) = (x, x) + (2y, -y) = (x + 2y, x - y) = (1, 4)$$

Now set corresponding components equal to each other to obtain

$$\begin{aligned} x + 2y &= 1 \\ x - y &= 4 \end{aligned}$$

Solve the system of equations to find $x = 3$ and $y = -1$.

MULTIPLICATION OF A ROW VECTOR AND A COLUMN VECTOR

9.10. Compute: (i) $(2, -3, 6)\begin{pmatrix} 8 \\ 2 \\ -3 \end{pmatrix}$, (ii) $(1, -1, 0, 5)\begin{pmatrix} 3 \\ -1 \\ 4 \end{pmatrix}$, (iii) $(3, -5, 2, 1)\begin{pmatrix} 4 \\ 1 \\ -2 \\ 5 \end{pmatrix}$.

(i) Multiply corresponding components and add:

$$(2, -3, 6)\begin{pmatrix} 8 \\ 2 \\ -3 \end{pmatrix} = 2 \cdot 8 + (-3) \cdot 2 + 6 \cdot (-3) = 16 - 6 - 18 = -8$$

(ii) The product is not defined since the vectors have different numbers of components.

(iii) Multiply corresponding components and add:

$$(3, -5, 2, 1)\begin{pmatrix}4\\1\\-2\\5\end{pmatrix} = 3\cdot4 + (-5)\cdot1 + 2\cdot(-2) + 1\cdot5 = 12 - 5 - 4 + 5 = 8$$

9.11. Let $u = (1, k, -3)$ and $v = \begin{pmatrix}2\\-5\\4\end{pmatrix}$. Determine k so that $u\cdot v = 0$.

Compute $u\cdot v$ and set it equal to 0:

$$u\cdot v = (1, k, -3)\begin{pmatrix}2\\-5\\4\end{pmatrix} = 1\cdot2 + (-5)\cdot k + (-3)\cdot4 = 2 - 5k - 12 = -10 - 5k = 0$$

The solution to the equation $-10 - 5k = 0$ is $k = -2$.

MISCELLANEOUS PROBLEMS

9.12. Prove Theorem 9.1(i): For any vectors u, v and w, $(u+v) + w = u + (v+w)$.

Let u_i, v_i and w_i be the ith components of u, v and w respectively. Then $u_i + v_i$ is the ith component of $u+v$ and so $(u_i+v_i)+w_i$ is the ith component of $(u+v)+w$. On the other hand, v_i+w_i is the ith component of $v+w$ and so $u_i+(v_i+w_i)$ is the ith component of $u+(v+w)$. But u_i, v_i and w_i are numbers for which the associative law holds, that is,

$$(u_i+v_i)+w_i = u_i + (v_i+w_i), \quad \text{for each } i$$

Accordingly, $(u+v)+w = u+(v+w)$ since all their corresponding components are equal.

9.13. Let 0 be the number zero and θ the zero vector. Show that, for any vector u, $0u = \theta$.

Method 1:

Let u_i be the ith component of u. Then $0u_i = 0$ is the ith component of $0u$. Since every component of $0u$ is zero, $0u$ is the zero vector, i.e. $0u = \theta$.

Method 2:

We use Theorem 9.1:

$$0u = (0+0)u = 0u + 0u$$

Now adding $-0u$ to both sides gives

$$\theta = 0u + (-0u) = 0u + [0u + (-0u)] = 0u + \theta = 0u$$

9.14. Prove Theorem 9.1(vi): $(k+k')u = ku + k'u$.

Observe that the first plus sign refers to the addition of the two scalars k and k' whereas the second plus sign refers to the vector addition of the two vectors ku and $k'u$.

Let u_i be the ith component of u. Then $(k+k')u_i$ is the ith component of $(k+k')u$. On the other hand, ku_i and $k'u_i$ are the ith components of the vectors ku and $k'u$, respectively, and so $ku_i + k'u_i$ is the ith component of the vector $ku + k'u$. But k, k' and u_i are numbers; hence

$$(k+k')u_i = ku_i + k'u_i, \quad \text{for each } i$$

Thus $(k+k')u = ku + k'u$, as corresponding components are equal.

9.15. The norm $\|u\|$ of a vector $u = (u_1, u_2, \ldots, u_n)$ is defined by

$$\|u\| = \sqrt{u_1^2 + u_2^2 + \cdots + u_n^2}$$

(i) Find $\|(3,-4,12)\|$.

(ii) Prove: $\|ku\| = |k|\,\|u\|$, for any real number k.

(i) $\|(3,-4,12)\| = \sqrt{3^2+(-4)^2+12^2} = \sqrt{9+16+144} = \sqrt{169} = 13$

(ii)
$$\begin{aligned}\|ku\| &= \|k(u_1,u_2,\ldots,u_n)\| = \|(ku_1,ku_2,\ldots,ku_n)\| \\ &= \sqrt{(ku_1)^2+(ku_2)^2+\cdots+(ku_n)^2} = \sqrt{k^2u_1^2+k^2u_2^2+\cdots+k^2u_n^2} \\ &= \sqrt{k^2(u_1^2+u_2^2+\cdots+u_n^2)} = \sqrt{k^2}\,\sqrt{u_1^2+u_2^2+\cdots+u_n^2} = |k|\,\|u\|\end{aligned}$$

Supplementary Problems

COLUMN VECTORS

9.16. Compute: (i) $\begin{pmatrix}2\\-4\\5\\-1\end{pmatrix}+\begin{pmatrix}-3\\4\\0\\-7\end{pmatrix}$ (ii) $\begin{pmatrix}1\\-2\\7\\6\end{pmatrix}+\begin{pmatrix}1\\-5\\8\end{pmatrix}$ (iii) $4\begin{pmatrix}6\\-3\\-5\\1\end{pmatrix}$ (iv) $-\begin{pmatrix}9\\-1\\-2\\-1\end{pmatrix}$

9.17. Compute: (i) $3\begin{pmatrix}0\\-2\\4\\-4\end{pmatrix}+5\begin{pmatrix}6\\-3\\-2\\1\end{pmatrix}$ (ii) $3\begin{pmatrix}-1\\0\\-5\\4\end{pmatrix}-2\begin{pmatrix}2\\-3\\1\\5\end{pmatrix}$ (iii) $-\begin{pmatrix}-5\\4\\-2\\1\end{pmatrix}-4\begin{pmatrix}3\\7\\-2\\-1\end{pmatrix}$

9.18. Compute: (i) $5\begin{pmatrix}1\\0\\-3\\-1\end{pmatrix}+2\begin{pmatrix}-2\\6\\-1\\-2\end{pmatrix}-4\begin{pmatrix}0\\-3\\-3\\1\end{pmatrix}$ (ii) $-2\begin{pmatrix}-3\\4\\1\\2\end{pmatrix}-6\begin{pmatrix}1\\-1\\-1\\-2\end{pmatrix}+3\begin{pmatrix}-3\\-1\\0\\3\end{pmatrix}$

9.19. Determine x and y if: (i) $\begin{pmatrix}x\\x+y\end{pmatrix}=\begin{pmatrix}y-2\\6\end{pmatrix}$ (ii) $x\begin{pmatrix}3\\2\end{pmatrix}=2\begin{pmatrix}y\\-1\end{pmatrix}$.

9.20. Determine x, y and z if $x\begin{pmatrix}1\\1\\1\end{pmatrix}+y\begin{pmatrix}1\\-1\\0\end{pmatrix}+z\begin{pmatrix}1\\0\\0\end{pmatrix}=\begin{pmatrix}3\\-1\\2\end{pmatrix}$.

9.21. Determine x and y if $x\begin{pmatrix}2\\y\end{pmatrix}=y\begin{pmatrix}1\\-2\end{pmatrix}$.

ROW VECTORS

9.22. Let $u=(2,-1,0,-3)$, $v=(1,-1,-1,3)$ and $w=(1,3,-2,2)$. Find:
(i) $3u$, (ii) $u+v$, (iii) $2u-3v$, (iv) $5u-3v-4w$, (v) $-u+2v-2w$.

9.23. Find x and y if $x(1,2)=-4(y,3)$.

9.24. Find x, y and z if $x(1,1,0)+y(2,0,-1)+z(0,1,1)=(-1,3,3)$.

MULTIPLICATION OF A ROW VECTOR AND A COLUMN VECTOR

9.25. Compute: (i) $(2,-3,-1)\begin{pmatrix}4\\1\\-2\end{pmatrix}$ (ii) $(1,-3,-2,4)\begin{pmatrix}-3\\2\\1\end{pmatrix}$ (iii) $(3,-1,2,0)\begin{pmatrix}1\\-1\\-2\\5\end{pmatrix}$

9.26. Determine k so that $(1, k, -2, -3)\begin{pmatrix} 3 \\ 2 \\ -4 \\ 1 \end{pmatrix} = 0.$

9.27. Prove: If $u = (u_1, \ldots, u_n)$ is a row vector such that $u \cdot v = 0$ for every column vector v with n components, then u is the zero vector.

MISCELLANEOUS PROBLEMS

9.28. Prove Theorem 9.1(v): For any vectors u and v and scalar k, $k(u+v) = ku + kv$.

9.29. Prove Theorem 9.1(vii): For any vector u and scalars k, k', $(kk')u = k(k'u)$.

9.30. Find (i) $\|(3, 5, -4)\|$, (ii) $\|(2, -3, 6, 4)\|$.

Answers to Supplementary Problems

9.16. (i) $\begin{pmatrix} -1 \\ 0 \\ 5 \\ -8 \end{pmatrix}$ (ii) Not defined. (iii) $\begin{pmatrix} 24 \\ -12 \\ -20 \\ 4 \end{pmatrix}$ (iv) $\begin{pmatrix} -9 \\ 1 \\ 2 \\ 1 \end{pmatrix}$

9.17. (i) $\begin{pmatrix} 30 \\ -21 \\ 2 \\ -7 \end{pmatrix}$ (ii) $\begin{pmatrix} -7 \\ 6 \\ -17 \\ 2 \end{pmatrix}$ (iii) $\begin{pmatrix} -7 \\ -32 \\ 10 \\ 3 \end{pmatrix}$

9.18. (i) $\begin{pmatrix} 1 \\ 24 \\ -5 \\ -13 \end{pmatrix}$ (ii) $\begin{pmatrix} -9 \\ -5 \\ 4 \\ 17 \end{pmatrix}$

9.19. (i) $x = 2,\ y = 4$ (ii) $x = -1,\ y = -3/2$

9.20. $x = 2,\ y = 3,\ z = -2$

9.21. $x = 0,\ y = 0$ or $x = -2,\ y = -4$

9.22. $3u = (6, -3, 0, -9)$, $u + v = (3, -2, -1, 0)$, $2u - 3v = (1, 1, 3, -15)$, $5u - 3v - 4w = (3, -14, 11, -32)$, $-u + 2v - 2w = (-2, -7, 2, 5)$

9.23. $x = -6,\ y = 3/2$

9.24. $x = 1,\ y = -1,\ z = 2$

9.25. (i) 7 (ii) Not defined (iii) 0

9.26. $k = -4$

9.30. (i) $5\sqrt{2}$, (ii) $\sqrt{65}$

Chapter 10

Matrices

MATRICES

A *matrix* is a rectangular array of numbers; the general form of a matrix with m rows and n columns is

$$\begin{pmatrix} a_{11} & a_{12} & a_{13} & \cdots & a_{1n} \\ a_{21} & a_{22} & a_{23} & \cdots & a_{2n} \\ \cdots & \cdots & \cdots & \cdots & \cdots \\ a_{m1} & a_{m2} & a_{m3} & \cdots & a_{mn} \end{pmatrix}$$

We denote such a matrix by

$$(a_{ij})_{m,n} \quad \text{or simply} \quad (a_{ij})$$

and call it an $m \times n$ matrix (read "m by n"). Note that the row and column of the element a_{ij} is indicated by its first and second subscript respectively.

Example 1.1: Consider the 2×3 matrix $\begin{pmatrix} 1 & -3 & 4 \\ 0 & 5 & -2 \end{pmatrix}$.

Its rows are $(1, -3, 4)$ and $(0, 5, -2)$ and its columns are $\begin{pmatrix} 1 \\ 0 \end{pmatrix}$, $\begin{pmatrix} -3 \\ 5 \end{pmatrix}$, and $\begin{pmatrix} 4 \\ -2 \end{pmatrix}$.

In this chapter, capital letters $A, B, \ldots$ denote matrices whereas lower case letters $a, b, \ldots$ denote numbers, which we call *scalars*. Two matrices A and B are equal, written $A = B$, if they have the same *shape*, i.e. the same number of rows and the same number of columns, and if corresponding elements are equal. Hence the equality of two $m \times n$ matrices is equivalent to a system of mn equalities, one for each pair of elements.

Example 1.2: The statement $\begin{pmatrix} x+y & 2z+w \\ x-y & z-w \end{pmatrix} = \begin{pmatrix} 3 & 5 \\ 1 & 4 \end{pmatrix}$

is equivalent to the system of equations

$$\begin{cases} x + y = 3 \\ x - y = 1 \\ 2z + w = 5 \\ z - w = 4 \end{cases}$$

The solution of the system of equations is $x = 2,\ y = 1,\ z = 3,\ w = -1$.

Remark: A matrix with one row is simply a row vector, and a matrix with one column is simply a column vector. Hence vectors are a special case of matrices.

MATRIX ADDITION

Let A and B be two matrices with the same shape, i.e. the same number of rows and of columns. The *sum* of A and B, written $A + B$, is the matrix obtained by adding corresponding elements from A and B:

$$\begin{pmatrix} a_{11} & a_{12} & \cdots & a_{1n} \\ a_{21} & a_{22} & \cdots & a_{2n} \\ \cdots & \cdots & \cdots & \cdots \\ a_{m1} & a_{m2} & \cdots & a_{mn} \end{pmatrix} + \begin{pmatrix} b_{11} & b_{12} & \cdots & b_{1n} \\ b_{21} & b_{22} & \cdots & b_{2n} \\ \cdots & \cdots & \cdots & \cdots \\ b_{m1} & b_{m2} & \cdots & b_{mn} \end{pmatrix} = \begin{pmatrix} a_{11}+b_{11} & a_{12}+b_{12} & \cdots & a_{1n}+b_{1n} \\ a_{21}+b_{21} & a_{22}+b_{22} & \cdots & a_{2n}+b_{2n} \\ \cdots & \cdots & \cdots & \cdots \\ a_{m1}+b_{m1} & a_{m2}+b_{m2} & \cdots & a_{mn}+b_{mn} \end{pmatrix}$$

Note that $A+B$ has the same shape as A and B. The sum of two matrices with different shapes is not defined.

Example 2.1: $\begin{pmatrix} 1 & -2 & 3 \\ 0 & 4 & 5 \end{pmatrix} + \begin{pmatrix} 3 & 0 & -6 \\ 2 & -3 & 1 \end{pmatrix} = \begin{pmatrix} 1+3 & -2+0 & 3+(-6) \\ 0+2 & 4+(-3) & 5+1 \end{pmatrix} = \begin{pmatrix} 4 & -2 & -3 \\ 2 & 1 & 6 \end{pmatrix}$

Example 2.2: The sum $\begin{pmatrix} 1 & -2 \\ 3 & 4 \end{pmatrix} + \begin{pmatrix} 0 & 5 & -2 \\ 1 & -3 & -1 \end{pmatrix}$

is not defined since the matrices have different shapes.

A matrix whose elements are all zero is called a *zero matrix* and is also denoted by 0. Part (iii) of the following theorem shows the similarity between the zero matrix and the scalar zero.

Theorem 10.1: For matrices A, B and C (with the same shape),

(i) $(A+B)+C = A+(B+C)$, i.e. addition is associative;

(ii) $A+B = B+A$, i.e. addition is commutative;

(iii) $A+0 = 0+A = A$.

SCALAR MULTIPLICATION

The product of a scalar k and a matrix A, written kA or Ak, is the matrix obtained by multiplying each element of A by k:

$$k\begin{pmatrix} a_{11} & a_{12} & \cdots & a_{1n} \\ a_{21} & a_{22} & \cdots & a_{2n} \\ \cdots & \cdots & \cdots & \cdots \\ a_{m1} & a_{m2} & \cdots & a_{mn} \end{pmatrix} = \begin{pmatrix} ka_{11} & ka_{12} & \cdots & ka_{1n} \\ ka_{21} & ka_{22} & \cdots & ka_{2n} \\ \cdots & \cdots & \cdots & \cdots \\ ka_{m1} & ka_{m2} & \cdots & ka_{mn} \end{pmatrix}$$

Note that A and kA have the same shape.

Example 3.1: $3\begin{pmatrix} 1 & -2 & 0 \\ 4 & 3 & -5 \end{pmatrix} = \begin{pmatrix} 3\cdot 1 & 3\cdot(-2) & 3\cdot 0 \\ 3\cdot 4 & 3\cdot 3 & 3\cdot(-5) \end{pmatrix} = \begin{pmatrix} 3 & -6 & 0 \\ 12 & 9 & -15 \end{pmatrix}$

We also introduce the following notation:

$$-A = (-1)A \quad \text{and} \quad A-B = A+(-B)$$

The next theorem follows directly from the above definition of scalar multiplication.

Theorem 10.2: For any scalars k_1 and k_2 and any matrices A and B (with the same shape),

(i) $(k_1k_2)A = k_1(k_2A)$

(ii) $k_1(A+B) = k_1A + k_1B$

(iii) $(k_1+k_2)A = k_1A + k_2A$

(iv) $1\cdot A = A$, and $0A = 0$

(v) $A+(-A) = (-A)+A = 0$.

Using (iii) and (iv) above, we also have that

$$A + A = 2A, \quad A + A + A = 3A, \quad \ldots$$

MATRIX MULTIPLICATION

Let A and B be matrices such that the number of columns of A is equal to the number of rows of B. Then the product of A and B, written AB, is the matrix with the same number of rows as A and of columns as B and whose element in the ith row and jth column is obtained by multiplying the ith row of A by the jth column of B:

$$\begin{pmatrix} a_{11} & \cdots & a_{1p} \\ \cdot & \cdots & \cdot \\ a_{i1} & \cdots & a_{ip} \\ \cdot & \cdots & \cdot \\ a_{m1} & \cdots & a_{mp} \end{pmatrix} \begin{pmatrix} b_{11} & \cdots & b_{1j} & \cdots & b_{1n} \\ \cdot & \cdots & \cdot & \cdots & \cdot \\ \cdot & \cdots & \cdot & \cdots & \cdot \\ \cdot & \cdots & \cdot & \cdots & \cdot \\ b_{p1} & \cdots & b_{pj} & \cdots & b_{pn} \end{pmatrix} = \begin{pmatrix} c_{11} & \cdots & c_{1n} \\ \cdot & \cdots & \cdot \\ \cdot & c_{ij} & \cdot \\ \cdot & \cdots & \cdot \\ c_{m1} & \cdots & c_{mn} \end{pmatrix}$$

where $c_{ij} = a_{i1}b_{1j} + a_{i2}b_{2j} + \cdots + a_{ip}b_{pj}$.

In other words, if $A = (a_{ij})$ is an $m \times p$ matrix and $B = (b_{ij})$ is a $p \times n$ matrix, then $AB = (c_{ij})$ is the $m \times n$ matrix for which

$$c_{ij} = a_{i1}b_{1j} + a_{i2}b_{2j} + \cdots + a_{ip}b_{pj} = \sum_{k=1}^{p} a_{ik}b_{kj}$$

If the number of columns of A is not equal to the number of rows of B, say A is $m \times p$ and B is $q \times n$ where $p \neq q$, then the product AB is not defined.

Example 4.1:

$$\begin{pmatrix} r & s \\ t & u \end{pmatrix} \begin{pmatrix} a_1 & a_2 & a_3 \\ b_1 & b_2 & b_3 \end{pmatrix} = \begin{pmatrix} ra_1 + sb_1 & ra_2 + sb_2 & ra_3 + sb_3 \\ ta_1 + ub_1 & ta_2 + ub_2 & ta_3 + ub_3 \end{pmatrix}$$

Example 4.2:

$$\begin{pmatrix} 1 & 2 \\ 3 & 4 \end{pmatrix} \begin{pmatrix} 1 & 1 \\ 0 & 2 \end{pmatrix} = \begin{pmatrix} 1\cdot 1 + 2\cdot 0 & 1\cdot 1 + 2\cdot 2 \\ 3\cdot 1 + 4\cdot 0 & 3\cdot 1 + 4\cdot 2 \end{pmatrix} = \begin{pmatrix} 1 & 5 \\ 3 & 11 \end{pmatrix}$$

$$\begin{pmatrix} 1 & 1 \\ 0 & 2 \end{pmatrix} \begin{pmatrix} 1 & 2 \\ 3 & 4 \end{pmatrix} = \begin{pmatrix} 1\cdot 1 + 1\cdot 3 & 1\cdot 2 + 1\cdot 4 \\ 0\cdot 1 + 2\cdot 3 & 0\cdot 2 + 2\cdot 4 \end{pmatrix} = \begin{pmatrix} 4 & 6 \\ 6 & 8 \end{pmatrix}$$

We see by the preceding example that matrices under the operation of matrix multiplication do not satisfy the commutative law, i.e. the products AB and BA of matrices need not be equal.

Matrix multiplication does, however, satisfy the following properties:

Theorem 10.3: (i) $(AB)C = A(BC)$

(ii) $A(B + C) = AB + AC$

(iii) $(B + C)A = BA + CA$

(iv) $k(AB) = (kA)B = A(kB)$, where k is a scalar.

We assume that the sums and products in the above theorem are defined.

Remark: In the special case where one of the factors of AB is a vector, then the product is also a vector:

$$(a_1, a_2, \ldots, a_p) \begin{pmatrix} b_{11} & b_{12} & \cdots & b_{1n} \\ b_{21} & b_{22} & \cdots & b_{2n} \\ \cdots & \cdots & \cdots & \cdots \\ b_{p1} & b_{p2} & \cdots & b_{pn} \end{pmatrix} = \left(\sum_{i=1}^{p} a_i b_{i1}, \ \sum_{i=1}^{p} a_i b_{i2}, \ \ldots, \ \sum_{i=1}^{p} a_i b_{in} \right)$$

$$\begin{pmatrix} a_{11} & a_{12} & \cdots & a_{1p} \\ a_{21} & a_{22} & \cdots & a_{2p} \\ \cdots & \cdots & \cdots & \cdots \\ a_{m1} & a_{m2} & \cdots & a_{mp} \end{pmatrix} \begin{pmatrix} b_1 \\ b_2 \\ \cdot \\ b_p \end{pmatrix} = \begin{pmatrix} a_{11}b_1 + \cdots + a_{1p}b_p \\ a_{21}b_1 + \cdots + a_{2p}b_p \\ \cdots\cdots\cdots\cdots \\ a_{m1}b_1 + \cdots + a_{mp}b_p \end{pmatrix}$$

Example 4.3:

$$(2, -3, 4)\begin{pmatrix} 1 & -3 \\ 5 & 0 \\ -2 & 4 \end{pmatrix} = (2\cdot 1 + (-3)\cdot 5 + 4\cdot(-2),\ 2\cdot(-3) + (-3)\cdot 0 + 4\cdot 4) = (-21, 10)$$

$$\begin{pmatrix} 1 & -3 \\ 5 & 0 \\ -2 & 4 \end{pmatrix}\begin{pmatrix} -1 \\ 2 \end{pmatrix} = \begin{pmatrix} 1\cdot(-1) + (-3)\cdot 2 \\ 5\cdot(-1) + 0\cdot 2 \\ (-2)\cdot(-1) + 4\cdot 2 \end{pmatrix} = \begin{pmatrix} -7 \\ -5 \\ 10 \end{pmatrix}$$

Example 4.4:

A system of linear equations, such as

$$\begin{cases} x + 2y - 3z = 4 \\ 5x - 6y + 8z = 8 \end{cases}$$

is equivalent to the matrix equation

$$\begin{pmatrix} 1 & 2 & -3 \\ 5 & -6 & 8 \end{pmatrix}\begin{pmatrix} x \\ y \\ z \end{pmatrix} = \begin{pmatrix} 4 \\ 8 \end{pmatrix}$$

That is, any solution to the system of equations is also a solution to the matrix equation, and vice versa.

SQUARE MATRICES

A matrix with the same number of rows as columns is called a *square* matrix. A square matrix with n rows and n columns is said to be of *order n*, and is called an *n-square matrix*. The main diagonal, or simply diagonal, of a square matrix $A = (a_{ij})$ is the numbers $a_{11}, a_{22}, \ldots, a_{nn}$.

Example 5.1:

The matrix

$$\begin{pmatrix} 1 & -2 & 0 \\ 0 & -4 & -1 \\ 5 & 3 & 2 \end{pmatrix}$$

is a square matrix of order 3. The numbers along the main diagonal are 1, −4 and 2.

The n-square matrix with 1's along the main diagonal and 0's elsewhere, e.g.,

$$\begin{pmatrix} 1 & 0 & 0 & 0 \\ 0 & 1 & 0 & 0 \\ 0 & 0 & 1 & 0 \\ 0 & 0 & 0 & 1 \end{pmatrix}$$

is called the unit matrix and will be denoted by I. The unit matrix I plays the same role in matrix multiplication as the number 1 does in the usual multiplication of numbers. Specifically,

Theorem 10.4: For any square matrix A,

$$AI = IA = A$$

ALGEBRA OF SQUARE MATRICES

Recall that not every two matrices can be added or multiplied. However, if we only consider square matrices of some given order n, then this inconvenience disappears. Specifically, any $n \times n$ matrix can be added to or multiplied by another $n \times n$ matrix, or multiplied by a scalar, and the result is again an $n \times n$ matrix.

In particular, if A is any n-square matrix, we can form powers of A:

$$A^2 = AA, \; A^3 = A^2A, \; \ldots \quad \text{and } A^0 = I$$

We can also form polynomials in A. That is, for any polynomial

$$f(x) \;=\; a_0 + a_1x + a_2x^2 + \cdots + a_nx^n$$

we define $f(A)$ to be the matrix

$$f(A) \;=\; a_0I + a_1A + a_2A^2 + \cdots + a_nA^n$$

In the case that $f(A)$ is the zero matrix, then A is said to be a *zero* or *root* of the polynomial $f(x)$.

Example 6.1: Let $A = \begin{pmatrix} 1 & 2 \\ 3 & -4 \end{pmatrix}$; then $A^2 = \begin{pmatrix} 7 & -6 \\ -9 & 22 \end{pmatrix}$.

If $f(x) = 2x^2 - 3x + 5$, then

$$f(A) \;=\; 2\begin{pmatrix} 7 & -6 \\ -9 & 22 \end{pmatrix} - 3\begin{pmatrix} 1 & 2 \\ 3 & -4 \end{pmatrix} + 5\begin{pmatrix} 1 & 0 \\ 0 & 1 \end{pmatrix} \;=\; \begin{pmatrix} 16 & -18 \\ -27 & 61 \end{pmatrix}$$

On the other hand, if $g(x) = x^2 + 3x - 10$ then

$$g(A) \;=\; \begin{pmatrix} 7 & -6 \\ -9 & 22 \end{pmatrix} + 3\begin{pmatrix} 1 & 2 \\ 3 & -4 \end{pmatrix} - 10\begin{pmatrix} 1 & 0 \\ 0 & 1 \end{pmatrix} = \begin{pmatrix} 0 & 0 \\ 0 & 0 \end{pmatrix}$$

Thus A is a zero of the polynomial $g(x)$.

TRANSPOSE

The transpose of a matrix A, written A^t, is the matrix obtained by writing the rows of A, in order, as columns:

$$\begin{pmatrix} a_1 & a_2 & \cdots & a_n \\ b_1 & b_2 & \cdots & b_n \\ \cdots & \cdots & \cdots & \cdots \\ c_1 & c_2 & \cdots & c_n \end{pmatrix}^t = \begin{pmatrix} a_1 & b_1 & \cdots & c_1 \\ a_2 & b_2 & \cdots & c_2 \\ \cdots & \cdots & \cdots & \cdots \\ a_n & b_n & \cdots & c_n \end{pmatrix}$$

Note that if A is an $m \times n$ matrix, then A^t is an $n \times m$ matrix.

Example 7.1:

$$\begin{pmatrix} 1 & 2 & 3 \\ 4 & -5 & 6 \end{pmatrix}^t = \begin{pmatrix} 1 & 4 \\ 2 & -5 \\ 3 & 6 \end{pmatrix}$$

The transpose operation on matrices satisfies the following properties.

Theorem 10.5: (i) $(A+B)^t = A^t + B^t$

(ii) $(A^t)^t = A$

(iii) $(kA)^t = kA^t$, for k a scalar

(iv) $(AB)^t = B^tA^t$.

Solved Problems

MATRIX ADDITION AND SCALAR MULTIPLICATION

10.1. Compute:

(i) $\begin{pmatrix} 1 & 2 & 3 \\ 4 & 5 & 6 \end{pmatrix} + \begin{pmatrix} 1 & -1 & 2 \\ 0 & 3 & -5 \end{pmatrix}$

(ii) $\begin{pmatrix} 1 & 2 & -3 \\ 0 & -4 & 1 \end{pmatrix} + \begin{pmatrix} 3 & 5 \\ 1 & -2 \end{pmatrix}$

(iii) $\begin{pmatrix} 1 & 2 & -3 & 4 \\ 0 & -5 & 1 & -1 \end{pmatrix} + \begin{pmatrix} 3 & -5 & 6 & -1 \\ 2 & 0 & -2 & -3 \end{pmatrix}$

(i) Add corresponding elements:

$$\begin{pmatrix} 1 & 2 & 3 \\ 4 & 5 & 6 \end{pmatrix} + \begin{pmatrix} 1 & -1 & 2 \\ 0 & 3 & -5 \end{pmatrix} = \begin{pmatrix} 1+1 & 2+(-1) & 3+2 \\ 4+0 & 5+3 & 6+(-5) \end{pmatrix} = \begin{pmatrix} 2 & 1 & 5 \\ 4 & 8 & 1 \end{pmatrix}$$

(ii) The sum is not defined since the matrices have different shapes.

(iii) Add corresponding elements:

$$\begin{pmatrix} 1 & 2 & -3 & 4 \\ 0 & -5 & 1 & -1 \end{pmatrix} + \begin{pmatrix} 3 & -5 & 6 & -1 \\ 2 & 0 & -2 & -3 \end{pmatrix} = \begin{pmatrix} 1+3 & 2+(-5) & (-3)+6 & 4+(-1) \\ 0+2 & (-5)+0 & 1+(-2) & (-1)+(-3) \end{pmatrix} = \begin{pmatrix} 4 & -3 & 3 & 3 \\ 2 & -5 & -1 & -4 \end{pmatrix}$$

10.2. Compute: (i) $3\begin{pmatrix} 2 & 4 \\ -3 & 1 \end{pmatrix}$ (ii) $-2\begin{pmatrix} 1 & 7 \\ 2 & -3 \\ 0 & -1 \end{pmatrix}$ (iii) $-\begin{pmatrix} 2 & -3 & 8 \\ 1 & -2 & -6 \end{pmatrix}$

(i) Multiply each element of the matrix by the scalar 3:

$$3\begin{pmatrix} 2 & 4 \\ -3 & 1 \end{pmatrix} = \begin{pmatrix} 3\cdot 2 & 3\cdot 4 \\ 3\cdot(-3) & 3\cdot 1 \end{pmatrix} = \begin{pmatrix} 6 & 12 \\ -9 & 3 \end{pmatrix}$$

(ii) Multiply each element of the matrix by the scalar -2:

$$-2\begin{pmatrix} 1 & 7 \\ 2 & -3 \\ 0 & -1 \end{pmatrix} = \begin{pmatrix} (-2)\cdot 1 & (-2)\cdot 7 \\ (-2)\cdot 2 & (-2)\cdot(-3) \\ (-2)\cdot 0 & (-2)\cdot(-1) \end{pmatrix} = \begin{pmatrix} -2 & -14 \\ -4 & 6 \\ 0 & 2 \end{pmatrix}$$

(iii) Multiply each element of the matrix by -1, or equivalently change the sign of each element in the matrix:

$$-\begin{pmatrix} 2 & -3 & 8 \\ 1 & -2 & -6 \end{pmatrix} = \begin{pmatrix} -2 & 3 & -8 \\ -1 & 2 & 6 \end{pmatrix}$$

10.3. Compute: $3\begin{pmatrix} 2 & -5 & 1 \\ 3 & 0 & -4 \end{pmatrix} - 2\begin{pmatrix} 1 & -2 & -3 \\ 0 & -1 & 5 \end{pmatrix} + 4\begin{pmatrix} 0 & 1 & -2 \\ 1 & -1 & -1 \end{pmatrix}$

First perform the scalar multiplication, and then the matrix addition:

$$3\begin{pmatrix} 2 & -5 & 1 \\ 3 & 0 & -4 \end{pmatrix} - 2\begin{pmatrix} 1 & -2 & -3 \\ 0 & -1 & 5 \end{pmatrix} + 4\begin{pmatrix} 0 & 1 & -2 \\ 1 & -1 & -1 \end{pmatrix}$$

$$= \begin{pmatrix} 6 & -15 & 3 \\ 9 & 0 & -12 \end{pmatrix} + \begin{pmatrix} -2 & 4 & 6 \\ 0 & 2 & -10 \end{pmatrix} + \begin{pmatrix} 0 & 4 & -8 \\ 4 & -4 & -4 \end{pmatrix}$$

$$= \begin{pmatrix} 6+(-2)+0 & -15+4+4 & 3+6+(-8) \\ 9+0+4 & 0+2+(-4) & -12+(-10)+(-4) \end{pmatrix} = \begin{pmatrix} 4 & -7 & 1 \\ 13 & -2 & -26 \end{pmatrix}$$

10.4. Find x, y, z and w if:

$$3\begin{pmatrix} x & y \\ z & w \end{pmatrix} = \begin{pmatrix} x & 6 \\ -1 & 2w \end{pmatrix} + \begin{pmatrix} 4 & x+y \\ z+w & 3 \end{pmatrix}$$

First write each side as a single matrix:

$$\begin{pmatrix} 3x & 3y \\ 3z & 3w \end{pmatrix} = \begin{pmatrix} x+4 & x+y+6 \\ z+w-1 & 2w+3 \end{pmatrix}$$

Set corresponding elements equal to each other to obtain the four linear equations

$$\begin{aligned} 3x &= x+4 \\ 3y &= x+y+6 \\ 3z &= z+w-1 \\ 3w &= 2w+3 \end{aligned} \qquad \text{or} \qquad \begin{aligned} 2x &= 4 \\ 2y &= 6+x \\ 2z &= w-1 \\ w &= 3 \end{aligned}$$

The solution of this system of equations is $x = 2$, $y = 4$, $z = 1$, $w = 3$.

MATRIX MULTIPLICATION

10.5. Let $(r \times s)$ denote a matrix with shape $r \times s$. When will the product of two matrices $(r \times s)(t \times u)$ be defined and what will be its shape?

The product $(r \times s)(t \times u)$ of an $r \times s$ matrix and a $t \times u$ matrix is defined if the inner numbers s and t are equal, i.e. $s = t$. The shape of the product will then be the outer numbers in the given order, i.e. $r \times u$.

10.6. Let $(r \times s)$ denote a matrix with shape $r \times s$. Find the shape of the following products if the product is defined:

(i) $(2\times 3)(3\times 4)$ (iv) $(5\times 2)(2\times 3)$
(ii) $(4\times 1)(1\times 2)$ (v) $(4\times 4)(3\times 3)$
(iii) $(1\times 2)(3\times 1)$ (vi) $(2\times 2)(2\times 4)$

In each case the product is defined if the inner numbers are equal, and then the product will have the shape of the outer numbers in the given order.

(i) The product is a 2×4 matrix.

(ii) The product is a 4×2 matrix.

(iii) The product is not defined since the inner numbers 2 and 3 are not equal.

(iv) The product is a 5×3 matrix.

(v) The product is not defined since the inner numbers 4 and 3 are not equal.

(vi) The product is a 2×4 matrix.

10.7. Let $A = \begin{pmatrix} 1 & 3 \\ 2 & -1 \end{pmatrix}$ and $B = \begin{pmatrix} 2 & 0 & -4 \\ 3 & -2 & 6 \end{pmatrix}$. Find (i) AB, (ii) BA.

(i) Now A is 2×2 and B is 2×3, so the product matrix AB is defined and is a 2×3 matrix. To obtain the elements in the first row of the product matrix AB, multiply the first row $(1, 3)$ of A by the columns $\begin{pmatrix} 2 \\ 3 \end{pmatrix}$, $\begin{pmatrix} 0 \\ -2 \end{pmatrix}$ and $\begin{pmatrix} -4 \\ 6 \end{pmatrix}$ of B, respectively:

$$\begin{pmatrix} 1 & 3 \\ 2 & -1 \end{pmatrix}\begin{pmatrix} 2 & 0 & -4 \\ 3 & -2 & 6 \end{pmatrix}$$

$$= \begin{pmatrix} 1\cdot 2 + 3\cdot 3 & 1\cdot 0 + 3\cdot(-2) & 1\cdot(-4) + 3\cdot 6 \end{pmatrix} = \begin{pmatrix} 11 & -6 & 14 \end{pmatrix}$$

To obtain the elements in the second row of the product matrix AB, multiply the second row $(2, -1)$ of A by the columns of B, respectively:

$$\begin{pmatrix} 1 & 3 \\ 2 & -1 \end{pmatrix}\begin{pmatrix} 2 & 0 & -4 \\ 3 & -2 & 6 \end{pmatrix}$$

$$= \begin{pmatrix} 11 & -6 & 14 \\ 2\cdot 2 + (-1)\cdot 3 & 2\cdot 0 + (-1)\cdot(-2) & 2\cdot(-4) + (-1)\cdot 6 \end{pmatrix} = \begin{pmatrix} 11 & -6 & 14 \\ 1 & 2 & -14 \end{pmatrix}$$

Thus
$$AB = \begin{pmatrix} 11 & -6 & 14 \\ 1 & 2 & -14 \end{pmatrix}$$

(ii) Now B is 2×3 and A is 2×2. Since the inner numbers 3 and 2 are not equal, the product BA is not defined.

10.8. Let $A = (2, 1)$ and $B = \begin{pmatrix} 1 & -2 & 0 \\ 4 & 5 & -3 \end{pmatrix}$. Find (i) AB, (ii) BA.

(i) Now A is 1×2 and B is 2×3, so the product AB is defined and is a 1×3 matrix, i.e. a row vector with 3 components.

To obtain the elements of the product AB, multiply the row of A by each column of B:

$$AB = (2,1)\begin{pmatrix} 1 & -2 & 0 \\ 4 & 5 & -3 \end{pmatrix} = (2\cdot 1 + 1\cdot 4,\ 2\cdot(-2) + 1\cdot 5,\ 2\cdot 0 + 1\cdot(-3)) = (6, 1, -3)$$

(ii) Now B is 2×3 and A is 1×2. Since the inner numbers 3 and 1 are not equal, the product BA is not defined.

10.9. Let $A = \begin{pmatrix} 2 & -1 \\ 1 & 0 \\ -3 & 4 \end{pmatrix}$ and $B = \begin{pmatrix} 1 & -2 & -5 \\ 3 & 4 & 0 \end{pmatrix}$. Find (i) AB, (ii) BA.

(i) Now A is 3×2 and B is 2×3, so the product AB is defined and is a 3×3 matrix. To obtain the first row of the product matrix AB, multiply the first row of A by each column of B, respectively:

$$\begin{pmatrix} 2 & -1 \\ 1 & 0 \\ -3 & 4 \end{pmatrix}\begin{pmatrix} 1 & -2 & -5 \\ 3 & 4 & 0 \end{pmatrix}$$

$$= \begin{pmatrix} 2\cdot 1 + (-1)\cdot 3 & 2\cdot(-2) + (-1)\cdot 4 & 2\cdot(-5) + (-1)\cdot 0 \\ & & \\ & & \end{pmatrix} = \begin{pmatrix} -1 & -8 & -10 \\ & & \\ & & \end{pmatrix}$$

To obtain the second row of the product matrix AB, multiply the second row of A by each column of B, respectively:

$$\begin{pmatrix} 2 & -1 \\ 1 & 0 \\ -3 & 4 \end{pmatrix}\begin{pmatrix} 1 & -2 & -5 \\ 3 & 4 & 0 \end{pmatrix}$$

$$= \begin{pmatrix} -1 & -8 & -10 \\ 1\cdot 1 + 0\cdot 3 & 1\cdot(-2) + 0\cdot 4 & 1\cdot(-5) + 0\cdot 0 \\ & & \end{pmatrix} = \begin{pmatrix} -1 & -8 & -10 \\ 1 & -2 & -5 \\ & & \end{pmatrix}$$

To obtain the third row of the product matrix AB, multiply the third row of A by each column of B, respectively:

$$\begin{pmatrix} 2 & -1 \\ 1 & 0 \\ -3 & 4 \end{pmatrix}\begin{pmatrix} 1 & -2 & -5 \\ 3 & 4 & 0 \end{pmatrix}$$

$$= \begin{pmatrix} -1 & -8 & -10 \\ 1 & -2 & -5 \\ (-3)\cdot 1 + 4\cdot 3 & (-3)\cdot(-2) + 4\cdot 4 & (-3)\cdot(-5) + 4\cdot 0 \end{pmatrix} = \begin{pmatrix} -1 & -8 & -10 \\ 1 & -2 & -5 \\ 9 & 22 & 15 \end{pmatrix}$$

Thus
$$AB = \begin{pmatrix} -1 & -8 & -10 \\ 1 & -2 & -5 \\ 9 & 22 & 15 \end{pmatrix}$$

(ii) Now B is 2×3 and A is 3×2, so the product BA is defined and is a 2×2 matrix. To obtain the first row of the product matrix BA, multiply the first row of B by each column of A, respectively:

$$\begin{pmatrix} 1 & -2 & -5 \\ 3 & 4 & 0 \end{pmatrix}\begin{pmatrix} 2 & -1 \\ 1 & 0 \\ -3 & 4 \end{pmatrix}$$

$$= \begin{pmatrix} 1\cdot 2 + (-2)\cdot 1 + (-5)\cdot(-3) & 1\cdot(-1) + (-2)\cdot 0 + (-5)\cdot 4 \\ & \end{pmatrix} = \begin{pmatrix} 15 & -21 \\ & \end{pmatrix}$$

To obtain the second row of the product matrix BA, multiply the second row of B by each column of A:

$$\begin{pmatrix} 1 & -2 & -5 \\ 3 & 4 & 0 \end{pmatrix}\begin{pmatrix} 2 & -1 \\ 1 & 0 \\ -3 & 4 \end{pmatrix}$$

$$= \begin{pmatrix} 15 & -21 \\ 3\cdot 2 + 4\cdot 1 + 0\cdot(-3) & 3\cdot(-1) + 4\cdot 0 + 0\cdot 4 \end{pmatrix} = \begin{pmatrix} 15 & -21 \\ 10 & -3 \end{pmatrix}$$

Thus
$$BA = \begin{pmatrix} 15 & -21 \\ 10 & -3 \end{pmatrix}$$

Remark: Observe that in this case both AB and BA are defined, but AB and BA are not equal; in fact, they do not even have the same shapes.

10.10. Let $A = \begin{pmatrix} 2 & -1 & 0 \\ 1 & 0 & -3 \end{pmatrix}$ and $B = \begin{pmatrix} 1 & -4 & 0 & 1 \\ 2 & -1 & 3 & -1 \\ 4 & 0 & -2 & 0 \end{pmatrix}$.

(i) Determine the shape of AB. (ii) Let c_{ij} denote the element in the ith row and jth column of the product matrix AB, that is, $AB = (c_{ij})$. Find: c_{23}, c_{14}, c_{21} and c_{12}.

(i) Since A is 2×3 and B is 3×4, the product AB is a 2×4 matrix.

(ii) Now c_{ij} is defined as the product of the ith row of A by the jth column of B. Hence:

$$c_{23} = (1, 0, -3)\begin{pmatrix} 0 \\ 3 \\ -2 \end{pmatrix} = 1\cdot 0 + 0\cdot 3 + (-3)\cdot(-2) = 0 + 0 + 6 = 6$$

$$c_{14} = (2, -1, 0)\begin{pmatrix} 1 \\ -1 \\ 0 \end{pmatrix} = 2\cdot 1 + (-1)\cdot(-1) + 0\cdot 0 = 2 + 1 + 0 = 3$$

$$c_{21} = (1, 0, -3)\begin{pmatrix} 1 \\ 2 \\ 4 \end{pmatrix} = 1\cdot 1 + 0\cdot 2 + (-3)\cdot 4 = 1 + 0 - 12 = -11$$

$$c_{12} = (2, -1, 0)\begin{pmatrix} -4 \\ -1 \\ 0 \end{pmatrix} = 2\cdot(-4) + (-1)\cdot(-1) + 0\cdot 0 = -8 + 1 + 0 = -7$$

10.11. Compute: (i) $\begin{pmatrix} 1 & 6 \\ -3 & 5 \end{pmatrix}\begin{pmatrix} 4 & 0 \\ 2 & -1 \end{pmatrix}$ (iii) $\begin{pmatrix} 1 \\ -6 \end{pmatrix}\begin{pmatrix} 1 & 6 \\ -3 & 5 \end{pmatrix}$

(ii) $\begin{pmatrix} 1 & 6 \\ -3 & 5 \end{pmatrix}\begin{pmatrix} 2 \\ -7 \end{pmatrix}$ (iv) $\begin{pmatrix} 1 \\ 6 \end{pmatrix}(3, 2)$ (v) $(2, -1)\begin{pmatrix} 1 \\ -6 \end{pmatrix}$

(i) The first factor is 2×2 and the second is 2×2, so the product is defined and is a 2×2 matrix:

$$\begin{pmatrix} 1 & 6 \\ -3 & 5 \end{pmatrix}\begin{pmatrix} 4 & 0 \\ 2 & -1 \end{pmatrix} = \begin{pmatrix} 1\cdot 4 + 6\cdot 2 & 1\cdot 0 + 6\cdot(-1) \\ (-3)\cdot 4 + 5\cdot 2 & (-3)\cdot 0 + 5\cdot(-1) \end{pmatrix} = \begin{pmatrix} 16 & -6 \\ -2 & -5 \end{pmatrix}$$

(ii) The first factor is 2×2 and the second is 2×1, so the product is defined and is a 2×1 matrix:

$$\begin{pmatrix} 1 & 6 \\ -3 & 5 \end{pmatrix}\begin{pmatrix} 2 \\ -7 \end{pmatrix} = \begin{pmatrix} 1 \cdot 2 + 6 \cdot (-7) \\ (-3) \cdot 2 + 5 \cdot (-7) \end{pmatrix} = \begin{pmatrix} -40 \\ -41 \end{pmatrix}$$

(iii) Now the first factor is 2×1 and the second is 2×2. Since the inner numbers 1 and 2 are distinct, the product is not defined..

(iv) Here the first factor is 2×1 and the second is 1×2, so the product is defined and is a 2×2 matrix:

$$\begin{pmatrix} 1 \\ 6 \end{pmatrix}(3, 2) = \begin{pmatrix} 1 \cdot 3 & 1 \cdot 2 \\ 6 \cdot 3 & 6 \cdot 2 \end{pmatrix} = \begin{pmatrix} 3 & 2 \\ 18 & 12 \end{pmatrix}$$

(v) The first factor is 1×2 and the second is 2×1, so the product is defined and is a 1×1 matrix which we frequently write as a scalar.

$$(2, -1)\begin{pmatrix} 1 \\ -6 \end{pmatrix} = (2 \cdot 1 + (-1) \cdot (-6)) = (8) = 8$$

SQUARE MATRICES

10.12. Let $A = \begin{pmatrix} 1 & 2 \\ 4 & -3 \end{pmatrix}$. Find (i) A^2, (ii) A^3, (iii) $f(A)$, where $f(x) = 2x^3 - 4x + 5$.
(iv) Show that A is a zero of the polynomial $g(x) = x^2 + 2x - 11$.

(i) $$A^2 = AA = \begin{pmatrix} 1 & 2 \\ 4 & -3 \end{pmatrix}\begin{pmatrix} 1 & 2 \\ 4 & -3 \end{pmatrix}$$

$$= \begin{pmatrix} 1 \cdot 1 + 2 \cdot 4 & 1 \cdot 2 + 2 \cdot (-3) \\ 4 \cdot 1 + (-3) \cdot 4 & 4 \cdot 2 + (-3) \cdot (-3) \end{pmatrix} = \begin{pmatrix} 9 & -4 \\ -8 & 17 \end{pmatrix}$$

(ii) $$A^3 = AA^2 = \begin{pmatrix} 1 & 2 \\ 4 & -3 \end{pmatrix}\begin{pmatrix} 9 & -4 \\ -8 & 17 \end{pmatrix}$$

$$= \begin{pmatrix} 1 \cdot 9 + 2 \cdot (-8) & 1 \cdot (-4) + 2 \cdot 17 \\ 4 \cdot 9 + (-3) \cdot (-8) & 4 \cdot (-4) + (-3) \cdot 17 \end{pmatrix} = \begin{pmatrix} -7 & 30 \\ 60 & -67 \end{pmatrix}$$

(iii) To find $f(A)$, first substitute A for x and $5I$ for the constant 5 in the given polynomial $f(x) = 2x^3 - 4x + 5$:

$$f(A) = 2A^3 - 4A + 5I = 2\begin{pmatrix} -7 & 30 \\ 60 & -67 \end{pmatrix} - 4\begin{pmatrix} 1 & 2 \\ 4 & -3 \end{pmatrix} + 5\begin{pmatrix} 1 & 0 \\ 0 & 1 \end{pmatrix}$$

Then multiply each matrix by its respective scalar:

$$= \begin{pmatrix} -14 & 60 \\ 120 & -134 \end{pmatrix} + \begin{pmatrix} -4 & -8 \\ -16 & 12 \end{pmatrix} + \begin{pmatrix} 5 & 0 \\ 0 & 5 \end{pmatrix}$$

Lastly, add the corresponding elements in the matrices:

$$= \begin{pmatrix} -14 - 4 + 5 & 60 - 8 + 0 \\ 120 - 16 + 0 & -134 + 12 + 5 \end{pmatrix} = \begin{pmatrix} -13 & 52 \\ 104 & -117 \end{pmatrix}$$

(iv) Now A is a zero of $g(x)$ if the matrix $g(A)$ is the zero matrix. Compute $g(A)$ as was done for $f(A)$, i.e. first substitute A for x and $11I$ for the constant 11 in $g(x) = x^2 + 2x - 11$:

$$g(A) = A^2 + 2A - 11I = \begin{pmatrix} 9 & -4 \\ -8 & 17 \end{pmatrix} + 2\begin{pmatrix} 1 & 2 \\ 4 & -3 \end{pmatrix} - 11\begin{pmatrix} 1 & 0 \\ 0 & 1 \end{pmatrix}$$

Then multiply each matrix by the scalar preceding it:

$$g(A) = \begin{pmatrix} 9 & -4 \\ -8 & 17 \end{pmatrix} + \begin{pmatrix} 2 & 4 \\ 8 & -6 \end{pmatrix} + \begin{pmatrix} -11 & 0 \\ 0 & -11 \end{pmatrix}$$

Lastly, add the corresponding elements in the matrices:

$$g(A) = \begin{pmatrix} 9+2-11 & -4+4+0 \\ -8+8+0 & 17-6-11 \end{pmatrix} = \begin{pmatrix} 0 & 0 \\ 0 & 0 \end{pmatrix}$$

Since $g(A) = 0$, A is a zero of the polynomial $g(x)$.

10.13. Let $B = \begin{pmatrix} 1 & 2 & 0 \\ 3 & -4 & 5 \\ 0 & -1 & 2 \end{pmatrix}$ and let $f(x) = x^2 - 4x + 3$. Find $f(B)$.

First compute B^2:

$$B^2 = BB = \begin{pmatrix} 1 & 2 & 0 \\ 3 & -4 & 5 \\ 0 & -1 & 2 \end{pmatrix}\begin{pmatrix} 1 & 2 & 0 \\ 3 & -4 & 5 \\ 0 & -1 & 2 \end{pmatrix}$$

$$= \begin{pmatrix} 1\cdot1+2\cdot3+0\cdot0 & 1\cdot2+2\cdot(-4)+0\cdot(-1) & 1\cdot0+2\cdot5+0\cdot2 \\ 3\cdot1+(-4)\cdot3+5\cdot0 & 3\cdot2+(-4)\cdot(-4)+5\cdot(-1) & 3\cdot0+(-4)\cdot5+5\cdot2 \\ 0\cdot1+(-1)\cdot3+2\cdot0 & 0\cdot2+(-1)\cdot(-4)+2\cdot(-1) & 0\cdot0+(-1)\cdot5+2\cdot2 \end{pmatrix}$$

$$= \begin{pmatrix} 7 & -6 & 10 \\ -9 & 17 & -10 \\ -3 & 2 & -1 \end{pmatrix}$$

Then

$$f(B) = B^2 - 4B + 3I = \begin{pmatrix} 7 & -6 & 10 \\ -9 & 17 & -10 \\ -3 & 2 & -1 \end{pmatrix} - 4\begin{pmatrix} 1 & 2 & 0 \\ 3 & -4 & 5 \\ 0 & -1 & 2 \end{pmatrix} + 3\begin{pmatrix} 1 & 0 & 0 \\ 0 & 1 & 0 \\ 0 & 0 & 1 \end{pmatrix}$$

$$= \begin{pmatrix} 7 & -6 & 10 \\ -9 & 17 & -10 \\ -3 & 2 & -1 \end{pmatrix} + \begin{pmatrix} -4 & -8 & 0 \\ -12 & 16 & -20 \\ 0 & 4 & -8 \end{pmatrix} + \begin{pmatrix} 3 & 0 & 0 \\ 0 & 3 & 0 \\ 0 & 0 & 3 \end{pmatrix}$$

$$= \begin{pmatrix} 7-4+3 & -6-8+0 & 10+0+0 \\ -9-12+0 & 17+16+3 & -10-20+0 \\ -3+0+0 & 2+4+0 & -1-8+3 \end{pmatrix} = \begin{pmatrix} 6 & -14 & 10 \\ -21 & 36 & -30 \\ -3 & 6 & -6 \end{pmatrix}$$

10.14. Let $A = \begin{pmatrix} 1 & 3 \\ 4 & -3 \end{pmatrix}$. Find a non-zero column vector $u = \begin{pmatrix} x \\ y \end{pmatrix}$ such that $Au = 3u$.

First set up the matrix equation $Au = 3u$:

$$\begin{pmatrix} 1 & 3 \\ 4 & -3 \end{pmatrix}\begin{pmatrix} x \\ y \end{pmatrix} = 3\begin{pmatrix} x \\ y \end{pmatrix}$$

Write each side as a single matrix (column vector):

$$\begin{pmatrix} x+3y \\ 4x-3y \end{pmatrix} = \begin{pmatrix} 3x \\ 3y \end{pmatrix}$$

Set corresponding elements equal to each other to obtain the system of equations

$$\begin{aligned} x + 3y &= 3x \\ 4x - 3y &= 3y \end{aligned} \qquad \text{or} \qquad \begin{aligned} -2x + 3y &= 0 \\ 4x - 6y &= 0 \end{aligned}$$

The two linear equations are the same (see Chapter 11) and there exist an infinite number of solutions. One such solution is $x = 3$, $y = 2$. In other words, the vector $u = \begin{pmatrix} 3 \\ 2 \end{pmatrix}$ is non-zero and has the property that $Au = 3u$.

TRANSPOSE

10.15. Find the transpose A^t of the matrix $A = \begin{pmatrix} 1 & 0 & 1 & 0 \\ 2 & 3 & 4 & 5 \\ 4 & 4 & 4 & 4 \end{pmatrix}$.

Rewrite the rows of A as the columns of A^t: $A^t = \begin{pmatrix} 1 & 2 & 4 \\ 0 & 3 & 4 \\ 1 & 4 & 4 \\ 0 & 5 & 4 \end{pmatrix}$.

10.16. Let A be an arbitrary matrix. Under what conditions is the product AA^t defined?

Suppose A is an $m \times n$ matrix; then A^t is $n \times m$. Thus the product AA^t is always defined. Observe that A^tA is also defined. Here AA^t is an $m \times m$ matrix, whereas A^tA is an $n \times n$ matrix.

10.17. Let $A = \begin{pmatrix} 1 & 2 & 0 \\ 3 & -1 & 4 \end{pmatrix}$. Find (i) AA^t, (ii) A^tA.

To obtain A^t, rewrite the rows of A as columns: $A^t = \begin{pmatrix} 1 & 3 \\ 2 & -1 \\ 0 & 4 \end{pmatrix}$. Then

$$AA^t = \begin{pmatrix} 1 & 2 & 0 \\ 3 & -1 & 4 \end{pmatrix}\begin{pmatrix} 1 & 3 \\ 2 & -1 \\ 0 & 4 \end{pmatrix}$$

$$= \begin{pmatrix} 1\cdot1+2\cdot2+0\cdot0 & 1\cdot3+2\cdot(-1)+0\cdot4 \\ 3\cdot1+(-1)\cdot2+4\cdot0 & 3\cdot3+(-1)\cdot(-1)+4\cdot4 \end{pmatrix} = \begin{pmatrix} 5 & 1 \\ 1 & 26 \end{pmatrix}$$

$$A^tA = \begin{pmatrix} 1 & 3 \\ 2 & -1 \\ 0 & 4 \end{pmatrix}\begin{pmatrix} 1 & 2 & 0 \\ 3 & -1 & 4 \end{pmatrix}$$

$$= \begin{pmatrix} 1\cdot1+3\cdot3 & 1\cdot2+3\cdot(-1) & 1\cdot0+3\cdot4 \\ 2\cdot1+(-1)\cdot3 & 2\cdot2+(-1)\cdot(-1) & 2\cdot0+(-1)\cdot4 \\ 0\cdot1+4\cdot3 & 0\cdot2+4\cdot(-1) & 0\cdot0+4\cdot4 \end{pmatrix} = \begin{pmatrix} 10 & -1 & 12 \\ -1 & 5 & -4 \\ 12 & -4 & 16 \end{pmatrix}$$

PROOFS

10.18. Prove Theorem 10.3(i): $(AB)C = A(BC)$.

Let $A = (a_{ij})$, $B = (b_{jk})$ and $C = (c_{kl})$. Furthermore, let $AB = S = (s_{ik})$ and $BC = T = (t_{jl})$. Then

$$s_{ik} = a_{i1}b_{1k} + a_{i2}b_{2k} + \cdots + a_{im}b_{mk} = \sum_{j=1}^{m} a_{ij}b_{jk}$$

$$t_{jl} = b_{j1}c_{1l} + b_{j2}c_{2l} + \cdots + b_{jn}c_{nl} = \sum_{k=1}^{n} b_{jk}c_{kl}$$

Now multiplying S by C, i.e. (AB) by C, the element in the ith row and lth column of the matrix $(AB)C$ is

$$s_{i1}c_{1l} + s_{i2}c_{2l} + \cdots + s_{in}c_{nl} = \sum_{k=1}^{n} s_{ik}c_{kl} = \sum_{k=1}^{n}\sum_{j=1}^{m} (a_{ij}b_{jk})c_{kl}$$

On the other hand, multiplying A by T, i.e. A by BC, the element in the ith row and lth column of the matrix $A(BC)$ is

$$a_{i1}t_{1l} + a_{i2}t_{2l} + \cdots + a_{im}t_{ml} = \sum_{j=1}^{m} a_{ik}t_{k} = \sum_{j=1}^{m}\sum_{k=1}^{n} a_{ij}(b_{jk}c_{kl})$$

Since the above sums are equal, the theorem is proven.

10.19. Prove Theorem 10.3(ii): $A(B+C) = AB + AC$.

Let $A = (a_{ij})$, $B = (b_{jk})$ and $C = (c_{jk})$. Furthermore, let $D = B + C = (d_{jk})$, $E = AB = (e_{ik})$ and $F = AC = (f_{ik})$. Then

$$d_{jk} = b_{jk} + c_{jk}$$

$$e_{ik} = a_{i1}b_{1k} + a_{i2}b_{2k} + \cdots + a_{im}b_{mk} = \sum_{j=1}^{m} a_{ij}b_{jk}$$

$$f_{ik} = a_{i1}c_{1k} + a_{i2}c_{2k} + \cdots + a_{im}c_{mk} = \sum_{j=1}^{m} a_{ij}c_{jk}$$

Hence the element in the ith row and kth column of the matrix $AB + AC$ is

$$e_{ik} + f_{ik} = \sum_{j=1}^{m} a_{ij}b_{jk} + \sum_{j=1}^{m} a_{ij}c_{jk} = \sum_{j=1}^{m} a_{ij}(b_{jk} + c_{jk})$$

On the other hand, the element in the ith row and kth column of the matrix $AD = A(B+C)$ is

$$a_{i1}d_{1k} + a_{i2}d_{2k} + \cdots + a_{im}d_{mk} = \sum_{j=1}^{m} a_{ij}d_{jk} = \sum_{j=1}^{m} a_{ij}(b_{jk} + c_{jk})$$

Thus $A(B+C) = AB + AC$ since the corresponding elements are equal.

10.20. Prove Theorem 10.5(iv): $(AB)^t = B^tA^t$.

Let $A = (a_{ij})$ and $B = (b_{jk})$. Then the element in the ith row and jth column of the matrix AB is

$$a_{i1}b_{1j} + a_{i2}b_{2j} + \cdots + a_{im}b_{mj} \qquad (1)$$

Thus (1) is the element which appears in the jth row and ith column of the transpose matrix $(AB)^t$.

On the other hand, the jth row of B^t consists of the elements from the jth column of B:

$$(b_{1j} \quad b_{2j} \quad \cdots \quad b_{mj}) \qquad (2)$$

Furthermore, the ith column of A^t consists of the elements from the ith row of A:

$$\begin{pmatrix} a_{i1} \\ a_{i2} \\ \cdot \\ \cdot \\ \cdot \\ a_{im} \end{pmatrix} \qquad (3)$$

Consequently, the element appearing in the jth row and ith column of the matrix B^tA^t is the product of (2) by (3) which gives (1). Thus $(AB)^t = B^tA^t$.

Supplementary Problems

MATRIX OPERATIONS

For Problems 10.21-23, let

$$A = \begin{pmatrix} 1 & -1 & 2 \\ 0 & 3 & 4 \end{pmatrix}, \quad B = \begin{pmatrix} 4 & 0 & -3 \\ -1 & -2 & 3 \end{pmatrix}, \quad C = \begin{pmatrix} 2 & -3 & 0 & 1 \\ 5 & -1 & -4 & 2 \\ -1 & 0 & 0 & 3 \end{pmatrix} \quad \text{and} \quad D = \begin{pmatrix} 2 \\ -1 \\ 3 \end{pmatrix}$$

10.21. Find: (i) $A + B$, (ii) $A + C$, (iii) $3A - 4B$.

10.22. Find: (i) AB, (ii) AC, (iii) AD, (iv) BC, (v) BD, (vi) CD.

10.23. Find: (i) A^t, (ii) A^tC, (iii) D^tA^t, (iv) B^tA, (v) D^tD, (vi) DD^t.

SQUARE MATRICES

10.24. Let $A = \begin{pmatrix} 2 & 2 \\ 3 & -1 \end{pmatrix}$. (i) Find A^2 and A^3. (ii) If $f(x) = x^3 - 3x^2 - 2x + 4$, find $f(A)$. (iii) If $g(x) = x^2 - x - 8$, find $g(A)$.

10.25. Let $B = \begin{pmatrix} 1 & 3 \\ 5 & 3 \end{pmatrix}$. (i) If $f(x) = 2x^2 - 4x + 3$, find $f(B)$. (ii) If $g(x) = x^2 - 4x - 12$, find $g(B)$. (iii) Find a non-zero column vector $u = \begin{pmatrix} x \\ y \end{pmatrix}$ such that $Bu = 6u$.

10.26. Matrices A and B are said to commute if $AB = BA$. Find all matrices $\begin{pmatrix} x & y \\ z & w \end{pmatrix}$ which commute with $\begin{pmatrix} 1 & 1 \\ 0 & 1 \end{pmatrix}$.

10.27. Let $A = \begin{pmatrix} 1 & 2 \\ 0 & 1 \end{pmatrix}$. Find A^n.

PROOFS

10.28. Prove Theorem 10.2(iii): $(k_1 + k_2)A = k_1A + k_2A$.

10.29. Prove Theorem 10.4(iii): $k(AB) = (kA)B = A(kB)$.

10.30. Prove Theorem 10.5(i): $(A + B)^t = A^t + B^t$.

Answers to Supplementary Problems

10.21. (i) $\begin{pmatrix} 5 & -1 & -1 \\ -1 & 1 & 7 \end{pmatrix}$ (ii) Not defined. (iii) $\begin{pmatrix} -13 & -3 & 18 \\ 4 & 17 & 0 \end{pmatrix}$

10.22. (i) Not defined. (iii) $\begin{pmatrix} 9 \\ 9 \end{pmatrix}$ (v) $\begin{pmatrix} -1 \\ 9 \end{pmatrix}$

(ii) $\begin{pmatrix} -5 & -2 & 4 & 5 \\ 11 & -3 & -12 & 18 \end{pmatrix}$ (iv) $\begin{pmatrix} 11 & -12 & 0 & -5 \\ -15 & 5 & 8 & 4 \end{pmatrix}$ (vi) Not defined.

10.23. (i) $\begin{pmatrix} 1 & 0 \\ -1 & 3 \\ 2 & 4 \end{pmatrix}$ (ii) Not defined. (iii) $(9, 9)$ (iv) $\begin{pmatrix} 4 & -7 & 4 \\ 0 & -6 & -8 \\ -3 & 12 & 6 \end{pmatrix}$ (v) 14 (vi) $\begin{pmatrix} 4 & -2 & 6 \\ -2 & 1 & -3 \\ 6 & -3 & 9 \end{pmatrix}$

10.24. (i) $A^2 = \begin{pmatrix} 10 & 2 \\ 3 & 7 \end{pmatrix}$, $A^3 = \begin{pmatrix} 26 & 18 \\ 27 & -1 \end{pmatrix}$ (ii) $f(A) = \begin{pmatrix} -4 & 8 \\ 12 & -16 \end{pmatrix}$ (iii) $g(A) = \begin{pmatrix} 0 & 0 \\ 0 & 0 \end{pmatrix}$.

10.25. (i) $f(B) = \begin{pmatrix} 31 & 12 \\ 20 & 39 \end{pmatrix}$ (ii) $g(B) = \begin{pmatrix} 0 & 0 \\ 0 & 0 \end{pmatrix}$ (iii) $u = \begin{pmatrix} 3 \\ 5 \end{pmatrix}$ or $\begin{pmatrix} 3k \\ 5k \end{pmatrix}$, $k \neq 0$.

10.26. Only matrices of the form $\begin{pmatrix} a & b \\ 0 & a \end{pmatrix}$ commute with $\begin{pmatrix} 1 & 1 \\ 0 & 1 \end{pmatrix}$.

10.27. $A^n = \begin{pmatrix} 1 & 2n \\ 0 & 1 \end{pmatrix}$

Chapter 11

Linear Equations

LINEAR EQUATION IN TWO UNKNOWNS

A linear equation in two unknowns x and y is of the form

$$ax + by = c$$

where a, b, c are real numbers. We seek a pair of numbers $\alpha = (k_1, k_2)$ which satisfies the equation, that is, for which

$$ak_1 + bk_2 = c$$

is a true statement. Solutions of the equation can be found by assigning arbitrary values to x and solving for y (or vice versa).

Example 1.1: Consider the equation

$$2x + y = 4$$

If we substitute $x = -2$ in the equation, we obtain

$$2 \cdot (-2) + y = 4 \quad \text{or} \quad -4 + y = 4 \quad \text{or} \quad y = 8$$

Hence $(-2, 8)$ is a solution. If we substitute $x = 3$ in the equation, we obtain

$$2 \cdot 3 + y = 4 \quad \text{or} \quad 6 + y = 4 \quad \text{or} \quad y = -2$$

Hence $(3, -2)$ is a solution. The table on the right lists six possible values for x and the *corresponding* values for y, i.e. six solutions of the equation.

x	y
-2	8
-1	6
0	4
1	2
2	0
3	-2

Now any solution $\alpha = (k_1, k_2)$ of the linear equation $ax + by = c$ determines a point in the Cartesian plane R^2. If a, b are not both zero, the solutions of the linear equation correspond precisely to the points on a straight line (whence the name linear equation).

Example 1.2: Consider the equation $2x + y = 4$ of the preceding example. On the right we have plotted the six solutions of the equation which appear in the table above. Note that they all lie on the same line. We call this line the graph of the equation since it corresponds precisely to the solution set of the equation.

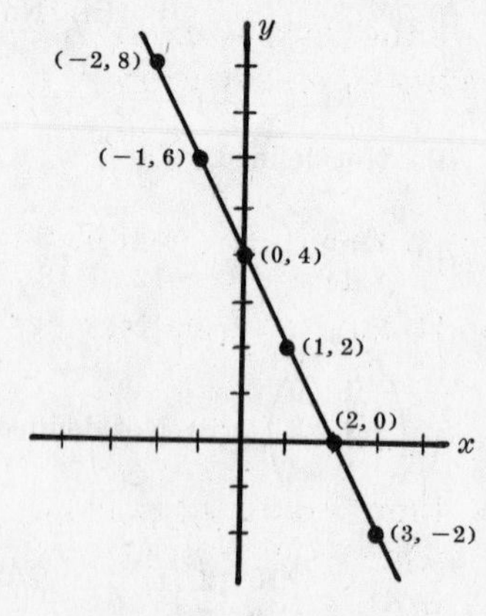

Graph of $2x + y = 4$

TWO LINEAR EQUATIONS IN TWO UNKNOWNS

We now consider a system of two linear equations in two unknowns x and y:

$$a_1x + b_1y = c_1$$
$$a_2x + b_2y = c_2$$

A pair of numbers which satisfies both equations is called a simultaneous solution of the given equations or a solution of the system of equations. There are three cases which can be described geometrically. (Here we assume that the coefficients of x and y in each equation are not both zero.)

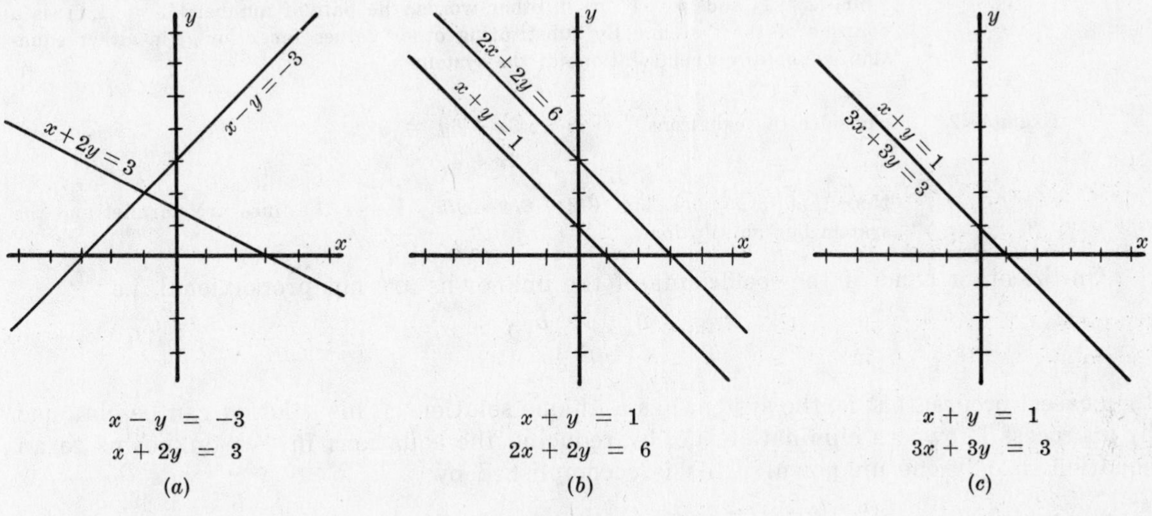

$$x - y = -3 \qquad x + y = 1 \qquad x + y = 1$$
$$x + 2y = 3 \qquad 2x + 2y = 6 \qquad 3x + 3y = 3$$

(a) (b) (c)

Fig. 10-1

1. *The system has exactly one solution.* Here the lines corresponding to the linear equations intersect in one point as shown in Fig. 10-1(a) above.

2. *The system has no solutions.* Here the lines corresponding to the linear equations are parallel as shown in Fig. 10-1(b) above.

3. *The system has an infinite number of solutions.* Here the lines corresponding to the linear equations coincide as shown in Fig. 10-1(c) above.

Now the special cases 2 and 3 can occur if and only if the coefficients of x and the coefficients of y are proportional:

$$\frac{a_1}{a_2} = \frac{b_1}{b_2}$$

There are then two possibilities:

(i) If the constant terms of the equations are in the same proportion, i.e.

$$\frac{a_1}{a_2} = \frac{b_1}{b_2} = \frac{c_1}{c_2}$$

then the lines are coincident. Hence the system has an infinite number of solutions which correspond to the solutions of either equation.

(ii) If the constant terms of the equations are not in the same proportion, i.e.

$$\frac{a_1}{a_2} = \frac{b_1}{b_2} \neq \frac{c_1}{c_2}$$

then the lines are parallel. Hence the system has no solution.

Example 2.1: Consider the system

$$3x + 6y = 9$$
$$2x + 4y = 6$$

Note that $3/2 = 6/4 = 9/6$. Hence the lines are coincident and the system has an infinite number of solutions which correspond to the solutions of either equation.

Particular solutions of the first equation and hence of the system can be obtained as shown in Example 1.1. For example, substitute $x = 1$ in the first equation to obtain

$$3 \cdot 1 + 6y = 9 \quad \text{or} \quad 3 + 6y = 9 \quad \text{or} \quad 6y = 6 \quad \text{or} \quad y = 1$$

Thus $x = 1$ and $y = 1$ or, in other words, the pair of numbers $\alpha = (1, 1)$ is a solution of the system. By substituting other values for x (or y) in either equation, we obtain other solutions of the system.

Example 2.2: Consider the equations

$$\begin{aligned} 3x + 6y &= 9 \\ 2x + 4y &= 5 \end{aligned}$$

Note that $3/2 = 6/4$ but $3/2 = 6/4 \neq 9/5$. Hence the lines are parallel and the system has no solution.

On the other hand, if the coefficients of the unknowns are not proportional, i.e.

$$\frac{a_1}{a_2} \neq \frac{b_1}{b_2}$$

then case 1 occurs; that is, the system has a unique solution. This solution can be obtained by a process known as elimination, i.e. by reducing the equations in two unknowns to an equation in only one unknown. This is accomplished by:

(i) multiplying each of the given equations by numbers such that the coefficients of one of the unknowns in the resulting equations are negatives of each other;

(ii) adding the resulting equations.

We illustrate this method in the following example.

Example 2.3: Consider the equations

$$\begin{aligned} (1) \qquad 3x + 2y &= 8 \\ (2) \qquad 2x - 5y &= -1 \end{aligned}$$

Note that $3/2 \neq 2/-5$, so the system has a unique solution. Multiply (*1*) by 2 and (*2*) by -3 and then add in order to "eliminate" x:

$$\begin{aligned} 2 \times (1)\text{:} \qquad 6x + 4y &= 16 \\ -3 \times (2)\text{:} \qquad -6x + 15y &= 3 \\ \hline \text{Addition:} \qquad 19y &= 19 \quad \text{or} \quad y = 1 \end{aligned}$$

Substitute $y = 1$ in (*1*) to obtain

$$3x + 2 \cdot 1 = 8 \quad \text{or} \quad 3x + 2 = 8 \quad \text{or} \quad 3x = 6 \quad \text{or} \quad x = 2$$

Thus $x = 2$ and $y = 1$ or, in other words, the pair $\alpha = (2, 1)$ is the unique solution to the given system.

GENERAL LINEAR EQUATION

We now consider a linear equation in any arbitrary number of unknowns, say $x_1, x_2, \ldots, x_n$. Such an equation is of the form

$$a_1x_1 + a_2x_2 + \cdots + a_nx_n = b$$

where $a_1, a_2, \ldots, a_n, b$ are real numbers. The numbers a_i are called the *coefficients* of x_i, and b is called the *constant* of the equation. An n-tuple of real numbers $\alpha = (k_1, k_2, \ldots, k_n)$, i.e. an n-component row vector, is a solution of the above equation if, on substituting k_i for x_i, the statement

$$a_1k_1 + a_2k_2 + \cdots + a_nk_n = b$$

is true. We then say that α satisfies the equation.

Example 3.1: Consider the equation $x + 2y - 4z + w = 3$

Now $\alpha = (3, 2, 1, 0)$ is a solution of the equation since

$$3 + 2 \cdot 2 - 4 \cdot 1 + 0 = 3 \quad \text{or} \quad 3 + 4 - 4 + 0 = 3 \quad \text{or} \quad 3 = 3$$

is a true statement. On the other hand, $\beta = (1, 1, 2, 2)$ is not a solution of the equation since

$$1 + 2 \cdot 1 - 4 \cdot 2 + 2 = 3 \quad \text{or} \quad 1 + 2 - 8 + 2 = 3 \quad \text{or} \quad -3 = 3$$

is not a true statement.

A linear equation is said to be *degenerate* if the coefficients of the unknowns are all zero. There are two cases:

(i) The constant is not zero, i.e. the equation is of the form

$$0x_1 + 0x_2 + \cdots + 0x_n = b, \quad \text{with } b \neq 0$$

Then there is no solution to the linear equation.

(ii) The constant is also zero, i.e. the equation is of the form

$$0x_1 + 0x_2 + \cdots + 0x_n = 0$$

Then every n-tuple of real numbers is a solution of the equation.

GENERAL SYSTEM OF LINEAR EQUATIONS

A system of m equations in n unknowns $x_1, x_2, \ldots, x_n$ is of the form

$$\begin{aligned} a_{11}x_1 + a_{12}x_2 + \cdots + a_{1n}x_n &= b_1 \\ a_{21}x_1 + a_{22}x_2 + \cdots + a_{2n}x_n &= b_2 \\ \cdots\cdots\cdots\cdots\cdots\cdots\cdots\cdots\cdots & \\ a_{m1}x_1 + a_{m2}x_2 + \cdots + a_{mn}x_n &= b_m \end{aligned} \qquad (1)$$

where the a_{ij}, b_i are real numbers. An n-tuple of numbers $\alpha = (k_1, k_2, \ldots, k_n)$ which satisfies all the equations is called a solution of the system.

We first consider the special case where all the above equations are degenerate, i.e. where every $a_{ij} = 0$. There are two cases:

(i) If one of the constants $b_i \neq 0$, i.e. the system has an equation of the form

$$0x_1 + 0x_2 + \cdots + 0x_n = b_i, \quad \text{with } b_i \neq 0$$

then this equation, and hence the system, has no solution.

(ii) If every constant $b_i = 0$, i.e. every equation in the system is of the form

$$0x_1 + 0x_2 + \cdots + 0x_n = 0$$

then each equation, and hence the system, has every n-tuple of real numbers as a solution.

Example 4.1: The system

$$\begin{aligned} 0x + 0y + 0z + 0w &= 0 \\ 0x + 0y + 0z + 0w &= 4 \\ 0x + 0y + 0z + 0w &= -1 \end{aligned}$$

has no solution since one of the constants on the right is not zero.

On the other hand, the system

$$\begin{aligned} 0x + 0y + 0z + 0w &= 0 \\ 0x + 0y + 0z + 0w &= 0 \\ 0x + 0y + 0z + 0w &= 0 \end{aligned}$$

has every 4-tuple $\alpha = (k_1, k_2, k_3, k_4)$ as a solution since all the constants on the right are zero.

In the usual case where the equations are not all degenerate, i.e. where one of the $a_{ij} \neq 0$, we reduce the system (1) to a simpler system which is equivalent to the original system, i.e. has the same solutions. This process of reduction, known as (Gauss) elimination, is as follows:

(i) Interchange equations and the position of the unknowns so that $a_{11} \neq 0$.

(ii) Multiply the first equation by the appropriate non-zero constant so that $a_{11} = 1$.

(iii) For each $i > 1$, multiply the first equation by $-a_{i1}$ and add it to the ith equation so that the first unknown is eliminated.

Then the system (1) is replaced by an equivalent system of the form

$$\begin{aligned} x_1 + a_{12}^* x_2 + a_{13}^* x_3 + \cdots + a_{1n}^* x_n &= b_1^* \\ a_{22}^* x_2 + a_{23}^* x_3 + \cdots + a_{2n}^* x_n &= b_2^* \\ \cdots\cdots\cdots\cdots\cdots\cdots\cdots\cdots\cdots & \\ a_{m2}^* x_2 + a_{m3}^* x_3 + \cdots + a_{mn}^* x_n &= b_m^* \end{aligned}$$

Example 4.2: Consider the system

$$\begin{aligned} 2x + 6y - z &= 4 \\ 3x - 2y - z &= 1 \\ 5x + 9y - 2z &= 12 \end{aligned}$$

Multiply the first equation by $\frac{1}{2}$ so that the system is replaced by

$$\begin{aligned} x + 3y - \tfrac{1}{2}z &= 2 \\ 3x - 2y - z &= 1 \\ 5x + 9y - 2z &= 12 \end{aligned}$$

Multiply the new first equation by -3 and add it to the second equation so that the system is replaced by

$$\begin{aligned} x + 3y - \tfrac{1}{2}z &= 2 \\ -11y + \tfrac{1}{2}z &= -5 \\ 5x + 9y - 2z &= 12 \end{aligned}$$

Lastly multiply the first equation by -5 and add it to the third equation so that the system is replaced by the following which is in the desired form:

$$\begin{aligned} x + 3y - \tfrac{1}{2}z &= 2 \\ -11y + \tfrac{1}{2}z &= -5 \\ -6y + \tfrac{1}{2}z &= 2 \end{aligned}$$

Example 4.3: Consider the system of the preceding example. Here we reduce the system to a simpler form without first changing the leading coefficient to 1. This method has the advantage of minimizing the number of fractions appearing.

Multiply the first equation by 3 and the second by -2 and then add to eliminate x from the second equation:

$$\begin{array}{ll} 3 \times \text{first:} & 6x + 18y - 3z = 12 \\ -2 \times \text{second:} & -6x + 4y + 2z = -2 \\ \hline \text{Addition:} & 22y - z = 10 \end{array}$$

Now multiply the first equation by 5 and the third equation by -2 to eliminate x from the third equation:

$$\begin{array}{ll} 5 \times \text{first:} & 10x + 30y - 5z = 20 \\ -2 \times \text{third:} & -10x - 18y + 4z = -24 \\ \hline \text{Addition:} & 12y - z = -4 \end{array}$$

Thus the system is equivalent to

$$\begin{aligned} 2x + 6y - z &= 4 \\ 22y - z &= 10 \\ 12y - z &= -4 \end{aligned} \qquad \text{or} \qquad \begin{aligned} x + 3y - \tfrac{1}{2}z &= 2 \\ 22y - z &= 10 \\ 12y - z &= -4 \end{aligned}$$

Continuing the above process, we obtain the following fundamental result:

Theorem 11.1: Every system of m equations in n unknowns can be reduced to an equivalent system of the following form:

$$\begin{aligned} x_1 + c_{12}x_2 + c_{13}x_3 + \cdots + c_{1r}x_r + \cdots + c_{1n}x_n &= d_1, \quad r \leq m \\ x_2 + c_{23}x_3 + \cdots + c_{2r}x_r + \cdots + c_{2n}x_n &= d_2 \\ x_3 + \cdots + c_{3r}x_r + \cdots + c_{3n}x_n &= d_3 \\ &\cdots \\ x_r + \cdots + c_{rn}x_n &= d_r \\ 0 &= d_{r+1} \\ &\cdots \\ 0 &= d_m \end{aligned} \tag{2}$$

Remark: A system of linear equations in the above form, where the leading coefficient is not zero, is said to be in *echelon* form.

The solutions of a system of equations in the echelon form (*2*) can easily be described and found.

(i) *Inconsistent Equations.* If the numbers $d_{r+1}, \ldots, d_m$ are not all zero, i.e. there is an equation of the form

$$0x_1 + 0x_2 + \cdots + 0x_n = d_k, \quad \text{with } d_k \neq 0$$

then the system is said to be *inconsistent*, and there are no solutions.

(ii) *Consistent Equations.* If the numbers $d_{r+1}, \ldots, d_m$ are all zero, then the system is said to be *consistent* and there does exist a solution. There are two cases:

(*a*) $r = n$, that is, there are as many non-zero equations as unknowns. Then we can successively solve uniquely for the unknowns $x_r, x_{r-1}, \ldots, x_1$ and so there exists a unique solution for the system.

(*b*) $r < n$, that is, there are more unknowns then there are non-zero equations. Then we can arbitrarily assign values to the unknowns $x_{r+1}, \ldots, x_n$ and then solve uniquely for the unknowns $x_r, \ldots, x_1$. Accordingly there exists an infinite number of solutions.

The following diagram shows the various cases:

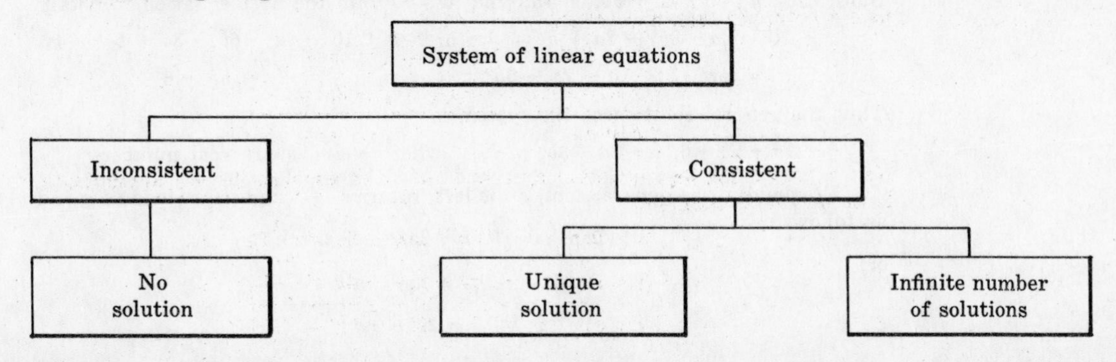

Example 4.4: Consider the system

$$\begin{aligned} x + 2y - 3z &= 4 \\ 0x + y + 2z &= 5 \\ 0x + 0y + 0z &= 3 \end{aligned} \qquad \text{or} \qquad \begin{aligned} x + 2y - 3z &= 4 \\ y + 2z &= 5 \\ 0 &= 3 \end{aligned}$$

Since the system has an equation of the form $0 = c$ with $c \neq 0$, the system is inconsistent and has no solution.

Example 4.5: Consider the system

$$\begin{aligned} x + 2y - 3z &= 4 \\ 0x + y + 2z &= 5 \\ 0x + 0y + z &= 1 \\ 0x + 0y + 0z &= 0 \end{aligned} \qquad \text{or} \qquad \begin{aligned} x + 2y - 3z &= 4 \\ y + 2z &= 5 \\ z &= 1 \\ 0 &= 0 \end{aligned} \qquad \text{or} \qquad \begin{aligned} x + 2y - 3z &= 4 \\ y + 2z &= 5 \\ z &= 1 \end{aligned}$$

Since there is no equation of the form $0 = c$, with $c \neq 0$, the system is consistent. Furthermore, since there are three unknowns and three non-zero equations, the system has a unique solution. Substituting $z = 1$ in the second equation we obtain

$$y + 2 \cdot 1 = 5 \quad \text{or} \quad y + 2 = 5 \quad \text{or} \quad y = 3$$

Substituting $z = 1$ and $y = 3$ in the first equation, we obtain

$$x + 2 \cdot 3 - 3 \cdot 1 = 4 \quad \text{or} \quad x + 6 - 3 = 4 \quad \text{or} \quad x + 3 = 4 \quad \text{or} \quad x = 1$$

Thus $x = 1$, $y = 3$ and $z = 1$ or, in other words, the ordered triple $(1, 3, 1)$ is the unique solution of the system.

Example 4.6: Consider the system

$$\begin{aligned} x + 2y - 3z + w &= 4 \\ 0x + y + 2z + 3w &= 5 \\ 0x + 0y + 0z + 0w &= 0 \end{aligned} \qquad \text{or} \qquad \begin{aligned} x + 2y - 3z + w &= 4 \\ y + 2z + 3w &= 5 \\ 0 &= 0 \end{aligned}$$

$$\text{or} \qquad \begin{aligned} x + 2y - 3z + w &= 4 \\ y + 2z + 3w &= 5 \end{aligned}$$

The system is consistent, and since there are more unknowns than non-zero equations the system has an infinite number of solutions. In fact, we can arbitrarily give values to z and w and solve for x and y. To obtain a particular solution substitute, say, $z = 1$ and $w = 2$ in the second equation to obtain

$$y + 2 \cdot 1 + 3 \cdot 2 = 5 \quad \text{or} \quad y + 2 + 6 = 5 \quad \text{or} \quad y + 8 = 5 \quad \text{or} \quad y = -3$$

Substitute $y = -3$, $z = 1$ and $w = 2$ in the first equation to obtain

$$x + 2 \cdot (-3) - 3 \cdot 1 + 2 = 4 \quad \text{or} \quad x - 6 - 3 + 2 = 4 \quad \text{or} \quad x - 7 = 4 \quad \text{or} \quad x = 11$$

Thus $x = 11$, $y = -3$, $z = 1$ and $w = 2$ or, in other words, the row vector $(11, -3, 1, 2)$ is a specific solution to the system.

To obtain the general solution to the system substitute, say, $z = a$ and $w = b$ in the second equation to obtain

$$y + 2a + 3b = 5 \quad \text{or} \quad y = 5 - 2a - 3b$$

Substitute $y = 5 - 2a - 3b$, $z = a$ and $w = b$ into the first equation to obtain

$$x + 2(5 - 2a - 3b) - 3a + b = 4 \quad \text{or} \quad x + 10 - 4a - 6b - 3a + b = 4$$

$$\text{or} \quad x + 10 - 7a - 5b = 4 \quad \text{or} \quad x = 7a + 5b - 6$$

Thus the general solution of the system is

$$(7a + 5b - 6,\ 5 - 2a - 3b,\ a,\ b), \quad \text{where } a \text{ and } b \text{ are real numbers}$$

Frequently, the general solution is left in terms of z and w (instead of a and b) as follows:

$$(7z + 5w - 6,\ 5 - 2z - 3w,\ z,\ w)$$

or,

$$\begin{aligned} x &= 7z + 5w - 6 \\ y &= 5 - 2z - 3w \end{aligned}$$

The previous remarks give us the next theorem.

Theorem 11.2: A system of linear equations belongs to exactly one of the following cases:

1. The system has no solution.
2. The system has a unique solution.
3. The system has an infinite number of solutions.

Recall that, in the special case of two equations in two unknowns, the above three cases correspond to the following cases described geometrically:

1. The two lines are parallel.
2. The two lines intersect in exactly one point.
3. The two lines are coincident.

HOMOGENEOUS SYSTEMS OF EQUATIONS

A system of linear equations is called homogeneous if it is of the form

$$\begin{aligned} a_{11}x_1 + a_{12}x_2 + \cdots + a_{1n}x_n &= 0 \\ a_{21}x_1 + a_{22}x_2 + \cdots + a_{2n}x_n &= 0 \\ \cdots\cdots\cdots\cdots\cdots\cdots\cdots\cdots& \\ a_{m1}x_1 + a_{m2}x_2 + \cdots + a_{mn}x_n &= 0 \end{aligned} \tag{3}$$

that is, if all the constants on the right are zero. By Theorem 11.1, the above system is equivalent to a system of the form

$$\begin{aligned} x_1 + c_{12}x_2 + \cdots + c_{1r}x_r + \cdots + c_{1n}x_n &= 0, \qquad r \leq m \\ x_2 + \cdots + c_{2r}x_r + \cdots + c_{2n}x_n &= 0 \\ &\vdots \\ x_r + \cdots + c_{rn}x_n &= 0 \\ 0 &= 0 \\ &\vdots \\ 0 &= 0 \end{aligned}$$

Clearly, the above system is consistent since the constants are all zero; in fact, the system (*3*) always has the *zero solution* $(0, 0, \ldots, 0)$ which is called the *trivial solution.* It has a non-trivial solution only if $r < n$. Hence if we originally begin with fewer equations than unknowns in (*3*), then $r < n$ and so the system always has a non-trivial solution. That is,

Theorem 11.3: A homogeneous system of linear equations with more unknowns than equations has a non-zero solution.

Example 5.1: The homogeneous system

$$\begin{aligned} x + 2y - 3z + w &= 0 \\ x - 3y + z - 2w &= 0 \\ 2x + y - 3z + 5w &= 0 \end{aligned}$$

has a non-trivial solution since there are more unknowns than equations.

Example 5.2: We reduce the following system to echelon form:

$$\begin{aligned} x + y - z &= 0 \\ 2x - 3y + z &= 0 \\ x - 4y + 2z &= 0 \end{aligned} \quad \text{to} \quad \begin{aligned} x + y - z &= 0 \\ -5y + 3z &= 0 \\ -5y + 3z &= 0 \end{aligned} \quad \text{to} \quad \begin{aligned} x + y - z &= 0 \\ y - \tfrac{3}{5}z &= 0 \\ 0 &= 0 \end{aligned}$$

The system has a non-trivial solution since we finally have only two non-zero equations in three unknowns. For example, choose $z = 5$; then $y = 3$ and $x = 2$. In other words $(2, 3, 5)$ is a particular solution.

Example 5.3: We reduce the following system to echelon form:

$$\begin{aligned} x + y - z &= 0 \\ 2x + 4y - z &= 0 \\ 3x + 2y + 2z &= 0 \end{aligned} \quad \text{to} \quad \begin{aligned} x + y - z &= 0 \\ 2y + z &= 0 \\ -y - z &= 0 \end{aligned} \quad \text{to} \quad \begin{aligned} x + y - z &= 0 \\ y + \tfrac{1}{2}z &= 0 \\ -\tfrac{1}{2}z &= 0 \end{aligned}$$

$$\text{to} \quad \begin{aligned} x + y - z &= 0 \\ y + \tfrac{1}{2}z &= 0 \\ z &= 0 \end{aligned}$$

Since in echelon form there are three non-zero equations in the three unknowns, the system has a unique solution, the zero solution $(0, 0, 0)$.

MATRICES AND LINEAR EQUATIONS

The system (*1*) of linear equations is equivalent to the matrix equation

$$\begin{pmatrix} a_{11} & a_{12} & \cdots & a_{1n} \\ a_{21} & a_{22} & \cdots & a_{2n} \\ \cdots & \cdots & \cdots & \cdots \\ a_{m1} & a_{m2} & \cdots & a_{mn} \end{pmatrix} \begin{pmatrix} x_1 \\ x_2 \\ \cdot \\ \cdot \\ \cdot \\ x_n \end{pmatrix} = \begin{pmatrix} b_1 \\ b_2 \\ \cdot \\ \cdot \\ \cdot \\ b_m \end{pmatrix}$$

in that every solution to (*1*) is a solution to the above matrix equation and vice versa. In other words, if $A = (a_{ij})$ is the matrix of coefficients, $X = (x_i)$ is the column vector of unknowns, and $B = (b_i)$ is the column vector of constants, then the system (*1*) is equivalent to the simple matrix equation

$$AX = B$$

In particular, the homogeneous system of linear equations (*3*) is equivalent to the matrix equation

$$AX = 0$$

Example 6.1: The following system of linear equations and the matrix equation are equivalent:

$$\begin{aligned} 2x + 3y - 4z &= 7 \\ x - 2y - 5z &= 3 \end{aligned}; \qquad \begin{pmatrix} 2 & 3 & -4 \\ 1 & -2 & -5 \end{pmatrix} \begin{pmatrix} x \\ y \\ z \end{pmatrix} = \begin{pmatrix} 7 \\ 3 \end{pmatrix}$$

In studying linear equations, it is usually simpler to use the language and theory of matrices.

Theorem 11.4: The general solution of a non-homogeneous system of linear equations $AX = B$ is obtained by adding the general solution of the homogeneous system $AX = 0$ to a particular solution of the non-homogeneous system $AX = B$.

Proof. Let α_1 be a fixed solution of $AX = B$. If β is any solution of the homogeneous system $AX = 0$, then

$$A(\alpha_1 + \beta) = A\alpha_1 + A\beta = B + 0 = B$$

That is, the sum $\alpha_1 + \beta$ is a solution to the non-homogeneous system $AX = B$.

On the other hand, suppose α_2 is a solution of $AX = B$ distinct from α_1. Then

$$A(\alpha_2 - \alpha_1) = A\alpha_2 - A\alpha_1 = B - B = 0$$

That is, the difference $\alpha_2 - \alpha_1$ is a solution of the homogeneous system $AX = 0$. But

$$\alpha_2 = \alpha_1 + (\alpha_2 - \alpha_1)$$

hence every solution of $AX = B$ can be obtained by adding a solution of $AX = 0$ to the particular solution α_1 of $AX = B$.

Theorem 11.5: If $\alpha_1, \alpha_2, \ldots, \alpha_n$ are solutions to the homogeneous system of linear equations $AX = 0$, then every linear combination of the α_i of the form

$$k_1\alpha_1 + k_2\alpha_2 + \cdots + k_n\alpha_n, \quad \text{where the } k_i \text{ are scalars,}$$

is also a solution of the homogeneous system $AX = 0$.

Proof. We are given that $A\alpha_1 = 0, \ldots, A\alpha_n = 0$. Hence

$$A(k_1\alpha_1 + \cdots + k_n\alpha_n) = k_1A\alpha_1 + \cdots + k_nA\alpha_n = k_10 + \cdots + k_n0 = 0$$

Accordingly, $k_1\alpha_1 + \cdots + k_n\alpha_n$ is also a solution of the homogeneous system $AX = 0$.

In particular, every multiple $k\alpha$ of any solution α of $AX = 0$ is also a solution of $AX = 0$.

Example 6.2: Consider the homogeneous system in Example 5.2:

$$\begin{aligned} x + y - z &= 0 \\ 2x - 3y + z &= 0 \\ x - 4y + 2z &= 0 \end{aligned}$$

As was noted, $\alpha = (2, 3, 5)$ is a solution of the above. Hence every multiple of α, such as $(4, 6, 10)$ and $(-6, -9, -15)$, is also a solution.

We now restate Theorem 11.2 using the language of matrix theory, and give an independent proof of the theorem.

Theorem 11.2: If $AX = B$ has more than one solution, then it has an infinite number of solutions.

Proof. Let α and β be distinct solutions to $AX = B$. Then the difference $\alpha - \beta$ is a non-trivial solution to the homogeneous system $AX = 0$:

$$A(\alpha - \beta) = A\alpha - A\beta = B - B = 0$$

Hence every multiple $k(\alpha - \beta)$ of $\alpha - \beta$ is a solution of $AX = 0$. Accordingly, for each scalar k,

$$\alpha + k(\alpha - \beta)$$

is a distinct solution to $AX = B$. Thus $AX = B$ has an infinite number of solutions as claimed.

VECTORS AND LINEAR EQUATIONS

The system (*1*) of linear equations is also equivalent to the vector equation

$$x_1\begin{pmatrix} a_{11} \\ a_{21} \\ \vdots \\ a_{m1} \end{pmatrix} + x_2\begin{pmatrix} a_{12} \\ a_{22} \\ \vdots \\ a_{m2} \end{pmatrix} + \cdots + x_n\begin{pmatrix} a_{1n} \\ a_{2n} \\ \vdots \\ a_{mn} \end{pmatrix} = \begin{pmatrix} b_1 \\ b_2 \\ \vdots \\ b_m \end{pmatrix}$$

In other words, if $\alpha_1, \alpha_2, \ldots, \alpha_n$ and β denote the above column vectors respectively, then the system (*1*) is equivalent to the vector equation

$$x_1\alpha_1 + x_2\alpha_2 + \cdots + x_n\alpha_n = \beta$$

If the above equation has a solution, then β is said to be a linear combination of the vectors α_i or is said to depend upon the vectors α_i. We restate this concept formally:

Definition: The vector β is a linear combination of the vectors $\alpha_1, \ldots, \alpha_n$ if there exist scalars $k_1, \ldots, k_n$ such that

$$\beta = k_1\alpha_1 + k_2\alpha_2 + \cdots + k_n\alpha_n$$

i.e. if there exists a solution to

$$\beta = x_1\alpha_1 + x_2\alpha_2 + \cdots + x_n\alpha_n$$

where the x_i are unknown scalars.

Remark: The above definition applies to both column vectors and row vectors, although our illustration was in terms of column vectors.

Example 7.1: Let

$$\beta = \begin{pmatrix} 2 \\ 3 \\ -4 \end{pmatrix}, \quad \alpha_1 = \begin{pmatrix} 1 \\ 1 \\ 1 \end{pmatrix}, \quad \alpha_2 = \begin{pmatrix} 1 \\ 1 \\ 0 \end{pmatrix} \quad \text{and} \quad \alpha_3 = \begin{pmatrix} 1 \\ 0 \\ 0 \end{pmatrix}$$

Then β does depend upon the vectors α_1, α_2 and α_3 since

$$\beta = -4\alpha_1 + 7\alpha_2 - \alpha_3$$

that is, the equation (or system)

$$\begin{pmatrix} 2 \\ 3 \\ -4 \end{pmatrix} = x\begin{pmatrix} 1 \\ 1 \\ 1 \end{pmatrix} + y\begin{pmatrix} 1 \\ 1 \\ 0 \end{pmatrix} + z\begin{pmatrix} 1 \\ 0 \\ 0 \end{pmatrix} \quad \text{or} \quad \begin{cases} 2 = x + y + z \\ 3 = x + y \\ -4 = x \end{cases}$$

has a solution $(-4, 7, -1)$.

VECTORS AND HOMOGENEOUS LINEAR EQUATIONS

The system (*3*) of homogeneous linear equations is equivalent to the vector equation

$$x_1\begin{pmatrix} a_{11} \\ a_{21} \\ \cdot \\ \cdot \\ \cdot \\ a_{m1} \end{pmatrix} + x_2\begin{pmatrix} a_{12} \\ a_{22} \\ \cdot \\ \cdot \\ \cdot \\ a_{m2} \end{pmatrix} + \cdots + x_n\begin{pmatrix} a_{1n} \\ a_{2n} \\ \cdot \\ \cdot \\ \cdot \\ a_{mn} \end{pmatrix} = \begin{pmatrix} 0 \\ 0 \\ \cdot \\ \cdot \\ \cdot \\ 0 \end{pmatrix}$$

That is, if $\alpha_1, \alpha_2, \ldots, \alpha_n$ denote the above column vectors respectively, then the homogeneous system (*3*) is equivalent to the vector equation

$$x_1\alpha_1 + x_2\alpha_2 + \cdots + x_n\alpha_n = 0$$

If the above equation has a non-zero solution, then the vectors $\alpha_1, \ldots, \alpha_n$ are said to be *dependent*; on the other hand, if the above equation has only the trivial (zero) solution, then the vectors are said to be independent. We restate this concept formally:

Definition: The vectors $\alpha_1, \alpha_2, \ldots, \alpha_n$ are dependent if there exists scalars $k_1, k_2, \ldots, k_n$, not all zero, such that

$$k_1\alpha_1 + k_2\alpha_2 + \cdots + k_n\alpha_n = 0$$

i.e. if there exists a non-trivial solution to

$$x_1\alpha_1 + x_2\alpha_2 + \cdots + x_n\alpha_n = 0$$

where the x_i are unknown scalars. Otherwise, the vectors are said to be independent.

Remark: The above definition applies to both column vectors and row vectors although our illustration was in terms of column vectors.

Example 8.1: The only solution to

$$x\begin{pmatrix}1\\1\\1\end{pmatrix} + y\begin{pmatrix}1\\1\\0\end{pmatrix} + z\begin{pmatrix}1\\0\\0\end{pmatrix} = \begin{pmatrix}0\\0\\0\end{pmatrix} \quad \text{or} \quad \begin{aligned} x+y+z &= 0\\ x+y &= 0\\ x &= 0\end{aligned}$$

is the trivial solution $x = 0$, $y = 0$ and $z = 0$. Hence the three vectors are independent.

Example 8.2: The vector equation (or: system of linear equations)

$$x\begin{pmatrix}1\\1\\1\end{pmatrix} + y\begin{pmatrix}2\\-1\\3\end{pmatrix} + z\begin{pmatrix}1\\-5\\3\end{pmatrix} = \begin{pmatrix}0\\0\\0\end{pmatrix} \quad \text{or} \quad \begin{aligned} x+2y+z &= 0\\ x-y-5z &= 0\\ x+3y+3z &= 0\end{aligned}$$

has the non-trivial solution $(3, -2, 1)$. Accordingly, the three vectors are dependent.

Solved Problems

LINEAR EQUATIONS IN TWO UNKNOWNS

11.1. Determine three distinct solutions of $2x - 3y = 14$ and plot its graph.

Choose any value for either unknown, say $x = -2$. Substitute $x = -2$ into the equation to obtain

$$2\cdot(-2) - 3y = 14 \quad \text{or} \quad -4 - 3y = 14 \quad \text{or} \quad -3y = 18 \quad \text{or} \quad y = -6$$

Thus $x = -2$ and $y = -6$ or, in other words, the pair $(-2, -6)$ is a solution.

Now substitute, say, $x = 0$ into the equation to obtain

$$2\cdot 0 - 3y = 14 \quad \text{or} \quad -3y = 14 \quad \text{or} \quad y = -14/3$$

Thus $(0, -14/3)$ is another solution.

Lastly, substitute, say, $y = 0$ into the equation to obtain

$$2x - 3\cdot 0 = 14 \quad \text{or} \quad 2x = 14 \quad \text{or} \quad x = 7$$

Hence $(7, 0)$ is still another solution.

Now plot the three solutions on the Cartesian plane R^2 as shown below; the line passing through these three points is the graph of the equation.

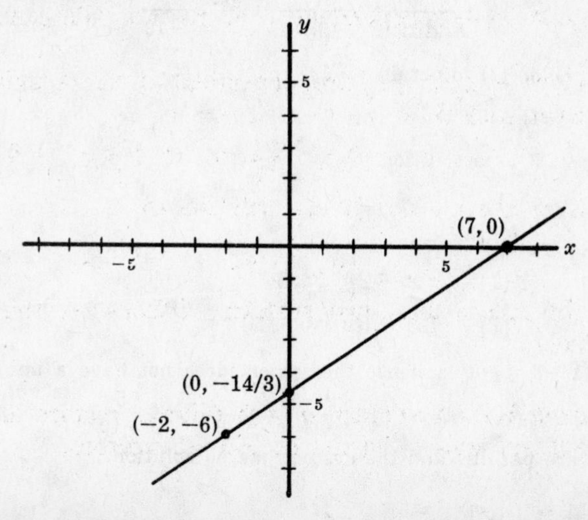

11.2. Solve the system: $\begin{aligned} 3x - 2y &= 7 \\ x + 2y &= 1 \end{aligned}$.

Since $3/1 \neq -2/2$, the system has a unique solution. Now the coefficients of y are already the negatives of each other; hence add both equations:

$$\begin{array}{lrcl} & 3x - 2y & = & 7 \\ & x + 2y & = & 1 \\ \hline \text{Addition:} & 4x & = & 8 \quad \text{or} \quad x = 2 \end{array}$$

Substitute $x = 2$ into the second equation to obtain

$$2 + 2y = 1 \quad \text{or} \quad 2y = -1 \quad \text{or} \quad y = -\tfrac{1}{2}$$

Thus $x = 2$ and $y = -\frac{1}{2}$ or, in other words, the pair $\alpha = (2, -\frac{1}{2})$ is the solution of the system.

Check your answer by substituting the solution back into both original equations:

$$3 \cdot 2 - 2 \cdot (-\tfrac{1}{2}) = 7 \quad \text{or} \quad 6 + 1 = 7 \quad \text{or} \quad 7 = 7$$

and

$$2 + 2 \cdot (-\tfrac{1}{2}) = 1 \quad \text{or} \quad 2 - 1 = 1 \quad \text{or} \quad 1 = 1$$

11.3. Solve the system: $\begin{aligned} (1)\ \ 2x + 5y &= 8 \\ (2)\ \ 3x - 2y &= -7 \end{aligned}$.

Note that $2/3 \neq 5/-2$; hence the system has a unique solution. To eliminate x, multiply (1) by 3 and (2) by -2 and then add:

$$\begin{array}{lrcl} 3 \times (1): & 6x + 15y & = & 24 \\ -2 \times (2): & -6x + \ \ 4y & = & 14 \\ \hline \text{Addition:} & 19y & = & 38 \quad \text{or} \quad y = 2 \end{array}$$

Substitute $y = 2$ into one of the original equations, say (1), to obtain

$$2x + 5 \cdot 2 = 8 \quad \text{or} \quad 2x + 10 = 8 \quad \text{or} \quad 2x = -2 \quad \text{or} \quad x = -1$$

Hence $x = -1$ and $y = 2$ or, in other words, the pair $(-1, 2)$ is the unique solution to the system.

Check your answer by substituting the solution back into both original equations:

$$\begin{array}{ll} (1): & 2 \cdot (-1) + 5 \cdot 2 = 8 \quad \text{or} \quad -2 + 10 = 8 \quad \text{or} \quad 8 = 8 \\ (2): & 3 \cdot (-1) - 2 \cdot 2 = -7 \quad \text{or} \quad -3 - 4 = -7 \quad \text{or} \quad -7 = -7 \end{array}$$

We could also solve the system by first eliminating y as follows. Multiply (1) by 2 and (2) by 5 and then add:

$$\begin{array}{lrcl} 2 \times (1): & 4x + 10y & = & 16 \\ 5 \times (2): & 15x - 10y & = & -35 \\ \hline \text{Addition:} & 19x & = & -19 \quad \text{or} \quad x = -1 \end{array}$$

Substitute $x = -1$ in (1) to obtain

$$2 \cdot (-1) + 5y = 8 \quad \text{or} \quad -2 + 5y = 8 \quad \text{or} \quad 5y = 10 \quad \text{or} \quad y = 2$$

Again we get $(-1, 2)$ as a solution.

11.4. Solve the system: $\begin{aligned} (1)\ \ \ \ \ x - 2y &= 5 \\ (2)\ \ -3x + 6y &= -10 \end{aligned}$.

Note that $1/-3 = -2/6$; hence the system does not have a unique solution. But

$$\frac{1}{-3} = \frac{-2}{6} \neq \frac{5}{-10}$$

Hence the lines are parallel and the system has no solution.

11.5. Solve the system: $\begin{array}{ll}(1) & 5x - 2y = 8 \\ (2) & 3x + 4y = 10\end{array}$.

Note that $5/3 \neq -2/4$; hence the system has a unique solution. To eliminate y, multiply (1) by 2 and add it to (2):

$$\begin{array}{lrcl} 2 \times (1): & 10x - 4y & = & 16 \\ (2): & 3x + 4y & = & 10 \\ \hline \text{Addition:} & 13x & = & 26 \quad \text{or} \quad x = 2 \end{array}$$

Substitute $x = 2$ in either original equation, say (2), to obtain

$$3 \cdot 2 + 4y = 10 \quad \text{or} \quad 6 + 4y = 10 \quad \text{or} \quad 4y = 4 \quad \text{or} \quad y = 1$$

Thus the pair $(2, 1)$ is the unique solution to the system.

Check the answer by substituting the solution back into both original equations:

$$\begin{array}{lllll} (1): & 5 \cdot 2 - 2 \cdot 1 = 8 \quad \text{or} & 10 - 2 = 8 \quad \text{or} & 8 = 8 \\ (2): & 3 \cdot 2 + 4 \cdot 1 = 10 \quad \text{or} & 6 + 4 = 10 \quad \text{or} & 10 = 10 \end{array}$$

11.6. Solve the system: $\begin{array}{ll}(1) & x - 2y = 5 \\ (2) & -3x + 6y = -15\end{array}$.

Note that
$$\frac{1}{-3} = \frac{-2}{6} = \frac{5}{-15}$$
Accordingly, the two lines are coincident. Hence the system has an infinite number of solutions which correspond to the solutions of either equation.

Particular solutions can be found as follows: Let $y = 1$ and substitute in (1) to obtain

$$x - 2 \cdot 1 = 5 \quad \text{or} \quad x - 2 = 5 \quad \text{or} \quad x = 7$$

Hence $(7, 1)$ is a particular solution. Let $y = 2$ and substitute in (1) to obtain

$$x - 2 \cdot 2 = 5 \quad \text{or} \quad x - 4 = 5 \quad \text{or} \quad x = 9$$

Then $(9, 2)$ is another specific solution of the system. And so forth.

The general solution to the system is obtained as follows. Let $y = a$ and substitute in (1) to obtain

$$x - 2y = 5 \quad \text{or} \quad x = 5 + 2a$$

Accordingly, the general solution to the system is $(5 + 2a, a)$, where a is any real number.

GENERAL SYSTEMS OF LINEAR EQUATIONS

11.7. Solve the following system in echelon form:

$$\begin{aligned} x + 2y - 3z + 4w &= 5 \\ 0x + y + 5z + 2w &= 1 \\ 0x + 0y + 0z + 0w &= 2 \end{aligned} \qquad \text{or} \qquad \begin{aligned} x + 2y - 3z + 4w &= 5 \\ y + 5z + 2w &= 1 \\ 0 &= 2 \end{aligned}$$

The system is inconsistent since it has an equation of the form $0 = c$ with $c \neq 0$. Hence the system has no solution.

11.8. Solve the following system in echelon form:

$$\begin{aligned} x + 3y - 2z &= 4 \\ 0x + y - 5z &= 2 \\ 0x + 0y + z &= -1 \\ 0x + 0y + 0z &= 0 \end{aligned} \qquad \text{or} \qquad \begin{aligned} x + 3y - 2z &= 4 \\ y - 5z &= 2 \\ z &= -1 \\ 0 &= 0 \end{aligned}$$

The system is consistent since it has no equation of the form $0=c$ with $c \neq 0$; hence the system has a solution. Furthermore, since there are three unknowns and three non-zero equations, the system has a unique solution.

Substitute $z=-1$ into the second equation to obtain

$$y - 5\cdot(-1) = 2 \quad \text{or} \quad y+5 = 2 \quad \text{or} \quad y = -3$$

Then substitute $y=-3$ and $z=-1$ into the first equation to obtain

$$x + 3\cdot(-3) - 2\cdot(-1) = 4 \quad \text{or} \quad x - 9 + 2 = 4 \quad \text{or} \quad x - 7 = 4 \quad \text{or} \quad x = 11$$

Thus $x=11$, $y=-3$ and $z=-1$ or, in other words, the ordered triple $(11, -3, -1)$ is the unique solution to the system.

11.9. Solve the following system in echelon form:

$$\begin{aligned} x + 2y - 3z + 4w &= 5 \\ 0x + y + 5z + 2w &= 1 \\ 0x + 0y + 0z + 0w &= 0 \end{aligned} \qquad \text{or} \qquad \begin{aligned} x + 2y - 3z + 4w &= 5 \\ y + 5z + 2w &= 1 \\ 0 &= 0 \end{aligned}$$

The system is consistent and so has a solution. Furthermore, since there are four unknowns and only two non-zero equations, the system has an infinite number of solutions which can be obtained by assigning arbitrary values to the unknowns z and w.

To find the general solution to the system let, say, $z=a$ and $w=b$, and substitute into the second equation to obtain

$$y + 5a + 2b = 1 \quad \text{or} \quad y = 1 - 5a - 2b$$

Now substitute $z=a$, $w=b$ and $y=1-5a-2b$ into the first equation to obtain

$$x + 2(1-5a-2b) - 3a + 4b = 5 \quad \text{or} \quad x + 2 - 10a - 4b - 3a + 4b = 5$$

$$\text{or} \quad x + 2 - 13a = 5 \quad \text{or} \quad x = 3 + 13a$$

Thus the general solution is

$$(3+13a,\ 1-5a-2b,\ a,\ b) \qquad \text{where } a \text{ and } b \text{ are real numbers}$$

To find a particular solution to the system let, say, $a=1$ and $b=2$ to obtain

$$(3 + 13\cdot 1,\ 1 - 5\cdot 1 - 2\cdot 2,\ 1,\ 2) \quad \text{or} \quad (16, -8, 1, 2)$$

Note. Some texts leave the general solution in terms of z and w as follows:

$$\begin{aligned} x &= 3 + 13z \\ y &= 1 - 5z - 2w \end{aligned}$$

or, $\quad (3+13z,\ 1-5z-2w,\ z,\ w) \quad$ where z and w are any real numbers

11.10. Solve the following system:

$$\begin{aligned} (1)\quad x + 2y - 4z &= -4 \\ (2)\quad 5x - 3y - 7z &= 6\,. \\ (3)\quad 3x - 2y + 3z &= 11 \end{aligned}$$

Reduce the system to echelon form by first eliminating x from the second and third equations. Multiply (1) by -5 and add to (2) to eliminate x from the second equation:

$$\begin{aligned} -5\times(1)\!:&\quad -5x - 10y + 20z = 20 \\ (2)\!:&\quad 5x - 3y - 7z = 6 \\ \hline \text{Addition:}&\quad -13y + 13z = 26 \quad \text{or} \quad y - z = -2 \end{aligned}$$

Now multiply (1) by -3 and add to (3) to eliminate x from the third equation:

$$\begin{aligned} -3\times(1)\!:&\quad -3x - 6y + 12z = 12 \\ (3)\!:&\quad 3x - 2y + 3z = 11 \\ \hline \text{Addition:}&\quad -8y + 15z = 23 \end{aligned}$$

Thus the original system is equivalent to the system

$$\begin{aligned} x + 2y - \ \ 4z &= -4 \\ y - \ \ z &= -2 \\ -8y + 15z &= 23 \end{aligned}$$

Next multiply the second equation by 8 and add to the third equation to eliminate y from the third equation:

$$\begin{array}{lrcl} 8 \times \text{second:} & 8y - \ \ 8z &=& -16 \\ \text{third:} & -8y + 15z &=& 23 \\ \hline \text{Addition:} & 7z &=& 7 \quad \text{or} \quad z = 1 \end{array}$$

Thus the original system is equivalent to the following system in echelon form:

$$\begin{aligned} x + 2y - 4z &= -4 \\ y - \ \ z &= -2 \\ z &= 1 \end{aligned}$$

The system is consistent since it has no equation of the form $0 = c$ with $c \neq 0$; hence the system has a solution. Furthermore, since there are three unknowns and also three non-zero equations, the system has a unique solution.

Substitute $z = 1$ into the second equation to obtain

$$y - 1 = -2 \quad \text{or} \quad y = -1$$

Now substitute $z = 1$ and $y = -1$ into the first equation to obtain

$$x + 2 \cdot (-1) - 4 \cdot 1 = -4 \quad \text{or} \quad x - 2 - 4 = -4 \quad \text{or} \quad x - 6 = -4 \quad \text{or} \quad x = 2$$

Thus $x = 2$, $y = -1$ and $z = 1$ or, in other words, the ordered triple $(2, -1, 1)$ is the unique solution to the system.

11.11. Solve the following system: $\begin{array}{lrcl} (1) & x + 2y - 3z &=& -1 \\ (2) & -3x + \ \ y - 2z &=& -7 \, . \\ (3) & 5x + 3y - 4z &=& 2 \end{array}$

Reduce to echelon form by first eliminating x from the second and third equations. Multiply (*1*) by 3 and add to (*2*) to eliminate x from the second equation:

$$\begin{array}{lrcl} 3 \times (1): & 3x + 6y - \ \ 9z &=& -3 \\ (2): & -3x + \ \ y - \ \ 2z &=& -7 \\ \hline \text{Addition:} & 7y - 11z &=& -10 \end{array}$$

Now multiply (*1*) by -5 and add to (*3*) to eliminate x from the third equation:

$$\begin{array}{lrcl} -5 \times (1): & -5x - 10y + 15z &=& 5 \\ (3): & 5x + \ \ 3y - \ \ 4z &=& 2 \\ \hline \text{Addition:} & -7y + 11z &=& 7 \end{array}$$

Thus the system is equivalent to the system

$$\begin{aligned} x + 2y - \ \ 3z &= -1 \\ 7y - 11z &= -10 \\ -7y + 11z &= 7 \end{aligned}$$

However, if we add the second equation to the third equation, we obtain

$$0x + 0y + 0z = -3 \quad \text{or} \quad 0 = -3$$

Thus the system is inconsistent since it gives rise to an equation $0 = c$ with $c \neq 0$; accordingly, the system has no solution.

11.12. Solve the following system:

$$\begin{aligned} (1)\quad x + 2y - 3z &= 6 \\ (2)\quad 2x - y + 4z &= 2 \\ (3)\quad 4x + 3y - 2z &= 14 \end{aligned}$$

Reduce the system to echelon form by first eliminating x from the second and third equations. Multiply (1) by -2 and add to (2):

$$\begin{array}{lrcl} -2 \times (1): & -2x - 4y + 6z & = & -12 \\ (2): & 2x - y + 4z & = & 2 \\ \hline \text{Addition:} & -5y + 10z & = & -10 \quad \text{or} \quad y - 2z = 2 \end{array}$$

Multiply (1) by -4 and add it to (3):

$$\begin{array}{lrcl} -4 \times (1): & -4x - 8y + 12z & = & -24 \\ (3): & 4x + 3y - 2z & = & 14 \\ \hline \text{Addition:} & -5y + 10z & = & -10 \quad \text{or} \quad y - 2z = 2 \end{array}$$

Thus the system is equivalent to

$$\begin{aligned} x + 2y - 3z &= 6 \\ y - 2z &= 2 \\ y - 2z &= 2 \end{aligned} \qquad \text{or simply} \qquad \begin{aligned} x + 2y - 3z &= 6 \\ y - 2z &= 2 \end{aligned}$$

(*Note.* Since the second and third equations are identical, we can disregard one of them. In fact, if we multiply the second equation by -1 and add to the third equation we obtain $0 = 0$.)

The system is now in echelon form. The system is consistent and since there are three unknowns but only two non-zero equations, the system has an infinite number of solutions which can be obtained by assigning arbitrary values to z.

To obtain the general solution to the system let, say, $z = a$. Then substitute $z = a$ into the second equation to obtain

$$y - 2a = 2 \quad \text{or} \quad y = 2 + 2a$$

Now substitute $z = a$ and $y = 2 + 2a$ into the first equation to obtain

$$x + 2(2 + 2a) - 3a = 6 \quad \text{or} \quad x + 4 + 4a - 3a = 6 \quad \text{or} \quad x + 4 + a = 6 \quad \text{or} \quad x = 2 - a$$

Thus the general solution to the system is

$$(2 - a,\ 2 + 2a,\ a) \qquad \text{where } a \text{ is any real number.}$$

To obtain a particular solution of the system let, say, $a = 1$ and substitute into the general solution to obtain $(2 - 1,\ 2 + 2 \cdot 1,\ 1)$ or $(1, 4, 1)$.

11.13. Solve the system:

$$\begin{aligned} (1)\quad 2x + y - 3z &= 1 \\ (2)\quad 3x - y - 4z &= 7 \\ (3)\quad 5x + 2y - 6z &= 5 \end{aligned}$$

Method 1.

Reduce to echelon form by first eliminating x from the second and third equations. Multiply (1) by -3 and (2) by 2 and then add to eliminate x from the second equation:

$$\begin{array}{lrcl} -3 \times (1): & -6x - 3y + 9z & = & -3 \\ 2 \times (2): & 6x - 2y - 8z & = & 14 \\ \hline \text{Addition:} & -5y + z & = & 11 \end{array}$$

Multiply (1) by -5 and (3) by 2 and then add to eliminate x from the third equation:

$$\begin{array}{lrcl} -5 \times (1): & -10x - 5y + 15z & = & -5 \\ 2 \times (3): & 10x + 4y - 12z & = & 10 \\ \hline \text{Addition:} & -y + 3z & = & 5 \quad \text{or} \quad y - 3z = -5 \end{array}$$

Thus the system is equivalent to the system

$$\begin{aligned} 2x + y - 3z &= 1 \\ -5y + z &= 11 \\ y - 3z &= -5 \end{aligned} \qquad \text{or, interchanging equations,} \qquad \begin{aligned} 2x + y - 3z &= 1 \\ y - 3z &= -5 \\ -5y + z &= 11 \end{aligned}$$

Now multiply the second equation by 5 and add to the third equation to eliminate y from the third equation:

$$\begin{array}{rr} 5 \times \text{second:} & 5y - 15z = -25 \\ \text{third:} & -5y + z = 11 \\ \hline \text{Addition:} & -14z = -14 \end{array} \quad \text{or } z = 1$$

Thus the original system is equivalent to the system

$$\begin{aligned} 2x + y - 3z &= 1 \\ y - 3z &= -5 \\ z &= 1 \end{aligned}$$

Substitute $z = 1$ into the second equation to obtain

$$y - 3 \cdot 1 = -5 \quad \text{or} \quad y - 3 = -5 \quad \text{or} \quad y = -2$$

Substitute $z = 1$ and $y = -2$ into the first equation to obtain

$$2x + (-2) - 3 \cdot 1 = 1 \quad \text{or} \quad 2x - 2 - 3 = 1 \quad \text{or} \quad 2x - 5 = 1 \quad \text{or} \quad 2x = 6 \quad \text{or} \quad x = 3$$

Thus $x = 3$, $y = -2$ and $z = 1$ is the unique solution to the system.

Method 2.

Consider y as the first unknown and eliminate y from the second and third equation. Add (*1*) to (*2*) to obtain

$$5x - 7z = 8$$

Multiply (*1*) by -2 and add to (*3*):

$$\begin{array}{rr} -2 \times (1): & -4x - 2y + 6z = -2 \\ (3): & 5x + 2y - 6z = 5 \\ \hline \text{Addition:} & x = 3 \end{array}$$

Thus the system is equivalent to

$$\begin{aligned} 2x + y - 3z &= 1 \\ 5x - 7z &= 8 \\ x &= 3 \end{aligned} \qquad \text{or} \qquad \begin{aligned} y - 3z + 2x &= 1 \\ -7z + 5x &= 8 \\ x &= 3 \end{aligned}$$

Substitute $x = 3$ into the second equation to obtain

$$-7z + 15 = 8 \quad \text{or} \quad -7z = -7 \quad \text{or} \quad z = 1$$

Substitute $x = 3$ and $z = 1$ into the first equation to obtain

$$y - 3 + 6 = 1 \quad \text{or} \quad y + 3 = 1 \quad \text{or} \quad y = -2$$

Thus $x = 3$, $y = -2$ and $z = 1$ is the unique solution of the system.

VECTORS AND LINEAR EQUATIONS

11.14. Reduce the following vector equation to an equivalent system of linear equations and solve:

$$\begin{pmatrix} 2 \\ -4 \\ 7 \end{pmatrix} = x \begin{pmatrix} 1 \\ 1 \\ 1 \end{pmatrix} + y \begin{pmatrix} 1 \\ 1 \\ 0 \end{pmatrix} + z \begin{pmatrix} 1 \\ 0 \\ 0 \end{pmatrix}$$

Multiply the vectors on the right by the unknown scalars and then add:

$$\begin{pmatrix} 2 \\ -4 \\ 7 \end{pmatrix} = x\begin{pmatrix} 1 \\ 1 \\ 1 \end{pmatrix} + y\begin{pmatrix} 1 \\ 1 \\ 0 \end{pmatrix} + z\begin{pmatrix} 1 \\ 0 \\ 0 \end{pmatrix} = \begin{pmatrix} x \\ x \\ x \end{pmatrix} + \begin{pmatrix} y \\ y \\ 0 \end{pmatrix} + \begin{pmatrix} z \\ 0 \\ 0 \end{pmatrix} = \begin{pmatrix} x+y+z \\ x+y \\ x \end{pmatrix}$$

Set the corresponding components of the vectors equal to each other to form the system

$$\begin{aligned} x+y+z &= 2 \\ x+y &= -4 \\ x &= 7 \end{aligned} \qquad \text{or} \qquad \begin{aligned} z+y+x &= 2 \\ y+x &= -4 \\ x &= 7 \end{aligned}$$

Substitute $x=7$ into the second equation to obtain

$$y+7 = -4 \quad \text{or} \quad y = -11$$

Substitute $x=7$ and $y=-11$ into the first equation to obtain

$$z-11+7 = 2 \quad \text{or} \quad z-4 = 2 \quad \text{or} \quad z = 6$$

Thus $x=7$, $y=-11$ and $z=6$ is the solution of the vector equation.

11.15. Write the vector $v = (1,-2,5)$ as a linear combination of the vectors $e_1 = (1,1,1)$, $e_2 = (1,2,3)$ and $e_3 = (2,-1,1)$.

We wish to express v as $v = xe_1 + ye_2 + ze_3$, with x, y and z as yet unknown. Thus we require:

$$\begin{aligned} (1,-2,5) &= x(1,1,1) + y(1,2,3) + z(2,-1,1) \\ &= (x,x,x) + (y,2y,3y) + (2z,-z,z) \\ &= (x+y+2z,\ x+2y-z,\ x+3y+z) \end{aligned}$$

Hence

$$\begin{aligned} x+y+2z &= 1 \\ x+2y-z &= -2 \\ x+3y+z &= 5 \end{aligned} \qquad \text{or} \qquad \begin{aligned} x+y+2z &= 1 \\ y-3z &= -3 \\ 2y-z &= 4 \end{aligned} \qquad \text{or} \qquad \begin{aligned} x+y+2z &= 1 \\ y-3z &= -3 \\ 5z &= 10 \end{aligned}$$

$$\text{or} \qquad \begin{aligned} x+y+2z &= 1 \\ y-3z &= -3 \\ z &= 2 \end{aligned}$$

Substitute $z=2$ into the second equation to obtain

$$y-3\cdot 2 = -3 \quad \text{or} \quad y-6 = -3 \quad \text{or} \quad y = 3$$

Substitute $z=2$ and $y=3$ into the first equation to obtain

$$x+3+2\cdot 2 = 1 \quad \text{or} \quad x+3+4 = 1 \quad \text{or} \quad x+7 = 1 \quad \text{or} \quad x = -6$$

Consequently, $$v = -6e_1 + 3e_2 + 2e_3$$

11.16. Write $v = \begin{pmatrix} 2 \\ 3 \\ -5 \end{pmatrix}$ as a linear combination of $e_1 = \begin{pmatrix} 1 \\ 2 \\ -3 \end{pmatrix}$, $e_2 = \begin{pmatrix} 2 \\ -1 \\ -4 \end{pmatrix}$ and $e_3 = \begin{pmatrix} 1 \\ 7 \\ -5 \end{pmatrix}$.

Set v as a linear combination of the e_i using unknowns x, y and z: $v = xe_1 + ye_2 + ze_3$.

$$\begin{pmatrix} 2 \\ 3 \\ -5 \end{pmatrix} = x\begin{pmatrix} 1 \\ 2 \\ -3 \end{pmatrix} + y\begin{pmatrix} 2 \\ -1 \\ -4 \end{pmatrix} + z\begin{pmatrix} 1 \\ 7 \\ -5 \end{pmatrix} = \begin{pmatrix} x \\ 2x \\ -3x \end{pmatrix} + \begin{pmatrix} 2y \\ -y \\ -4y \end{pmatrix} + \begin{pmatrix} z \\ 7z \\ -5z \end{pmatrix} = \begin{pmatrix} x+2y+z \\ 2x-y+7z \\ -3x-4y-5z \end{pmatrix}$$

Hence

$$\begin{aligned} x + 2y + z &= 2 \\ 2x - y + 7z &= 3 \\ -3x - 4y - 5z &= -5 \end{aligned} \quad \text{or} \quad \begin{aligned} x + 2y + z &= 2 \\ -5y + 5z &= -1 \\ 2y - 2z &= 1 \end{aligned} \quad \text{or} \quad \begin{aligned} x + 2y + z &= 2 \\ y - z &= \tfrac{1}{2} \\ -5y + 5z &= -1 \end{aligned}$$

$$\text{or} \quad \begin{aligned} x + 2y + z &= 2 \\ y - z &= \tfrac{1}{2} \\ 0 &= \tfrac{3}{2} \end{aligned}$$

The system is inconsistent and so has no solution. Accordingly, v cannot be written as a linear combination of the vectors e_1, e_2 and e_3.

DEPENDENCE AND INDEPENDENCE

11.17. **Determine whether the vectors $(1,1,-1)$, $(2,-3,1)$ and $(8,-7,1)$ are dependent or independent.**

First set a linear combination of the vectors equal to the zero vector:

$$(0,0,0) = x(1,1,-1) + y(2,-3,1) + z(8,-7,1)$$

Then reduce the vector equation to an equivalent system of homogeneous linear equations:

$$\begin{aligned} (0,0,0) &= (x, x, -x) + (2y, -3y, y) + (8z, -7z, z) \\ &= (x + 2y + 8z,\ x - 3y - 7z,\ -x + y + z) \end{aligned}$$

or,

$$\begin{aligned} x + 2y + 8z &= 0 \\ x - 3y - 7z &= 0 \\ -x + y + z &= 0 \end{aligned}$$

Lastly, reduce the system to echelon form. Eliminate x from the second and third equation to obtain

$$\begin{aligned} x + 2y + 8z &= 0 \\ -5y - 15z &= 0 \\ 3y + 9z &= 0 \end{aligned} \quad \text{or} \quad \begin{aligned} x + 2y + 8z &= 0 \\ y + 3z &= 0 \\ y + 3z &= 0 \end{aligned} \quad \text{or} \quad \begin{aligned} x + 2y + 8z &= 0 \\ y + 3z &= 0 \\ 0 &= 0 \end{aligned}$$

Since there are three unknowns but only two non-zero equations, the system has a non-trivial solution. Thus the original vectors are dependent.

Remark: We do not need to solve the system to determine dependence or independence; we only need to know if a non-zero solution exists.

11.18. **Determine whether the vectors $\begin{pmatrix}1\\-2\\-3\end{pmatrix}$, $\begin{pmatrix}2\\3\\-1\end{pmatrix}$ and $\begin{pmatrix}3\\2\\1\end{pmatrix}$ are dependent or independent.**

First set a linear combination of the vectors equal to the zero vector; and then obtain an equivalent system of homogeneous linear equations:

$$\begin{pmatrix}0\\0\\0\end{pmatrix} = x\begin{pmatrix}1\\-2\\-3\end{pmatrix} + y\begin{pmatrix}2\\3\\-1\end{pmatrix} + z\begin{pmatrix}3\\2\\1\end{pmatrix} = \begin{pmatrix}x\\-2x\\-3x\end{pmatrix} + \begin{pmatrix}2y\\3y\\-y\end{pmatrix} + \begin{pmatrix}3z\\2z\\z\end{pmatrix} = \begin{pmatrix}x + 2y + 3z\\-2x + 3y + 2z\\-3x - y + z\end{pmatrix}$$

or,

$$\begin{aligned} x + 2y + 3z &= 0 \\ -2x + 3y + 2z &= 0 \\ -3x - y + z &= 0 \end{aligned}$$

Eliminate x from the second and third equations to obtain

$$\begin{aligned} x + 2y + 3z &= 0 \\ 7y + 8z &= 0 \\ 5y + 10z &= 0 \end{aligned} \quad \text{or} \quad \begin{aligned} x + 2y + 3z &= 0 \\ y + 2z &= 0 \\ 7y + 8z &= 0 \end{aligned}$$

Eliminate y from the third equation to obtain

$$\begin{aligned} x + 2y + 3z &= 0 \\ y + 2z &= 0 \\ -6z &= 0 \end{aligned} \qquad \text{or} \qquad \begin{aligned} x + 2y + 3z &= 0 \\ y + 2z &= 0 \\ z &= 0 \end{aligned}$$

In echelon form the system has three unknowns and three non-zero equations; hence the system has the unique solution $x=0$, $y=0$, $z=0$, i.e. the zero solution. Hence the original vectors are independent.

11.19. Show that the vectors $\begin{pmatrix} 1 \\ -2 \\ 3 \\ -4 \end{pmatrix}, \begin{pmatrix} 1 \\ 2 \\ 1 \\ 5 \end{pmatrix}, \begin{pmatrix} 3 \\ -7 \\ 0 \\ 2 \end{pmatrix}, \begin{pmatrix} 2 \\ 0 \\ -6 \\ -5 \end{pmatrix}$ and $\begin{pmatrix} -8 \\ 1 \\ -7 \\ 4 \end{pmatrix}$ are dependent.

Reduce the problem to a homogeneous system of equations by setting a linear combination of the vectors equal to the zero vector:

$$\begin{pmatrix} 0 \\ 0 \\ 0 \\ 0 \end{pmatrix} = x\begin{pmatrix} 1 \\ -2 \\ 3 \\ -4 \end{pmatrix} + y\begin{pmatrix} 1 \\ 2 \\ 1 \\ 5 \end{pmatrix} + z\begin{pmatrix} 3 \\ -7 \\ 0 \\ 2 \end{pmatrix} + v\begin{pmatrix} 2 \\ 0 \\ -6 \\ -5 \end{pmatrix} + w\begin{pmatrix} -8 \\ 1 \\ -7 \\ 4 \end{pmatrix}$$

$$= \begin{pmatrix} x \\ -2x \\ 3x \\ -4x \end{pmatrix} + \begin{pmatrix} y \\ 2y \\ y \\ 5y \end{pmatrix} + \begin{pmatrix} 3z \\ -7z \\ 0 \\ 2z \end{pmatrix} + \begin{pmatrix} 2v \\ 0 \\ -6v \\ -5v \end{pmatrix} + \begin{pmatrix} -8w \\ w \\ -7w \\ 4w \end{pmatrix}$$

$$= \begin{pmatrix} x + y + 3z + 2v - 8w \\ -2x + 2y - 7z + w \\ 3x + y - 6v - 7w \\ -4x + 5y + 2z - 5v + 4w \end{pmatrix}$$

or,

$$\begin{aligned} x + y + 3z + 2v - 8w &= 0 \\ -2x + 2y - 7z \qquad + w &= 0 \\ 3x + y \qquad - 6v - 7w &= 0 \\ -4x + 5y + 2z - 5v + 4w &= 0 \end{aligned}$$

Now the system of homogeneous equations has five unknowns but only four equations. Hence, by Theorem 11.3, the system has a non-trivial solution and so the vectors are dependent.

11.20. Let $v_1, v_2, \ldots, v_m$ be n-component vectors. Show that if the number of vectors is greater than the number of components, i.e. $m > n$, then the vectors are dependent.

Note, as in the preceding problem, that the vector equation

$$x_1v_1 + x_2v_2 + \cdots + x_mv_m = 0$$

gives rise to a homogeneous system of linear equations in which the number of unknowns equals the number of vectors and the number of equations equals the number of components. Hence if the number of vectors is greater than the number of components, then the system has a non-zero solution by Theorem 11.3 and so the vectors are dependent.

Supplementary Problems

LINEAR EQUATIONS IN TWO UNKNOWNS

11.21. Solve:

(i) $\begin{cases} 2x - 5y = 1 \\ 3x + 2y = 11 \end{cases}$ (ii) $\begin{cases} 2x + 3y = 1 \\ 5x + 7y = 3 \end{cases}$ (iii) $\begin{cases} 3x - 2y = 7 \\ x + 3y = -16 \end{cases}$

11.22. Solve:

(i) $\begin{cases} 2x + 4y = 10 \\ 3x + 6y = 15 \end{cases}$ (ii) $\begin{cases} 4x - 2y = 5 \\ -6x + 3y = 1 \end{cases}$ (iii) $\begin{cases} 2x - 3y = 5t \\ 3x + y = 2t \end{cases}$

11.23. Solve:

(i) $\begin{cases} \dfrac{2x}{3} + \dfrac{y}{2} = 8 \\ \dfrac{x}{6} - \dfrac{y}{4} = -1 \end{cases}$ (ii) $\begin{cases} \dfrac{2x-1}{3} + \dfrac{y+2}{4} = 4 \\ \dfrac{x+3}{2} - \dfrac{x-y}{3} = 3 \end{cases}$ (iii) $\begin{cases} 2x - 4 = 3y \\ 5y - x = 5 \end{cases}$

GENERAL SYSTEMS OF LINEAR EQUATIONS

11.24. Solve:

(i) $\begin{cases} 2x + y - 3z = 5 \\ 3x - 2y + 2z = 5 \\ 5x - 3y - z = 16 \end{cases}$ (ii) $\begin{cases} 2x + 3y - 2z = 5 \\ x - 2y + 3z = 2 \\ 4x - y + 4z = 1 \end{cases}$ (iii) $\begin{cases} x + 2y + 3z = 3 \\ 2x + 3y + 8z = 4 \\ 3x + 2y + 17z = 1 \end{cases}$

11.25. Solve:

(i) $\begin{cases} 2x + 3y = 3 \\ x - 2y = 5 \\ 3x + 2y = 7 \end{cases}$ (ii) $\begin{cases} x + 2y - 3z + 2w = 2 \\ 2x + 5y - 8z + 6w = 5 \\ 3x + 4y - 5z + 2w = 4 \end{cases}$ (iii) $\begin{cases} x + 2y - z + 3w = 3 \\ 2x + 4y + 4z + 3w = 9 \\ 3x + 6y - z + 8w = 10 \end{cases}$

11.26. Solve:

(i) $\begin{matrix} x + 2y + 2z = 2 \\ 3x - 2y - z = 5 \\ 2x - 5y + 3z = -4 \\ x + 4y + 6z = 0 \end{matrix}$ (ii) $\begin{matrix} x + 5y + 4z - 13w = 3 \\ 3x - y + 2z + 5w = 2 \\ 2x + 2y + 3z - 4w = 1 \end{matrix}$

VECTORS AND EQUATIONS

11.27. Write $v = (9, 5, 7)$ as a linear combination of the vectors $e_1 = (1, 1, 1)$, $e_2 = (1, -1, 0)$ and $e_3 = (2, 0, 1)$.

11.28. Determine whether the vectors $(1, -2, 3, 1)$, $(3, 2, 1, -2)$ and $(1, 6, -5, -4)$ are dependent or independent.

11.29. Determine whether the vectors $\begin{pmatrix} 1 \\ 1 \\ -1 \end{pmatrix}, \begin{pmatrix} 2 \\ 1 \\ 0 \end{pmatrix}$ and $\begin{pmatrix} -1 \\ 1 \\ 2 \end{pmatrix}$ are dependent or independent.

11.30. Prove: If the vectors e_1, e_2 and e_3 are independent, then any vector v can be written in at most one way as a linear combination of the vectors e_1, e_2 and e_3.

Answers to Supplementary Problems

11.21. (i) (3, 1), or $x = 3$, $y = 1$ (ii) (2, −1), or $x = 2$, $y = -1$ (iii) (−1, −5), or $x = -1$, $y = -5$

11.22. (i) $(5 - 2a, a)$, or $x = 5 - 2a$, $y = a$ (ii) no solution (iii) $(t, -t)$, or $x = t$, $y = -t$

11.23. (i) (6, 8) (ii) (5, 2) (iii) (5, 2)

11.24. (i) (1, −3, −2) (ii) no solution (iii) $(-1 - 7a, 2 + 2a, a)$, or $\begin{cases} x = -1 - 7z \\ y = 2 + 2z \end{cases}$

11.25. (i) (3, −1) (ii) $(-a + 2b, 1 + 2a - 2b, a, b)$, or $\begin{cases} x = -z + 2w \\ y = 1 + 2z - 2w \end{cases}$

(iii) $(6 - 2a - 5b, a, b, 2b - 1)$, or $\begin{cases} x = 6 - 2y - 5z \\ w = 2z - 1 \end{cases}$

11.26. (i) (2, 1, −1) (ii) no solution

11.27. $v = 2e_1 - 3e_2 + 5e_3$

11.28. Dependent

11.29. Independent

11.30. Proof: Suppose $v = a_1e_1 + a_2e_2 + a_3e_3$ and $v = b_1e_1 + b_2e_2 + b_3e_3$. Subtracting, we obtain

$$0 = v - v = (a_1 - b_1)e_1 + (a_2 - b_2)e_2 + (a_3 - b_3)e_3$$

Since the vectors are independent, the coefficients must be zero, i.e.

$$a_1 - b_1 = 0, \quad a_2 - b_2 = 0 \quad \text{and} \quad a_3 - b_3 = 0$$

Accordingly, $a_1 = b_1$, $a_2 = b_2$ and $a_3 = b_3$.

Chapter 12

Determinants of Order Two and Three

INTRODUCTION

To every square matrix there is assigned a specific number called the determinant of the matrix. This determinant function, first discovered in the investigation of systems of linear equations, has many interesting properties in its own right.

In this chapter we define and investigate the determinant of square matrices of order two and three, including its application to linear equations. Determinants of matrices of higher order are beyond the scope of this text.

DETERMINANTS OF ORDER ONE

Write $\det(A)$ or $|A|$ for the determinant of the matrix A; it is a number assigned to square matrices only.

The determinant of a 1×1 matrix (a) is the number a itself: $\det(a) = a$.

Observe that in the linear equation in one unknown x,

$$ax = b$$

we can view a as a 1×1 matrix (a) applied to the one-component vector x. If $\det(a) \neq 0$, i.e. if $a \neq 0$, then the equation has the unique solution $x = \frac{1}{a}b$. However, if $\det(a) = 0$, i.e. if $a = 0$, then the equation has no solution if $b \neq 0$, and every number is a solution if $b = 0$; hence the solution is not unique.

As we shall subsequently see, the preceding result holds true in general, i.e. $\det(A) \neq 0$ is a necessary and sufficient condition for the linear equation $Ax = b$ to have a unique solution.

DETERMINANTS OF ORDER TWO

The determinant of the 2×2 matrix $\begin{pmatrix} a & b \\ c & d \end{pmatrix}$ is denoted and defined as follows:

$$\begin{vmatrix} a & b \\ c & d \end{vmatrix} = ad - bc$$

Example 1.1:

$$\begin{vmatrix} 5 & 4 \\ 2 & 3 \end{vmatrix} = 5 \cdot 3 - 4 \cdot 2 = 15 - 8 = 7 \qquad \begin{vmatrix} 2 & 1 \\ -4 & 6 \end{vmatrix} = 2 \cdot 6 - 1 \cdot (-4) = 12 + 4 = 16$$

We emphasize that a square array of numbers enclosed by straight lines is not a matrix but denotes the number that the determinant function assigns to the enclosed array of numbers, i.e. to the enclosed square matrix.

A very important property of the determinant function is that it is multiplicative; that is,

Theorem 12.1: The determinant of a product of matrices is the product of the determinants of the matrices: $\det(AB) = \det(A)\det(B)$.

Remark: The diagram on the right may help the reader obtain the determinant of a 2×2 matrix. Here the arrow slanting downward connects the two numbers in the plus term of the determinant $ad - bc$, and the arrow slanting upward connects the two numbers in the minus term of the determinant $ad - bc$.

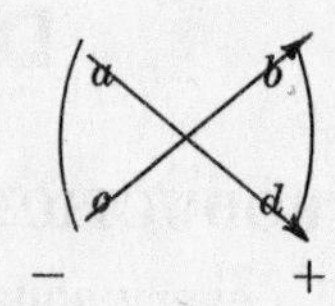

LINEAR EQUATIONS IN TWO UNKNOWNS AND DETERMINANTS

Consider two linear equations in two unknowns:

$$\begin{aligned} a_1x + b_1y &= c_1 \\ a_2x + b_2y &= c_2 \end{aligned}$$

Let us solve the system by eliminating y. Multiply the first equation by b_2 and the second equation by $-b_1$, and then add:

$$\begin{array}{lrcl} b_2 \times \text{first:} & a_1b_2x + b_1b_2y &=& b_2c_1 \\ -b_1 \times \text{second:} & -a_2b_1x - b_1b_2y &=& -b_1c_2 \\ \hline \text{Addition:} & (a_1b_2 - a_2b_1)x &=& b_2c_1 - b_1c_2 \end{array}$$

Now the system has a unique solution if and only if the coefficient of x in this last equation is not zero, i.e.

$$D = \begin{vmatrix} a_1 & b_1 \\ a_2 & b_2 \end{vmatrix} = a_1b_2 - a_2b_1 \neq 0$$

Observe that D is the determinant of the matrix of coefficients of the system of equations. In this case, where $D \neq 0$, we can uniquely solve for x and y as quotients of determinants as follows:

$$x = \frac{N_x}{D} = \frac{b_2c_1 - b_1c_2}{a_1b_2 - a_2b_1} = \frac{\begin{vmatrix} c_1 & b_1 \\ c_2 & b_2 \end{vmatrix}}{\begin{vmatrix} a_1 & b_1 \\ a_2 & b_2 \end{vmatrix}} \quad \text{and} \quad y = \frac{N_y}{D} = \frac{a_1c_2 - a_2c_1}{a_1b_2 - a_2b_1} = \frac{\begin{vmatrix} a_1 & c_1 \\ a_2 & c_2 \end{vmatrix}}{\begin{vmatrix} a_1 & b_1 \\ a_2 & b_2 \end{vmatrix}}$$

Here D, the determinant of the matrix of coefficients, appears in the denominator of both quotients. The numerators N_x and N_y of the quotients for x and y, respectively, can be obtained by substituting the column of constant terms in place of the column of coefficients of the given unknown in the matrix of coefficients.

Example 2.1: Solve using determinants:

$$\begin{aligned} 2x - 3y &= 7 \\ 3x + 5y &= 1 \end{aligned}$$

The determinant D of the matrix of coefficients is

$$D = \begin{vmatrix} 2 & -3 \\ 3 & 5 \end{vmatrix} = 2 \cdot 5 - 3 \cdot (-3) = 10 + 9 = 19$$

Since $D \neq 0$, the system has a unique solution. To obtain the numerator N_x replace, in the matrix of coefficients, the coefficients of x by the constant terms:

$$N_x = \begin{vmatrix} 7 & -3 \\ 1 & 5 \end{vmatrix} = 7 \cdot 5 - 1 \cdot (-3) = 35 + 3 = 38$$

To obtain the numerator N_y replace, in the matrix of coefficients, the coefficients of y by the constant terms:

$$N_y = \begin{vmatrix} 2 & 7 \\ 3 & 1 \end{vmatrix} = 2 \cdot 1 - 3 \cdot 7 = 2 - 21 = -19$$

Thus the unique solution of the system is

$$x = \frac{N_x}{D} = \frac{38}{19} = 2 \quad \text{and} \quad y = \frac{N_y}{D} = \frac{-19}{19} = -1$$

Example 2.2: Solve using determinants: $\begin{cases} 2x = 5 + y \\ 3 + 2y + 3x = 0 \end{cases}$.

First arrange the equations in standard form:

$$\begin{aligned} 2x - y &= 5 \\ 3x + 2y &= -3 \end{aligned}$$

The determinant D of the matrix of coefficients is

$$D = \begin{vmatrix} 2 & -1 \\ 3 & 2 \end{vmatrix} = 2 \cdot 2 - 3 \cdot (-1) = 4 + 3 = 7$$

Since $D \neq 0$, the system has a unique solution. Now,

$$N_x = \begin{vmatrix} 5 & -1 \\ -3 & 2 \end{vmatrix} = 5 \cdot 2 - (-3)(-1) = 10 - 3 = 7$$

and $$N_y = \begin{vmatrix} 2 & 5 \\ 3 & -3 \end{vmatrix} = 2 \cdot (-3) - 3 \cdot 5 = -6 - 15 = -21$$

Thus the unique solution of the system is

$$x = \frac{N_x}{D} = \frac{7}{7} = 1 \quad \text{and} \quad y = \frac{N_y}{D} = \frac{-21}{7} = -3$$

Example 2.3: Solve using determinants: $\begin{cases} 2x - 4y = 7 \\ 3x - 6y = 5 \end{cases}$

The determinant D of the matrix of coefficients is

$$D = \begin{vmatrix} 2 & -4 \\ 3 & -6 \end{vmatrix} = 2 \cdot (-6) - 3 \cdot (-4) = -12 + 12 = 0$$

Since $D = 0$, the system does not have a unique solution, and we cannot solve the system by determinants. (The discussion of the previous chapter shows that the system has no solution since $2/3 = -4/-6 \neq 7/5$.)

DETERMINANTS OF ORDER THREE

The determinant of a 3 by 3 matrix is defined as follows:

$$\begin{vmatrix} a_1 & b_1 & c_1 \\ a_2 & b_2 & c_2 \\ a_3 & b_3 & c_3 \end{vmatrix} = a_1b_2c_3 + a_2b_3c_1 + a_3b_1c_2 - a_1b_3c_2 - a_2b_1c_3 - a_3b_2c_1$$

This may be written as

$$a_1(b_2c_3 - b_3c_2) - b_1(a_2c_3 - a_3c_2) + c_1(a_2b_3 - a_3b_2)$$

or

$$a_1\begin{vmatrix} b_2 & c_2 \\ b_3 & c_3 \end{vmatrix} - b_1\begin{vmatrix} a_2 & c_2 \\ a_3 & c_3 \end{vmatrix} + c_1\begin{vmatrix} a_2 & b_2 \\ a_3 & b_3 \end{vmatrix}$$

which is a linear combination of three determinants of order two whose coefficients (with alternating signs) are the first row of the given matrix. Note that each 2×2 matrix can be obtained by deleting, in the original matrix, the row and column containing its coefficient:

$$a_1\begin{vmatrix} a_1 & b_1 & c_1 \\ a_2 & b_2 & c_2 \\ a_3 & b_3 & c_3 \end{vmatrix} - b_1\begin{vmatrix} a_1 & b_1 & c_1 \\ a_2 & b_2 & c_2 \\ a_3 & b_3 & c_3 \end{vmatrix} + c_1\begin{vmatrix} a_1 & b_1 & c_1 \\ a_2 & b_2 & c_2 \\ a_3 & b_3 & c_3 \end{vmatrix}$$

Example 3.1:

Compute the determinant of $A = \begin{pmatrix} 2 & 3 & -4 \\ 0 & -4 & 2 \\ 1 & -1 & 5 \end{pmatrix}$.

$$\begin{aligned} \det(A) &= 2\begin{vmatrix} 2 & 3 & -4 \\ 0 & -4 & 2 \\ 1 & -1 & 5 \end{vmatrix} - 3\begin{vmatrix} 2 & 3 & -4 \\ 0 & -4 & 2 \\ 1 & -1 & 5 \end{vmatrix} + (-4)\begin{vmatrix} 2 & 3 & -4 \\ 0 & -4 & 2 \\ 1 & -1 & 5 \end{vmatrix} \\ &= 2\begin{vmatrix} -4 & 2 \\ -1 & 5 \end{vmatrix} - 3\begin{vmatrix} 0 & 2 \\ 1 & 5 \end{vmatrix} + (-4)\begin{vmatrix} 0 & -4 \\ 1 & -1 \end{vmatrix} \\ &= 2(-20+2) - 3(0-2) - 4(0+4) = -36 + 6 - 16 = -46 \end{aligned}$$

Example 3.2:

$$\begin{aligned} \begin{vmatrix} 2 & 1 & 3 \\ -5 & 6 & -1 \\ 4 & 0 & -2 \end{vmatrix} &= 2\begin{vmatrix} 6 & -1 \\ 0 & -2 \end{vmatrix} - 1\begin{vmatrix} -5 & -1 \\ 4 & -2 \end{vmatrix} + 3\begin{vmatrix} -5 & 6 \\ 4 & 0 \end{vmatrix} \\ &= 2(-12-0) - (10+4) + 3(0-24) = -24 - 14 - 72 = -110 \end{aligned}$$

Remark: Problems 12.8 and 12.18 show that the determinant of a 3×3 matrix can be expressed as a linear combination of three determinants of order two with coefficients from any row or from any column, and not just from the first row as exhibited here.

LINEAR EQUATIONS IN THREE UNKNOWNS AND DETERMINANTS

Consider three linear equations in the three unknowns x, y and z:

$$\begin{aligned} a_1x + b_1y + c_1z &= d_1 \\ a_2x + b_2y + c_2z &= d_2 \\ a_3x + b_3y + c_3z &= d_3 \end{aligned}$$

Now the above system has a unique solution if and only if the determinant of the matrix of coefficients is not zero:

$$D = \begin{vmatrix} a_1 & b_1 & c_1 \\ a_2 & b_2 & c_2 \\ a_3 & b_3 & c_3 \end{vmatrix} \neq 0$$

In this case, the unique solution of the system can be expressed as quotients of determinants,

$$x = \frac{N_x}{D}, \quad y = \frac{N_y}{D}, \quad z = \frac{N_z}{D}$$

where the denominator D in each quotient is the determinant of the matrix of coefficients, as above, and the numerators N_x, N_y and N_z are obtained by replacing the column of coefficients of the unknown in the matrix of coefficients by the column of constant terms:

$$N_x = \begin{vmatrix} d_1 & b_1 & c_1 \\ d_2 & b_2 & c_2 \\ d_3 & b_3 & c_3 \end{vmatrix}, \quad N_y = \begin{vmatrix} a_1 & d_1 & c_1 \\ a_2 & d_2 & c_2 \\ a_3 & d_3 & c_3 \end{vmatrix}, \quad N_z = \begin{vmatrix} a_1 & b_1 & d_1 \\ a_2 & b_2 & d_2 \\ a_3 & b_3 & d_3 \end{vmatrix}$$

We emphasize that if the determinant D of the matrix of coefficients is zero then the system has either no solution or an infinite number of solutions.

Example 4.1:

Solve the following system by determinants:

$$\begin{aligned} 2x + y - z &= 3 \\ x + y + z &= 1 \\ x - 2y - 3z &= 4 \end{aligned}$$

The determinant D of the matrix of coefficients is obtained as follows:

$$D = \begin{vmatrix} 2 & 1 & -1 \\ 1 & 1 & 1 \\ 1 & -2 & -3 \end{vmatrix} = 2\begin{vmatrix} 1 & 1 \\ -2 & -3 \end{vmatrix} - 1\begin{vmatrix} 1 & 1 \\ 1 & -3 \end{vmatrix} + (-1)\begin{vmatrix} 1 & 1 \\ 1 & -2 \end{vmatrix}$$

$$= 2(-3+2) - 1(-3-1) - 1(-2-1) = -2 + 4 + 3 = 5$$

Since $D \neq 0$, the system has a unique solution. We evaluate N_x, N_y and N_z, the numerators for x, y and z respectively:

$$N_x = \begin{vmatrix} 3 & 1 & -1 \\ 1 & 1 & 1 \\ 4 & -2 & -3 \end{vmatrix} = 3(-3+2) - 1(-3-4) - 1(-2-4) = -3 + 7 + 6 = 10$$

$$N_y = \begin{vmatrix} 2 & 3 & -1 \\ 1 & 1 & 1 \\ 1 & 4 & -3 \end{vmatrix} = 2(-3-4) - 3(-3-1) - 1(4-1) = -14 + 12 - 3 = -5$$

$$N_z = \begin{vmatrix} 2 & 1 & 3 \\ 1 & 1 & 1 \\ 1 & -2 & 4 \end{vmatrix} = 2(4+2) - 1(4-1) + 3(-2-1) = 12 - 3 - 9 = 0$$

Thus the unique solution is

$$x = \frac{N_x}{D} = \frac{10}{5} = 2, \quad y = \frac{N_y}{D} = \frac{-5}{5} = -1, \quad z = \frac{N_z}{D} = \frac{0}{5} = 0$$

INVERTIBLE MATRICES

A square matrix A is said to be *invertible* if there exists a matrix B with the property that

$$AB = BA = I, \text{ the identity matrix}$$

Such a matrix B is unique; for

$$AB_1 = B_1A = I \text{ and } AB_2 = B_2A = I \quad \text{implies} \quad B_1 = B_1I = B_1(AB_2) = (B_1A)B_2 = IB_2 = B_2.$$

We call such a matrix B the *inverse* of A and denote it by A^{-1}. Observe that the above relation is symmetric; that is, if B is the inverse of A, then A is also the inverse of B.

Example 5.1:

$$\begin{pmatrix} 2 & 5 \\ 1 & 3 \end{pmatrix}\begin{pmatrix} 3 & -5 \\ -1 & 2 \end{pmatrix} = \begin{pmatrix} 2\cdot 3 + 5\cdot(-1) & 2\cdot(-5) + 5\cdot 2 \\ 1\cdot 3 + 3\cdot(-1) & 1\cdot(-5) + 3\cdot 2 \end{pmatrix} = \begin{pmatrix} 1 & 0 \\ 0 & 1 \end{pmatrix}$$

$$\begin{pmatrix} 3 & -5 \\ -1 & 2 \end{pmatrix}\begin{pmatrix} 2 & 5 \\ 1 & 3 \end{pmatrix} = \begin{pmatrix} 3\cdot 2 + (-5)\cdot 1 & 3\cdot 5 + (-5)\cdot 3 \\ (-1)\cdot 2 + 2\cdot 1 & (-1)\cdot 5 + 2\cdot 3 \end{pmatrix} = \begin{pmatrix} 1 & 0 \\ 0 & 1 \end{pmatrix}$$

Thus $\begin{pmatrix} 2 & 5 \\ 1 & 3 \end{pmatrix}$ and $\begin{pmatrix} 3 & -5 \\ -1 & 2 \end{pmatrix}$ are inverses.

It is known that $AB = I$ if and only if $BA = I$ (Problem 12.17); hence it is necessary to test only one product to determine whether two given matrices are inverses, as in the next example.

Example 5.2:

$$\begin{pmatrix} 1 & 0 & 2 \\ 2 & -1 & 3 \\ 4 & 1 & 8 \end{pmatrix}\begin{pmatrix} -11 & 2 & 2 \\ -4 & 0 & 1 \\ 6 & -1 & -1 \end{pmatrix}$$

$$= \begin{pmatrix} 1\cdot(-11) + 0\cdot(-4) + 2\cdot 6 & 1\cdot 2 + 0\cdot 0 + 2\cdot(-1) & 1\cdot 2 + 0\cdot 1 + 2\cdot(-1) \\ 2\cdot(-11) + (-1)\cdot(-4) + 3\cdot 6 & 2\cdot 2 + (-1)\cdot 0 + 3\cdot(-1) & 2\cdot 2 + (-1)\cdot 1 + 3\cdot(-1) \\ 4\cdot(-11) + 1\cdot(-4) + 8\cdot 6 & 4\cdot 2 + 1\cdot 0 + 8\cdot(-1) & 4\cdot 2 + 1\cdot 1 + 8\cdot(-1) \end{pmatrix}$$

$$= \begin{pmatrix} 1 & 0 & 0 \\ 0 & 1 & 0 \\ 0 & 0 & 1 \end{pmatrix}$$

Thus the two matrices are invertible and are inverses of each other.

Example 5.3:

Find the inverse of $\begin{pmatrix} 1 & 2 \\ 3 & 4 \end{pmatrix}$. We seek scalars x, y, z and w for which

$$\begin{pmatrix} 1 & 2 \\ 3 & 4 \end{pmatrix}\begin{pmatrix} x & y \\ z & w \end{pmatrix} = \begin{pmatrix} 1 & 0 \\ 0 & 1 \end{pmatrix} \quad \text{or} \quad \begin{pmatrix} x + 2z & y + 2w \\ 3x + 4z & 3y + 4w \end{pmatrix} = \begin{pmatrix} 1 & 0 \\ 0 & 1 \end{pmatrix}$$

or which satisfy $\begin{cases} x + 2z = 1 \\ 3x + 4z = 0 \end{cases}$ $\begin{cases} y + 2w = 0 \\ 3y + 4w = 1 \end{cases}$

The solutions of the systems are $x = -2$, $z = 3/2$ and $y = 1$, $w = -1/2$. Thus the inverse of the given matrix is $\begin{pmatrix} -2 & 1 \\ \frac{3}{2} & -\frac{1}{2} \end{pmatrix}$. We test our answer:

$$\begin{pmatrix} 1 & 2 \\ 3 & 4 \end{pmatrix}\begin{pmatrix} -2 & 1 \\ \frac{3}{2} & -\frac{1}{2} \end{pmatrix} = \begin{pmatrix} 1\cdot(-2) + 2\cdot\frac{3}{2} & 1\cdot 1 + 2\cdot(-\frac{1}{2}) \\ 3\cdot(-2) + 4\cdot\frac{3}{2} & 3\cdot 1 + 4\cdot(-\frac{1}{2}) \end{pmatrix} = \begin{pmatrix} 1 & 0 \\ 0 & 1 \end{pmatrix}$$

INVERTIBLE MATRICES AND DETERMINANTS

We now calculate the inverse of a general 2×2 matrix $A = \begin{pmatrix} a & b \\ c & d \end{pmatrix}$. We seek scalars x, y, z and w such that

$$\begin{pmatrix} a & b \\ c & d \end{pmatrix}\begin{pmatrix} x & y \\ z & w \end{pmatrix} = \begin{pmatrix} 1 & 0 \\ 0 & 1 \end{pmatrix} \quad \text{or} \quad \begin{pmatrix} ax + bz & ay + bw \\ cx + dz & cy + dw \end{pmatrix} = \begin{pmatrix} 1 & 0 \\ 0 & 1 \end{pmatrix}$$

which reduces to solving the following two systems of linear equations in two unknowns:

$$\begin{cases} ax + bz = 1 \\ cx + dz = 0 \end{cases} \qquad \begin{cases} ay + bw = 0 \\ cy + dw = 1 \end{cases}$$

Note first that one equation in each system is homogeneous and has $(0,0)$ as a solution, whereas the other equation does not have $(0,0)$ as a solution. In other words, the lines in each system are not coincident and so each system has either a unique solution or no solution. Furthermore, the matrix of coefficients in each system is identical to the original matrix A. Accordingly, the systems have a solution (and so A has an inverse) if and only if the determinant of A is not zero. We formally state this result (which holds for matrices of all orders):

Theorem 12.2: A matrix is invertible if and only if its determinant is not zero.

Usually a matrix is said to be *singular* if its determinant is zero, and *nonsingular* otherwise. Hence the preceding theorem can be restated as follows:

Theorem 12.3: A matrix is invertible if and only if it is nonsingular.

Now if the determinant of the above matrix A is not zero, then we can uniquely solve for the unknowns x, y, z and w as follows:

$$x = \frac{d}{ad-bc} = \frac{d}{|A|}, \quad y = \frac{-b}{ad-bc} = \frac{-b}{|A|}, \quad z = \frac{-c}{ad-bc} = \frac{-c}{|A|} \quad \text{and} \quad w = \frac{a}{ad-bc} = \frac{a}{|A|}$$

Accordingly,

$$A^{-1} = \begin{pmatrix} a & b \\ c & d \end{pmatrix}^{-1} = \begin{pmatrix} d/|A| & -b/|A| \\ -c/|A| & a/|A| \end{pmatrix} = \frac{1}{|A|}\begin{pmatrix} d & -b \\ -c & a \end{pmatrix}$$

In other words, we can obtain the inverse of a 2×2 matrix, with determinant not zero, by (i) interchanging the elements on the main diagonal, (ii) taking the negative of the other elements, and (iii) dividing each element by the determinant of the original matrix.

Example 6.1: Find the inverse of the matrix $A = \begin{pmatrix} 2 & 3 \\ 4 & 5 \end{pmatrix}$.

Now $\det(A) = |A| = 2 \cdot 5 - 4 \cdot 3 = 10 - 12 = -2$. Hence the matrix does have an inverse. To obtain A^{-1}, interchange the elements 2 and 5 on the main diagonal and take the negative of the other elements to obtain the matrix

$$\begin{pmatrix} 5 & -3 \\ -4 & 2 \end{pmatrix}$$

Now divide each element by -2, the determinant of A, to obtain A^{-1}:

$$A^{-1} = \begin{pmatrix} -\frac{5}{2} & \frac{3}{2} \\ 2 & -1 \end{pmatrix}$$

Solved Problems

DETERMINANTS OF ORDER TWO

12.1. Compute the determinant of each matrix:

$$\text{(i)} \begin{pmatrix} 3 & -2 \\ 4 & 5 \end{pmatrix} \quad \text{(ii)} \begin{pmatrix} -1 & 6 \\ 0 & 4 \end{pmatrix} \quad \text{(iii)} \begin{pmatrix} a-b & b \\ b & a+b \end{pmatrix} \quad \text{(iv)} \begin{pmatrix} a-b & a \\ a & a+b \end{pmatrix}$$

(i) $\begin{vmatrix} 3 & -2 \\ 4 & 5 \end{vmatrix} = 3 \cdot 5 - (-2) \cdot 4 = 15 + 8 = 23$

(ii) $\begin{vmatrix} -1 & 6 \\ 0 & 4 \end{vmatrix} = -1 \cdot 4 - 6 \cdot 0 = -4$

(iii) $\begin{vmatrix} a-b & b \\ b & a+b \end{vmatrix} = (a-b)(a+b) - b \cdot b = a^2 - b^2 - b^2 = a^2 - 2b^2$

(iv) $\begin{vmatrix} a-b & a \\ a & a+b \end{vmatrix} = (a-b)(a+b) - a \cdot a = a^2 - b^2 - a^2 = -b^2$

12.2. Solve for x and y, using determinants:

(i) $\begin{aligned} 2x + y &= 7 \\ 3x - 5y &= 4 \end{aligned}$ (ii) $\begin{aligned} ax - 2by &= c \\ 3ax - 5by &= 2c \end{aligned}$, where $ab \neq 0$

(i) $D = \begin{vmatrix} 2 & 1 \\ 3 & -5 \end{vmatrix} = -10 - 3 = -13, \quad N_x = \begin{vmatrix} 7 & 1 \\ 4 & -5 \end{vmatrix} = -35 - 4 = -39,$

$$N_y = \begin{vmatrix} 2 & 7 \\ 3 & 4 \end{vmatrix} = 8 - 21 = -13$$

$$x = \frac{N_x}{D} = \frac{-39}{-13} = 3 \quad \text{and} \quad y = \frac{N_y}{D} = \frac{-13}{-13} = 1$$

(ii) $D = \begin{vmatrix} a & -2b \\ 3a & -5b \end{vmatrix} = -5ab + 6ab = ab, \quad N_x = \begin{vmatrix} c & -2b \\ 2c & -5b \end{vmatrix} = -5bc + 4bc = -bc,$

$$N_y = \begin{vmatrix} a & c \\ 3a & 2c \end{vmatrix} = 2ac - 3ac = -ac$$

$$x = \frac{N_x}{D} = \frac{-bc}{ab} = -\frac{c}{a}, \quad y = \frac{N_y}{D} = \frac{-ac}{ab} = -\frac{c}{b}$$

12.3. Determine those values of k for which $\begin{vmatrix} k & k \\ 4 & 2k \end{vmatrix} = 0$.

$$\begin{vmatrix} k & k \\ 4 & 2k \end{vmatrix} = 2k^2 - 4k = 0 \quad \text{or} \quad 2k(k-2) = 0$$

Hence $2k = 0$ or $k = 0$, and $k - 2 = 0$ or $k = 2$. That is, if $k = 0$ or $k = 2$, the determinant is zero.

12.4. Prove Theorem 12.1 (in the case of 2×2 matrices): The determinant of a product of matrices is the product of the determinants of the matrices, i.e. $\det(AB) = \det(A)\det(B)$.

Let $A = \begin{pmatrix} a & b \\ c & d \end{pmatrix}$ and $B = \begin{pmatrix} r & s \\ t & u \end{pmatrix}$, and so $AB = \begin{pmatrix} ar + bt & as + bu \\ cr + dt & cs + du \end{pmatrix}$. Hence

$$\begin{aligned} \det(AB) &= (ar + bt)(cs + du) - (as + bu)(cr + dt) \\ &= acrs + adru + bcst + bdtu - acrs - adst - bcru - bdtu \\ &= adru + bcst - adst - bcru \end{aligned}$$

and $\det(A)\det(B) = (ad - bc)(ru - st) = adru - adst - bcru + bcst$

Thus $\det(AB) = \det(A)\det(B)$.

DETERMINANTS OF ORDER THREE

12.5. Find the determinant of each matrix:

$$\text{(i)} \begin{pmatrix} 1 & 2 & 3 \\ 4 & -2 & 3 \\ 0 & 5 & -1 \end{pmatrix} \quad \text{(ii)} \begin{pmatrix} 4 & -1 & -2 \\ 0 & 2 & -3 \\ 5 & 2 & 1 \end{pmatrix} \quad \text{(iii)} \begin{pmatrix} 2 & -3 & 4 \\ 1 & 2 & -3 \\ -1 & -2 & 5 \end{pmatrix}$$

(i)
$$\begin{vmatrix} 1 & 2 & 3 \\ 4 & -2 & 3 \\ 0 & 5 & -1 \end{vmatrix} = 1\begin{vmatrix} 1 & 2 & 3 \\ 4 & -2 & 3 \\ 0 & 5 & -1 \end{vmatrix} - 2\begin{vmatrix} 1 & 2 & 3 \\ 4 & -2 & 3 \\ 0 & 5 & -1 \end{vmatrix} + 3\begin{vmatrix} 1 & 2 & 3 \\ 4 & -2 & 3 \\ 0 & 5 & -1 \end{vmatrix}$$

$$= 1\begin{vmatrix} -2 & 3 \\ 5 & -1 \end{vmatrix} - 2\begin{vmatrix} 4 & 3 \\ 0 & -1 \end{vmatrix} + 3\begin{vmatrix} 4 & -2 \\ 0 & 5 \end{vmatrix}$$

$$= 1(2-15) - 2(-4+0) + 3(20+0) = -13 + 8 + 60 = 55$$

(ii)
$$\begin{vmatrix} 4 & -1 & -2 \\ 0 & 2 & -3 \\ 5 & 2 & 1 \end{vmatrix} = 4\begin{vmatrix} 2 & -3 \\ 2 & 1 \end{vmatrix} - (-1)\begin{vmatrix} 0 & -3 \\ 5 & 1 \end{vmatrix} + (-2)\begin{vmatrix} 0 & 2 \\ 5 & 2 \end{vmatrix}$$

$$= 4(2+6) + 1(0+15) - 2(0-10) = 32 + 15 + 20 = 67$$

(iii)
$$\begin{vmatrix} 2 & -3 & 4 \\ 1 & 2 & -3 \\ -1 & -2 & 5 \end{vmatrix} = 2\begin{vmatrix} 2 & -3 \\ -2 & 5 \end{vmatrix} - (-3)\begin{vmatrix} 1 & -3 \\ -1 & 5 \end{vmatrix} + 4\begin{vmatrix} 1 & 2 \\ -1 & -2 \end{vmatrix}$$

$$= 2(10-6) + 3(5-3) + 4(-2+2) = 8 + 6 + 0 = 14$$

12.6. Solve, using determinants:
$$\begin{cases} 3y + 2x = z + 1 \\ 3x + 2z = 8 - 5y \\ 3z - 1 = x - 2y \end{cases}$$

First arrange the equation in standard form with the unknowns appearing in columns:

$$\begin{aligned} 2x + 3y - z &= 1 \\ 3x + 5y + 2z &= 8 \\ x - 2y - 3z &= -1 \end{aligned}$$

Compute the determinant D of the matrix of coefficients:

$$D = \begin{vmatrix} 2 & 3 & -1 \\ 3 & 5 & 2 \\ 1 & -2 & -3 \end{vmatrix} = 2(-15+4) - 3(-9-2) - (-6-5) = -22 + 33 + 11 = 22$$

Since $D \neq 0$, the system has a unique solution. To compute N_x, N_y and N_z, replace the coefficients of the unknown in the matrix of coefficients by the constant terms:

$$N_x = \begin{vmatrix} 1 & 3 & -1 \\ 8 & 5 & 2 \\ -1 & -2 & -3 \end{vmatrix} = 1(-15+4) - 3(-24+2) - 1(-16+5) = -11 + 66 + 11 = 66$$

$$N_y = \begin{vmatrix} 2 & 1 & -1 \\ 3 & 8 & 2 \\ 1 & -1 & -3 \end{vmatrix} = 2(-24+2) - 1(-9-2) - 1(-3-8) = -44 + 11 + 11 = -22$$

$$N_z = \begin{vmatrix} 2 & 3 & 1 \\ 3 & 5 & 8 \\ 1 & -2 & -1 \end{vmatrix} = 2(-5+16) - 3(-3-8) + 1(-6-5) = 22 + 33 - 11 = 44$$

Hence $\quad x = \frac{N_x}{D} = \frac{66}{22} = 3, \quad y = \frac{N_y}{D} = \frac{-22}{22} = -1, \quad z = \frac{N_z}{D} = \frac{44}{22} = 2$

12.7. Consider the 3×3 matrix $\begin{pmatrix} a_1 & b_1 & c_1 \\ a_2 & b_2 & c_2 \\ a_3 & b_3 & c_3 \end{pmatrix}$. Show that the diagram appearing below, where the first two columns are rewritten to the right of the matrix, can be used to obtain the determinant of the given matrix:

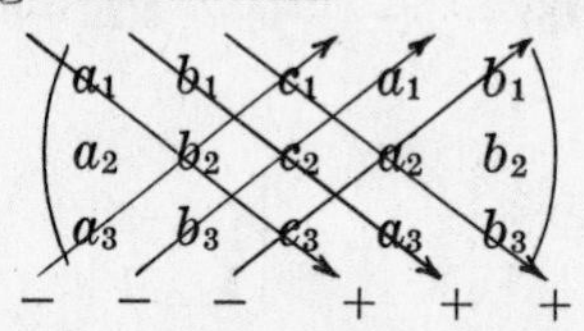

Form the product of each of the three numbers joined by an arrow slanting downward, and precede each product by a plus sign as follows:

$$+ a_1b_2c_3 + b_1c_2a_3 + c_1a_2b_3$$

Now form the product of each of the three numbers joined by an arrow slanting upward, and precede each product by a minus sign as follows:

$$- a_3b_2c_1 - b_3c_2a_1 - c_3a_2b_1$$

Then the determinant of the given matrix is precisely the sum of the above two expressions:

$$\begin{vmatrix} a_1 & b_1 & c_1 \\ a_2 & b_2 & c_2 \\ a_3 & b_3 & c_3 \end{vmatrix} = a_1b_2c_3 + b_1c_2a_3 + c_1a_2b_3 - a_3b_2c_1 - b_3c_2a_1 - c_3a_2b_1$$

12.8. We represented the determinant of a 3×3 matrix A as a linear combination of determinants of order two with coefficients from the first row of A. Show that $\det(A)$ can also be represented as a linear combination of determinants of order two with coefficients from (i) the second row of A, (ii) the third row of A.

(i) $$\begin{vmatrix} a_1 & b_1 & c_1 \\ a_2 & b_2 & c_2 \\ a_3 & b_3 & c_3 \end{vmatrix} = a_1b_2c_3 + a_2b_3c_1 + a_3b_1c_2 - a_1b_3c_2 - a_2b_1c_3 - a_3b_2c_1$$

$$= -a_2(b_1c_3 - b_3c_1) + b_2(a_1c_3 - a_3c_1) - c_2(a_1b_3 - a_3b_1)$$

$$= -a_2\begin{vmatrix} b_1 & c_1 \\ b_3 & c_3 \end{vmatrix} + b_2\begin{vmatrix} a_1 & c_1 \\ a_3 & c_3 \end{vmatrix} - c_2\begin{vmatrix} a_1 & b_1 \\ a_3 & b_3 \end{vmatrix}$$

$$= -a_2\begin{vmatrix} a_1 & b_1 & c_1 \\ a_2 & b_2 & c_2 \\ a_3 & b_3 & c_3 \end{vmatrix} + b_2\begin{vmatrix} a_1 & b_1 & c_1 \\ a_2 & b_2 & c_2 \\ a_3 & b_3 & c_3 \end{vmatrix} - c_2\begin{vmatrix} a_1 & b_1 & c_1 \\ a_2 & b_2 & c_2 \\ a_3 & b_3 & c_3 \end{vmatrix}$$

(ii) $$= a_3(b_1c_2 - b_2c_1) - b_3(a_1c_2 - a_2c_1) + c_3(a_1b_2 - a_2b_1)$$

$$= a_3\begin{vmatrix} b_1 & c_1 \\ b_2 & c_2 \end{vmatrix} - b_3\begin{vmatrix} a_1 & c_1 \\ a_2 & c_2 \end{vmatrix} + c_3\begin{vmatrix} a_1 & b_1 \\ a_2 & b_2 \end{vmatrix}$$

$$= a_3\begin{vmatrix} a_1 & b_1 & c_1 \\ a_2 & b_2 & c_2 \\ a_3 & b_3 & c_3 \end{vmatrix} - b_3\begin{vmatrix} a_1 & b_1 & c_1 \\ a_2 & b_2 & c_2 \\ a_3 & b_3 & c_3 \end{vmatrix} + c_3\begin{vmatrix} a_1 & b_1 & c_1 \\ a_2 & b_2 & c_2 \\ a_3 & b_3 & c_3 \end{vmatrix}$$

Observe that the 2×2 matrix accompanying any coefficient can be obtained by deleting, in the original matrix, the row and column containing the coefficient. Furthermore, the signs accompanying the coefficients form a checkerboard pattern in the original matrix:

$$\begin{pmatrix} + & - & + \\ - & + & - \\ + & - & + \end{pmatrix}$$

Remark: We can also expand the above determinant in terms of the elements of any column in an analogous way (Problem 12.18).

INVERTIBLE MATRICES

12.9. Find the inverse of $\begin{pmatrix} 3 & 5 \\ 2 & 3 \end{pmatrix}$.

Method 1.

We seek scalars x, y, z and w for which

$$\begin{pmatrix} 3 & 5 \\ 2 & 3 \end{pmatrix}\begin{pmatrix} x & y \\ z & w \end{pmatrix} = \begin{pmatrix} 1 & 0 \\ 0 & 1 \end{pmatrix} \quad \text{or} \quad \begin{pmatrix} 3x+5z & 3y+5w \\ 2x+3z & 2y+3w \end{pmatrix} = \begin{pmatrix} 1 & 0 \\ 0 & 1 \end{pmatrix}$$

or which satisfy

$$\begin{cases} 3x+5z = 1 \\ 2x+3z = 0 \end{cases} \quad \text{and} \quad \begin{cases} 3y+5w = 0 \\ 2y+3w = 1 \end{cases}$$

To solve the first system, multiply the first equation by 2 and the second equation by -3 and then add:

$$\begin{array}{lrcl} 2 \times \text{first:} & 6x + 10z & = & 2 \\ -3 \times \text{second:} & -6x - 9z & = & 0 \\ \hline \text{Addition:} & z & = & 2 \end{array}$$

Substitute $z=2$ into the first equation to obtain

$$3x + 5\cdot 2 = 1 \quad \text{or} \quad 3x + 10 = 1 \quad \text{or} \quad 3x = -9 \quad \text{or} \quad x = -3$$

To solve the second system, multiply the first equation by 2 and the second equation by -3 and then add:

$$\begin{array}{lrcl} 2 \times \text{first:} & 6y + 10w & = & 0 \\ -3 \times \text{second:} & -6y - 9w & = & -3 \\ \hline \text{Addition:} & w & = & -3 \end{array}$$

Substitute $w=-3$ in the first equation to obtain

$$3y + 5\cdot(-3) = 0 \quad \text{or} \quad 3y - 15 = 0 \quad \text{or} \quad 3y = 15 \quad \text{or} \quad y = 5$$

Thus the inverse of the given matrix is $\begin{pmatrix} -3 & 5 \\ 2 & -3 \end{pmatrix}$.

Method 2.

We use the general formula for the inverse of a 2×2 matrix. First find the determinant of the given matrix:

$$\begin{vmatrix} 3 & 5 \\ 2 & 3 \end{vmatrix} = 3\cdot 3 - 2\cdot 5 = 9 - 10 = -1$$

Now interchange the elements on the main diagonal of the given matrix and take the negative of the other elements to obtain

$$\begin{pmatrix} 3 & -5 \\ -2 & 3 \end{pmatrix}$$

Lastly, divide each element of this matrix by the determinant of the given matrix, that is, by -1:

$$\begin{pmatrix} -3 & 5 \\ 2 & -3 \end{pmatrix}$$

The above is the required inverse.

12.10. Find the inverse of $\begin{pmatrix} 2 & 3 & -1 \\ 1 & 2 & 1 \\ -1 & -1 & 3 \end{pmatrix}$.

We seek nine scalars $x_1, x_2, x_3, y_1, y_2, y_3, z_1, z_2, z_3$ so that

$$\begin{pmatrix} 2 & 3 & -1 \\ 1 & 2 & 1 \\ -1 & -1 & 3 \end{pmatrix}\begin{pmatrix} x_1 & x_2 & x_3 \\ y_1 & y_2 & y_3 \\ z_1 & z_2 & z_3 \end{pmatrix} = \begin{pmatrix} 1 & 0 & 0 \\ 0 & 1 & 0 \\ 0 & 0 & 1 \end{pmatrix}$$

or

$$\begin{pmatrix} 2x_1 + 3y_1 - z_1 & 2x_2 + 3y_2 - z_2 & 2x_3 + 3y_3 - z_3 \\ x_1 + 2y_1 + z_1 & x_2 + 2y_2 + z_2 & x_3 + 2y_3 + z_3 \\ -x_1 - y_1 + 3z_1 & -x_2 - y_2 + 3z_2 & -x_3 - y_3 + 3z_3 \end{pmatrix} = \begin{pmatrix} 1 & 0 & 0 \\ 0 & 1 & 0 \\ 0 & 0 & 1 \end{pmatrix}$$

or which satisfy

$$\begin{cases} 2x_1 + 3y_1 - z_1 = 1 \\ x_1 + 2y_1 + z_1 = 0 \\ -x_1 - y_1 + 3z_1 = 0 \end{cases} \qquad \begin{cases} 2x_2 + 3y_2 - z_2 = 0 \\ x_2 + 2y_2 + z_2 = 1 \\ -x_2 - y_2 + 3z_2 = 0 \end{cases} \qquad \begin{cases} 2x_3 + 3y_3 - z_3 = 0 \\ x_3 + 2y_3 + z_3 = 0 \\ -x_3 - y_3 + 3z_3 = 1 \end{cases}$$

We solve each system by determinants. Observe that the matrix of coefficients in each system is the original matrix. Its determinant D is

$$D = \begin{vmatrix} 2 & 3 & -1 \\ 1 & 2 & 1 \\ -1 & -1 & 3 \end{vmatrix} = 2(6+1) - 3(3+1) - 1(-1+2) = 14 - 12 - 1 = 1$$

Since $D \neq 0$, each system has a solution and so the given matrix has an inverse. The solution of each of the three systems follows:

$$x_1 = 7,\ y_1 = -4,\ z_1 = 1; \quad x_2 = -8,\ y_2 = 5,\ z_2 = -1; \quad x_3 = 5,\ y_3 = -3,\ z_3 = 1$$

Thus the inverse of the given matrix is $\begin{pmatrix} 7 & -8 & 5 \\ -4 & 5 & -3 \\ 1 & -1 & 1 \end{pmatrix}$.

Supplementary Problems

12.11. Find the determinant of each matrix:

(i) $\begin{pmatrix} 2 & 5 \\ 4 & 1 \end{pmatrix}$ (ii) $\begin{pmatrix} 6 & 1 \\ 3 & -2 \end{pmatrix}$ (iii) $\begin{pmatrix} 4 & -5 \\ 0 & 2 \end{pmatrix}$ (iv) $\begin{pmatrix} 1 & 0 \\ 0 & 1 \end{pmatrix}$ (v) $\begin{pmatrix} -2 & 8 \\ -5 & -2 \end{pmatrix}$ (vi) $\begin{pmatrix} 4 & 9 \\ 5 & -3 \end{pmatrix}$

12.12. Solve each system using determinants:

(i) $\begin{cases} 3x + 5y = 8 \\ 4x - 2y = 1 \end{cases}$ (ii) $\begin{cases} 2x - 3y = -1 \\ 4x + 7y = -1 \end{cases}$ (iii) $\begin{cases} 5x = 2y - 7 \\ 3y = 4 + 4x \end{cases}$

12.13. Find the determinant of each matrix:

(i) $\begin{pmatrix} 2 & 1 & 1 \\ 0 & 5 & -2 \\ 1 & -3 & 4 \end{pmatrix}$ (ii) $\begin{pmatrix} 3 & -2 & -4 \\ 2 & 5 & -1 \\ 0 & 6 & 1 \end{pmatrix}$ (iii) $\begin{pmatrix} -2 & -1 & 4 \\ 6 & -3 & -2 \\ 4 & 1 & 2 \end{pmatrix}$ (iv) $\begin{pmatrix} 7 & 6 & 5 \\ 1 & 2 & 1 \\ 3 & -2 & 1 \end{pmatrix}$

12.14. Solve each system using determinants:

(i) $\begin{cases} 2x - 5y + 2z = 7 \\ x + 2y - 4z = 3 \\ 3x - 4y - 6z = 5 \end{cases}$ (ii) $\begin{cases} 2z + 3 = y + 3x \\ x - 3z = 2y + 1 \\ 3y + z = 2 - 2x \end{cases}$

12.15. Find the inverse of each matrix: (i) $\begin{pmatrix} 3 & 2 \\ 7 & 5 \end{pmatrix}$ (ii) $\begin{pmatrix} 2 & -3 \\ 1 & 3 \end{pmatrix}$

12.16. Find the inverse of $\begin{pmatrix} -1 & 2 & -3 \\ 2 & 1 & 0 \\ 4 & -2 & 5 \end{pmatrix}$. (*Note*: There will be 9 unknowns.)

12.17. Prove: $AB = I$ if and only if $BA = I$, where A and B are square matrices and I is the identity matrix.

12.18. Express the determinant of a 3×3 matrix as a linear combination of determinants of order two with coefficients from (i) the first column, (ii) the second column, (iii) the third column.

Answers to Supplementary Problems

12.11. (i) -18, (ii) -15, (iii) 8, (iv) 1, (v) 44, (vi) -57

12.12. (i) $x = \frac{21}{26}$, $y = \frac{29}{26}$ (ii) $x = -\frac{5}{13}$, $y = \frac{1}{13}$ (iii) $x = -\frac{13}{7}$, $y = -\frac{8}{7}$

12.13. (i) 21, (ii) −11, (iii) 100, (iv) 0

12.14. (i) $x = 5,\ y = 1,\ z = 1$ (ii) $x = z + y,\ y = -z$. The determinant of this system = 0.

12.15. (i) $\begin{pmatrix} 5 & -2 \\ -7 & 3 \end{pmatrix}$ (ii) $\begin{pmatrix} 1/3 & 1/3 \\ -1/9 & 2/9 \end{pmatrix}$

12.16. $\begin{pmatrix} -5 & 4 & -3 \\ 10 & -7 & 6 \\ 8 & -6 & 5 \end{pmatrix}$

12.18.

$$\begin{vmatrix} a_1 & b_1 & c_1 \\ a_2 & b_2 & c_2 \\ a_3 & b_3 & c_3 \end{vmatrix} = a_1\begin{vmatrix} b_2 & c_2 \\ b_3 & c_3 \end{vmatrix} - a_2\begin{vmatrix} b_1 & c_1 \\ b_3 & c_3 \end{vmatrix} + a_3\begin{vmatrix} b_1 & c_1 \\ b_2 & c_2 \end{vmatrix}$$

$$= -b_1\begin{vmatrix} a_2 & c_2 \\ a_3 & c_3 \end{vmatrix} + b_2\begin{vmatrix} a_1 & c_1 \\ a_3 & c_3 \end{vmatrix} - b_3\begin{vmatrix} a_1 & c_1 \\ a_2 & c_2 \end{vmatrix}$$

$$= c_1\begin{vmatrix} a_2 & b_2 \\ a_3 & b_3 \end{vmatrix} - c_2\begin{vmatrix} a_1 & b_1 \\ a_3 & b_3 \end{vmatrix} + c_3\begin{vmatrix} a_1 & b_1 \\ a_2 & b_2 \end{vmatrix}$$

Chapter 13

The Binomial Coefficients and Theorem

FACTORIAL NOTATION

The product of the positive integers from 1 to n inclusive is denoted by $n!$ (read "n factorial"):

$$n! = 1 \cdot 2 \cdot 3 \cdot \cdots \cdot (n-2)(n-1)n$$

We can also define $n!$ recursively as follows:

$$1! = 1 \quad \text{and} \quad n! = n \cdot (n-1)!$$

It is also convenient to define $0! = 1$.

Example 1.1: $2! = 1 \cdot 2 = 2$, $3! = 1 \cdot 2 \cdot 3 = 6$, $4! = 1 \cdot 2 \cdot 3 \cdot 4 = 24$,

$5! = 5 \cdot 4! = 5 \cdot 24 = 120$, $6! = 6 \cdot 5! = 6 \cdot 120 = 720$

Example 1.2: $\frac{8!}{6!} = \frac{8 \cdot 7 \cdot 6!}{6!} = 8 \cdot 7 = 56 \qquad 12 \cdot 11 \cdot 10 = \frac{12 \cdot 11 \cdot 10 \cdot 9!}{9!} = \frac{12!}{9!}$

$$\frac{12 \cdot 11 \cdot 10}{1 \cdot 2 \cdot 3} = 12 \cdot 11 \cdot 10 \cdot \frac{1}{3!} = \frac{12!}{3!\,9!}$$

Example 1.3:

$$n(n-1)\cdots(n-r+1) = \frac{n(n-1)\cdots(n-r+1)(n-r)(n-r-1)\cdots 3 \cdot 2 \cdot 1}{(n-r)(n-r-1)\cdots 3 \cdot 2 \cdot 1} = \frac{n!}{(n-r)!}$$

$$\frac{n(n-1)\cdots(n-r+1)}{1 \cdot 2 \cdot 3 \cdots (r-1)r} = n(n-1)\cdots(n-r+1) \cdot \frac{1}{r!} = \frac{n!}{(n-r)!} \cdot \frac{1}{r!} = \frac{n!}{r!\,(n-r)!}$$

BINOMIAL COEFFICIENTS

The symbol $\binom{n}{r}$ (read "$n\,C\,r$"), where r and n are positive integers with $r \leq n$, is defined as follows:

$$\binom{n}{r} = \frac{n(n-1)(n-2)\cdots(n-r+1)}{1 \cdot 2 \cdot 3 \cdots (r-1)r}$$

These numbers are called the *binomial coefficients* in view of the theorem in the following section.

Example 2.1:

$$\binom{8}{2} = \frac{8 \cdot 7}{1 \cdot 2} = 28 \qquad \binom{9}{4} = \frac{9 \cdot 8 \cdot 7 \cdot 6}{1 \cdot 2 \cdot 3 \cdot 4} = 126$$

$$\binom{10}{3} = \frac{10 \cdot 9 \cdot 8}{1 \cdot 2 \cdot 3} = 120 \qquad \binom{13}{1} = \frac{13}{1} = 13$$

$$\binom{12}{5} = \frac{12 \cdot 11 \cdot 10 \cdot 9 \cdot 8}{1 \cdot 2 \cdot 3 \cdot 4 \cdot 5} = 792$$

Note that $\binom{n}{r}$ has exactly r factors in both the numerator and the denominator.

Observe, by Example 1.3, that

$$\binom{n}{r} = \frac{n(n-1)\cdots(n-r+1)}{1 \cdot 2 \cdot 3 \cdots (r-1)r} = \frac{n!}{r!\,(n-r)!}$$

But $n-(n-r) = r$; hence we have the following important relation:

Theorem 13.1: $\binom{n}{n-r} = \binom{n}{r}$ or, in other words, if $a+b=n$ then $\binom{n}{a} = \binom{n}{b}$.

Example 2.2:

Compute $\binom{10}{7}$. By definition,

$$\binom{10}{7} = \frac{10 \cdot 9 \cdot 8 \cdot 7 \cdot 6 \cdot 5 \cdot 4}{1 \cdot 2 \cdot 3 \cdot 4 \cdot 5 \cdot 6 \cdot 7} = 120$$

On the other hand, $10-7=3$ and so we can also compute $\binom{10}{7}$ as follows:

$$\binom{10}{7} = \binom{10}{3} = \frac{10 \cdot 9 \cdot 8}{1 \cdot 2 \cdot 3} = 120$$

Observe that the second method saves space and time.

Example 2.3:

Compute $\binom{14}{12}$ and $\binom{20}{17}$. Now $14-12=2$ and $20-17=3$; hence

$$\binom{14}{12} = \binom{14}{2} = \frac{14 \cdot 13}{1 \cdot 2} = 91 \qquad \binom{20}{17} = \binom{20}{3} = \frac{20 \cdot 19 \cdot 18}{1 \cdot 2 \cdot 3} = 1140$$

Remark: Using the above formula for $\binom{n}{r}$ and the fact that $0! = 1$, we can extend the definition of $\binom{n}{r}$ to the following cases:

$$\binom{n}{0} = \frac{n!}{0!\,n!} = 1 \quad \text{and, in particular,} \quad \binom{0}{0} = \frac{0!}{0!\,0!} = 1$$

BINOMIAL THEOREM

The Binomial Theorem, which can be proved by mathematical induction, gives the general expression for the expansion of $(a+b)^n$.

Theorem 13.2 (Binomial Theorem):

$$\begin{aligned}(a+b)^n &= a^n + na^{n-1}b + \frac{n(n-1)}{1 \cdot 2}a^{n-2}b^2 + \frac{n(n-1)(n-2)}{1 \cdot 2 \cdot 3}a^{n-3}b^3 + \cdots \\ &\quad + \frac{n(n-1)}{1 \cdot 2}a^2b^{n-2} + nab^{n-1} + b^n \\ &= a^n + \binom{n}{1}a^{n-1}b + \binom{n}{2}a^{n-2}b^2 + \binom{n}{3}a^{n-3}b^3 + \cdots \\ &\quad + \binom{n}{2}a^2b^{n-2} + \binom{n}{1}ab^{n-1} + b^n \\ &= \sum_{r=0}^{n} \binom{n}{r} a^{n-r}b^r\end{aligned}$$

Example 3.1:

$$\begin{aligned}(a+b)^6 &= a^6 + 6a^5b + \frac{6 \cdot 5}{1 \cdot 2}a^4b^2 + \frac{6 \cdot 5 \cdot 4}{1 \cdot 2 \cdot 3}a^3b^3 + \frac{6 \cdot 5}{1 \cdot 2}a^2b^4 + 6ab^5 + b^6 \\ &= a^6 + 6a^5b + 15a^4b^2 + 20a^3b^3 + 15a^2b^4 + 6ab^5 + b^6\end{aligned}$$

$$(a+b)^7 = a^7 + 7a^6b + \frac{7\cdot 6}{1\cdot 2}a^5b^2 + \frac{7\cdot 6\cdot 5}{1\cdot 2\cdot 3}a^4b^3 + \frac{7\cdot 6\cdot 5}{1\cdot 2\cdot 3}a^3b^4 + \frac{7\cdot 6}{1\cdot 2}a^2b^5 + 7ab^6 + b^7$$

$$= a^7 + 7a^6b + 21a^5b^2 + 35a^4b^3 + 35a^3b^4 + 21a^2b^5 + 7ab^6 + b^7$$

The following properties of the expansion of $(a+b)^n$ should be observed:

(i) There are $n+1$ terms.

(ii) The sum of the exponents of a and b in every term is n.

(iii) The exponents of a decrease term by term from n to 0; the exponents of b increase term by term from 0 to n.

(iv) The coefficient of any term is $\binom{n}{k}$ where k is the exponent of either a or b. (This property follows from Theorem 13.1.)

(v) The coefficients of terms equidistant from the ends are equal.

Example 3.2:

Expand $(2x+3y^2)^5$ and simplify.

Observe that a and b in the Binomial Theorem correspond in this case to $2x$ and $3y^2$ respectively. Thus:

$$(2x+3y^2)^5 = (2x)^5 + 5(2x)^4(3y^2) + \frac{5\cdot 4}{1\cdot 2}(2x)^3(3y^2)^2 + \frac{5\cdot 4}{1\cdot 2}(2x)^2(3y^2)^3 + 5(2x)(3y^2)^4 + (3y^2)^5$$

$$= 2^5x^5 + 5\cdot 2^4x^4\cdot 3y^2 + 10\cdot 2^3x^3\cdot 3^2y^4 + 10\cdot 2^2x^2\cdot 3^3y^6 + 5\cdot 2x\cdot 3^4y^8 + 3^5y^{10}$$

$$= 32x^5 + 240x^4y^2 + 720x^3y^4 + 1080x^2y^6 + 810xy^8 + 243y^{10}$$

PASCAL'S TRIANGLE

The coefficients of the successive powers of $a+b$ can be arranged in a triangular array of numbers, called Pascal's triangle, as follows:

$$(a+b)^0 = 1$$
$$(a+b)^1 = a + b$$
$$(a+b)^2 = a^2 + 2ab + b^2$$
$$(a+b)^3 = a^3 + 3a^2b + 3ab^2 + b^3$$
$$(a+b)^4 = a^4 + 4a^3b + 6a^2b^2 + 4ab^3 + b^4$$
$$(a+b)^5 = a^5 + 5a^4b + 10a^3b^2 + 10a^2b^3 + 5ab^4 + b^5$$
$$(a+b)^6 = a^6 + 6a^5b + 15a^4b^2 + 20a^3b^3 + 15a^2b^4 + 6ab^5 + b^6$$
. . . .

1
1 1
1 2 1
1 3 3 1
1 4 6 4 1
1 5 (10) 10 5 1
1 6 (15) (20) 15 6 1
. . . .

Pascal's triangle has the following interesting properties.

(i) The first number and the last number in each row is 1.

(ii) Every other number in the array can be obtained by adding the two numbers appearing directly above it. For example: $10 = 4+6$, $15 = 5+10$, $20 = 10+10$.

Now since the numbers appearing in Pascal's triangle are the binomial coefficients, we can rewrite the triangular array of numbers as follows:

$$\binom{0}{0}$$

$$\binom{1}{0} \qquad \binom{1}{0}$$

$$\binom{2}{0} \qquad \binom{2}{1} \qquad \binom{2}{2}$$

$$\binom{3}{0} \qquad \binom{3}{1} \qquad \binom{3}{2} \qquad \binom{3}{3}$$

$$\binom{4}{0} \qquad \binom{4}{1} \qquad \binom{4}{2} \qquad \binom{4}{3} \qquad \binom{4}{4}$$

....

$$\binom{n}{0} \qquad \binom{n}{1} \cdots \binom{n}{r-1} \qquad \binom{n}{r} \cdots \binom{n}{n-1} \qquad \binom{n}{n}$$

$$\binom{n+1}{0} \qquad \binom{n+1}{1} \qquad \cdots \qquad \binom{n+1}{r} \qquad \cdots \qquad \binom{n+1}{n} \qquad \binom{n+1}{n+1}$$

....

In order to establish the above property (ii) of Pascal's triangle, it is necessary to prove the following theorem.

Theorem 13.3: $\binom{n+1}{r} = \binom{n}{r-1} + \binom{n}{r}$

In the next example we prove a special case of this theorem; the general proof of Theorem 13.3 is given as a solved problem.

Example 4.1:

We prove $\binom{12}{7} = \binom{11}{6} + \binom{11}{7}$ which is Theorem 13.3 when $n = 11$ and $r = 7$. Now

$$\binom{11}{6} + \binom{11}{7} = \frac{11!}{6!\,5!} + \frac{11!}{7!\,4!}$$

Multiply the first fraction by $\frac{7}{7}$ and the second fraction by $\frac{5}{5}$ to obtain the same denominator, and then add:

$$\binom{11}{6} + \binom{11}{7} = \frac{7 \cdot 11!}{7 \cdot 6!\,5!} + \frac{5 \cdot 11!}{7! \cdot 5 \cdot 4!} = \frac{7 \cdot 11!}{7!\,5!} + \frac{5 \cdot 11!}{7!\,5!}$$

$$= \frac{7 \cdot 11! + 5 \cdot 11!}{7!\,5!} = \frac{(7+5) \cdot 11!}{7!\,5!} = \frac{12 \cdot 11!}{7!\,5!} = \frac{12!}{7!\,5!} = \binom{12}{7}$$

The general proof of Theorem 13.3 is very similar to the proof given above.

MULTINOMIAL COEFFICIENTS

Let $n_1, n_2, \ldots, n_r$ be non-negative integers such that $n_1 + n_2 + \cdots + n_r = n$. Then the expression $\binom{n}{n_1, n_2, \ldots, n_r}$ is defined as follows:

$$\binom{n}{n_1, n_2, \ldots, n_r} = \frac{n!}{n_1!\,n_2! \cdots n_r!}$$

These numbers are called the *multinomial coefficients* in view of the following theorem which generalizes the binomial theorem.

Theorem 13.4: $$(a_1 + a_2 + \cdots + a_r)^n = \sum_{n_1+n_2+\cdots+n_r=n} \binom{n}{n_1, n_2, \ldots, n_r} a_1^{n_1} a_2^{n_2} \cdots a_r^{n_r}$$

Example 5.1:

$$\binom{7}{2, 3, 2} = \frac{7!}{2!\,3!\,2!} = \frac{7 \cdot 6 \cdot 5 \cdot 4 \cdot 3 \cdot 2 \cdot 1}{2 \cdot 1 \cdot 3 \cdot 2 \cdot 1 \cdot 2 \cdot 1} = 210$$

$$\binom{8}{4, 2, 2, 0} = \frac{8!}{4!\,2!\,2!\,0!} = \frac{8 \cdot 7 \cdot 6 \cdot 5 \cdot 4 \cdot 3 \cdot 2 \cdot 1}{4 \cdot 3 \cdot 2 \cdot 1 \cdot 2 \cdot 1 \cdot 2 \cdot 1 \cdot 1} = 420$$

Example 5.2:

Expand $(a + b + c)^3$. Consider all triplets of non-negative integers (n_1, n_2, n_3) for which $n_1 + n_2 + n_3 = 3$:

$$\begin{aligned}(a+b+c)^3 &= \binom{3}{3,0,0} a^3b^0b^0 + \binom{3}{2,1,0} a^2b^1c^0 + \binom{3}{2,0,1} a^2b^0c^1 \\ &\quad + \binom{3}{1,1,1} a^1b^1c^1 + \binom{3}{0,3,0} a^0b^3c^0 + \binom{3}{0,2,1} a^0b^2c^1 \\ &\quad + \binom{3}{1,2,0} a^1b^2c^0 + \binom{3}{0,0,3} a^0b^0c^3 + \binom{3}{1,0,2} a^1b^0c^2 + \binom{3}{0,1,2} a^0b^1c^2 \\ &= a^3 + 3a^2b + 3a^2c + 6abc + b^3 + 3b^2c + 3ab^2 + c^3 + 3ac^2 + 3bc^2\end{aligned}$$

Solved Problems

FACTORIAL

13.1. Compute 4!, 5!, 6!, 7! and 8!.

$4! = 1 \cdot 2 \cdot 3 \cdot 4 = 24$

$5! = 1 \cdot 2 \cdot 3 \cdot 4 \cdot 5 = 5 \cdot 4! = 5 \cdot 24 = 120$

$6! = 1 \cdot 2 \cdot 3 \cdot 4 \cdot 5 \cdot 6 = 6 \cdot 5! = 6 \cdot 120 = 720$

$7! = 7 \cdot 6! = 7 \cdot 720 = 5040$

$8! = 8 \cdot 7! = 8 \cdot 5040 = 40{,}320$

13.2. Compute: (i) $\dfrac{13!}{11!}$, (ii) $\dfrac{7!}{10!}$.

(i) $\dfrac{13!}{11!} = \dfrac{13 \cdot 12 \cdot 11 \cdot 10 \cdot 9 \cdot 8 \cdot 7 \cdot 6 \cdot 5 \cdot 4 \cdot 3 \cdot 2 \cdot 1}{11 \cdot 10 \cdot 9 \cdot 8 \cdot 7 \cdot 6 \cdot 5 \cdot 4 \cdot 3 \cdot 2 \cdot 1} = 13 \cdot 12 = 156$

or $\dfrac{13!}{11!} = \dfrac{13 \cdot 12 \cdot 11!}{11!} = 13 \cdot 12 = 156$

(ii) $\dfrac{7!}{10!} = \dfrac{7!}{10 \cdot 9 \cdot 8 \cdot 7!} = \dfrac{1}{10 \cdot 9 \cdot 8} = \dfrac{1}{720}$

13.3. Write in terms of factorials: (i) $27 \cdot 26$, (ii) $\dfrac{1}{14 \cdot 13 \cdot 12}$, (iii) 56.

(i) $27 \cdot 26 = \dfrac{27 \cdot 26 \cdot 25!}{25!} = \dfrac{27!}{25!}$

(ii) $\dfrac{1}{14 \cdot 13 \cdot 12} = \dfrac{11!}{14 \cdot 13 \cdot 12 \cdot 11!} = \dfrac{11!}{14!}$

(iii) $56 = \dfrac{56 \cdot 55!}{55!} = \dfrac{56!}{55!}$

or $56 = 8 \cdot 7 = \dfrac{8 \cdot 7 \cdot 6!}{6!} = \dfrac{8!}{6!}$

13.4. Simplify: (i) $\dfrac{n!}{(n-1)!}$, (ii) $\dfrac{(n+2)!}{n!}$, (iii) $\dfrac{(n+1)!}{(n-1)!}$, (iv) $\dfrac{(n-1)!}{(n+2)!}$.

(i) $\dfrac{n!}{(n-1)!} = \dfrac{n(n-1)(n-2)\cdots 3\cdot 2\cdot 1}{(n-1)(n-2)\cdots 3\cdot 2\cdot 1} = n$ or, simply, $\dfrac{n!}{(n-1)!} = \dfrac{n(n-1)!}{(n-1)!} = n$

(ii) $\dfrac{(n+2)!}{n!} = \dfrac{(n+2)(n+1)n(n-1)(n-2)\cdots 3\cdot 2\cdot 1}{n(n-1)(n-2)\cdots 3\cdot 2\cdot 1} = (n+2)(n+1) = n^2+3n+2$

or, simply, $\dfrac{(n+2)!}{n!} = \dfrac{(n+2)(n+1)\cdot n!}{n!} = (n+2)(n+1) = n^2+3n+2$

(iii) $\dfrac{(n+1)!}{(n-1)!} = \dfrac{(n+1)n(n-1)(n-2)\cdots 3\cdot 2\cdot 1}{(n-1)(n-2)\cdots 3\cdot 2\cdot 1} = (n+1)\cdot n = n^2+n$

or, simply, $\dfrac{(n+1)!}{(n-1)!} = \dfrac{(n+1)\cdot n\cdot (n-1)!}{(n-1)!} = (n+1)\cdot n = n^2+n$

(iv) $\dfrac{(n-1)!}{(n+2)!} = \dfrac{(n-1)!}{(n+2)(n+1)\cdot n\cdot (n-1)!} = \dfrac{1}{(n+2)(n+1)\cdot n} = \dfrac{1}{n^3+3n^2+2n}$

BINOMIAL COEFFICIENTS

13.5. Compute: (i) $\dbinom{16}{3}$, (ii) $\dbinom{12}{4}$, (iii) $\dbinom{15}{5}$.

Recall that there are as many factors in the numerator as in the denominator.

(i) $\dbinom{16}{3} = \dfrac{16\cdot 15\cdot 14}{1\cdot 2\cdot 3} = 560$ (iii) $\dbinom{15}{5} = \dfrac{15\cdot 14\cdot 13\cdot 12\cdot 11}{1\cdot 2\cdot 3\cdot 4\cdot 5} = 3003$

(ii) $\dbinom{12}{4} = \dfrac{12\cdot 11\cdot 10\cdot 9}{1\cdot 2\cdot 3\cdot 4} = 495$

13.6. Compute: (i) $\dbinom{8}{5}$, (ii) $\dbinom{9}{7}$, (iii) $\dbinom{10}{6}$.

(i) $\dbinom{8}{5} = \dfrac{8\cdot 7\cdot 6\cdot 5\cdot 4}{1\cdot 2\cdot 3\cdot 4\cdot 5} = 56$

Note that $8-5=3$; hence we could also compute $\dbinom{8}{5}$ as follows:

$$\binom{8}{5} = \binom{8}{3} = \frac{8\cdot 7\cdot 6}{1\cdot 2\cdot 3} = 56$$

(ii) Now $9-7=2$; hence $\dbinom{9}{7} = \dbinom{9}{2} = \dfrac{9\cdot 8}{1\cdot 2} = 36$.

(iii) Now $10-6=4$; hence $\dbinom{10}{6} = \dbinom{10}{4} = \dfrac{10\cdot 9\cdot 8\cdot 7}{1\cdot 2\cdot 3\cdot 4} = 210$.

13.7. Prove: $\dbinom{17}{6} = \dbinom{16}{5} + \dbinom{16}{6}$.

Now $\dbinom{16}{5} + \dbinom{16}{6} = \dfrac{16!}{5!\,11!} + \dfrac{16!}{6!\,10!}$. Multiply the first fraction by $\dfrac{6}{6}$ and the second by $\dfrac{11}{11}$ to obtain the same denominator in both fractions; and then add:

$$\binom{16}{5} + \binom{16}{6} = \frac{6\cdot 16!}{6\cdot 5!\cdot 11!} + \frac{11\cdot 16!}{6!\cdot 11\cdot 10!} = \frac{6\cdot 16!}{6!\cdot 11!} + \frac{11\cdot 16!}{6!\cdot 11!}$$

$$= \frac{6\cdot 16! + 11\cdot 16!}{6!\cdot 11!} = \frac{(6+11)\cdot 16!}{6!\cdot 11!} = \frac{17\cdot 16!}{6!\cdot 11!} = \frac{17!}{6!\cdot 11!} = \binom{17}{6}$$

13.8. Prove Theorem 13.3: $\binom{n+1}{r} = \binom{n}{r-1} + \binom{n}{r}$.

(The technique in this proof is similar to that of the preceding problem and Example 4.1.)

Now $\binom{n}{r-1} + \binom{n}{r} = \frac{n!}{(r-1)!\cdot(n-r+1)!} + \frac{n!}{r!\cdot(n-r)!}$. To obtain the same denominator in both fractions, multiply the first fraction by $\frac{r}{r}$ and the second fraction by $\frac{n-r+1}{n-r+1}$. Hence

$$\binom{n}{r-1} + \binom{n}{r} = \frac{r\cdot n!}{r\cdot(r-1)!\cdot(n-r+1)!} + \frac{(n-r+1)\cdot n!}{r!\cdot(n-r+1)\cdot(n-r)!}$$

$$= \frac{r\cdot n!}{r!\,(n-r+1)!} + \frac{(n-r+1)\cdot n!}{r!\,(n-r+1)!}$$

$$= \frac{r\cdot n! + (n-r+1)\cdot n!}{r!\,(n-r+1)!} = \frac{[r+(n-r+1)]\cdot n!}{r!\,(n-r+1)!}$$

$$= \frac{(n+1)\,n!}{r!\,(n-r+1)!} = \frac{(n+1)!}{r!\,(n-r+1)!} = \binom{n+1}{r}$$

BINOMIAL THEOREM

13.9. Expand and simplify: $(x+3y)^3$.

$$(x+3y)^3 = (x)^3 + \frac{3}{1}(x)^2(3y) + \frac{3}{1}(x)(3y)^2 + (3y)^3 = x^3 + 9x^2y + 27xy^2 + 27y^3$$

13.10. Expand and simplify: $(2x-y)^4$.

$$(2x-y)^4 = (2x)^4 + \frac{4}{1}(2x)^3(-y) + \frac{4\cdot 3}{1\cdot 2}(2x)^2(-y)^2 + \frac{4}{1}(2x)(-y)^3 + (-y)^4$$

$$= 16x^4 - 32x^3y + 24x^2y^2 - 8xy^3 + y^4$$

13.11. Expand and simplify: $(2x+y^2)^5$.

$$(2x+y^2)^5 = (2x)^5 + \frac{5}{1}(2x)^4(y^2) + \frac{5\cdot 4}{1\cdot 2}(2x)^3(y^2)^2 + \frac{5\cdot 4}{1\cdot 2}(2x)^2(y^2)^3 + \frac{5}{1}(2x)(y^2)^4 + (y^2)^5$$

$$= 32x^5 + 80x^4y^2 + 80x^3y^4 + 40x^2y^6 + 10xy^8 + y^{10}$$

13.12. Expand and simplify: $(x^2-2y)^6$.

$$(x^2-2y)^6 = (x^2)^6 + \frac{6}{1}(x^2)^5(-2y) + \frac{6\cdot 5}{1\cdot 2}(x^2)^4(-2y)^2 + \frac{6\cdot 5\cdot 4}{1\cdot 2\cdot 3}(x^2)^3(-2y)^3$$

$$+ \frac{6\cdot 5}{1\cdot 2}(x^2)^2(-2y)^4 + \frac{6}{1}(x^2)(-2y)^5 + (-2y)^6$$

$$= x^{12} - 12x^{10}y + 60x^8y^2 - 160x^6y^3 + 240x^4y^4 - 192x^2y^5 + 64y^6$$

13.13. Given $(a+b)^6 = a^6 + 6a^5b + 15a^4b^2 + 20a^3b^3 + 15a^2b^4 + 6ab^5 + b^6$. Expand $(a+b)^7$ and $(a+b)^8$.

We could expand $(a+b)^7$ and $(a+b)^8$ using the binomial theorem. However, given the expansion of $(a+b)^6$, it is simpler to use Pascal's triangle to obtain the coefficients of $(a+b)^7$ and $(a+b)^8$. First write down the coefficients of $(a+b)^6$ and then compute the next two rows of the triangle:

1 6 15 20 15 6 1

1 7 21 (35) 35 21 7 1

1 8 (28) 56 70 (56) 28 8 1

Recall that each number is obtained by adding the two numbers above, e.g. $35 = 15+20$, $28 = 7+21$, $56 = 35+21$. From the above diagram we obtain

$$(a+b)^7 = a^7 + 7a^6b + 21a^5b^2 + 35a^4b^3 + 35a^3b^4 + 21a^2b^5 + 7ab^6 + b^7$$

$$(a+b)^8 = a^8 + 8a^7b + 28a^6b^2 + 56a^5b^3 + 70a^4b^4 + 56a^3b^5 + 28a^2b^6 + 8ab^7 + b^8$$

13.14. (i) How does $(a-b)^n$ differ from $(a+b)^n$? (ii) Use Problem 13.11 to obtain $(2x-y^2)^5$.

(i) The expansion of $(a-b)^n$ will be identical to the expansion of $(a+b)^n$ except that the signs in $(a-b)^n$ will alternate beginning with plus, i.e. terms with even powers of b will be plus and terms with odd powers of b will be minus. This follows from the fact that $(-b)^r = b^r$ if r is even and $(-b)^r = -b^r$ if r is odd.

(ii) By Problem 13.11,

$$(2x+y^2)^5 = 32x^5 + 80x^4y^2 + 80x^3y^4 + 40x^2y^6 + 10xy^8 + y^{10}$$

To obtain $(2x-y^2)^5$, just alternate the signs in $(2x+y^2)^5$:

$$(2x-y^2)^5 = 32x^5 - 80x^4y^2 + 80x^3y^4 - 40x^2y^6 + 10xy^8 - y^{10}$$

13.15. Obtain and simplify the term in the expansion of $(2x^2-y^3)^8$ which contains x^{10}.

Method 1.

The general term in the expansion of $(2x^2-y^3)^8$ is

$$\binom{8}{r}(2x^2)^{8-r}(-y^3)^r = \binom{8}{r}2^{8-r}x^{16-2r}(-y^3)^r$$

Hence the term containing x^{10} has $x^{16-2r} = x^{10}$ or $16-2r = 10$ or $r=3$, that is, is the term

$$\binom{8}{3}(2x^2)^5(-y^3)^3 = \binom{8}{3}2^5x^{10}(-y^9) = -\frac{8\cdot 7\cdot 6}{1\cdot 2\cdot 3}32x^{10}y^9 = -1792x^{10}y^9$$

Method 2.

Each term in the expansion contains a binomial coefficient, a power of $2x^2$ and a power of $-y^3$; hence first write down the following:

$$(\quad)(2x^2)(-y^3)$$

To obtain x^{10}, the power of $2x^2$ must be 5; hence insert the exponent 5 to obtain

$$(\quad)(2x^2)^5(-y^3)$$

Now the sum of the exponents must add up to 8, the exponent of $(2x^2-y^3)^8$; hence the exponent of $(-y^3)$ must be 3. Insert the 3 to obtain

$$(\quad)(2x^2)^5(-y^3)^3$$

Last, the binomial coefficient must contain 8 above and either exponent below. Thus the required term is

$$\binom{8}{3}(2x^2)^5(-y^3)^3 = -1792x^{10}y^9$$

13.16. Prove: $2^4 = 16 = \binom{4}{0} + \binom{4}{1} + \binom{4}{2} + \binom{4}{3} + \binom{4}{4}$.

Expand $(1+1)^4$ using the binomial theorem:

$$\begin{aligned} 2^4 = (1+1)^4 &= \binom{4}{0}1^4 + \binom{4}{1}1^3 1^1 + \binom{4}{2}1^2 1^2 + \binom{4}{3}1^1 1^3 + \binom{4}{4}1^4 \\ &= \binom{4}{0} + \binom{4}{1} + \binom{4}{2} + \binom{4}{3} + \binom{4}{4} \end{aligned}$$

13.17. Prove the Binomial Theorem 13.2: $(a+b)^n = \sum_{r=0}^{n} \binom{n}{r} a^{n-r} b^r$.

The theorem is true for $n = 1$, since

$$\sum_{r=0}^{1} \binom{1}{r} a^{1-r} b^r = \binom{1}{0} a^1 b^0 + \binom{1}{1} a^0 b^1 = a + b = (a+b)^1$$

We assume the theorem holds for $(a+b)^n$ and prove it is true for $(a+b)^{n+1}$.

$$\begin{aligned} (a+b)^{n+1} &= (a+b)(a+b)^n \\ &= (a+b)\left[a^n + \binom{n}{1}a^{n-1}b + \cdots + \binom{n}{r-1}a^{n-r+1}b^{r-1}\right. \\ &\qquad \left. + \binom{n}{r}a^{n-r}b^r + \cdots + \binom{n}{1}ab^{n-1} + b^n\right] \end{aligned}$$

Now the term in the product which contains b^r is obtained from

$$\begin{aligned} b\left[\binom{n}{r-1}a^{n-r+1}b^{r-1}\right] + a\left[\binom{n}{r}a^{n-r}b^r\right] &= \binom{n}{r-1}a^{n-r+1}b^r + \binom{n}{r}a^{n-r+1}b^r \\ &= \left[\binom{n}{r-1} + \binom{n}{r}\right]a^{n-r+1}b^r \end{aligned}$$

But, by Theorem 13.3, $\binom{n}{r-1} + \binom{n}{r} = \binom{n+1}{r}$. Thus the term containing b^r is $\binom{n+1}{r}a^{n-r+1}b^r$. Note that $(a+b)(a+b)^n$ is a polynomial of degree $n+1$ in b. Consequently,

$$(a+b)^{n+1} = (a+b)(a+b)^n = \sum_{r=0}^{n+1} \binom{n+1}{r} a^{n-r+1} b^r$$

which was to be proved.

MULTINOMIAL COEFFICIENTS

13.18. Compute: (i) $\binom{6}{3,2,1}$, (ii) $\binom{8}{4,2,2,0}$, (iii) $\binom{10}{5,3,2,2}$.

(i) $\binom{6}{3,2,1} = \dfrac{6!}{3!\,2!\,1!} = \dfrac{6\cdot5\cdot4\cdot3\cdot2\cdot1}{3\cdot2\cdot1\cdot2\cdot1\cdot1} = 60$

(ii) $\binom{8}{4,2,2,0} = \dfrac{8!}{4!\,2!\,2!\,0!} = \dfrac{8\cdot7\cdot6\cdot5\cdot4\cdot3\cdot2\cdot1}{4\cdot3\cdot2\cdot1\cdot2\cdot1\cdot2\cdot1\cdot1} = 420$

(iii) The expression $\binom{10}{5,3,2,2}$ has no meaning since $5+3+2+2 \neq 10$.

13.19. Show that $\binom{n}{n_1, n_2} = \binom{n}{n_1} = \binom{n}{n_2}$.

Observe that the expression $\binom{n}{n_1, n_2}$ implicitly implies that $n_1 + n_2 = n$ or $n_2 = n - n_1$. Hence

$$\binom{n}{n_2} = \binom{n}{n_1} = \frac{n!}{n_1!\,(n-n_1)!} = \frac{n!}{n_1!\,n_2!} = \binom{n}{n_1, n_2}$$

13.20. Find the term in the expansion of $(2x^3 - 3xy^2 + z^2)^6$ which contains x^{11} and y^4.

The general term of the expansion is

$$\binom{6}{a, b, c}(2x^3)^a(-3xy^2)^b(z^2)^c = \binom{6}{a, b, c}2^a x^{3a}(-3)^b x^b y^{2b} z^{2c}$$
$$= \binom{6}{a, b, c}2^a(-3)^b x^{3a+b} y^{2b} z^{2c}$$

Thus the term containing x^{11} and y^4 has $3a + b = 11$ and $2b = 4$ or, $b = 2$ and $a = 3$. Also, since $a + b + c = 6$, we have $c = 1$. Substituting in the above gives

$$\binom{6}{3, 2, 1}2^3(-3)^2 x^{11} y^4 z^2 = -\frac{6!}{3!\,2!\,1!}\,8\cdot 9\, x^{11} y^4 z^2 = -4320x^{11}y^4z^2$$

Supplementary Problems

FACTORIAL

13.21. Compute: (i) 9!, (ii) 10!, (iii) 11!

13.22. Compute: (i) $\frac{16!}{14!}$, (ii) $\frac{14!}{11!}$, (iii) $\frac{8!}{10!}$, (iv) $\frac{10!}{13!}$.

13.23. Write in terms of factorials: (i) $24\cdot 23\cdot 22\cdot 21$, (ii) $\frac{1}{10\cdot 11\cdot 12}$, (iii) 42.

13.24. Simplify: (i) $\frac{(n+1)!}{n!}$, (ii) $\frac{n!}{(n-2)!}$, (iii) $\frac{(n-1)!}{(n+2)!}$, (iv) $\frac{(n-r+1)!}{(n-r-1)!}$.

BINOMIAL AND MULTINOMIAL COEFFICIENTS

13.25. Compute: (i) $\binom{5}{2}$, (ii) $\binom{7}{3}$, (iii) $\binom{14}{2}$, (iv) $\binom{6}{4}$, (v) $\binom{20}{17}$, (vi) $\binom{18}{15}$.

13.26. Show (without using Theorem 13.3) that $\binom{12}{8} = \binom{11}{7} + \binom{11}{8}$.

13.27. Compute: (i) $\binom{9}{3, 5, 1}$, (ii) $\binom{7}{3, 2, 2, 0}$, (iii) $\binom{6}{2, 2, 1, 1, 0}$.

BINOMIAL AND MULTINOMIAL THEOREMS

13.28. Expand and simplify: (i) $(2x + y^2)^3$, (ii) $(x^2 - 3y)^4$, (iii) $(\frac{1}{2}a + 2b)^5$, (iv) $(2a^2 - b)^6$.

13.29. The eighth row of Pascal's triangle is as follows:

1 8 28 56 70 56 28 8 1

Compute the ninth and tenth rows of the triangle.

13.30. Show that $\binom{n}{0} + \binom{n}{1} + \binom{n}{2} + \binom{n}{3} + \cdots + \binom{n}{n} = 2^n$.

13.31. Show that $\binom{n}{0} - \binom{n}{1} + \binom{n}{2} - \binom{n}{3} + \cdots \pm \binom{n}{n} = 0$.

13.32. Find the term in the expansion of $(2x^2 - \frac{1}{2}y^3)^8$ which contains x^8.

13.33. Find the term in the expansion of $(3xy^2 + z^2)^7$ which contains y^6.

13.34. Find the term in the expansion of $(2x^2 - y^3 + \frac{1}{2}z)^7$ which contains x^4 and z^4.

13.35. Find the term in the expression of $(xy - y^2 + 2z)^6$ which contains x^3 and y^5.

Answers to Supplementary Problems

13.21. (i) 362,880 (ii) 3,628,800 (iii) 39,916,800

13.22. (i) 240 (ii) 2184 (iii) 1/90 (iv) 1/1716

13.23. (i) 24!/20! (ii) 9!/12! (iii) 42!/41! or 7!/5!

13.24. (i) $n+1$ (ii) $n(n-1) = n^2 - n$ (iii) $1/[n(n+1)(n+2)]$ (iv) $(n-r)(n-r+1)$

13.25. (i) 10 (ii) 35 (iii) 91 (iv) 15 (v) 1140 (vi) 816

13.27. (i) 504 (ii) 210 (iii) 180

13.28. (i) $8x^3 + 12x^2y^2 + 6xy^4 + y^6$

(ii) $x^8 - 12x^6y + 54x^4y^2 - 108x^2y^3 + 81y^4$

(iii) $a^5/32 + 5a^4b/8 + 5a^3b^2 + 20a^2b^3 + 40ab^4 + 32b^5$

(iv) $64a^{12} - 192a^{10}b + 240a^8b^2 - 160a^6b^3 + 60a^4b^4 - 12a^2b^5 + b^6$

13.29.

1 8 28 56 70 56 28 8 1

1 9 36 84 126 126 84 36 9 1

1 10 45 120 210 252 210 120 45 10 1

13.30. *Hint.* Expand $(1+1)^n$.

13.31. *Hint.* Expand $(1-1)^n$.

13.32. $70x^8y^{12}$

13.33. $945x^3y^6z^8$

13.34. $-(105/4)x^4y^3z^4$

13.35. $-240x^3y^5z^2$

Chapter 14

Permutations, Ordered Samples

FUNDAMENTAL PRINCIPLE OF COUNTING

Combinatorial analysis, which includes the study of permutations, combinations and partitions, is concerned with determining the number of logical possibilities of some event without necessarily enumerating each case. The following basic principle of counting is used throughout.

Fundamental Principle of Counting: If some event can occur in n_1 different ways, and if, following this event, a second event can occur in n_2 different ways, and, following this second event, a third event can occur in n_3 different ways, ..., then the number of ways the events can occur in the order indicated is $n_1 \cdot n_2 \cdot n_3 \cdots$.

Example 1.1: Suppose a license plate contains two letters followed by three digits with the first digit not zero. How many different license plates can be printed?

Each letter can be printed in 26 different ways, the first digit in 9 ways and each of the other two digits in 10 ways. Hence

$$26 \cdot 26 \cdot 9 \cdot 10 \cdot 10 = 608{,}400$$

different plates can be printed.

Example 1.2: In how many ways can an organization containing 26 members elect a president, treasurer and secretary (assuming no person is elected to more than one position)?

The president can be elected in 26 different ways; following this, the treasurer can be elected in 25 different ways (since the person chosen president is not eligible to be treasurer); and, following this, the secretary can be elected in 24 different ways. Thus, by the above principle of counting, there are

$$26 \cdot 25 \cdot 24 = 15{,}600$$

different ways in which the organization can elect its officers.

PERMUTATIONS

Any arrangement of a set of n objects in a given order is called a *permutation* of the objects (taken all at a time). Any arrangement of any $r \leq n$ of these objects in a given order is called an *r-permutation* or a *permutation of the n objects taken r at a time.*

Example 2.1: Consider the set of letters a, b, c and d. Then:

(i) *bdca, dcba* and *acdb* are permutations of the 4 letters (taken all at a time);

(ii) *bad, adb, cbd* and *bca* are permutations of the 4 letters taken 3 at a time;

(iii) *ad, cb, da* and *bd* are permutations of the 4 letters taken 2 at a time.

The number of permutations of n objects taken r at a time is denoted by

$$P(n, r), \quad {}_nP_r, \quad P_{n,r} \quad \text{or} \quad P^n_r$$

We shall use $P(n, r)$. Before we derive the general formula for $P(n, r)$ we consider a particular case.

Example 2.2: How many permutations are there of 6 objects, say a, b, c, d, e and f, taken three at a time? In other words, we want to find the number of "three letter words" using the above six letters without repetitions.

Let the general three letter word be represented by the following three boxes:

Now the first letter can be chosen in 6 different ways; following this, the second letter can be chosen in 5 different ways; and, following this, the last letter can be chosen in 4 different ways. Write each number in its appropriate box as follows:

6	5	4

Thus by the fundamental principle of counting there are $6 \cdot 5 \cdot 4 = 120$ possible three letter words without repetitions from the six letters, or there are 120 permutations of 6 objects taken 3 at a time:

$$P(6, 3) = 120$$

The derivation of the formula for the number of permutations of n objects taken r at a time, or the number of r-permutations of n objects, $P(n, r)$, follows the procedure in the preceding example. The first element in an r-permutation of n objects can be chosen in n different ways; following this, the second element in the permutation can be chosen in $n-1$ ways; and, following this, the third element in the permutation can be chosen in $n-2$ ways. Continuing in this manner, we have that the rth (last) element in the r-permutation can be chosen in $n-(r-1) = n-r+1$ ways. Thus, by the fundamental principle of counting, we have

Theorem 14.1: $P(n, r) = n(n-1)(n-2) \cdots (n-r+1)$.

Recall (see Example 1.3 of Chapter 13) that

$$n(n-1)(n-2)\cdots(n-r+1) = \frac{n(n-1)(n-2)\cdots(n-r+1)\cdot(n-r)!}{(n-r)!} = \frac{n!}{(n-r)!}$$

In other words,

Corollary 14.2: $P(n, r) = \dfrac{n!}{(n-r)!}$.

In the special case in which $r = n$, we have

$$P(n, n) = n(n-1)(n-2)\cdots 3\cdot 2\cdot 1 = n!$$

In other words,

Corollary 14.3: There are $n!$ permutations of n objects (taken all at a time).

Example 2.3: How many permutations are there of 3 objects, say a, b and c?

By Corollary 14.3 there are $3! = 1 \cdot 2 \cdot 3 = 6$ permutations of the letters a, b and c. These are *abc, acb, bac, bca, cab, cba*.

PERMUTATIONS WITH REPETITIONS

Frequently we want to find the number of permutations of objects some of which are alike, as illustrated in the examples below. The general formula is as follows:

Theorem 14.4: The number of permutations of n objects of which n_1 are alike, n_2 are alike, ..., n_r are alike is

$$\frac{n!}{n_1!\, n_2! \cdots n_r!}$$

We indicate the proof of the above theorem by a particular example. Suppose we want to form all possible 5 letter words using the letters from the word "DADDY". Now there are $5! = 120$ permutations of the objects D_1, A, D_2, D_3, Y, where the three D's are distinguished. Observe that the following six permutations

$$D_1D_2D_3AY,\ D_2D_1D_3AY,\ D_3D_1D_2AY,\ D_1D_3D_2AY,\ D_2D_3D_1AY, \text{ and } D_3D_2D_1AY$$

produce the same word when the subscripts are removed. The 6 comes from the fact that there are $3! = 3\cdot2\cdot1 = 6$ different ways of placing the three D's in the first three positions in the permutation. This is true for each of the three possible positions in which the D's can appear. Accordingly there are

$$\frac{5!}{3!} = \frac{120}{6} = 20$$

different 5 letter words that can be formed using the letters from the word "DADDY".

Example 3.1: How many 7 letter words can be formed using the letters of the word "BENZENE"? We seek the number of permutations of 7 objects of which 3 are alike (the three E's), and 2 are alike (the two N's). By Theorem 14.4, there are

$$\frac{7!}{3!\,2!} = \frac{7\cdot6\cdot5\cdot4\cdot3\cdot2\cdot1}{3\cdot2\cdot1\cdot2\cdot1} = 420$$

such words.

Example 3.2: How many different signals, each consisting of 8 flags hung in a vertical line, can be formed from a set of 4 indistinguishable red flags, 3 indistinguishable white flags, and a blue flag? We seek the number of permutations of 8 objects of which 4 are alike and 3 are alike. There are

$$\frac{8!}{4!\,3!} = \frac{8\cdot7\cdot6\cdot5\cdot4\cdot3\cdot2\cdot1}{4\cdot3\cdot2\cdot1\cdot3\cdot2\cdot1} = 280$$

different signals.

ORDERED SAMPLES

Many problems in combinatorial analysis and, in particular, probability, are concerned with choosing a ball from an urn containing n balls (or a card from a deck, or a person from a population). When we choose one ball after the other from the urn, say r times, we call each choice an ordered sample of size r. We consider two cases:

(i) *Sampling with replacement.* Here the ball is replaced in the urn before the next ball is chosen. Now since there are n different ways to choose each ball, by the fundamental principle of counting there are

$$\overbrace{n\cdot n\cdot n\cdots n}^{r \text{ times}} = n^r$$

different ordered samples with replacement of size r.

(ii) *Sampling without replacement.* Here the ball is not replaced in the urn after it is chosen. Thus there are no repetitions in the ordered sample. In other words, an ordered sample of size r without replacement is simply an r-permutation of the objects in the urn. Accordingly, there are

$$P(n, r) = n(n-1)(n-2)\cdots(n-r+1) = \frac{n!}{(n-r)!}$$

different ordered samples of size r without replacement from a population of n objects.

Example 4.1: In how many ways can one choose three cards in succession from a deck of 52 cards (i) with replacement, (ii) without replacement? If each card is replaced in the deck before the next card is chosen, then each card can be chosen in 52 different ways. Hence there are

$$52 \cdot 52 \cdot 52 = 52^3 = 140{,}608$$

different ordered samples of size 3 with replacement.

In case there is no replacement, then the first card can be chosen in 52 different ways; the second card can be chosen in 51 different ways; and, lastly, the third card can be chosen in 50 different ways. Thus there are

$$52 \cdot 51 \cdot 50 = 132{,}600$$

different ordered samples of size 3 without replacement.

Solved Problems

14.1. There are 4 bus lines between A and B; and 3 bus lines between B and C.

(i) In how many ways can a man travel by bus from A to C by way of B?

(ii) In how many ways can a man travel roundtrip by bus from A to C by way of B?

(iii) In how many ways can a man travel roundtrip by bus from A to C by way of B, if he doesn't want to use a bus line more than once?

(i) There are 4 ways to go from A to B and 3 ways to go from B to C; hence there are $4 \cdot 3 = 12$ ways to go from A to C by way of B.

(ii) **Method 1.**

There are 12 ways to go from A to C by way of B, and 12 ways to return. Hence there are $12 \cdot 12 = 144$ ways to travel roundtrip.

Method 2.

The man will travel from A to B to C to B to A. Enter these letters with connecting arrows as follows:

$$A \rightarrow B \rightarrow C \rightarrow B \rightarrow A$$

Above each arrow write the number of ways that part of the trip can be traveled; 4 ways from A to B, 3 ways from B to C, 3 ways from C to B, and 4 ways from B to A:

$$A \xrightarrow{4} B \xrightarrow{3} C \xrightarrow{3} B \xrightarrow{4} A$$

Thus there are $4 \cdot 3 \cdot 3 \cdot 4 = 144$ ways to travel roundtrip.

(iii) The man will travel from A to B to C to B to A. Enter these letters with connecting arrows as follows:

$$A \rightarrow B \rightarrow C \rightarrow B \rightarrow A$$

The man can travel 4 ways from A to B and 3 ways from B to C; but he can only travel 2 ways from C to B and 3 ways from B to A since he doesn't want to use a bus line more than once. Enter these numbers above the corresponding arrows as follows:

$$A \xrightarrow{4} B \xrightarrow{3} C \xrightarrow{2} B \xrightarrow{3} A$$

Thus there are $4 \cdot 3 \cdot 2 \cdot 3 = 72$ ways to travel roundtrip without using the same bus line more than once.

14.2. If repetitions are not permitted, (i) how many 3 digit numbers can be formed from the six digits 2, 3, 5, 6, 7 and 9? (ii) How many of these are less than 400? (iii) How many are even? (iv) How many are odd? (v) How many are multiples of 5?

In each case draw three boxes $\boxed{}\,\boxed{}\,\boxed{}$ to represent an arbitrary number, and then write in each box the number of digits that can be placed there.

(i) The box on the left can be filled in 6 ways; following this, the middle box can be filled in 5 ways; and, lastly, the box on the right can be filled in 4 ways: $\boxed{6}\,\boxed{5}\,\boxed{4}$. Thus there are $6 \cdot 5 \cdot 4 = 120$ numbers.

(ii) The box on the left can be filled in only 2 ways, by 2 or 3, since each number must be less than 400; the middle box can be filled in 5 ways; and, lastly, the box on the right can be filled in 4 ways: $\boxed{2}\,\boxed{5}\,\boxed{4}$. Thus there are $2 \cdot 5 \cdot 4 = 40$ numbers.

(iii) The box on the right can be filled in only 2 ways, by 2 or 6, since the numbers must be even; the box on the left can then be filled in 5 ways; and, lastly, the middle box can be filled in 4 ways: $\boxed{5}\,\boxed{4}\,\boxed{2}$. Thus there are $5 \cdot 4 \cdot 2 = 40$ numbers.

(iv) The box on the right can be filled in only 4 ways, by 3, 5, 7 or 9, since the numbers must be odd; the box on the left can then be filled in 5 ways; and, lastly, the box in the middle can be filled in 4 ways: $\boxed{5}\,\boxed{4}\,\boxed{4}$. Thus there are $5 \cdot 4 \cdot 4 = 80$ numbers.

(v) The box on the right can be filled in only 1 way, by 5, since the numbers must be multiples of 5; the box on the left can then be filled in 5 ways; and, lastly, the box in the middle can be filled in 4 ways: $\boxed{5}\,\boxed{4}\,\boxed{1}$. Thus there are $5 \cdot 4 \cdot 1 = 20$ numbers.

14.3. Solve the preceding problem if repetitions are permitted.

(i) Each box can be filled in 6 ways: $\boxed{6}\,\boxed{6}\,\boxed{6}$. Thus there are $6 \cdot 6 \cdot 6 = 216$ numbers.

(ii) The box on the left can still be filled in only two ways, by 2 or 3, and each of the others in 6 ways: $\boxed{2}\,\boxed{6}\,\boxed{6}$. Thus there are $2 \cdot 6 \cdot 6 = 72$ numbers.

(iii) The box on the right can still be filled in only 2 ways, by 2 or 6, and each of the others in 6 ways: $\boxed{6}\,\boxed{6}\,\boxed{2}$. Thus there are $6 \cdot 6 \cdot 2 = 72$ numbers.

(iv) The box on the right can still be filled in only 4 ways, by 3, 5, 7 or 9, and each of the other boxes in 6 ways: $\boxed{6}\,\boxed{6}\,\boxed{4}$. Thus there are $6 \cdot 6 \cdot 4 = 144$ numbers.

(v) The box on the left can still be filled in only 1 way, by 5, and each of the other boxes in 6 ways: $\boxed{6}\,\boxed{6}\,\boxed{1}$. Thus there are $6 \cdot 6 \cdot 1 = 36$ numbers.

14.4. In how many ways can a party of 7 persons arrange themselves (i) in a row of 7 chairs? (ii) around a circular table?

(i) The seven persons can arrange themselves in a row in $7 \cdot 6 \cdot 5 \cdot 4 \cdot 3 \cdot 2 \cdot 1 = 7!$ ways.

(ii) One person can sit at any place in the circular table. The other six persons can then arrange themselves in $6 \cdot 5 \cdot 4 \cdot 3 \cdot 2 \cdot 1 = 6!$ ways around the table.

This is an example of a *circular permutation*. In general, n objects can be arranged in a circle in $(n-1)(n-2) \cdots 3 \cdot 2 \cdot 1 = (n-1)!$ ways.

14.5. (i) In how many ways can 3 boys and 2 girls sit in a row? (ii) In how many ways can they sit in a row if the boys and girls are each to sit together? (iii) In how many ways can they sit in a row if just the girls are to sit together?

(i) The five persons can sit in a row in $5 \cdot 4 \cdot 3 \cdot 2 \cdot 1 = 5! = 120$ ways.

(ii) There are 2 ways to distribute them according to sex: BBBGG or GGBBB. In each case the boys can sit in $3 \cdot 2 \cdot 1 = 3! = 6$ ways, and the girls can sit in $2 \cdot 1 = 2! = 2$ ways. Thus, altogether, there are $2 \cdot 3! \cdot 2! = 2 \cdot 6 \cdot 2 = 24$ ways.

(iii) There are 4 ways to distribute them according to sex: GGBBB, BGGBB, BBGGB, BBBGG. Note that each way corresponds to the number, 0, 1, 2 or 3, of boys sitting to the left of the girls. In each case, the boys can sit in $3!$ ways, and the girls in $2!$ ways. Thus, altogether, there are $4 \cdot 3! \cdot 2! = 4 \cdot 6 \cdot 2 = 48$ ways.

14.6. Solve the preceding problem in the case of r boys and s girls. (Answers are to be left in factorials.)

(i) The $r+s$ persons can sit in a row in $(r+s)!$ ways.

(ii) There are still 2 ways to distribute them according to sex, the boys on the left or the girls on the left. In each case the boys can sit in $r!$ ways and the girls in $s!$ ways. Thus, altogether, there are $2 \cdot r! \cdot s!$ ways.

(iii) There are $r+1$ ways to distribute them according to sex, each way corresponding to the number, $0, 1, \ldots, r$, of boys sitting to the left of the girls. In each case the boys can sit in $r!$ ways and the girls in $s!$ ways. Thus, altogether, there are $(r+1) \cdot r! \cdot s!$ ways.

14.7. How many distinct permutations can be formed from all the letters of each word: (i) them, (ii) that, (iii) radar, (iv) unusual, (v) sociological.

(i) $4! = 24$, since there are 4 letters and no repetition.

(ii) $\dfrac{4!}{2!} = 12$, since there are 4 letters of which 2 are t.

(iii) $\dfrac{5!}{2!\,2!} = 30$, since there are 5 letters of which 2 are r and 2 are a.

(iv) $\dfrac{7!}{3!} = 840$, since there are 7 letters of which 3 are u.

(v) $\dfrac{12!}{3!\,2!\,2!\,2!}$, since there are 12 letters of which 3 are o, 2 are c, 2 are i, and 2 are l.

14.8. How many different signals, each consisting of 6 flags hung in a vertical line, can be formed from 4 identical red flags and 2 identical blue flags?

This problem concerns permutations with repetitions. There are $\dfrac{6!}{4!\,2!} = 15$ signals since there are 6 flags of which 4 are red and 2 are blue.

14.9. In how many ways can 4 mathematics books, 3 history books, 3 chemistry books and 2 sociology books be arranged on a shelf so that all books of the same subject are together?

First the books must be arranged on the shelf in 4 units according to subject matter: ☐ ☐ ☐ ☐. The box on the left can be filled by any of the four subjects; the next by any 3 subjects remaining; the next by any 2 subjects remaining; and the box on the right by the last subject: [4] [3] [2] [1]. Thus there are $4 \cdot 3 \cdot 2 \cdot 1 = 4!$ ways to arrange the books on the shelf according to subject matter.

In each of the above cases, the mathematics books can be arranged in $4!$ ways, the history books in $3!$ ways, the chemistry books in $3!$ ways, and the sociology books in $2!$ ways. Thus, altogether, there are $4!\,4!\,3!\,3!\,2! = 41{,}472$ arrangements.

14.10. Solve the preceding problem if there are r subjects, $A_1, A_2, \ldots, A_r$, containing $s_1, s_2, \ldots, s_r$ books, respectively.

Since there are r subjects, the books can be arranged on the shelf in $r!$ ways according to subject matter. In each case, the books in subject A_1 can be arranged in $s_1!$ ways, the books in subject A_2 in $s_2!$ ways, ..., the books in subject A_r in $s_r!$ ways. Thus, altogether, there are $r!\,s_1!\,s_2!\cdots s_r!$ arrangements.

14.11. (i) In how many ways can 3 Americans, 4 Frenchmen, 4 Danes and 2 Italians be seated in a row so that those of the same nationality sit together?

(ii) Solve the same problem if they sit at a round table.

(i) The 4 nationalities can be arranged in a row in $4!$ ways. In each case, the 3 Americans can be seated in $3!$ ways, the 4 Frenchmen in $4!$ ways, the 4 Danes in $4!$ ways, and the 2 Italians in $2!$ ways. Thus, altogether, there are $4!\,3!\,4!\,4!\,2! = 165{,}888$ arrangements.

(ii) The 4 nationalities can be arranged in a circle in $3!$ ways (see Problem 14.4 on circular permutations). In each case, the 3 Americans can be seated in $3!$ ways, the 4 Frenchmen in $4!$ ways, the 4 Danes in $4!$ ways, and the 2 Italians in $2!$ ways. Thus, altogether, there are $3!\,3!\,4!\,4!\,2! = 41{,}472$ arrangements.

14.12. Find the total number of positive integers that can be formed from the digits 1, 2, 3 and 4 if no digit is repeated in any one integer.

Note that no integer can contain more than 4 digits. Let s_1, s_2, s_3 and s_4 denote the number of integers containing 1, 2, 3 and 4 digits respectively. We compute each s_i separately.

Since there are 4 digits, there are 4 integers containing exactly one digit, i.e. $s_1 = 4$. Also, since there are 4 digits, there are $4 \cdot 3 = 12$ integers containing two digits, i.e. $s_2 = 12$. Similarly, there are $4 \cdot 3 \cdot 2 = 24$ integers containing three digits and $4 \cdot 3 \cdot 2 \cdot 1 = 24$ integers containing four digits, i.e. $s_3 = 24$ and $s_4 = 24$. Thus, altogether, there are $s_1 + s_2 + s_3 + s_4 = 4 + 12 + 24 + 24 = 64$ integers.

14.13. Suppose an urn contains 8 balls. Find the number of ordered samples of size 3 (i) with replacement, (ii) without replacement.

(i) Each ball in the ordered sample can be chosen in 8 ways; hence there are $8 \cdot 8 \cdot 8 = 8^3 = 512$ samples with replacement.

(ii) The first ball in the ordered sample can be chosen in 8 ways, the next in 7 ways, and the last in 6 ways. Thus there are $8 \cdot 7 \cdot 6 = 496$ samples without replacement.

14.14. Find n if (i) $P(n, 2) = 72$, (ii) $P(n, 4) = 42\,P(n, 2)$, (iii) $2\,P(n, 2) + 50 = P(2n, 2)$.

(i) $P(n,2) = n(n-1) = n^2 - n$; hence $n^2 - n = 72$ or $n^2 - n - 72 = 0$ or $(n-9)(n+8) = 0$. Since n must be positive, the only answer is $n = 9$.

(ii) $P(n,4) = n(n-1)(n-2)(n-3)$ and $P(n,2) = n(n-1)$. Hence

$$n(n-1)(n-2)(n-3) = 42n(n-1) \quad \text{or, if } n \neq 0, \neq 1, \quad (n-2)(n-3) = 42$$

$$\text{or} \quad n^2 - 5n + 6 = 42 \quad \text{or} \quad n^2 - 5n - 36 = 0 \quad \text{or} \quad (n-9)(n+4) = 0$$

Since n must be positive, the only answer is $n = 9$.

(iii) $P(n,2) = n(n-1) = n^2 - n$ and $P(2n,2) = 2n(2n-1) = 4n^2 - 2n$. Hence

$$2(n^2 - n) + 50 = 4n^2 - 2n \quad \text{or} \quad 2n^2 - 2n + 50 = 4n^2 - 2n \quad \text{or} \quad 50 = 2n^2 \quad \text{or} \quad n^2 = 25$$

Since n must be positive, the only answer is $n = 5$.

14.15. Find the relationship between $\binom{n}{r}$ and $P(n, r)$.

Recall that $\binom{n}{r} = \dfrac{n!}{r!\,(n-r)!}$ and $P(n,r) = \dfrac{n!}{(n-r)!}$. Multiply $P(n,r)$ by $\dfrac{r!}{r!}$ to obtain

$$P(n,r) = \frac{n!}{(n-r)!} = \frac{r!}{r!}\cdot\frac{n!}{(n-r)!} = r!\,\frac{n!}{r!\,(n-r)!} = r!\binom{n}{r}$$

Supplementary Problems

14.16. (i) How many automobile license plates can be made if each plate contains 2 different letters followed by 3 different digits? (ii) Solve the problem if the first digit cannot be 0.

14.17. There are 6 roads between A and B and 4 roads between B and C.

(i) In how many ways can one drive from A to C by way of B?

(ii) In how many ways can one drive roundtrip from A to C by way of B?

(iii) In how many ways can one drive roundtrip from A to C without using the same road more than once?

14.18. Find the number of ways in which 6 people can ride a toboggan if one of three must drive.

14.19. (i) Find the number of ways in which five persons can sit in a row.

(ii) How many ways are there if two of the persons insist on sitting next to one another?

14.20. Solve the preceding problem if they sit around a circular table.

14.21. Find the number of ways in which a judge can award first, second and third places in a contest with ten contestants.

14.22. (i) Find the number of four letter words that can be formed from the letters of the word HISTORY. (ii) How many of them contain only consonants? (iii) How many of them begin and end in a consonant? (iv) How many of them begin with a vowel? (v) How many contain the letter Y? (vi) How many begin with T and end in a vowel? (vii) How many begin with T and also contain S? (viii) How many contain both vowels?

14.23. How many different signals, each consisting of 8 flags hung in a vertical line, can be formed from 4 red flags, 2 blue flags and 2 green flags?

14.24. Find the number of permutations that can be formed from all the letters of each word: (i) queue, (ii) committee, (iii) proposition, (iv) baseball.

14.25. (i) Find the number of ways in which 4 boys and 4 girls can be seated in a row if the boys and girls are to have alternate seats.

(ii) Find the number of ways if they sit alternately and if one boy and one girl are to sit in adjacent seats.

(iii) Find the number of ways if they sit alternately and if one boy and one girl must not sit in adjacent seats.

14.26. Solve the preceding problem if they sit around a circular table.

14.27. An urn contains 10 balls. Find the number of ordered samples (i) of size 3 with replacement, (ii) of size 3 without replacement, (iii) of size 4 with replacement, (iv) of size 5 without replacement.

14.28. Find the number of ways in which 5 large books, 4 medium-size books and 3 small books can be placed on a shelf so that all books of the same size are together.

14.29. Consider all positive integers with 3 different digits. (Note that 0 cannot be the first digit.) (i) How many are greater than 700? (ii) How many are odd? (iii) How many are even? (iv) How many are divisible by 5?

14.30. (i) Find the number of distinct permutations that can be formed from all of the letters of the word ELEVEN. (ii) How many of them begin and end with E? (iii) How many of them have the 3 E's together? (iv) How many begin with E and end with N?

Answers to Supplementary Problems

14.16. (i) $26 \cdot 25 \cdot 10 \cdot 9 \cdot 8 = 468{,}000$ (ii) $26 \cdot 25 \cdot 9 \cdot 9 \cdot 8 = 421{,}200$

14.17. (i) $6 \cdot 4 = 24$ (ii) $6 \cdot 4 \cdot 4 \cdot 6 = 24 \cdot 24 = 576$ (iii) $6 \cdot 4 \cdot 3 \cdot 5 = 360$

14.18. $3 \cdot 5 \cdot 4 \cdot 3 \cdot 2 \cdot 1 = 360$

14.19. (i) $5! = 120$ (ii) $4 \cdot 2! \cdot 3! = 48$

14.20. (i) $4! = 24$ (ii) $2!\,3! = 12$

14.21. $10 \cdot 9 \cdot 8 = 720$

14.22. (i) $7 \cdot 6 \cdot 5 \cdot 4 = 840$ (iii) $5 \cdot 5 \cdot 4 \cdot 4 = 400$ (v) $4 \cdot 6 \cdot 5 \cdot 4 = 480$ (vii) $1 \cdot 3 \cdot 5 \cdot 4 = 60$

(ii) $5 \cdot 4 \cdot 3 \cdot 2 = 120$ (iv) $2 \cdot 6 \cdot 5 \cdot 4 = 240$ (vi) $1 \cdot 5 \cdot 4 \cdot 2 = 40$ (viii) $4 \cdot 3 \cdot 5 \cdot 4 = 240$

14.23. $\frac{8!}{4!\,2!\,2!} = 420$

14.24. (i) $\frac{5!}{2!\,2!} = 30$ (ii) $\frac{9!}{2!\,2!\,2!} = 45{,}360$ (iii) $\frac{11!}{2!\,3!\,2!} = 1{,}663{,}200$ (iv) $\frac{8!}{2!\,2!\,2!} = 5040$

14.25. (i) $2 \cdot 4! \cdot 4! = 1152$ (ii) $2 \cdot 7 \cdot 3! \cdot 3! = 504$ (iii) $1152 - 504 = 648$

14.26. (i) $3! \cdot 4! = 144$ (ii) $2 \cdot 3! \cdot 3! = 72$ (iii) $144 - 72 = 72$

14.27. (i) $10 \cdot 10 \cdot 10 = 1000$ (iii) $10 \cdot 10 \cdot 10 \cdot 10 = 10{,}000$

(ii) $10 \cdot 9 \cdot 8 = 720$ (iv) $10 \cdot 9 \cdot 8 \cdot 7 \cdot 6 = 30{,}240$

14.28. $3!\,5!\,4!\,3! = 103{,}680$

14.29. (i) $3 \cdot 9 \cdot 8 = 216$ (ii) $8 \cdot 8 \cdot 5 = 320$

(iii) $9 \cdot 8 \cdot 1 = 72$ end in 0, and $8 \cdot 8 \cdot 4 = 256$ end in the other even digits; hence, altogether, $72 + 256 = 328$ are even.

(iv) $9 \cdot 8 \cdot 1 = 72$ end in 0, and $8 \cdot 8 \cdot 1 = 64$ end in 5; hence, altogether, $72 + 64 = 136$ are divisible by 5.

14.30. (i) $\frac{6!}{3!} = 120$ (ii) $4! = 24$ (iii) $4 \cdot 3! = 24$ (iv) $\frac{4!}{2!} = 12$

Chapter 15

Combinations, Ordered Partitions

COMBINATIONS

Suppose we have a collection of n objects. A *combination* of these n objects taken r at a time is any selection of r of the objects where order doesn't count. In other words, an *r-combination* of a set of n objects is any subset of r elements.

Example 1.1: The combinations of the letters a, b, c, d taken 3 at a time are

$$\{a, b, c\},\ \{a, b, d\},\ \{a, c, d\},\ \{b, c, d\} \quad \text{or simply} \quad abc,\ abd,\ acd,\ bcd$$

Observe that the following combinations are equal:

$$abc,\ acb,\ bac,\ bca,\ cab \text{ and } cba$$

That is, each denotes the same set $\{a, b, c\}$.

The number of combinations of n objects taken r at a time is denoted by $C(n, r)$. The symbols ${}_nC_r$, $C_{n,r}$ and C_r^n also appear in various texts. Before we give the general formula for $C(n, r)$, we consider a special case.

Example 1.2: How many combinations are there of the four objects, a, b, c and d, taken 3 at a time?

Each combination consisting of 3 objects determines $3! = 6$ permutations of the objects in the combination:

Combinations	Permutations
abc	*abc, acb, bac, bca, cab, cba*
abd	*abd, adb, bad, bda, dab, dba*
acd	*acd, adc, cad, cda, dac, dca*
bcd	*bcd, bdc, cbd, cdb, dbc, dcb*

Thus the number of combinations multiplied by $3!$ equals the number of permutations:

$$C(4, 3) \cdot 3! = P(4, 3) \quad \text{or} \quad C(4, 3) = \frac{P(4, 3)}{3!}$$

But $P(4, 3) = 4 \cdot 3 \cdot 2 = 24$ and $3! = 6$; hence $C(4, 3) = 4$ as noted in Example 1.1.

Since any combination of n objects taken r at a time determines $r!$ permutations of the objects in the combination, we can conclude that

$$P(n, r) = r!\, C(n, r)$$

Thus we obtain

Theorem 15.1: $C(n, r) = \dfrac{P(n, r)}{r!} = \dfrac{n!}{r!\,(n-r)!}$

Recall that the binomial coefficient $\binom{n}{r}$ was defined to be $\frac{n!}{r!\,(n-r)!}$; hence

$$C(n, r) = \binom{n}{r}$$

We shall use $C(n,r)$ and $\binom{n}{r}$ interchangeably.

Example 1.3: How many committees of 3 can be formed from 8 people? Each committee is, essentially, a combination of the 8 people taken 3 at a time. Thus

$$C(8, 3) = \binom{8}{3} = \frac{8 \cdot 7 \cdot 6}{1 \cdot 2 \cdot 3} = 56$$

different committees can be formed.

Example 1.4: A farmer buys 3 cows, 2 pigs and 4 hens from a man who has 6 cows, 5 pigs and 8 hens. How many choices does the farmer have?

The farmer can choose the cows in $\binom{6}{3}$ ways, the pigs in $\binom{5}{2}$ ways, and the hens in $\binom{8}{4}$ ways. Hence altogether he can choose the animals in

$$\binom{6}{3}\binom{5}{2}\binom{8}{4} = \frac{6 \cdot 5 \cdot 4}{1 \cdot 2 \cdot 3} \cdot \frac{5 \cdot 4}{1 \cdot 2} \cdot \frac{8 \cdot 7 \cdot 6 \cdot 5}{1 \cdot 2 \cdot 3 \cdot 4} = 20 \cdot 10 \cdot 70 = 14{,}000 \text{ ways}$$

PARTITIONS AND CROSS-PARTITIONS

Recall (Page 60) that a *partition* of a set X is a subdivision of X into subsets which are disjoint and whose union is X, i.e. such that each $a \in X$ belongs to one and only one of the subsets. In other words, the collection $\{A_1, A_2, \ldots, A_m\}$ of subsets of X is a partition of X iff

(i) $X = A_1 \cup A_2 \cup \cdots \cup A_m$; (ii) for any A_i, A_j, either $A_i = A_j$ or $A_i \cap A_j = \emptyset$

The subsets of a partition are called *cells.*

Example 2.1: Consider the following classes of subsets of $X = \{1, 2, \ldots, 8, 9\}$:

(i) $[\{1,3,5\}, \{2,6\}, \{4,8,9\}]$

(ii) $[\{1,3,5\}, \{2,4,6,8\}, \{5,7,9\}]$

(iii) $[\{1,3,5\}, \{2,4,6,8\}, \{7,9\}]$

Then (i) is not a partition of X, since $7 \in X$ but 7 does not belong to any cell in (i). Also (ii) is not a partition of X since $5 \in X$ and 5 belongs to both $\{1,3,5\}$ and $\{5,7,9\}$. On the other hand, (iii) is a partition of X since each element of X belongs to exactly one cell.

Suppose $\{A_1, A_2, \ldots, A_r\}$ and $\{B_1, B_2, \ldots, B_s\}$ are both partitions of the same set X. Then the class of intersections $\{A_i \cap B_j\}$ form a new partition of X called the *cross-partition.*

Example 2.2: Let $\{A, B, C, D\}$ be the partition of the undergraduate students at a university into freshmen, sophomores, juniors and seniors, respectively; and let $\{M, F\}$ be the partition of the students into males and females, respectively. The cross-partition consists of the following:

$A \cap M$: male freshmen	$A \cap F$: female freshmen
$B \cap M$: male sophomores	$B \cap F$: female sophomores
$C \cap M$: male juniors	$C \cap F$: female juniors
$D \cap M$: male seniors	$D \cap F$: female seniors

ORDERED PARTITIONS

Suppose an urn A contains seven balls numbered 1 through 7. We compute the number of ways we can draw, first, 2 balls from the urn, then 3 balls from the urn, and lastly 2 balls from the urn. In other words, we want to compute the number of *ordered partitions*

$$(A_1, A_2, A_3)$$

of the set of 7 balls into cells A_1 containing 2 balls, A_2 containing 3 balls and A_3 containing 2 balls. We call these ordered partitions since we distinguish between

$$(\{1,2\}, \{3,4,5\}, \{6,7\}) \quad \text{and} \quad (\{6,7\}, \{3,4,5\}, \{1,2\})$$

each of which determines the same partition of A.

Now we begin with 7 balls in the urn, so there are $\binom{7}{2}$ ways of drawing the first 2 balls, i.e. of determining the first cell A_1; following this, there are 5 balls left in the urn and so there are $\binom{5}{3}$ ways of drawing the 3 balls, i.e. of determining the second cell A_2; finally, there are 2 balls left in the urn and so there are $\binom{2}{2}$ ways of determining the last cell A_3. Hence there are

$$\binom{7}{2}\binom{5}{3}\binom{2}{2} = \frac{7\cdot 6}{1\cdot 2}\cdot\frac{5\cdot 4\cdot 3}{1\cdot 2\cdot 3}\cdot\frac{2\cdot 1}{1\cdot 2} = 210$$

different ordered partitions of A into cells A_1 containing 2 balls, A_2 containing 3 balls, and A_3 containing 2 balls.

We state the above result in its general form as a theorem.

Theorem 15.2: Let A contain n elements and let $n_1, n_2, \ldots, n_r$ be positive integers with $n_1 + n_2 + \cdots + n_r = n$. Then there exist

$$\binom{n}{n_1}\binom{n-n_1}{n_2}\binom{n-n_1-n_2}{n_3}\cdots\binom{n-n_1-n_2-\cdots-n_{r-1}}{n_r}$$

different ordered partitions of A of the form $(A_1, A_2, \ldots, A_r)$ where A_1 contains n_1 elements, A_2 contains n_2 elements, ..., and A_r contains n_r elements.

Now observe that

$$\binom{7}{2}\binom{5}{3}\binom{2}{2} = \frac{7!}{2!5!}\cdot\frac{5!}{3!2!}\cdot\frac{2!}{2!0!} = \frac{7!}{2!3!2!}$$

since each numerator after the first is cancelled by the second term in the denominator of the previous factor. Similarly,

$$\binom{n}{n_1}\binom{n-n_1}{n_2}\binom{n-n_1-n_2}{n_3}\cdots\binom{n_r}{n_r}$$

$$= \frac{n!}{n_1!(n-n_1)!}\cdot\frac{(n-n_1)!}{n_2!(n-n_1-n_2)!}\cdot\frac{(n-n_1-n_2)!}{n_3!(n-n_1-n_2-n_3)!}\cdots\frac{n_r!}{n_r!}$$

$$= \frac{n!}{n_1!\,n_2!\,n_3!\cdots n_r!}$$

The above computation together with the preceding theorem gives us our next theorem.

Theorem 15.3: Let A contain n elements and let $n_1, n_2, \ldots, n_r$ be positive integers with $n_1 + n_2 + \cdots + n_r = n$. Then there exist

$$\frac{n!}{n_1!\, n_2!\, n_3! \cdots n_r!}$$

different ordered partitions of A of the form $(A_1, A_2, \ldots, A_r)$ where A_1 contains n_1 elements, A_2 contains n_2 elements, ..., and A_r contains n_r elements.

Example 3.1: In how many ways can 9 toys be divided between 4 children if the youngest child is to receive 3 toys and each of the others 2 toys?

We wish to find the number of ordered partitions of the 9 toys into 4 cells containing 3, 2, 2 and 2 toys respectively. By Theorem 15.3, there are

$$\frac{9!}{3!\,2!\,2!\,2!} = 2520$$

such ordered partitions.

Solved Problems

COMBINATIONS

15.1. In how many ways can a committee consisting of 3 men and 2 women be chosen from 7 men and 5 women?

The 3 men can be chosen from the 7 men in $\binom{7}{3}$ ways, and the 2 women can be chosen from the 5 women in $\binom{5}{2}$ ways. Hence the committee can be chosen in $\binom{7}{3}\binom{5}{2} = \frac{7 \cdot 6 \cdot 5}{1 \cdot 2 \cdot 3} \cdot \frac{5 \cdot 4}{1 \cdot 2} = 350$ ways.

15.2. A bag contains 6 white balls and 5 black balls. Find the number of ways 4 balls can be drawn from the bag if (i) they can be of any color, (ii) 2 must be white and 2 black, (iii) they must all be of the same color.

(i) The 4 balls (of any color) can be chosen from the 11 balls in the urn in $\binom{11}{4} = \frac{11 \cdot 10 \cdot 9 \cdot 8}{1 \cdot 2 \cdot 3 \cdot 4} =$ 330 ways.

(ii) The 2 white balls can be chosen in $\binom{6}{2}$ ways, and the 2 black balls can be chosen in $\binom{5}{2}$ ways.

Thus there are $\binom{6}{2}\binom{5}{2} = \frac{6 \cdot 5}{1 \cdot 2} \cdot \frac{5 \cdot 4}{1 \cdot 2} = 150$ ways of drawing 2 white balls and 2 black balls.

(iii) There are $\binom{6}{4} = 15$ ways of drawing 4 white balls, and $\binom{5}{4} = 5$ ways of drawing 4 black balls. Thus there are $15 + 5 = 20$ ways of drawing 4 balls of the same color.

15.3. A delegation of 4 students is selected each year from a college to attend the National Student Association annual meeting. (i) In how many ways can the delegation be chosen if there are 12 eligible students? (ii) In how many ways if two of the eligible students will not attend the meeting together? (iii) In how many ways if two of the eligible students are married and will only attend the meeting together?

(i) The 4 students can be chosen from the 12 students in $\binom{12}{4} = \frac{12 \cdot 11 \cdot 10 \cdot 9}{1 \cdot 2 \cdot 3 \cdot 4} = 495$ ways.

(ii) Let A and B denote the students who will not attend the meeting together.

Method 1.

If neither A nor B is included, then the delegation can be chosen in $\binom{10}{4} = \frac{10 \cdot 9 \cdot 8 \cdot 7}{1 \cdot 2 \cdot 3 \cdot 4} =$ 210 ways. If either A or B, but not both, is included, then the delegation can be chosen in $2 \cdot \binom{10}{3} = 2 \cdot \frac{10 \cdot 9 \cdot 8}{1 \cdot 2 \cdot 3} = 240$ ways. Thus, altogether, the delegation can be chosen in $210 + 240 = 450$ ways.

Method 2.

If A and B are both included, then the other 2 members of the delegation can be chosen in $\binom{10}{2} = 45$ ways. Thus there are $495 - 45 = 450$ ways the delegation can be chosen if A and B are not both included.

(iii) Let C and D denote the married students. If C and D do not go, then the delegation can be chosen in $\binom{10}{4} = 210$ ways. If both C and D go, then the delegation can be chosen in $\binom{10}{2} = 45$ ways. Altogether, the delegation can be chosen in $210 + 45 = 255$ ways.

15.4. There are 12 points $A, B, \ldots$ in a given plane, no three on the same line. (i) How many lines are determined by the points? (ii) How many of these lines pass through the point A? (iii) How many triangles are determined by the points? (iv) How many of these triangles contain the point A as a vertex?

(i) Since two points determine a line, there are $\binom{12}{2} = \frac{12 \cdot 11}{1 \cdot 2} = 66$ lines.

(ii) To determine a line through A, one other point must be chosen; hence there are 11 lines through A.

(iii) Since three points determine a triangle, there are $\binom{12}{3} = \frac{12 \cdot 11 \cdot 10}{1 \cdot 2 \cdot 3} = 220$ triangles.

(iv) **Method 1.**

To determine a triangle with vertex A, two other points must be chosen; hence there are $\binom{11}{2} = \frac{11 \cdot 10}{1 \cdot 2} = 55$ triangles with A as a vertex.

Method 2.

There are $\binom{11}{3} = \frac{11 \cdot 10 \cdot 9}{1 \cdot 2 \cdot 3} = 165$ triangles without A as a vertex. Thus $220 - 165 = 55$ of the triangles do have A as a vertex.

15.5. A student is to answer 8 out of 10 questions on an exam. (i) How many choices has he? (ii) How many if he must answer the first 3 questions? (iii) How many if he must answer at least 4 of the first 5 questions?

(i) The 8 questions can be selected in $\binom{10}{8} = \binom{10}{2} = \frac{10 \cdot 9}{1 \cdot 2} = 45$ ways.

(ii) If he answers the first 3 questions, then he can choose the other 5 questions from the last 7 questions in $\binom{7}{5} = \binom{7}{2} = \frac{7 \cdot 6}{1 \cdot 2} = 21$ ways.

(iii) If he answers all the first 5 questions, then he can choose the other 3 questions from the last 5 in $\binom{5}{3} = 10$ ways. On the other hand, if he answers only 4 of the first 5 questions, then he can choose these 4 in $\binom{5}{4} = \binom{5}{1} = 5$ ways, and he can choose the other 4 questions from the last 5 in $\binom{5}{4} = \binom{5}{1} = 5$ ways; hence he can choose the 8 questions in $5 \cdot 5 =$ 25 ways. Thus he has a total of 35 choices.

15.6. How many diagonals has an octagon?

An octagon has 8 sides and 8 vertices. Any two vertices determine either a side or a diagonal. Thus there are $\binom{8}{2} = \frac{8 \cdot 7}{1 \cdot 2} = 28$ sides plus diagonals. But there are 8 sides; hence there are $28 - 8 = 20$ diagonals.

15.7. How many diagonals has a regular polygon with n sides?

The regular polygon with n sides has, also, n vertices. Any two vertices determine either a side or a diagonal. Thus there are $\binom{n}{2} = \frac{n(n-1)}{1 \cdot 2} = \frac{n(n-1)}{2}$ sides plus diagonals. But there are n sides; hence there are

$$\frac{n(n-1)}{2} - n = \frac{n^2 - n}{2} - \frac{2n}{2} = \frac{n^2 - 3n}{2} = \frac{n(n-3)}{2} \quad \text{diagonals}$$

15.8. Which regular polygon has the same number of diagonals as sides?

The regular polygon with n sides has, by the preceding problem, $\frac{n^2 - 3n}{2}$ diagonals. Thus we seek the polygon of n sides for which

$$\frac{n^2 - 3n}{2} = n \quad \text{or} \quad n^2 - 3n = 2n \quad \text{or} \quad n^2 - 5n = 0 \quad \text{or} \quad n(n-5) = 0$$

Since n must be a positive integer, the only answer is $n = 5$. In other words, the pentagon is the only regular polygon with the same number of diagonals as sides.

15.9. A man is dealt a poker hand (5 cards) from an ordinary playing deck. In how many ways can he be dealt (i) a spade flush (5 spades), (ii) an ace high spade flush (5 spades with the ace), (iii) a flush (5 of the same suit), (iv) an ace high full house (3 aces with another pair), (v) a full house (3 of one kind and 2 of another), (vi) 3 aces (without the fourth ace or another pair), (vii) 3 of a kind (without another pair), (viii) two pair?

(i) The 5 spades can be dealt from the 13 spades in $\binom{13}{5} = \frac{13 \cdot 12 \cdot 11 \cdot 10 \cdot 9}{1 \cdot 2 \cdot 3 \cdot 4 \cdot 5} = 1287$ ways.

(ii) If he receives the spade ace, then he can be dealt the other 4 spades from the remaining 12 spades in $\binom{12}{4} = \frac{12 \cdot 11 \cdot 10 \cdot 9}{1 \cdot 2 \cdot 3 \cdot 4} = 495$ ways.

(iii) There are 4 suits, and a flush from each suit can be dealt (see (i)) in $\binom{13}{5}$ ways; hence a flush can be dealt in $4 \cdot \binom{13}{5} = 5148$ ways.

(iv) The 3 aces can be dealt from the 4 aces in $\binom{4}{3} = \binom{4}{1} = 4$ ways. The pair can be selected in 12 ways, and the 2 cards of this pair can be dealt in $\binom{4}{2}$ ways. Thus the ace high full house can be dealt in $4 \cdot 12 \cdot \binom{4}{2} = 4 \cdot 12 \cdot \frac{4 \cdot 3}{1 \cdot 2} = 288$ ways.

(v) One kind can be selected in 13 ways and 3 of this kind can be dealt in $\binom{4}{3}$ ways; another kind can be selected in 12 ways and 2 of this kind can be dealt in $\binom{4}{2}$ ways. Thus a full house can be dealt in

$$13 \cdot \binom{4}{3} \cdot 12 \cdot \binom{4}{2} = 3744 \quad \text{ways}$$

(Observe, by (iii), that a flush is more common than a full house.)

(vi) The 3 aces can be selected from 4 aces in $\binom{4}{3}$ ways. The 2 other cards can be selected from the remaining 12 kinds in $\binom{12}{2}$ ways; and 1 card in each kind can be dealt in 4 ways. Thus a poker hand of 3 aces can be dealt in

$$\binom{4}{3} \cdot \binom{12}{2} \cdot 4 \cdot 4 = 4224 \quad \text{ways}$$

(vii) One kind can be selected in 13 ways; and 3 of this kind can be dealt in $\binom{4}{3}$ ways. The 2 other kinds can be selected from the remaining 12 kinds in $\binom{12}{2}$ ways; and 1 card in each kind can be dealt in 4 ways. Thus a poker hand of 3 of a kind can be dealt in

$$13 \cdot \binom{4}{3} \cdot \binom{12}{2} \cdot 4 \cdot 4 = 54{,}912 \quad \text{ways}$$

(viii) Two kinds (for the pairs) can be selected from the 13 kinds in $\binom{13}{2}$ ways; and 2 of each kind can be dealt in $\binom{4}{2}$ ways. Another kind can be selected in 11 ways; and 1 of this kind in 4 ways. Thus a poker hand of two pair can be dealt in

$$\binom{13}{2} \cdot \binom{4}{2} \cdot \binom{4}{2} \cdot 11 \cdot 4 = 123{,}552 \quad \text{ways}$$

15.10. Consider 4 vowels (including a) and 8 consonants (including b).

(i) How many 5 letter "words" containing 2 different vowels and 3 different consonants can be formed from the letters?

(ii) How many of them contain b?

(iii) How many of them begin with b?

(iv) How many of them begin with a?

(v) How many of them begin with a and contain b?

(i) The 2 vowels can be selected from the 4 vowels in $\binom{4}{2}$ ways, and the 3 consonants can be selected from the 8 consonants in $\binom{8}{3}$ ways. Furthermore, each 5 letters can be arranged in a row (as a "word") in 5! ways. Thus we can form

$$\binom{4}{2} \cdot \binom{8}{3} \cdot 5! = 6 \cdot 56 \cdot 120 = 40{,}320 \quad \text{words}$$

(ii) The 2 vowels can still be selected in $\binom{4}{2}$ ways. However, since b is one of the consonants, the other 2 consonants can be selected from the remaining 7 consonants in $\binom{7}{2}$ ways. Again each 5 letters can be arranged as a word in 5! ways. Thus we can form

$$\binom{4}{2} \cdot \binom{7}{2} \cdot 5! = 6 \cdot 21 \cdot 120 = 15{,}120 \quad \text{words containing } b$$

(iii) The 2 vowels can be selected in $\binom{4}{2}$ ways, and the other 2 consonants can be selected in $\binom{7}{2}$ ways. The 4 letters can be arranged, following b, in 4! ways. Thus we can form

$$\binom{4}{2} \cdot \binom{7}{2} \cdot 4! = 6 \cdot 21 \cdot 24 = 3024 \quad \text{words beginning with } b$$

(iv) The other vowel can be chosen in 3 ways, and the 3 consonants can be chosen in $\binom{8}{3}$ ways. The 4 letters can be arranged, following a, in $4!$ ways. Thus we form $3 \cdot \binom{8}{3} \cdot 4! = 3 \cdot 56 \cdot 24 = 4032$ words beginning with a.

(v) The other vowel can be chosen in 3 ways, and the 2 other consonants can be chosen in $\binom{7}{2}$ ways. The 4 letters can be arranged, following a, in $4!$ ways. Thus we can form $3 \cdot \binom{7}{2} \cdot 4! = 3 \cdot 21 \cdot 24 = 1512$ words beginning with a and containing b.

15.11. How many committees of 5 with a given chairman can be selected from 12 persons?

The chairman can be chosen in 12 ways and, following this, the other 4 on the committee can be chosen from the 11 remaining in $\binom{11}{4}$ ways. Thus there are $12 \cdot \binom{11}{4} = 12 \cdot 330 = 3960$ such committees.

15.12. Find the number of subsets of a set X containing n elements.

Method 1.

The number of subsets of X with $r \leq n$ elements is given by $\binom{n}{r}$. Hence, altogether, there are

$$\binom{n}{0} + \binom{n}{1} + \binom{n}{2} + \cdots + \binom{n}{n-1} + \binom{n}{n}$$

subsets of X. The above sum (see Problem 13.30, Page 151) is equal to 2^n, i.e. there are 2^n subsets of X.

Method 2.

There are two possibilities for each element of X: either it belongs to the subset or it doesn't; hence there are

$$\overbrace{2 \cdot 2 \cdot \cdots \cdot 2}^{n \text{ times}} = 2^n$$

ways to form a subset of X, i.e. there are 2^n different subsets of X.

15.13. In how many ways can a teacher choose one or more students from six eligible students?

Method 1.

By the preceding problem, there are $2^6 = 64$ subsets of the set consisting of the six students. However, the empty set must be deleted since one or more students are chosen. Accordingly, there are $2^6 - 1 = 64 - 1 = 63$ ways to choose the students.

Method 2.

Either 1, 2, 3, 4, 5 or 6 students are chosen. Hence the number of choices is

$$\binom{6}{1} + \binom{6}{2} + \binom{6}{3} + \binom{6}{4} + \binom{6}{5} + \binom{6}{6} = 6 + 15 + 20 + 15 + 6 + 1 = 63$$

15.14. In how many ways can three or more persons be selected from 12 persons?

There are $2^{12} - 1 = 4096 - 1 = 4095$ ways of choosing one or more of the twelve persons. Now there are $\binom{12}{1} + \binom{12}{2} = 12 + 66 = 78$ ways of choosing one or two of the twelve persons. Hence there are $4096 - 78 = 4018$ ways of choosing three or more.

PARTITIONS AND CROSS-PARTITIONS

15.15. Which of the following are partitions of $X = \{a, b, c, d, e, f, g\}$?

(i) $[\{a, b, c\}, \{d\}, \{f, g\}]$

(ii) $[\{a, b, c\}, \{c, d, e\}, \{f, g\}]$

(iii) $[\{a, b, c\}, \{d, e\}, \{f, g\}]$

(i) This is not a partition of X since $e \in X$, but e does not belong to any cell.

(ii) This is not a partition of X since $c \in X$ and belongs to both $\{a, b, c\}$ and $\{c, d, e\}$.

(iii) This is a partition of X. Each element of x belongs to exactly one cell.

15.16. Let $\mathcal{A} = \{A_1, A_2, \ldots, A_m\}$ and $\mathcal{B} = \{B_1, B_2, \ldots, B_n\}$ be partitions of the same set X. Show that the cross-partition $\mathcal{C} = \{A_i \cap B_j : A_i \in \mathcal{A}, B_j \in \mathcal{B}\}$ is also a partition of X.

Let $x \in X$. Then x belongs to some A_{i_0} and to some B_{j_0}, since $\mathcal{A}$ and $\mathcal{B}$ are partitions of X. Thus $x \in A_{i_0} \cap B_{j_0}$ and so belongs to a member of the cross-partition.

On the other hand, suppose $x \in A_{i_0} \cap B_{j_0}$ and $A_{i_1} \cap B_{j_1}$. Then $x \in A_{i_0}$ and A_{i_1}, whence $A_{i_0} = A_{i_1}$ since $\mathcal{A}$ is a partition of X. Similarly, $B_{j_0} = B_{j_1}$. Accordingly, $A_{i_0} \cap B_{j_0} = A_{i_1} \cap B_{j_1}$ and so $\mathcal{C}$ is a partition of X.

15.17. Let $X = \{1, 2, 3, 4, 5, 6, 7, 8\}$. Find the cross-partition of each pair of partitions of X:

(i) $[\{1, 2, 3, 4\}, \{5, 6, 7, 8\}]$ and $[\{1, 2, 8\}, \{3, 4, 5, 6, 7\}]$

(ii) $[\{1, 2, 3, 4\}, \{5, 6, 7, 8\}]$ and $[\{1, 2\}, \{3, 4, 5\}, \{6, 7, 8\}]$

To obtain the cross-partition, intersect each set of one partition with each set of the other.

(i) $\{1, 2, 3, 4\} \cap \{1, 2, 8\} = \{1, 2\}$; $\quad \{5, 6, 7, 8\} \cap \{1, 2, 8\} = \{8\}$

$\{1, 2, 3, 4\} \cap \{3, 4, 5, 6, 7\} = \{3, 4\}$; $\quad \{5, 6, 7, 8\} \cap \{3, 4, 5, 6, 7\} = \{5, 6, 7\}$

Thus the cross-partition is $[\{1, 2\}, \{3, 4\}, \{8\}, \{5, 6, 7\}]$.

(ii) $\{1, 2, 3, 4\} \cap \{1, 2\} = \{1, 2\}$; $\quad \{5, 6, 7, 8\} \cap \{1, 2\} = \emptyset$

$\{1, 2, 3, 4\} \cap \{3, 4, 5\} = \{3, 4\}$; $\quad \{5, 6, 7, 8\} \cap \{3, 4, 5\} = \{5\}$

$\{1, 2, 3, 4\} \cap \{6, 7, 8\} = \emptyset$; $\quad \{5, 6, 7, 8\} \cap \{6, 7, 8\} = \{6, 7, 8\}$

Thus the cross-partition is

$[\{1, 2\}, \{3, 4\}, \{5\}, \{6, 7, 8\}, \emptyset]$ or, simply, $[\{1, 2\}, \{3, 4\}, \{5\}, \{6, 7, 8\}]$

Remark: We do not distinguish partitions which differ by the empty set.

15.18. Let X denote the domain of a proposition $P(p, q, r)$ of three variables:

$$X = \{\text{TTT, TTF, TFT, TFF, FTT, FTF, FFT, FFF}\}$$

(i) Partition X according to the number of T's appearing in each element.

(ii) For the proposition $P = (p \rightarrow q) \wedge r$, partition X into $\mathcal{T}(P)$ and $(\mathcal{T}(P))^c$. (Recall that $\mathcal{T}(P)$ is the truth set of P.)

(iii) For the proposition $Q = \sim p \rightarrow (q \wedge r)$, partition X into $\mathcal{T}(Q)$ and $(\mathcal{T}(Q))^c$.

(iv) Find the cross-partition of the partitions of X in (ii) and (iii).

(i) Combine the elements of X with the same number of T's to form the cells:

$$[\{\text{FFF}\}, \{\text{TFF, FTF, FFT}\}, \{\text{TTF, TFT, FTT}\}, \{\text{TTT}\}]$$

(ii) and (iii). Construct the truth tables of $P = (p \to q) \wedge r$ and $Q = \sim p \to (q \wedge r)$:

p	q	r	$p \to q$	P $(p \to q) \wedge r$	$\sim p$	$q \wedge r$	Q $\sim p \to (q \wedge r)$
T	T	T	T	T	F	T	T
T	T	F	T	F	F	F	T
T	F	T	F	F	F	F	T
T	F	F	F	F	F	F	T
F	T	T	T	T	T	T	T
F	T	F	T	F	T	F	F
F	F	T	T	T	T	F	F
F	F	F	T	F	T	F	F

Note that P is true in Cases (lines) 1, 5 and 7, and Q is true in Cases 1, 2, 3, 4 and 5. Thus

$$\mathcal{T}(P) = \{\text{TTT, FTT, FFT}\} \quad \text{and} \quad \mathcal{T}(Q) = \{\text{TTT, TTF, TFT, TFF, FTT}\}$$

and so the required partitions are

$$[\{\text{TTT, FTT, FFT}\}, \{\text{TTF, TFT, TFF, FTF, FFF}\}]$$

and

$$[\{\text{TTT, TTF, TFT, TFF, FTT}\}, \{\text{FTF, FFT, FFF}\}]$$

(iv) The cross-partition is obtained by intersecting each set of one partition with each set of the other:

$$\{\text{TTT, FTT, FFT}\} \cap \{\text{TTT, TTF, TFT, TFF, FTT}\} = \{\text{TTT, FTT}\}$$

$$\{\text{TTT, FTT, FFT}\} \cap \{\text{FTF, FFT, FFF}\} = \{\text{FFT}\}$$

$$\{\text{TTF, TFT, TFF, FTF, FFF}\} \cap \{\text{TTT, TTF, TFT, TFF, FTT}\} = \{\text{TTF, TFT, TFF}\}$$

$$\{\text{TTF, TFT, TFF, FTF, FFF}\} \cap \{\text{FTF, FFT, FFF}\} = \{\text{FTF, FFF}\}$$

Thus the cross-partition is

$$[\{\text{TTT, FTT}\}, \{\text{FFT}\}, \{\text{TTF, TFT, TFF}\}, \{\text{FTF, FFF}\}]$$

ORDERED AND UNORDERED PARTITIONS

15.19. In how many ways can 7 toys be divided among 3 children if the youngest gets 3 toys and each of the others gets 2?

We seek the number of ordered partitions of 7 objects into cells containing 3, 2 and 2 objects, respectively. By Theorem 15.3, there are $\frac{7!}{3!\,2!\,2!} = 210$ such partitions.

15.20. There are 12 students in a class. In how many ways can the 12 students take 3 different tests if 4 students are to take each test?

Method 1.

We seek the number of ordered partitions of the 12 students into cells containing 4 students each. By Theorem 15.3, there are $\frac{12!}{4!\,4!\,4!} = 34{,}650$ such partitions.

Method 2.

There are $\binom{12}{4}$ ways to choose 4 students to take the first test; following this, there are $\binom{8}{4}$ ways to choose 4 students to take the second test. The remaining students take the third test. Thus, altogether, there are $\binom{12}{4} \cdot \binom{8}{4} = 495 \cdot 70 = 34{,}650$ ways for the students to take the tests.

15.21. In how many ways can 12 students be partitioned into 3 teams, A_1, A_2 and A_3, so that each team contains 4 students?

Method 1.

Observe that each partition $\{A_1, A_2, A_3\}$ of the students can be arranged in $3! = 6$ ways as an ordered partition. Since (see the preceding problem) there are $\frac{12!}{4!\,4!\,4!} = 34{,}650$ such ordered partitions, there are $34{,}650/6 = 5775$ (unordered) partitions.

Method 2.

Let A denote one of the students. Then there are $\binom{11}{3}$ ways to choose 3 other students to be on the same team as A. Now let B denote a student who is not on the same team as A; then there are $\binom{7}{3}$ ways to choose 3 students of the remaining students to be on the same team as B. The remaining 4 students constitute the third team. Thus, altogether, there are $\binom{11}{3} \cdot \binom{7}{3} = 165 \cdot 35 = 5775$ ways to partition the students.

15.22. In how many ways can 6 students be partitioned into (i) 2 teams containing 3 students each, (ii) 3 teams containing 2 students each?

(i) **Method 1.**

There are $\frac{6!}{3!\,3!} = 20$ ordered partitions into 2 cells containing 3 students each. Since each unordered partition determines $2! = 2$ ordered partitions, there are $20/2 = 10$ unordered partitions.

Method 2.

Let A denote one of the students; then there are $\binom{5}{2} = 10$ ways to choose 2 other students to be on the same team as A. The other 3 students constitute the other team. In other words, there are 10 ways to partition the students.

(ii) **Method 1.**

There are $\frac{6!}{2!\,2!\,2!} = 90$ ordered partitions into 3 cells containing 2 students each. Since each unordered partition determines $3! = 6$ ordered partitions, there are $90/6 = 15$ unordered partitions.

Method 2.

Let A denote one of the students. Then there are 5 ways to choose the other student to be on the same team as A. Let B denote a student who isn't on the same team as A; then there are 3 ways to choose another student to be on the same team as B. The remaining 2 students constitute the other team. Thus, altogether, there are $5 \cdot 3 = 15$ ways to partition the students.

15.23. In how many ways can a class X with 10 students be partitioned into 4 teams A_1, A_2, B_1, B_2 where A_1 and A_2 contain 2 students each and B_1 and B_2 contain 3 students each?

Method 1.

There are $\frac{10!}{2!\,2!\,3!\,3!} = 25{,}200$ ordered partitions of X into 4 cells containing 2, 2, 3 and 3 students, respectively. However, each unordered partition $\{A_1, A_2, B_1, B_2\}$ of X determines $2! \cdot 2! = 4$ ordered partitions of X. Thus, altogether, there are $25{,}200/4 = 6300$ unordered partitions.

Method 2.

There are $\binom{10}{4}$ ways to choose 4 students who will be on the teams A_1 and A_2, and there are 3 ways in which each 4 students can be partitioned into 2 teams of 2 students each. On the other hand, there are 10 ways (see preceding problem) to partition the remaining 6 students into 2 teams containing 3 students each. Thus, altogether, there are $\binom{10}{4} \cdot 3 \cdot 10 = 210 \cdot 3 \cdot 10 = 6300$ ways to partition the students.

15.24. In how many ways can 9 students be partitioned into three teams containing 4, 3 and 2 students, respectively?

Since all the cells contain different numbers of students, the number of unordered partitions equals the number of ordered partitions which, by Theorem 15.3, is $\frac{9!}{4!\,3!\,2!} = 1260$.

15.25. (i) In how many ways can a set X containing 10 elements be partitioned into two cells?

(ii) In how many ways can 10 students be partitioned into two teams?

(i) **Method 1.**

Each subset A of X determines an ordered partition (A, A^c) of X (where A^c is the complement of A), and so there are $2^{10} = 1024$ such ordered partitions. However, each unordered partition $\{A, B\}$ determines two ordered partitions, (A, B) and (B, A), and so there are $1024/2 = 512$ unordered partitions.

Method 2.

Let x denote one of the elements in X. Now $X - \{x\}$ contains 9 elements, and so there are $2^9 = 512$ ways of choosing a subset of $X - \{x\}$ to be in the same cell as x. In other words, there are 512 such partitions. (Observe that $\{X, \emptyset\}$ is one of these partitions.)

(ii) Here we assume that each team must contain at least one student, hence we do not allow the partition for which one team contains 10 students, and the other team none. Accordingly, there are $512 - 1 = 511$ possible partitions.

Supplementary Problems

COMBINATIONS

15.26. A class contains 9 boys and 3 girls. (i) In how many ways can the teacher choose a committee of 4? (ii) How many of them will contain at least one girl? (iii) How many of them will contain exactly one girl?

15.27. A woman has 11 close friends. (i) In how many ways can she invite 5 of them to dinner? (ii) In how many ways if two of the friends are married and will not attend separately? (iii) In how many ways if two of them are not on speaking terms and will not attend together?

15.28. A woman has 11 close friends of whom 6 are also women. (i) In how many ways can she invite 3 or more of them to a party? (ii) In how many ways can she invite 3 or more of them if she wants the same number of men as women (including herself)?

15.29. There are 10 points $A, B, \ldots$ in a plane, no three on the same line. (i) How many lines are determined by the points? (ii) How many of these lines do not pass through A or B? (iii) How many triangles are determined by the points? (iv) How many of these triangles contain the point A? (v) How many of these triangles contain the side AB?

15.30. A student is to answer 10 out of 13 questions on an exam. (i) How many choices has he? (ii) How many if he must answer the first two questions? (iii) How many if he must answer the first or second question but not both? (iv) How many if he must answer exactly 3 of the first 5 questions? (v) How many if he must answer at least 3 of the first 5 questions?

15.31. How many diagonals has a (i) hexagon, (ii) decagon?

15.32. Which regular polygon has (i) twice as many diagonals as sides, (ii) three times as many diagonals as sides?

15.33. (i) How many triangles are determined by the vertices of an octagon?

(ii) How many if the sides of the octagon are not to be sides of any triangle?

15.34. How many triangles are determined by the vertices of a regular polygon with n sides if the sides of the polygon are not to be sides of any triangle?

15.35. A man is dealt a poker hand (5 cards) from an ordinary playing deck. In how many ways can he be dealt (i) a straight flush, (ii) four of a kind, (iii) a straight, (iv) a pair of aces, (v) two of a kind (a pair)?

15.36. The English alphabet has 26 letters of which 5 are vowels.

(i) How many 5 letter words containing 3 different consonants and 2 different vowels can be formed?

(ii) How many of them contain the letter b?

(iii) How many of them contain the letters b and c?

(iv) How many of them begin with b and contain the letter c?

(v) How many of them begin with b and end with c?

(vi) How many of them contain the letters a and b?

(vii) How many of them begin with a and contain b?

(viii) How many of them begin with b and contain a?

(ix) How many of them begin with a and end with b?

(x) How many of them contain the letters a, b and c?

PARTITIONS AND CROSS-PARTITIONS

15.37. Which of the following are partitions of $X = \{a, e, i, o, u\}$?

(i) $[\{a, e, i\}, \{o\}, \{u\}]$

(ii) $[\{a, i, u\}, \{e\}, \{o, u\}]$

(iii) $[\{a, e\}, \{o\}, \{u\}]$

(iv) $[\{a, e\}, \{i\}, \{o\}, \{a, e\}, \{u\}]$

(v) $[\{a, e, i\}, \{o, u\}, \emptyset]$

(vi) $[\emptyset, \{a\}, \{e, u\}, \emptyset, \{i, o\}]$

15.38. Find the cross-partition of each pair of partitions of $X = \{1, 2, 3, 4, 5, 6\}$:

(i) $[\{1, 2, 3\}, \{4, 5, 6\}]$ and $[\{1, 3, 5\}, \{2, 4, 6\}]$.

(ii) $[\{1, 2, 3\}, \{4, 5, 6\}]$ and $[\{1, 2\}, \{3, 4, 5\}, \{6\}]$.

15.39. Let X denote the domain of a proposition $P(p, q, r)$ of three variables:

$$X = \{\text{TTT, TTF, TFT, TFF, FTT, FTF, FFT, FFF}\}$$

(i) Partition X according to the number of F's appearing in each element.

(ii) For the proposition $P = (p \vee q) \wedge r$, partition X into $\mathcal{T}(P)$ and $(\mathcal{T}(P))^c$.

(iii) Partition X into $\mathcal{T}(q)$ and $\mathcal{T}(\sim q)$.

(iv) Find the cross-partition of the partitions in (ii) and (iii).

15.40. For any proposition $P = P(p, q, \ldots)$, show that $[\mathcal{T}(P), \mathcal{T}(\sim P)]$ forms a partition of the domain of P.

15.41. Let the partition $\{A_1, A_2, \ldots, A_r\}$ be a refinement of the partition $\{B_1, B_2, \ldots, B_s\}$, that is, each A_i is a subset of some B_j. Describe the cross-partition.

15.42. Show that a cross-partition is a refinement (see preceding problem) of each of the original partitions.

ORDERED AND UNORDERED PARTITIONS

15.43. In how many ways can 9 toys be divided evenly among 3 children?

15.44. In how many ways can 9 students be evenly divided into three teams?

15.45. In how many ways can 10 students be divided into three teams, one containing 4 students and the others 3?

15.46. There are 12 balls in an urn. In how many ways can 3 balls be drawn from the urn, four times in succession?

15.47. In how many ways can a club with 12 members be partitioned into three committees containing 5, 4 and 3 members respectively?

15.48. In how many ways can n students be partitioned into two teams containing at least one student?

15.49. In how many ways can 14 men be partitioned into 6 committees where 2 of the committees contain 3 men and the others 2?

15.50. In how many ways can a set X with 3 elements be partitioned into (i) three ordered cells, (ii) three (unordered) cells?

15.51. In how many ways can a set X with 4 elements be partitioned into (i) three ordered cells, (ii) three (unordered) cells?

15.52. In bridge, 13 cards are dealt to each of four players who are called North, South, East and West. The distribution of the cards is called a bridge hand. (i) How many bridge hands are there? (ii) In how many of them will one player be dealt all four aces? (iii) In how many of them will each player be dealt an ace? (iv) In how many of them will North be dealt 8 spades and South the other 5 spades? (v) In how many of them will North and South have, together, all four aces? (*Remark.* Leave answers in factorial notation.)

Answers to Supplementary Problems

15.26. (i) $\binom{12}{4} = 495$, (ii) $\binom{12}{4} - \binom{9}{4} = 369$, (iii) $3 \cdot \binom{9}{3} = 252$

15.27. (i) $\binom{11}{5} = 462$, (ii) $\binom{9}{3} + \binom{9}{5} = 210$, (iii) $2 \cdot \binom{9}{4} = 252$

15.28. (i) $2^{11} - 1 - \binom{11}{1} - \binom{11}{2} = 1981$ or $\binom{11}{3} + \binom{11}{4} + \cdots + \binom{11}{11} = 1981$

(ii) $\binom{5}{5}\binom{6}{4} + \binom{5}{4}\binom{6}{3} + \binom{5}{3}\binom{6}{2} + \binom{5}{2}\binom{6}{1} = 325$

15.29. (i) $\binom{10}{2} = 45$, (ii) $\binom{8}{2} = 28$, (iii) $\binom{10}{3} = 120$, (iv) $\binom{9}{2} = 36$, (v) 8

15.30. (i) $\binom{13}{10} = \binom{13}{3} = 286$

(ii) $\binom{11}{8} = \binom{11}{3} = 165$

(iii) $2 \cdot \binom{11}{9} = 2 \cdot \binom{11}{2} = 110$

(iv) $\binom{5}{3}\binom{8}{7} = 80$

(v) $\binom{5}{3}\binom{8}{7} + \binom{5}{4}\binom{8}{6} + \binom{5}{5}\binom{8}{5} = 276$

15.31. (i) $\binom{6}{2} - 6 = 9$, (ii) $\binom{10}{2} - 10 = 35$

15.32. *Hint.* By Problem 15.7, the regular polygon with n sides has $(n^2 - 3n)/2$ diagonals.
(i) $(n^2 - 3n)/2 = 2n$ or $n = 7$, (ii) $(n^2 - 3n)/2 = 3n$ or $n = 9$

15.33. (i) $\binom{8}{3} = 56$, (ii) $\binom{8}{3} - 8\binom{4}{1} - 8 = 16$

15.34. $\binom{n}{3} - n\binom{n-4}{1} - n \;=\; \frac{n}{6}(n-5)(n-4)$

15.35. (i) $4 \cdot 10 = 40$, (ii) $13 \cdot 48 = 624$, (iii) $10 \cdot 4^5 - 40 = 10{,}200$. (We subtract the number of straight flushes.) (iv) $\binom{4}{2}\binom{12}{3} \cdot 4^3 = 84{,}480$, (v) $13 \cdot \binom{4}{2}\binom{12}{3} \cdot 4^3 = 1{,}098{,}240$

15.36. (i) $\binom{21}{3}\binom{5}{2} \cdot 5! = 1{,}596{,}000$

(ii) $\binom{20}{2}\binom{5}{2} \cdot 5! = 228{,}000$

(iii) $19 \cdot \binom{5}{2} \cdot 5! = 22{,}800$

(iv) $19 \cdot \binom{5}{2} \cdot 4! = 4560$

(v) $19 \cdot \binom{5}{2} \cdot 3! = 1140$

(vi) $4 \cdot \binom{20}{2} \cdot 5! = 91{,}200$

(vii) $4 \cdot \binom{20}{2} \cdot 4! = 18{,}240$

(viii) 18,240

(ix) $4 \cdot \binom{20}{2} \cdot 3! = 4456$

(x) $4 \cdot 19 \cdot 5! = 9120$

15.37. (i) yes, (ii) no, (iii) no, (iv) yes, (v) yes, (vi) yes

15.38. (i) $[\{1,3\}, \{2\}, \{5\}, \{4,6\}]$, (ii) $[\{1,2\}, \{3\}, \emptyset, \{4,5\}, \{6\}]$

15.39. (i) [{TTT}, {TTF, TFT, FTT}, {TFF, FTF, FFT}, {FFF}]

(ii) [{TTT, TFT, FTT}, {TTF, TFF, FTF, FFT, FFF}]

(iii) [{TTT, TTF, FTT, FTF}, {TFT, TFF, FFT, FFF}]

(iv) [{TTT, FTT}, {TFT}, {TTF, FTF}, {TFF, FFT, FFF}]

15.41. $\{A_1, A_2, \ldots, A_r\}$, i.e. the refinement partition.

15.42. $A_i \cap B_j \subset A_i$ and $A_i \cap B_j \subset B_j$.

15.43. $\frac{9!}{3!\,3!\,3!} = 1680$

15.44. $1680/3! = 280$ or $\binom{8}{2}\binom{5}{2} = 280$

15.45. $\frac{10!}{4!\,3!\,3!} \cdot \frac{1}{2!} = 2100$ or $\binom{10}{4}\binom{5}{2} = 2100$

15.46. $\frac{12!}{3!\,3!\,3!\,3!} = 369{,}600$

15.47. $\frac{12!}{5!\,4!\,3!} = 27{,}720$

15.48. $2^{n-1} - 1$

15.49. $\frac{14!}{3!\,3!\,2!\,2!\,2!\,2!} \cdot \frac{1}{2!\,4!} = 3{,}153{,}150$

15.50. (i) $3^3 = 27$ (Each element can be placed in any of the three cells.)

(ii) The number of elements in the three cells can be distributed as follows:

(*a*) $[\{3\}, \{0\}, \{0\}]$, (*b*) $[\{2\}, \{1\}, \{0\}]$, (*c*) $[\{1\}, \{1\}, \{1\}]$

Thus the number of partitions is $1+3+1 = 5$.

15.51. (i) $3^4 = 81$. (ii) The number of elements in the three cells can be distributed as follows: (*a*) $[\{4\}, \{0\}, \{0\}]$, (*b*) $[\{3\}, \{1\}, \{0\}]$, (*c*) $[\{2\}, \{2\}, \{0\}]$, (*d*) $[\{2\}, \{1\}, \{1\}]$. Thus the number of partitions is $1+4+3+6 = 14$.

15.52. (i) $\frac{52!}{13!\,13!\,13!\,13!}$ (ii) $4 \cdot \frac{48!}{9!\,13!\,13!\,13!}$ (iii) $4! \cdot \frac{48!}{12!\,12!\,12!\,12!}$ (iv) $\binom{13}{8}\frac{39!}{5!\,8!\,13!\,13!}$

(v) $2 \cdot \frac{48!}{9!\,13!\,13!\,13!} + 2 \cdot 4 \cdot \frac{48!}{10!\,12!\,13!\,13!} + 2 \cdot 3 \cdot \frac{48!}{11!\,11!\,13!\,13!} \;=\; 2300\,\frac{48!}{11!\,13!\,13!\,13!}$

Chapter 16

Tree Diagrams

TREE DIAGRAMS

A tree diagram is a device used to enumerate all the logical possibilities of a sequence of events where each event can occur in a finite number of ways. The construction of tree diagrams is illustrated in the following examples.

Example 1: Find the product set $A \times B \times C$ where $A = \{1, 2\}$, $B = \{a, b, c\}$ and $C = \{3, 4\}$.

The tree diagram follows:

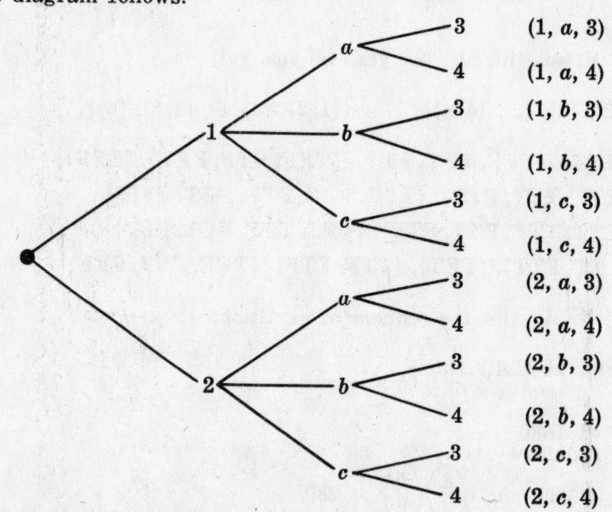

Observe that the tree is constructed from left to right, and that the number of branches at each point corresponds to the number of ways the next event can occur.

Example 2: Mark and Eric are to play a tennis tournament. The first person to win two games in a row or who wins a total of three games wins the tournament. The following diagram gives the various ways the tournament can occur.

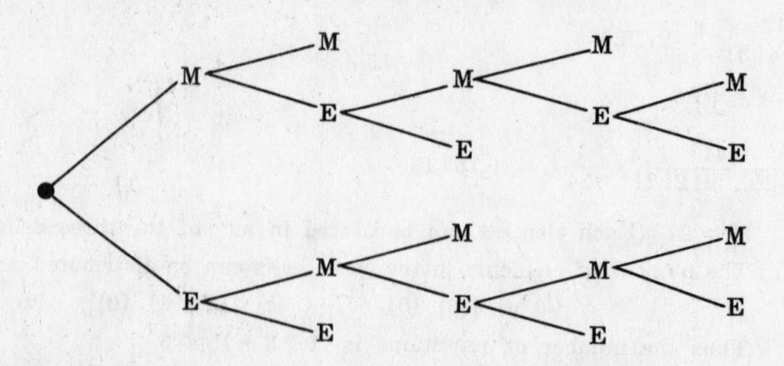

Observe that there are 10 endpoints which correspond to the 10 ways that the tournament can occur:

MM, MEMM, MEMEM, MEMEE, MEE, EMM, EMEMM, EMEME, EMEE, EE

The path from the beginning of the tree to the endpoint describes who won which game in the individual tournament.

Example 3: A man has time to play roulette at most five times. At each play he wins or loses a dollar. The man begins with one dollar and will stop playing before the five times if he loses all his money or if he wins three dollars, i.e. if he has four dollars. The tree diagram describes the way the betting can occur. Each number in the diagram denotes the number of dollars he has at that point.

The betting can occur in 11 different ways. Note that he will stop betting before the five times are up in only three of the cases.

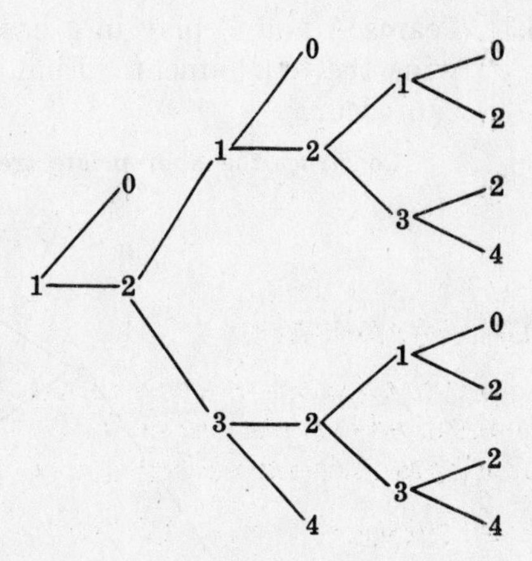

Solved Problems

16.1. Construct the tree diagram for the number of permutations of $\{a, b, c\}$.

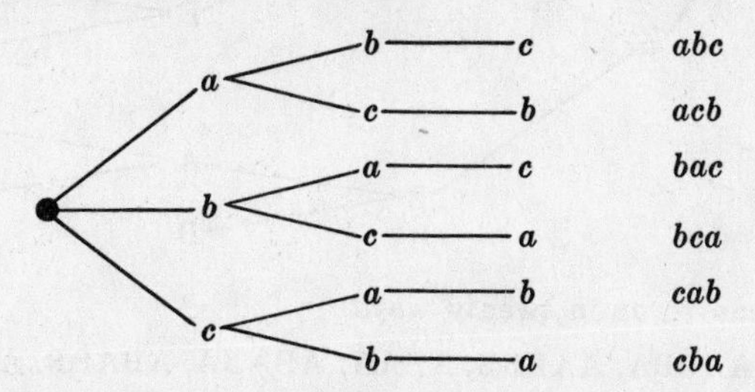

The six permutations are listed on the right of the diagram.

16.2. Find the product set $\{1,2,3\} \times \{2,4\} \times \{2,3,4\}$ by constructing the appropriate tree diagram.

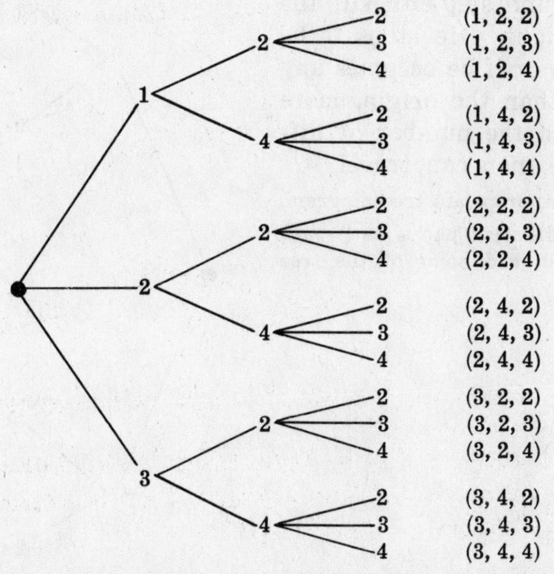

The eighteen elements of the product set are listed to the right of the tree diagram.

16.3. Teams A and B play in a basketball tournament. The team that first wins 3 games wins the tournament. Find the number of possible ways in which the tournament can occur.

Construct the appropriate tree diagram:

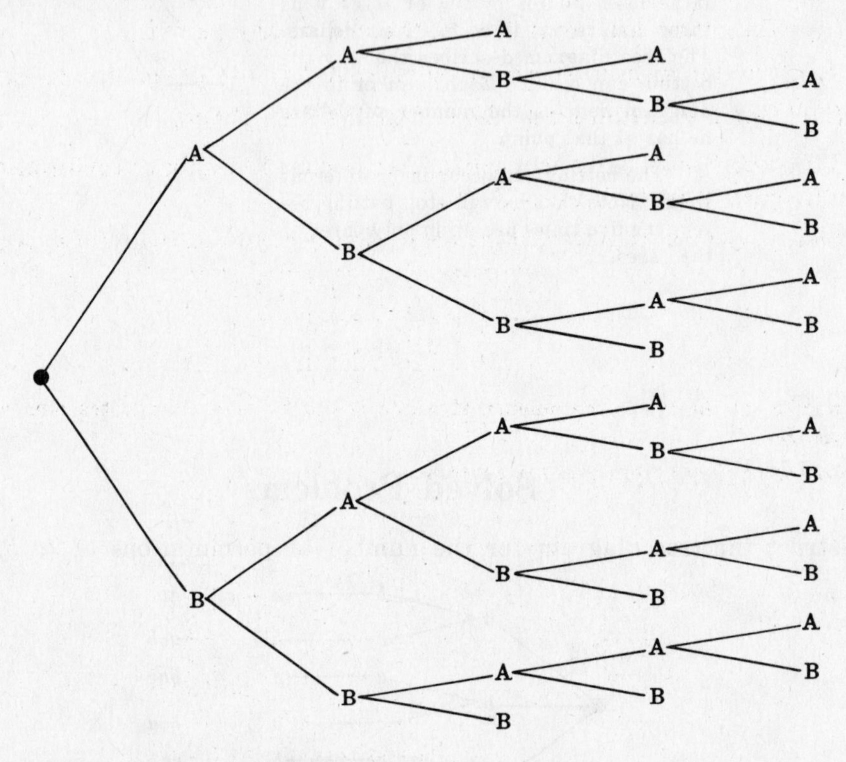

The tournament can occur in twenty ways:

AAA, AABA, AABBA, AABBB, ABAA, ABABA, ABABB, ABBAA, ABBAB, ABBB
BAAA, BAABA, BAABB, BABAA, BABAB, BABB, BBAAA, BBAAB, BBAB, BBB

16.4. A man is at the origin on the x-axis and takes a one unit step either to the left or to the right. He stops if he reaches 3 or -3, or if he occupies any position, other than the origin, more than once. Find the number of different paths the man can travel.

Construct the appropriate tree diagram.

There are 14 different paths, each path corresponding to an end point of the tree diagram.

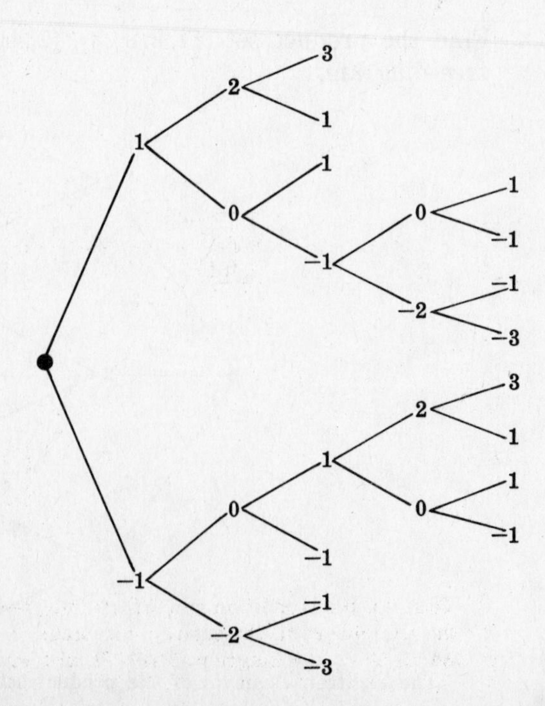

16.5. Teams A and B play a hockey game consisting of three periods. The number of goals (points) scored in each period satisfies the following conditions:

(i) The total number of points scored by both teams in any one period is one or two.

(ii) The teams are never tied at the end of any period.

(iii) Neither team is ahead by more than 2 points at the end of any period.

(iv) Team A is leading at the end of the first period.

Find the number of ways the scoring can occur if the above conditions are satisfied. In how many ways will team B win?

Note first, by condition (i), that the scoring in any one period can occur in only five ways:

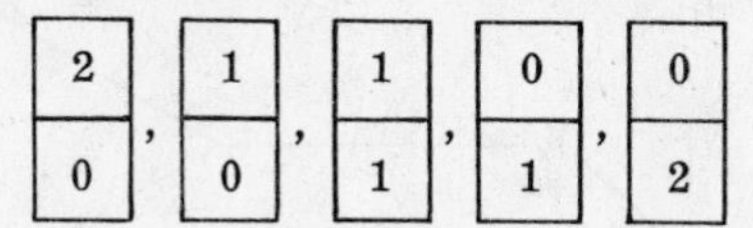

The number at the top is the number of points scored by team A, and the number at the bottom by team B.

Construct the appropriate tree diagram:

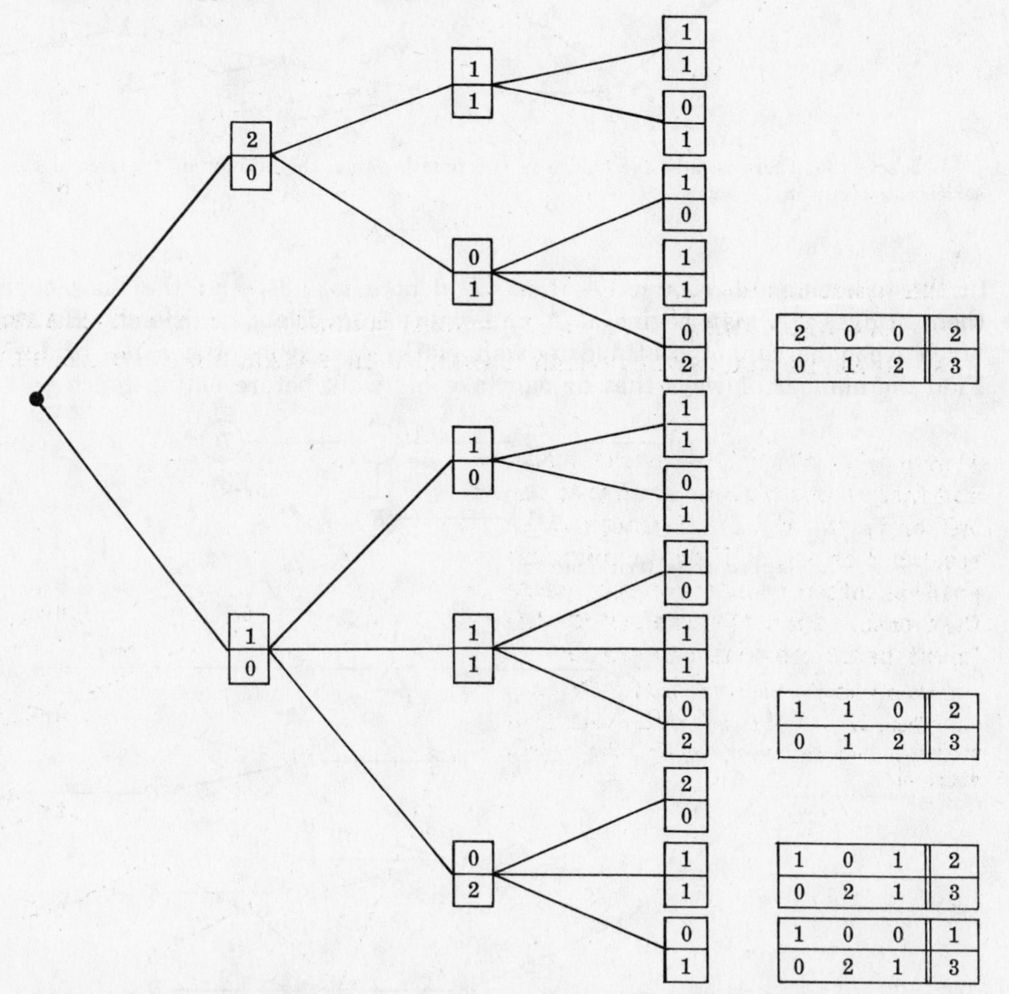

Observe that condition (iv) allows only two of the above five scores to occur in the first period. By the tree diagram, the scoring can occur in only 13 ways, and in 4 of these which are exhibited on the right of the diagram, Team B will win.

16.6. Teams A and B play in baseball's world series. (Here, the team that first wins 4 games wins the series.) Find the number of ways the series can occur if (i) A wins the first game and (ii) the team that wins the second game also wins the fourth game.

Construct the appropriate tree diagram:

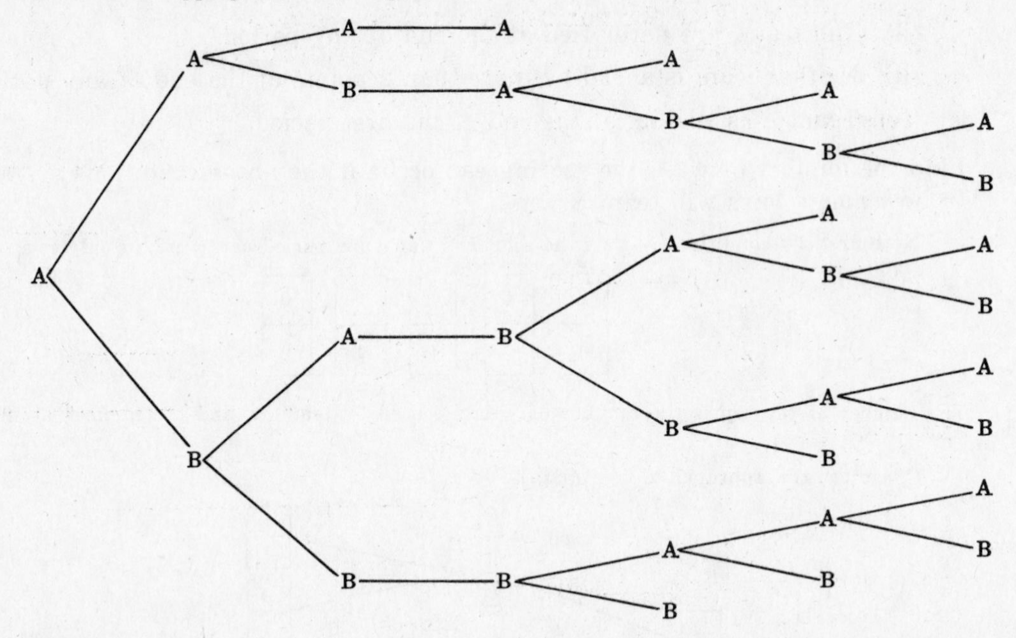

Observe that there is only one choice in the fourth game, the winner of the second game. The series can occur in 15 ways.

16.7. In the following diagram let $A, B, \ldots, F$ denote islands, and the lines connecting them bridges. A man begins at A and walks from island to island. He stops for lunch when he cannot continue to walk without crossing the same bridge twice. Find the number of ways that he can take his walk before eating lunch.

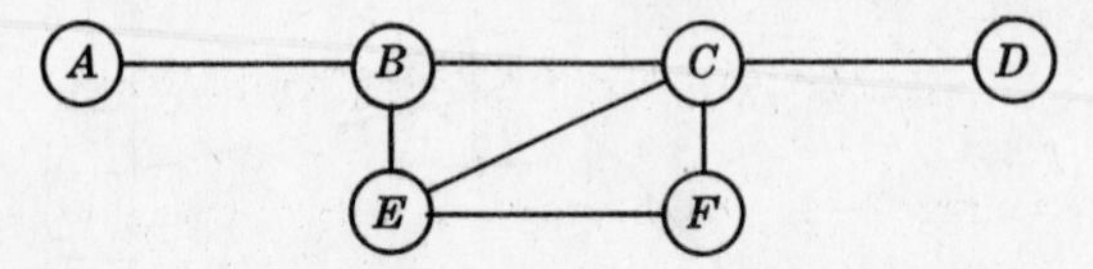

Construct the appropriate tree diagram:

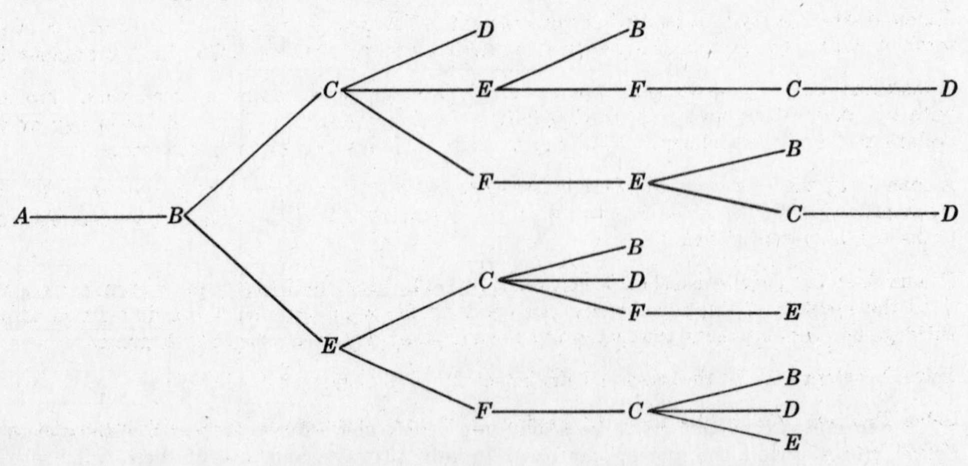

There are eleven ways to take his walk. Observe that he must eat his lunch at either B, D or E.

16.8. Consider the adjacent diagram with nine points $A, B, C, R, S, T, X, Y, Z$. A man begins at X and is allowed to move horizontally or vertically, one step at a time. He stops when he cannot continue to walk without reaching the same point more than once. Find the number of ways he can take his walk, if he first moves from X to R. (By symmetry, the total number of ways is twice this.)

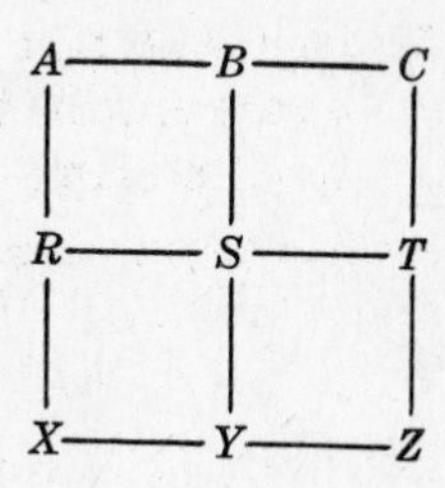

Construct the appropriate tree diagram:

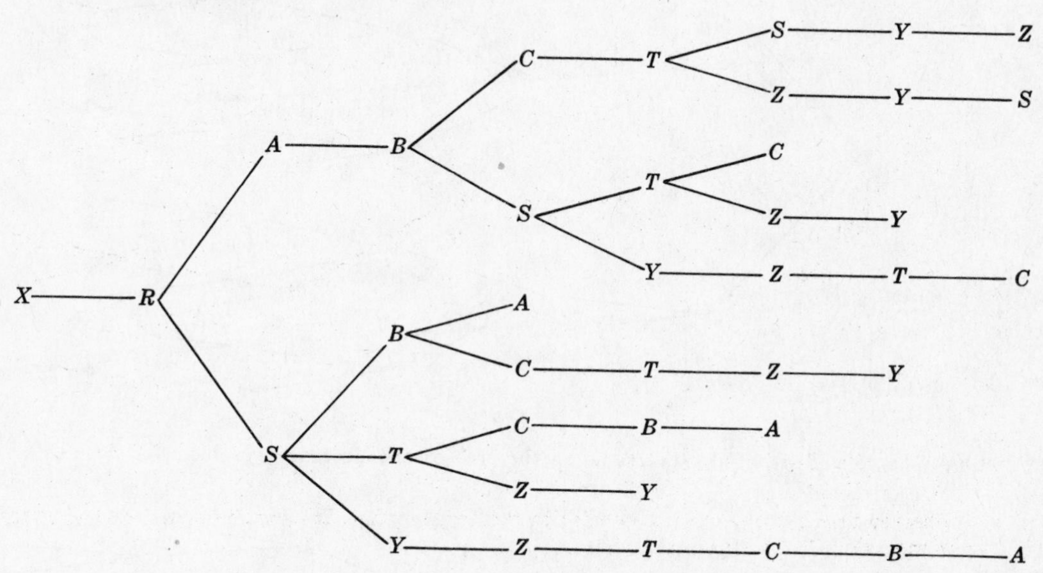

There are 10 different trips. (Note that in only 4 of them are all nine points covered.)

Supplementary Problems

16.9. Construct the tree diagram for the number of permutations of $\{a, b, c, d\}$.

16.10. Find the product set $\{a, b, c\} \times \{a, c\} \times \{b\} \times \{b, c\}$ by constructing the appropriate tree diagram.

16.11. Teams A and B play in a basketball tournament. The first team that wins two games in a row or a total of four games wins the tournament. Find the number of ways the tournament can occur.

16.12. A man has time to play roulette five times. He wins or loses a dollar at each play. The man begins with two dollars and will stop playing before the five times if he loses all his money or wins three dollars (i.e. has five dollars). Find the number of ways the playing can occur.

16.13. A man is at the origin on the x-axis and takes a unit step either to the left or to the right. He stops after 5 steps or if he reaches 3 or -2. Construct the tree diagram to describe all possible paths the man can travel.

16.14. Teams A and B play in baseball's world series. (The team that first wins 4 games wins the series.) Find the number of ways the series can occur if the team that wins the first game also wins the third game, and the team that wins the second game also wins the fourth game.

16.15. Solve Problem 16.7 if the man (i) begins at E, (ii) begins at F.

16.16. Solve Problem 16.8 if the man (i) begins at Y and first moves to S, (ii) begins at Y and first moves to Z.

16.17. Solve Problem 16.5 if condition (ii) is replaced by the following: (ii) The game does not end in a tie.

Answers to Supplementary Problems

16.12.

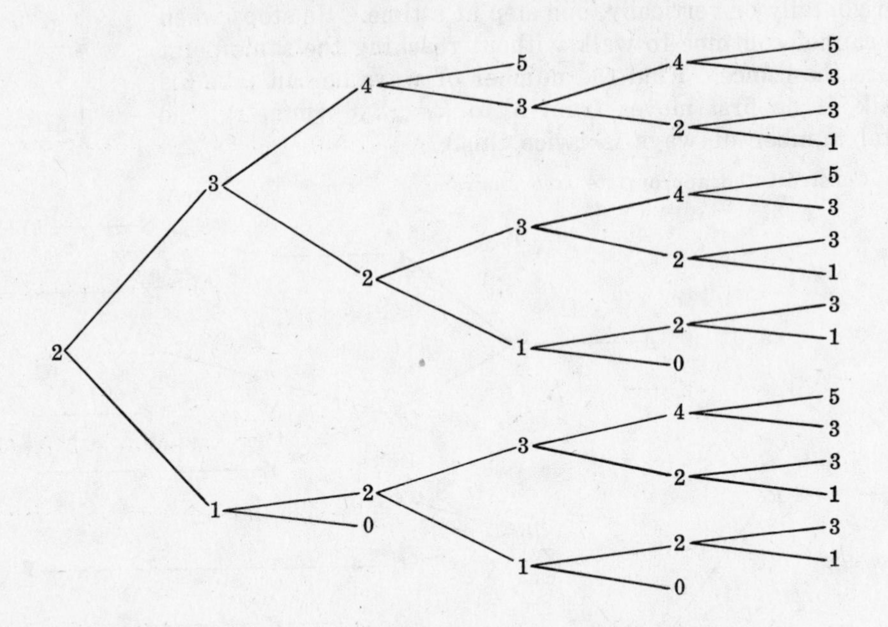

16.13. *Hint.* The tree is essentially the same as the tree of the preceding problem.

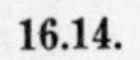
16.14.

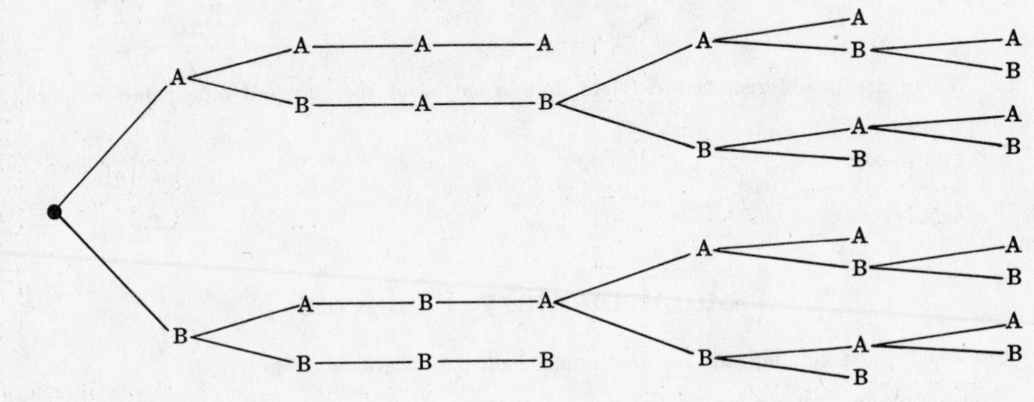

16.15. (i)

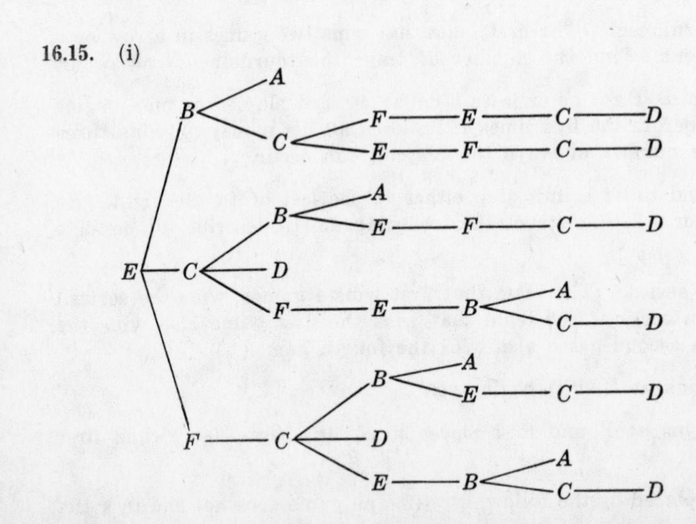

(ii)

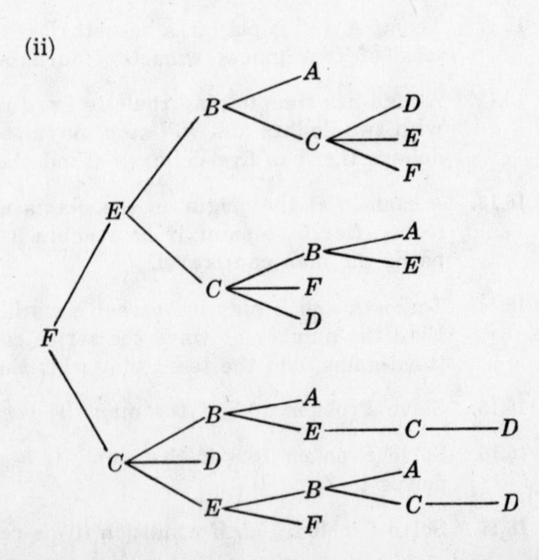

16.16. (i)

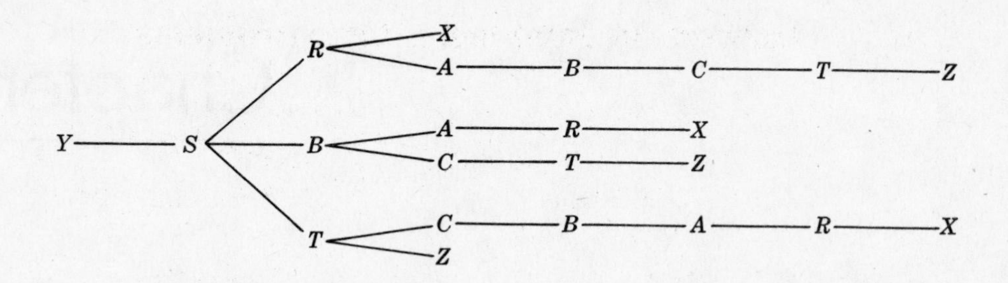

(ii)

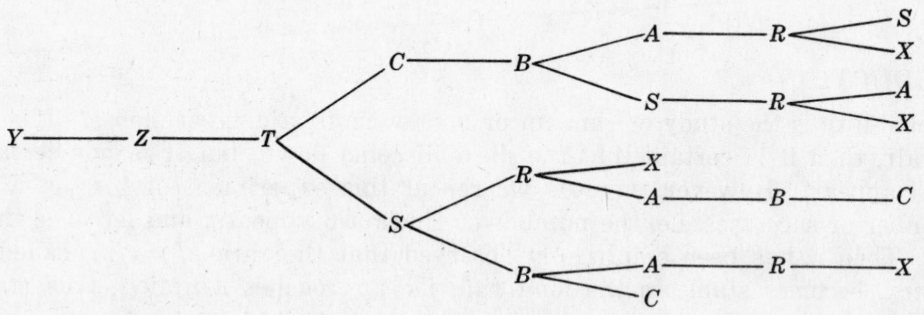

A R S
X
C B
S R A
X
Y Z T
R X
A B C
S
A R X
B
C

16.17. The tree is the same as in Problem 16.5, except that a tie is now possible at the end of the second period. Hence add the following branches onto the first period scores of $\begin{array}{|c|}2\\\hline 0\end{array}$ and $\begin{array}{|c|}1\\\hline 0\end{array}$, respectively:

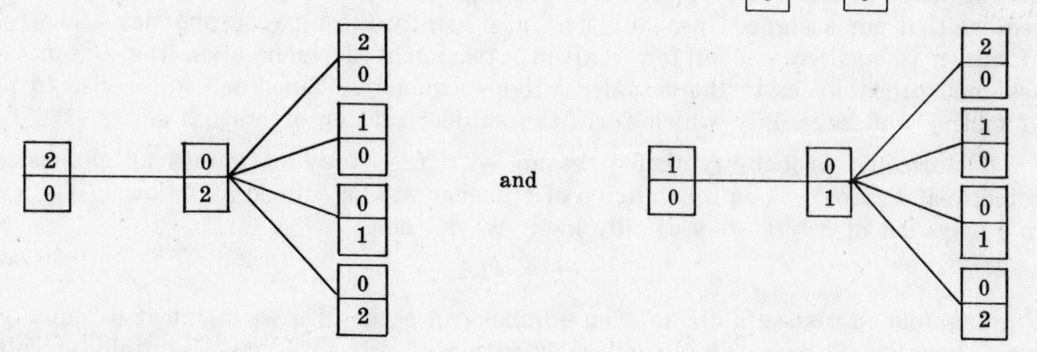

Thus altogether there are $13+8 = 21$ ways that the scoring can occur. Team B will win in $4+4 = 8$ ways.

Chapter 17

Probability

INTRODUCTION

Probability is the study of random or nondeterministic experiments. If a die is tossed in the air, then it is certain that the die will come down, but it is not certain that, say, a 6 will appear. However, suppose we repeat this experiment of tossing a die; let s be the number of successes, i.e. the number of times a 6 appears, and let n be the number of tosses. Then it has been empirically observed that the ratio $f = s/n$, called the *relative frequency*, becomes stable in the long run, i.e. approaches a limit. This stability is the basis of probability theory.

In probability theory, we define a mathematical model of the above phenomenon by assigning "probabilities" (or: the limit values of the relative frequencies) to each possible outcome of an experiment. Furthermore, since the relative frequency of each outcome is non-negative and the sum of the relative frequencies of all possible outcomes is unity, we require that our assigned "probabilities" also satisfy these two properties. The reliability of our mathematical model for a given experiment depends upon the closeness of the assigned probabilities to the actual relative frequency. This then gives rise to problems of testing and reliability which form the subject matter of statistics.

Historically, probability theory began with the study of games of chance, such as roulette and cards. The probability p of an event A was defined as follows: if A can occur in s ways out of a total of n equally likely ways, then

$$p = P(A) = \frac{s}{n}$$

For example, in tossing a die an even number can occur in 3 ways out of 6 "equally likely" ways; hence $p = \frac{3}{6} = \frac{1}{2}$. This classical definition of probability is essentially circular since the idea of "equally likely" is the same as that of "with equal probability" which has not been defined. The modern treatment of probability theory is purely axiomatic. In particular, we require that the probabilities of our events satisfy the three axioms listed in Theorem 17.1. However, those spaces consisting of n equally likely elementary events, called equiprobable spaces, are an important class of finite probability spaces and shall provide us with many examples.

SAMPLE SPACE AND EVENTS

The set S of all possible outcomes of some given experiment is called the *sample space*. A particular outcome, i.e. an element in S, is called a *sample point* or *sample*. An *event* A is a set of outcomes or, in other words, a subset of the sample space S. In particular, the set $\{a\}$ consisting of a single sample $a \in S$ is an event and called an *elementary event*. Furthermore, the empty set $\emptyset$ and S itself are subsets of S and so are events; $\emptyset$ is sometimes called the *impossible event*.

Since an event is a set, we can combine events to form new events using the various set operations:

(i) $A \cup B$ is the event that occurs iff A occurs *or* B occurs (or both);

(ii) $A \cap B$ is the event that occurs iff A occurs *and* B occurs;

(iii) A^c, the complement of A, also written $\bar{A}$, is the event that occurs iff A does *not* occur.

Two events A and B are called *mutually exclusive* if they are disjoint, i.e. $A \cap B = \emptyset$. In other words, A and B are mutually exclusive iff they cannot occur simultaneously.

Example 1.1: Experiment: Toss a die and observe the number that appears on top. Then the sample space consists of the six possible numbers:

$$S = \{1, 2, 3, 4, 5, 6\}$$

Let A be the event that an even number occurs, B that an odd number occurs and C that a prime number occurs:

$$A = \{2, 4, 6\}, \quad B = \{1, 3, 5\}, \quad C = \{2, 3, 5\}$$

Then:

$A \cup C = \{2, 3, 4, 5, 6\}$ is the event that an even or a prime number occurs;

$B \cap C = \{3, 5\}$ is the event that an odd prime number occurs;

$C^c = \{1, 4, 6\}$ is the event that a prime number does not occur.

Note that A and B are mutually exclusive: $A \cap B = \emptyset$; in other words, an even number and an odd number cannot occur simultaneously.

Example 1.2: Experiment: Toss a coin 3 times and observe the sequence of heads (H) and tails (T) that appears. The sample space S consists of eight elements:

$$S = \{HHH, HHT, HTH, HTT, THH, THT, TTH, TTT\}$$

Let A be the event that two or more heads appear consecutively, and B that all the tosses are the same:

$$A = \{HHH, HHT, THH\} \quad \text{and} \quad B = \{HHH, TTT\}$$

Then $A \cap B = \{HHH\}$ is the elementary event in which only heads appear. The event that 5 heads appear is the empty set $\emptyset$.

Example 1.3: Experiment: Toss a coin until a head appears and then count the number of times the coin was tossed. The sample space of this experiment is $S = \{1, 2, 3, \ldots, \infty\}$. Here ∞ refers to the case when a head never appears and so the coin is tossed an infinite number of times. This is an example of a sample space which is *countably infinite.*

Example 1.4: Experiment: Let a pencil drop, head first, into a rectangular box and note the point on the bottom of the box that the pencil first touches. Here S consists of all the points on the bottom of the box. Let the rectangular area on the right represent these points. Let A and B be the events that the pencil drops into the corresponding areas illustrated on the right. This is an example of a sample space which is not finite nor even countably infinite, i.e. which is non-denumerable. For theoretical reasons every subset of this space cannot be considered to be an event.

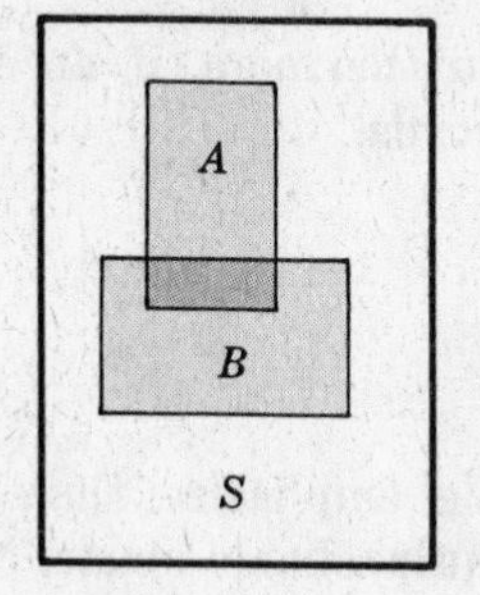

The sample spaces in the preceding two examples, as noted, are not finite. The theory concerning such sample spaces, especially the non-denumerable case, lies beyond the scope of this text. Thus, unless otherwise stated, all our sample spaces S shall be finite.

FINITE PROBABILITY SPACES

Let S be a finite sample space: $S = \{a_1, a_2, \ldots, a_n\}$. A finite probability space is obtained by assigning to each point $a_i \in S$ a real number p_i, called the *probability* of a_i, satisfying the following properties:

(i) each p_i is non-negative, $p_i \geq 0$

(ii) the sum of the p_i is one, $p_1 + p_2 + \cdots + p_n = 1$.

The *probability* of any event A, written $P(A)$, is then defined to be the sum of the probabilities of the points in A. For notational convenience we write $P(a_i)$ for $P(\{a_i\})$.

Example 2.1: Let three coins be tossed and the number of heads observed; then the sample space is $S = \{0, 1, 2, 3\}$. We obtain a probability space by the following assignment

$$P(0) = \tfrac{1}{8}, \quad P(1) = \tfrac{3}{8}, \quad P(2) = \tfrac{3}{8} \quad \text{and} \quad P(3) = \tfrac{1}{8}$$

since each probability is non-negative and the sum of the probabilities is 1. Let A be the event that at least one head appears and let B be the event that all heads or all tails appear:

$$A = \{1, 2, 3\} \quad \text{and} \quad B = \{0, 3\}$$

Then, by definition,

$$P(A) = P(1) + P(2) + P(3) = \tfrac{3}{8} + \tfrac{3}{8} + \tfrac{1}{8} = \tfrac{7}{8}$$

and $$P(B) = P(0) + P(3) = \tfrac{1}{8} + \tfrac{1}{8} = \tfrac{1}{4}$$

Example 2.2: Three horses A, B and C are in a race; A is twice as likely to win as B and B is twice as likely to win as C. What are their respective probabilities of winning, i.e. $P(A)$, $P(B)$ and $P(C)$?

Let $P(C) = p$; since B is twice as likely to win as C, $P(B) = 2p$; and since A is twice as likely to win as B, $P(A) = 2P(B) = 2(2p) = 4p$. Now the sum of the probabilities must be 1; hence

$$p + 2p + 4p = 1 \quad \text{or} \quad 7p = 1 \quad \text{or} \quad p = \tfrac{1}{7}$$

Accordingly,

$$P(A) = 4p = \tfrac{4}{7}, \quad P(B) = 2p = \tfrac{2}{7}, \quad P(C) = p = \tfrac{1}{7}$$

Question: What is the probability that B or C wins, i.e. $P(\{B, C\})$? By definition

$$P(\{B, C\}) = P(B) + P(C) = \tfrac{2}{7} + \tfrac{1}{7} = \tfrac{3}{7}$$

EQUIPROBABLE SPACES

Frequently, the physical characteristics of an experiment suggest that the various outcomes of the sample space be assigned equal probabilities. Such a finite probability space S, where each sample point has the same probability, will be called an *equiprobable space*. In particular, if S contains n points then the probability of each point is $1/n$. Furthermore, if an event A contains r points then its probability is $r \cdot \dfrac{1}{n} = \dfrac{r}{n}$. In other words,

$$P(A) = \frac{\text{number of elements in } A}{\text{number of elements in } S}$$

or

$$P(A) = \frac{\text{number of ways that the event } A \text{ can occur}}{\text{number of ways that the sample space } S \text{ can occur}}$$

We emphasize that the above formula for $P(A)$ can only be used with respect to an equiprobable space, and cannot be used in general.

The expression "at random" will be used only with respect to an equiprobable space; formally, the statement "choose a point at random from a set S" shall mean that S is an equiprobable space, i.e. that each sample point in S has the same probability.

Example 3.1: Let a card be selected at random from an ordinary deck of 52 cards. Let

$$A = \{\text{the card is a spade}\}$$

and $$B = \{\text{the card is a face card, i.e. a jack, queen or king}\}$$

We compute $P(A)$, $P(B)$ and $P(A \cap B)$. Since we have an equiprobable space,

$$P(A) = \frac{\text{number of spades}}{\text{number of cards}} = \frac{13}{52} = \frac{1}{4} \qquad P(B) = \frac{\text{number of face cards}}{\text{number of cards}} = \frac{12}{52} = \frac{3}{13}$$

$$P(A \cap B) = \frac{\text{number of spade face cards}}{\text{number of cards}} = \frac{3}{52}$$

Example 3.2: Let 2 items be chosen at random from a lot containing 12 items of which 4 are defective. Let

$$A = \{\text{both items are defective}\} \quad \text{and} \quad B = \{\text{both items are non-defective}\}$$

Find $P(A)$ and $P(B)$. Now

S can occur in $\binom{12}{2} = 66$ ways, the number of ways that 2 items can be chosen from 12 items;

A can occur in $\binom{4}{2} = 6$ ways, the number of ways that 2 defective items can be chosen from 4 defective items;

B can occur in $\binom{8}{2} = 28$ ways, the number of ways that 2 non-defective items can be chosen from 8 non-defective items.

Accordingly, $P(A) = \frac{6}{66} = \frac{1}{11}$ and $P(B) = \frac{28}{66} = \frac{14}{33}$.

THEOREMS ON FINITE PROBABILITY SPACES

The following theorem follows directly from the fact that the probability of an event is the sum of the probabilities of its points.

Theorem 17.1: The probability function P defined on the class of all events in a finite probability space satisfies the following axioms:

[P₁] For every event A, $P(A) \geq 0$.

[P₂] $P(S) = 1$.

[P₃] If events A and B are mutually exclusive, then

$$P(A \cup B) = P(A) + P(B)$$

Using mathematical induction, **[P₃]** can be generalized as follows:

Corollary 17.2: If $A_1, A_2, \ldots, A_r$ are pairwise mutually exclusive events, then

$$P(A_1 \cup A_2 \cup \cdots \cup A_r) = P(A_1) + P(A_2) + \cdots + P(A_r)$$

Using only the properties of Theorem 17.1, we can prove (see Problems 4.19-4.22).

Theorem 17.3: If $\emptyset$ is the empty set, and A and B are arbitrary events, then:

(i) $P(\emptyset) = 0$;

(ii) $P(A^c) = 1 - P(A)$;

(iii) $P(A \setminus B) = P(A) - P(A \cap B)$, i.e. $P(A \cap B^c) = P(A) - P(A \cap B)$;

(iv) $A \subset B$ implies $P(A) \leq P(B)$.

Property (ii) of the preceding theorem is very useful in many applications.

Example 4.1: Refer to Example 3.2, where 2 items are chosen at random from 12 items of which 4 are defective. Find the probability p that at least one item is defective.

Now $$C = \{\text{at least one item is defective}\}$$

is the complement of $B = \{\text{both items are non-defective}\}$

i.e. $C = B^c$; and we found that $P(B) = \frac{14}{33}$. Hence

$$p = P(C) = P(B^c) = 1 - P(B) = 1 - \tfrac{14}{33} = \tfrac{19}{33}$$

The *odds* that an event with probability p occurs is defined to be the ratio $p : (1-p)$. Thus the odds that at least one item is defective is $\frac{19}{33} : \frac{14}{33}$ or $19:14$ which is read "19 to 14".

Observe that axiom $[\mathbf{P}_3]$ gives the probability of a union of events in the case that the events are mutually exclusive, i.e. disjoint. The general formula follows:

Theorem 17.4: For any events A and B, $P(A \cup B) = P(A) + P(B) - P(A \cap B)$.

Corollary 17.5: For any events A, B and C,

$$P(A \cup B \cup C) = P(A) + P(B) + P(C) - P(A \cap B) - P(A \cap C) - P(B \cap C) + P(A \cap B \cap C)$$

Example 4.2: Of 100 students, 30 are taking mathematics, 20 are taking music and 10 are taking both mathematics and music. If a student is selected at random, find the probability p that he is taking mathematics or music.

Let A = {students taking mathematics} and B = {students taking music}. Then $A \cap B$ = {students taking mathematics and music} and $A \cup B$ = {students taking mathematics or music}.

Since the space is equiprobable, then $P(A) = \frac{30}{100} = \frac{3}{10}$, $P(B) = \frac{20}{100} = \frac{1}{5}$, $P(A \cap B) = \frac{10}{100} = \frac{1}{10}$. Thus by Theorem 17.4,

$$p = P(A \cup B) = P(A) + P(B) - P(A \cap B) = \tfrac{3}{10} + \tfrac{1}{5} - \tfrac{1}{10} = \tfrac{2}{5}$$

CLASSICAL BIRTHDAY PROBLEM

The classical birthday problem concerns the probability p that n people have distinct birthdays. In solving this problem, we assume that a person's birthday can fall on any day with the same probability.

Since there are n people and 365 different days, there are 365^n ways in which the n people can have their birthdays. On the other hand, if the n persons are to have distinct birthdays, then the first person can be born on any of the 365 days, the second person can be born on the remaining 364 days, the third person can be born on the remaining 363 days, etc. Thus there are $365 \cdot 364 \cdot 363 \cdots (365 - n + 1)$ ways the n persons can have distinct birthdays. Accordingly,

$$p = \frac{365 \cdot 364 \cdot 363 \cdots (365 - n + 1)}{365^n} = \frac{365}{365} \cdot \frac{364}{365} \cdot \frac{363}{365} \cdots \frac{365 - n + 1}{365}$$

It can be shown that for $n \geq 23$, $p < \frac{1}{2}$; in other words, amongst 23 or more people it is more likely that at least two of them have the same birthday than that they all have distinct birthdays.

Solved Problems

SAMPLE SPACES AND EVENTS

17.1. Let A and B be events. Find an expression and exhibit the Venn diagram for the event that: (i) A but not B occurs, i.e. only A occurs; (ii) either A or B, but not both occurs, i.e. exactly one of the two events occurs.

(i) Since A but not B occurs, shade the area of A outside of B as in Figure (*a*) below. Note that B^c, the complement of B, occurs since B does not occur; hence A and B^c occurs. In other words, the event is $A \cap B^c$.

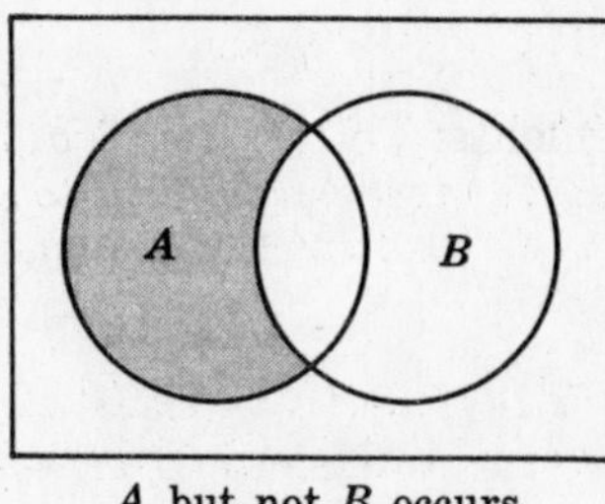

A but not B occurs.
(a)

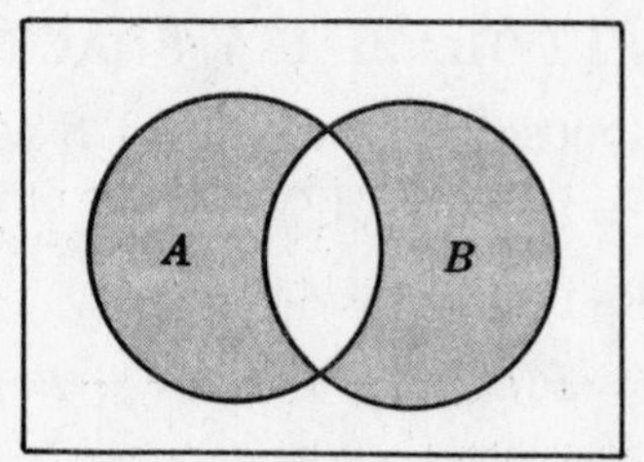

Either A or B, but not both, occurs.
(b)

(ii) Since A or B but not both occurs, shade the area of A and B except where they intersect as in Figure (b) above. The event is equivalent to A but not B occurs or B but not A occurs. Now, as in (i), A but not B is the event $A \cap B^c$, and B but not A is the event $B \cap A^c$. Thus the given event is $(A \cap B^c) \cup (B \cap A^c)$.

17.2. Let A, B and C be events. Find an expression and exhibit the Venn diagram for the event that (i) A and B but not C occurs, (ii) only A occurs.

(i) Since A and B but not C occurs, shade the intersection of A and B which lies outside of C, as in Figure (a) below. The event is $A \cap B \cap C^c$.

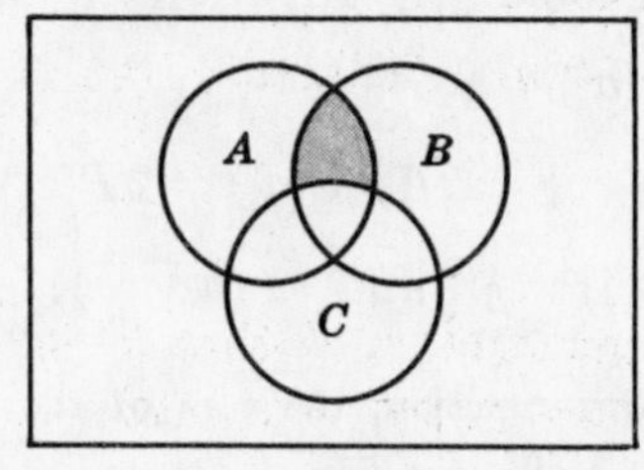

A and B but not C occurs.
(a)

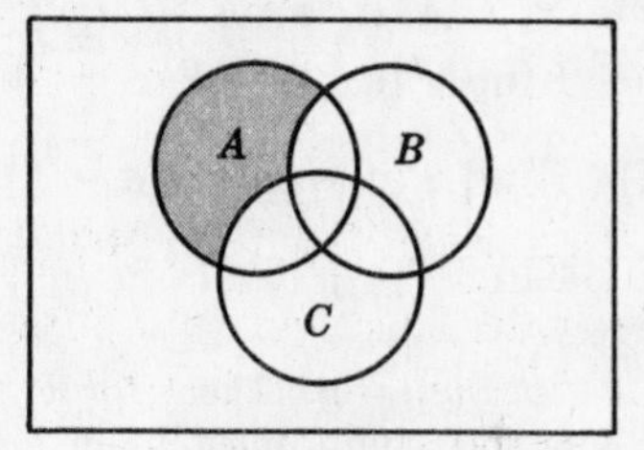

Only A occurs.
(b)

(ii) Since only A is to occur, shade the area of A which lies outside of B and of C, as in Figure (b) above. The event is $A \cap B^c \cap C^c$.

17.3. Let a coin and a die be tossed; let the sample space S consist of the twelve elements:

$$S = \{H1, H2, H3, H4, H5, H6, T1, T2, T3, T4, T5, T6\}$$

(i) Express explicitly the following events: A = {heads and an even number appears}, B = {a prime number appears}, C = {tails and an odd number appears}.

(ii) Express explicitly the event that: (a) A or B occurs, (b) B and C occurs, (c) only B occurs.

(iii) Which of the events A, B and C are mutually exclusive?

(i) To obtain A, choose those elements of S consisting of an H and an even number: $A = \{H2, H4, H6\}$.

To obtain B, choose those points in S involving a prime number: $B = \{H2, H3, H5, T2, T3, T5\}$.

To obtain C, choose those points in S consisting of a T and an odd number: $C = \{T1, T3, T5\}$.

(ii) (a) A or $B = A \cup B = \{H2, H4, H6, H3, H5, T2, T3, T5\}$

(b) B and $C = B \cap C = \{T3, T5\}$

(c) Choose those elements of B which do not lie in A or C: $B \cap A^c \cap C^c = \{H3, H5, T2\}$.

(iii) A and C are mutually exclusive since $A \cap C = \emptyset$.

FINITE PROBABILITY SPACES

17.4. Suppose a sample space S consists of 4 elements: $S = \{a_1, a_2, a_3, a_4\}$. Which function defines a probability space on S?

(i) $P(a_1) = \frac{1}{2}$, $P(a_2) = \frac{1}{3}$, $P(a_3) = \frac{1}{4}$, $P(a_4) = \frac{1}{5}$.

(ii) $P(a_1) = \frac{1}{2}$, $P(a_2) = \frac{1}{4}$, $P(a_3) = -\frac{1}{4}$, $P(a_4) = \frac{1}{2}$.

(iii) $P(a_1) = \frac{1}{2}$, $P(a_2) = \frac{1}{4}$, $P(a_3) = \frac{1}{8}$, $P(a_4) = \frac{1}{8}$.

(iv) $P(a_1) = \frac{1}{2}$, $P(a_2) = \frac{1}{4}$, $P(a_3) = \frac{1}{4}$, $P(a_4) = 0$.

(i) Since the sum of the values on the sample points is greater than one, $\frac{1}{2} + \frac{1}{3} + \frac{1}{4} + \frac{1}{5} = \frac{77}{60}$, the function does not define a probability space on S.

(ii) Since $P(a_3) = -\frac{1}{4}$, a negative number, the function does not define a probability space on S.

(iii) Since each value is non-negative, and the sum of the values is one, $\frac{1}{2} + \frac{1}{4} + \frac{1}{8} + \frac{1}{8} = 1$, the function does define a probability space on S.

(iv) The values are non-negative and add up to one; hence the function does define a probability space on S.

17.5. Let $S = \{a_1, a_2, a_3, a_4\}$, and let P be a probability function on S.

(i) Find $P(a_1)$ if $P(a_2) = \frac{1}{3}$, $P(a_3) = \frac{1}{6}$, $P(a_4) = \frac{1}{9}$.

(ii) Find $P(a_1)$ and $P(a_2)$ if $P(a_3) = P(a_4) = \frac{1}{4}$ and $P(a_1) = 2\,P(a_2)$.

(iii) Find $P(a_1)$ if $P(\{a_2, a_3\}) = \frac{2}{3}$, $P(\{a_2, a_4\}) = \frac{1}{2}$ and $P(a_2) = \frac{1}{3}$.

(i) Let $P(a_1) = p$. Then, for P to be a probability function, the sum of the probabilities on the sample points must be one: $p + \frac{1}{3} + \frac{1}{6} + \frac{1}{9} = 1$ or $p = \frac{7}{18}$.

(ii) Let $P(a_2) = p$, then $P(a_1) = 2p$. Hence $2p + p + \frac{1}{4} + \frac{1}{4} = 1$ or $p = \frac{1}{6}$. Thus $P(a_2) = \frac{1}{6}$ and $P(a_1) = \frac{1}{3}$.

(iii) Let $P(a_1) = p$.

$$P(a_3) = P(\{a_2, a_3\}) - P(a_2) = \tfrac{2}{3} - \tfrac{1}{3} = \tfrac{1}{3}$$

$$P(a_4) = P(\{a_2, a_4\}) - P(a_2) = \tfrac{1}{2} - \tfrac{1}{3} = \tfrac{1}{6}$$

Then $p + \frac{1}{3} + \frac{1}{3} + \frac{1}{6} = 1$ or $p = \frac{1}{6}$, that is, $P(a_1) = \frac{1}{6}$.

17.6. A coin is weighted so that heads is twice as likely to appear as tails. Find $P(\mathrm{T})$ and $P(\mathrm{H})$.

Let $P(\mathrm{T}) = p$; then $P(\mathrm{H}) = 2p$. Now set the sum of the probabilities equal to one: $p + 2p = 1$ or $p = \frac{1}{3}$. Thus $P(\mathrm{T}) = p = \frac{1}{3}$ and $P(\mathrm{H}) = 2p = \frac{2}{3}$.

17.7. Two men, m_1 and m_2, and three women, w_1, w_2 and w_3, are in a chess tournament. Those of the same sex have equal probabilities of winning, but each man is twice as likely to win as any woman. (i) Find the probability that a woman wins the tournament. (ii) If m_1 and w_1 are married, find the probability that one of them wins the tournament.

Set $P(w_1) = p$; then $P(w_2) = P(w_3) = p$ and $P(m_1) = P(m_2) = 2p$. Next set the sum of the probabilities of the five sample points equal to one: $p + p + p + 2p + 2p = 1$ or $p = \frac{1}{7}$.

We seek (i) $P(\{w_1, w_2, w_3\})$ and (ii) $P(\{m_1, w_1\})$. Then, by definition,

$$P(\{w_1, w_2, w_3\}) = P(w_1) + P(w_2) + P(w_3) = \tfrac{1}{7} + \tfrac{1}{7} + \tfrac{1}{7} = \tfrac{3}{7}$$

$$P(\{m_1, w_1\}) = P(m_1) + P(w_1) = \tfrac{2}{7} + \tfrac{1}{7} = \tfrac{3}{7}$$

17.8. Let a die be weighted so that the probability of a number appearing when the die is tossed is proportional to the given number (e.g. 6 has twice the probability of appearing as 3). Let $A = \{\text{even number}\}$, $B = \{\text{prime number}\}$, $C = \{\text{odd number}\}$.

(i) Describe the probability space, i.e. find the probability of each sample point.

(ii) Find $P(A)$, $P(B)$ and $P(C)$.

(iii) Find the probability that: (*a*) an even or prime number occurs; (*b*) an odd prime number occurs; (*c*) A but not B occurs.

(i) Let $P(1) = p$. Then $P(2) = 2p$, $P(3) = 3p$, $P(4) = 4p$, $P(5) = 5p$ and $P(6) = 6p$. Since the sum of the probabilities must be one, we obtain $p + 2p + 3p + 4p + 5p + 6p = 1$ or $p = 1/21$. Thus

$$P(1) = \tfrac{1}{21}, \quad P(2) = \tfrac{2}{21}, \quad P(3) = \tfrac{1}{7}, \quad P(4) = \tfrac{4}{21}, \quad P(5) = \tfrac{5}{21}, \quad P(6) = \tfrac{2}{7}$$

(ii) $P(A) = P(\{2, 4, 6\}) = \tfrac{4}{7}$, $\quad P(B) = P(\{2, 3, 5\}) = \tfrac{10}{21}$, $\quad P(C) = P(\{1, 3, 5\}) = \tfrac{3}{7}$.

(iii) (*a*) The event that an even or prime number occurs is $A \cup B = \{2, 4, 6, 3, 5\}$, or that 1 does not occur. Thus $P(A \cup B) = 1 - P(1) = \tfrac{20}{21}$.

(*b*) The event that an odd prime number occurs is $B \cap C = \{3, 5\}$. Thus $P(B \cap C) = P(\{3, 5\}) = \tfrac{8}{21}$.

(*c*) The event that A but not B occurs is $A \cap B^c = \{4, 6\}$. Hence $P(A \cap B^c) = P(\{4, 6\}) = \tfrac{10}{21}$.

17.9. Find the probability p of an event if the odds that it will occur are $a : b$, that is, "a to b".

The odds that an event with probability p occurs is the ratio $p : (1 - p)$. Hence

$$\frac{p}{1-p} = \frac{a}{b} \quad \text{or} \quad bp = a - ap \quad \text{or} \quad ap + bp = a \quad \text{or} \quad p = \frac{a}{a+b}$$

17.10. Find the probability p of an event if the odds that it will occur are "3 to 2".

$\frac{p}{1-p} = \frac{3}{2}$ from which $p = \frac{3}{5}$. We can also use the formula of the preceding problem to obtain the answer directly: $p = \frac{a}{a+b} = \frac{3}{3+2} = \frac{3}{5}$.

EQUIPROBABLE SPACES

17.11. Determine the probability p of each event:

(i) an even number appears in the toss of a fair die;

(ii) a king appears in drawing a single card from an ordinary deck of 52 cards;

(iii) at least one tail appears in the toss of three fair coins;

(iv) a white marble appears in drawing a single marble from an urn containing 4 white, 3 red and 5 blue marbles.

(i) The event can occur in three ways (a 2, 4 or 6) out of 6 equally likely cases; hence $p = \frac{3}{6} = \frac{1}{2}$.

(ii) There are 4 kings among the 52 cards; hence $p = \frac{4}{52} = \frac{1}{13}$.

(iii) If we consider the coins distinguished, then there are 8 equally likely cases: HHH, HHT, HTH, HTT, THH, THT, TTH, TTT. Only the first case is not favorable to the given event; hence $p = \frac{7}{8}$.

(iv) There are $4 + 3 + 5 = 12$ marbles, of which 4 are white; hence $p = \frac{4}{12} = \frac{1}{3}$.

17.12. Two cards are drawn at random from an ordinary deck of 52 cards. Find the probability p that (i) both are spades, (ii) one is a spade and one is a heart.

There are $\binom{52}{2} = 1326$ ways to draw 2 cards from 52 cards.

(i) There are $\binom{13}{2} = 78$ ways to draw 2 spades from 13 spades; hence

$$p = \frac{\text{number of ways 2 spades can be drawn}}{\text{number of ways 2 cards can be drawn}} = \frac{78}{1326} = \frac{1}{17}$$

(ii) Since there are 13 spades and 13 hearts, there are $13 \cdot 13 = 169$ ways to draw a spade and a heart; hence $p = \frac{169}{1326} = \frac{13}{102}$.

17.13. Three light bulbs are chosen at random from 15 bulbs of which 5 are defective. Find the probability p that (i) none is defective, (ii) exactly one is defective, (iii) at least one is defective.

There are $\binom{15}{3} = 455$ ways to choose 3 bulbs from the 15 bulbs.

(i) Since there are $15 - 5 = 10$ non-defective bulbs, there are $\binom{10}{3} = 120$ ways to choose 3 non-defective bulbs. Thus $p = \frac{120}{455} = \frac{24}{91}$.

(ii) There are 5 defective bulbs and $\binom{10}{2} = 45$ different pairs of non-defective bulbs; hence there are $5 \cdot 45 = 225$ ways to choose 3 bulbs of which one is defective. Thus $p = \frac{225}{455} = \frac{45}{91}$.

(iii) The event that at least one is defective is the complement of the event that none are defective which has, by (i), probability $\frac{24}{91}$. Hence $p = 1 - \frac{24}{91} = \frac{67}{91}$.

17.14. Two cards are selected at random from 10 cards numbered 1 to 10. Find the probability p that the sum is odd if (i) the two cards are drawn together, (ii) the two cards are drawn one after the other without replacement, (iii) the two cards are drawn one after the other with replacement.

(i) There are $\binom{10}{2} = 45$ ways to select 2 cards out of 10. The sum is odd if one number is odd and the other is even. There are 5 even numbers and 5 odd numbers; hence there are $5 \cdot 5 = 25$ ways of choosing an even and an odd number. Thus $p = \frac{25}{45} = \frac{5}{9}$.

(ii) There are $10 \cdot 9 = 90$ ways to draw two cards one after the other without replacement. There are $5 \cdot 5 = 25$ ways to draw an even number and then an odd number, and $5 \cdot 5 = 25$ ways to draw an odd number and then an even number; hence $p = \frac{25+25}{90} = \frac{50}{90} = \frac{5}{9}$.

(iii) There are $10 \cdot 10 = 100$ ways to draw two cards one after the other with replacement. As in (ii), there are $5 \cdot 5 = 25$ ways to draw an even number and then an odd number, and $5 \cdot 5 = 25$ ways to draw an odd number and then an even number; hence $p = \frac{25+25}{100} = \frac{50}{100} = \frac{1}{2}$.

17.15. An ordinary deck of 52 cards is dealt out at random to 4 persons (called North, South, East and West) with each person receiving 13 cards. Find the probability p of each event (answers may be left in combinatorial and factorial notation):

(i) North receives all 4 aces.

(ii) North receives no aces.

(iii) North receives 5 spades, 5 clubs, 2 hearts and 1 diamond.

(iv) North and South receive between them all the spades.

(v) North and South receive between them all the aces and kings.

(vi) Each person receives an ace.

(i) North can be dealt 13 cards in $\binom{52}{13}$ ways. If he receives the 4 aces, then he can be dealt the other 9 cards from the remaining 48 cards in $\binom{48}{9}$ ways. Thus

$$p = \frac{\binom{48}{9}}{\binom{52}{13}} = \frac{\frac{48!}{9!\,39!}}{\frac{52!}{13!\,39!}} = \frac{13 \cdot 12 \cdot 11 \cdot 10}{52 \cdot 51 \cdot 50 \cdot 49} = \frac{11}{4165}$$

(ii) North can be dealt 13 cards in $\binom{52}{13}$ ways. If he receives no aces, then he can be dealt 13 cards from the remaining 48 cards in $\binom{48}{13}$ ways. Thus

$$p = \frac{\binom{48}{13}}{\binom{52}{13}} = \frac{\frac{48!}{13!\,35!}}{\frac{52!}{13!\,39!}} = \frac{39 \cdot 38 \cdot 37 \cdot 36}{52 \cdot 51 \cdot 50 \cdot 49}$$

(iii) North can be dealt 13 cards in $\binom{52}{13}$ ways. He can be dealt 5 spades in $\binom{13}{5}$ ways, 5 clubs in $\binom{13}{5}$ ways, 2 hearts in $\binom{13}{2}$ ways, and 1 diamond in $\binom{13}{1} = 13$ ways. Hence $p = \frac{\binom{13}{5}\binom{13}{5}\binom{13}{2}\binom{13}{1}}{\binom{52}{13}}$.

(iv) North and South can be dealt 26 cards between them in $\binom{52}{26}$ ways. If they receive the 13 spades, then they can be dealt 13 other cards from the remaining 39 cards in $\binom{39}{13}$ ways. Hence $p = \binom{39}{13} / \binom{52}{13}$.

(v) North and South can be dealt 26 cards between them in $\binom{52}{26}$ ways. If they receive the 4 aces and the 4 kings, then they can be dealt the other $26 - 8 = 18$ cards from the remaining $52 - 8 = 44$ cards in $\binom{44}{18}$ ways. Thus $p = \binom{44}{18} / \binom{52}{26}$.

(vi) The 52 cards can be distributed among the 4 persons in $\frac{52!}{13!\,13!\,13!\,13!}$ ways (an ordered partition). Now each person can get an ace in $4!$ ways. The remaining 48 cards can be distributed among the 4 persons in $\frac{48!}{12!\,12!\,12!\,12!}$ ways. Thus

$$p = \frac{4!\frac{48!}{12!\,12!\,12!\,12!}}{\frac{52!}{13!\,13!\,13!\,13!}} = \frac{4!\,13^4}{52 \cdot 51 \cdot 50 \cdot 49} = \frac{13^3}{17 \cdot 25 \cdot 49}$$

17.16. Six married couples are standing in a room.

(i) If 2 people are chosen at random, find the probability p that (*a*) they are married, (*b*) one is male and one is female.

(ii) If 4 people are chosen at random, find the probability p that (*a*) 2 married couples are chosen, (*b*) no married couple is among the 4, (*c*) exactly one married couple is among the 4.

(iii) If the 12 people are divided into six pairs, find the probability p that (*a*) each pair is married, (*b*) each pair contains a male and a female.

(i) There are $\binom{12}{2} = 66$ ways to choose 2 people from the 12 people.

(a) There are 6 married couples; hence $p = \frac{6}{66} = \frac{1}{11}$.

(b) There are 6 ways to choose a male and 6 ways to choose a female; hence $p = \frac{6\cdot 6}{66} = \frac{6}{11}$.

(ii) There are $\binom{12}{4} = 495$ ways to choose 4 people from the 12 people.

(a) There are $\binom{6}{2} = 15$ ways to choose 2 couples from the 6 couples; hence $p = \frac{15}{495} = \frac{1}{33}$.

(b) The 4 persons come from 4 different couples. There are $\binom{6}{4} = 15$ ways to choose 4 couples from the 6 couples, and there are 2 ways to choose one person from each couple. Hence $p = \frac{2\cdot 2\cdot 2\cdot 2\cdot 15}{495} = \frac{16}{33}$.

(c) This event is mutually disjoint from the preceding two events (which are also mutually disjoint) and at least one of these events must occur. Hence $p + \frac{1}{33} + \frac{16}{33} = 1$ or $p = \frac{16}{33}$.

(iii) There are $\frac{12!}{2!\,2!\,2!\,2!\,2!\,2!} = \frac{12!}{2^6}$ ways to partition the 12 people into 6 ordered cells with 2 people in each.

(a) The 6 couples can be placed into the 6 ordered cells in 6! ways. Hence $p = \frac{6!}{12!/2^6} = \frac{1}{10,395}$.

(b) The six men can be placed one each into the 6 cells in 6! ways, and the 6 women can be placed one each into the 6 cells in 6! ways. Hence $p = \frac{6!\,6!}{12!/2^6} = \frac{16}{231}$.

17.17. If 5 balls are placed in 5 cells at random, find the probability p that exactly one cell is empty.

There are 5^5 ways to place the 5 balls in the 5 cells. If exactly one cell is empty, then another cell contains 2 balls and each of the remaining 3 cells contains one ball. There are 5 ways to select the empty cell, then 4 ways to select the cell containing 2 balls, and $\binom{5}{2} = 10$ ways to select 2 balls to go into this cell. Finally there are 3! ways to distribute the remaining 3 balls among the remaining 3 cells. Thus $p = \frac{5\cdot 4\cdot 10\cdot 3!}{5^5} = \frac{48}{125}$.

17.18. A class contains 10 men and 20 women of which half the men and half the women have brown eyes. Find the probability p that a person chosen at random is a man or has brown eyes.

Let A = {person is a man} and B = {person has brown eyes}. We seek $P(A \cup B)$.

Then $P(A) = \frac{10}{30} = \frac{1}{3}$, $P(B) = \frac{15}{30} = \frac{1}{2}$, $P(A \cap B) = \frac{5}{30} = \frac{1}{6}$. Thus, by Theorem 17.4,

$$p = P(A \cup B) = P(A) + P(B) - P(A \cap B) = \tfrac{1}{3} + \tfrac{1}{2} - \tfrac{1}{6} = \tfrac{2}{3}$$

PROOFS OF THEOREMS

17.19. Prove Theorem 17.3(i): If $\emptyset$ is the empty set, then $P(\emptyset) = 0$.

Let A be any event; then A and $\emptyset$ are disjoint and $A \cup \emptyset = A$.

Thus $P(A) = P(A \cup \emptyset) = P(A) + P(\emptyset)$, from which our result follows.

17.20. Prove Theorem 17.3(ii): For any event A, $P(A^c) = 1 - P(A)$.

The sample space S can be decomposed into the mutually exclusive events A and A^c: $S = A \cup A^c$.

Thus $1 = P(S) = P(A \cup A^c) = P(A) + P(A^c)$, from which our result follows.

17.21. Prove Theorem 17.3(iii):

$$P(A \setminus B) = P(A) - P(A \cap B)$$

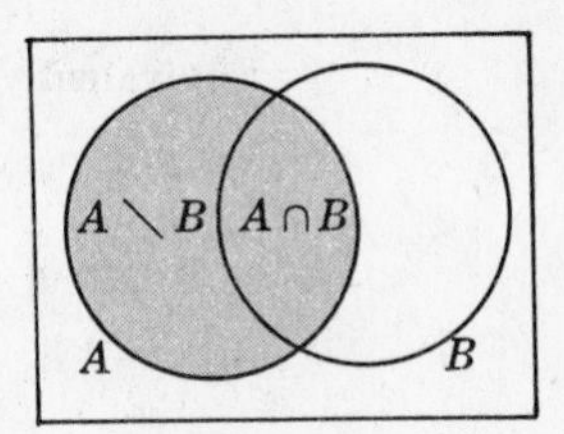

A is shaded.

A can be decomposed into the mutually exclusive events $A \setminus B$ and $A \cap B$: $A = (A \setminus B) \cup (A \cap B)$. Thus

$$P(A) = P(A \setminus B) + P(A \cap B)$$

from which our result follows.

17.22. Prove Theorem 17.3(iv):

$$\text{If } A \subset B, \text{ then } P(A) \leq P(B).$$

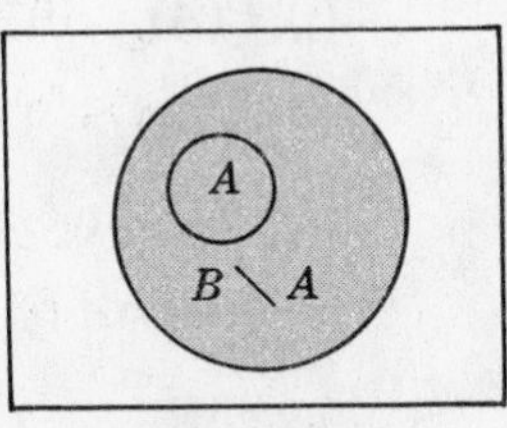

B is shaded.

If $A \subset B$, then B can be decomposed into the mutually exclusive events A and $B \setminus A$. Thus

$$P(B) = P(A) + P(B \setminus A)$$

The result now follows from the fact that $P(B \setminus A) \geq 0$.

17.23. Prove Theorem 17.4: For any events A and B,

$$P(A \cup B) = P(A) + P(B) - P(A \cap B)$$

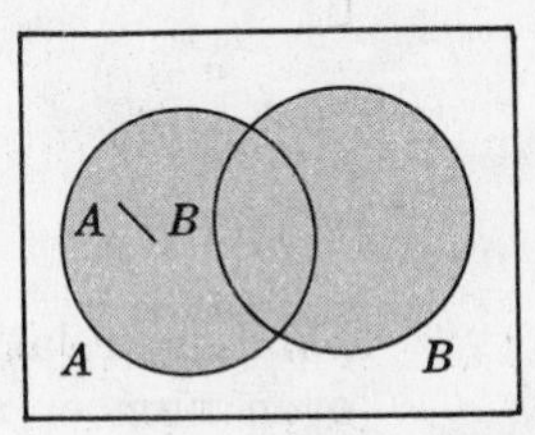

$A \cup B$ is shaded.

Note that $A \cup B$ can be decomposed into the mutually exclusive events $A \setminus B$ and B: $A \cup B = (A \setminus B) \cup B$. Thus, using Theorem 17.3(iii), we have

$$\begin{aligned} P(A \cup B) &= P(A \setminus B) + P(B) \\ &= P(A) - P(A \cap B) + P(B) \\ &= P(A) + P(B) - P(A \cap B) \end{aligned}$$

17.24. Prove Corollary 17.5: For any events A, B and C,

$$P(A \cup B \cup C) = P(A) + P(B) + P(C) - P(A \cap B) - P(A \cap C) - P(B \cap C) + P(A \cap B \cap C)$$

Let $D = B \cup C$. Then $A \cap D = A \cap (B \cup C) = (A \cap B) \cup (A \cap C)$ and

$$P(A \cap D) = P(A \cap B) + P(A \cap C) - P(A \cap B \cap A \cap C) = P(A \cap B) + P(A \cap C) - P(A \cap B \cap C)$$

Thus

$$\begin{aligned} P(A \cup B \cup C) &= P(A \cup D) = P(A) + P(D) - P(A \cap D) \\ &= P(A) + P(B) + P(C) - P(B \cap C) - [P(A \cap B) + P(A \cap C) - P(A \cap B \cap C)] \\ &= P(A) + P(B) + P(C) - P(B \cap C) - P(A \cap B) - P(A \cap C) + P(A \cap B \cap C) \end{aligned}$$

MISCELLANEOUS PROBLEMS

17.25. Let A and B be events with $P(A) = \frac{3}{8}$, $P(B) = \frac{1}{2}$ and $P(A \cap B) = \frac{1}{4}$. Find (i) $P(A \cup B)$, (ii) $P(A^c)$ and $P(B^c)$, (iii) $P(A^c \cap B^c)$, (iv) $P(A^c \cup B^c)$, (v) $P(A \cap B^c)$, (vi) $P(B \cap A^c)$.

(i) $P(A \cup B) = P(A) + P(B) - P(A \cap B) = \frac{3}{8} + \frac{1}{2} - \frac{1}{4} = \frac{5}{8}$

(ii) $P(A^c) = 1 - P(A) = 1 - \frac{3}{8} = \frac{5}{8}$ and $P(B^c) = 1 - P(B) = 1 - \frac{1}{2} = \frac{1}{2}$

(iii) Using De Morgan's Law, $(A \cup B)^c = A^c \cap B^c$, we have

$$P(A^c \cap B^c) = P((A \cup B)^c) = 1 - P(A \cup B) = 1 - \frac{5}{8} = \frac{3}{8}$$

(iv) Using De Morgan's Law, $(A \cap B)^c = A^c \cup B^c$, we have

$$P(A^c \cup B^c) = P((A \cap B)^c) = 1 - P(A \cap B) = 1 - \frac{1}{4} = \frac{3}{4}$$

Equivalently,

$$P(A^c \cup B^c) = P(A^c) + P(B^c) - P(A^c \cap B^c) = \tfrac{5}{8} + \tfrac{1}{2} - \tfrac{3}{8} = \tfrac{3}{4}$$

(v) $P(A \cap B^c) = P(A \setminus B) = P(A) - P(A \cap B) = \tfrac{3}{8} - \tfrac{1}{4} = \tfrac{1}{8}$

(vi) $P(B \cap A^c) = P(B) - P(A \cap B) = \tfrac{1}{2} - \tfrac{1}{4} = \tfrac{1}{4}$

17.26. Let A and B be events with $P(A \cup B) = \tfrac{3}{4}$, $P(A^c) = \tfrac{2}{3}$ and $P(A \cap B) = \tfrac{1}{4}$. Find (i) $P(A)$, (ii) $P(B)$, (iii) $P(A \cap B^c)$.

(i) $P(A) = 1 - P(A^c) = 1 - \tfrac{2}{3} = \tfrac{1}{3}$

(ii) Substitute in $P(A \cup B) = P(A) + P(B) - P(A \cap B)$ to obtain $\tfrac{3}{4} = \tfrac{1}{3} + P(B) - \tfrac{1}{4}$ or $P(B) = \tfrac{2}{3}$.

(iii) $P(A \cap B^c) = P(A) - P(A \cap B) = \tfrac{1}{3} - \tfrac{1}{4} = \tfrac{1}{12}$

17.27. A die is tossed 100 times. The following table lists the six numbers and frequency with which each number appeared:

Number	1	2	3	4	5	6
Frequency	14	17	20	18	15	16

Find the relative frequency f of the event (i) a 3 appears, (ii) a 5 appears, (iii) an even number appears, (iv) a prime appears.

The relative frequency $f = \dfrac{\text{number of successes}}{\text{total number of trials}}$.

(i) $f = \frac{20}{100} = .20$ (ii) $f = \frac{15}{100} = .15$ (iii) $f = \frac{17+18+16}{100} = .51$ (iv) $f = \frac{17+20+15}{100} = .52$

17.28. Let $S = \{a_1, a_2, \ldots, a_s\}$ and $T = \{b_1, b_2, \ldots, b_t\}$ be finite probability spaces. Let the number $p_{ij} = P(a_i)P(b_j)$ be assigned to the ordered pair (a_i, b_j) in the product set $S \times T = \{(s,t): s \in S,\ t \in T\}$. Show that the p_{ij} define a probability space on $S \times T$, i.e. that the p_{ij} are non-negative and add up to one. (This is called the *product probability space*. We emphasize that this is not the only probability function that can be defined on the product set $S \times T$.)

Since $P(a_i), P(b_j) \geq 0$, for each i and each j, $p_{ij} = P(a_i)P(b_j) \geq 0$. Furthermore,

$$\begin{aligned} & p_{11} + p_{12} + \cdots + p_{1t} + p_{21} + p_{22} + \cdots + p_{2t} + \cdots + p_{s1} + p_{s2} + \cdots + p_{st} \\ &= P(a_1)P(b_1) + \cdots + P(a_1)P(b_t) + \cdots + P(a_s)P(b_1) + \cdots + P(a_s)P(b_t) \\ &= P(a_1)[P(b_1) + \cdots + P(b_t)] + \cdots + P(a_s)[P(b_1) + \cdots + P(b_t)] \\ &= P(a_1)\cdot 1 + \cdots + P(a_s)\cdot 1 = P(a_1) + \cdots + P(a_s) = 1 \end{aligned}$$

Supplementary Problems

SAMPLE SPACES AND EVENTS

17.29. Let A and B be events. Find an expression and exhibit the Venn diagram for the event that (i) A or not B occurs, (ii) neither A nor B occurs.

17.30. Let A, B and C be events. Find an expression and exhibit the Venn diagram for the event that (i) exactly one of the three events occurs, (ii) at least two of the events occurs, (iii) none of the events occurs, (iv) A or B, but not C, occurs.

17.31. Let a penny, a dime and a die be tossed.

(i) Describe a suitable sample space S.

(ii) Express explicitly the following events: A = {two heads and an even number appears}, B = {a 2 appears}, C = {exactly one head and a prime number appears}.

(iii) Express explicitly the event that (*a*) A and B occur, (*b*) only B occurs, (*c*) B or C occur.

FINITE PROBABILITY SPACES

17.32. Which function defines a probability space on $S = \{a_1, a_2, a_3\}$?

(i) $P(a_1) = \frac{1}{4}$, $P(a_2) = \frac{1}{3}$, $P(a_3) = \frac{1}{2}$ (iii) $P(a_1) = \frac{1}{6}$, $P(a_2) = \frac{1}{3}$, $P(a_3) = \frac{1}{2}$

(ii) $P(a_1) = \frac{2}{3}$, $P(a_2) = -\frac{1}{3}$, $P(a_3) = \frac{2}{3}$ (iv) $P(a_1) = 0$, $P(a_2) = \frac{1}{3}$, $P(a_3) = \frac{2}{3}$

17.33. Let P be a probability function on $S = \{a_1, a_2, a_3\}$. Find $P(a_1)$ if (i) $P(a_2) = \frac{1}{3}$ and $P(a_3) = \frac{1}{4}$, (ii) $P(a_1) = 2\,P(a_2)$ and $P(a_3) = \frac{1}{4}$, (iii) $P(\{a_2, a_3\}) = 2\,P(a_1)$, (iv) $P(a_3) = 2\,P(a_2)$ and $P(a_2) = 3\,P(a_1)$.

17.34. Let A and B be events with $P(A \cup B) = \frac{7}{8}$, $P(A \cap B) = \frac{1}{4}$ and $P(A^c) = \frac{5}{8}$. Find $P(A)$, $P(B)$ and $P(A \cap B^c)$.

17.35. Let A and B be events with $P(A) = \frac{1}{2}$, $P(A \cup B) = \frac{3}{4}$ and $P(B^c) = \frac{5}{8}$. Find $P(A \cap B)$, $P(A^c \cap B^c)$, $P(A^c \cup B^c)$ and $P(B \cap A^c)$.

17.36. A coin is weighted so that heads is three times as likely to appear as tails. Find $P(\text{H})$ and $P(\text{T})$.

17.37. Three students A, B and C are in a swimming race. A and B have the same probability of winning and each is twice as likely to win as C. Find the probability that B or C wins.

17.38. A die is weighted so that the even numbers have the same chance of appearing, the odd numbers have the same chance of appearing, and each even number is twice as likely to appear as any odd number. Find the probability that (i) an even number appears, (ii) a prime number appears, (iii) an odd number appears, (iv) an odd prime number appears.

17.39. Find the probability of an event if the odds that it will occur are (i) 2 to 1, (ii) 5 to 11.

17.40. In a swimming race, the odds that A will win are 2 to 3 and the odds that B will win are 1 to 4. Find the probability p and the odds that A or B win the race.

EQUIPROBABLE SPACES

17.41. A class contains 5 freshmen, 4 sophomores, 8 juniors and 3 seniors. A student is chosen at random to represent the class. Find the probability that the student is (i) a sophomore, (ii) a senior, (iii) a junior or senior.

17.42. One card is selected at random from 50 cards numbered 1 to 50. Find the probability that the number on the card is (i) divisible by 5, (ii) prime, (iii) ends in the digit 2.

17.43. Of 10 girls in a class, 3 have blue eyes. If two of the girls are chosen at random, what is the probability that (i) both have blue eyes, (ii) neither has blue eyes, (iii) at least one has blue eyes?

17.44. Three bolts and three nuts are put in a box. If two parts are chosen at random, find the probability that one is a bolt and one a nut.

17.45. Ten students, $A, B, \ldots$, are in a class. If a committee of 3 is chosen at random from the class, find the probability that (i) A belongs to the committee, (ii) B belongs to the committee, (iii) A and B belong to the committee, (iv) A or B belongs to the committee.

17.46. A class consists of 6 girls and 10 boys. If a committee of 3 is chosen at random from the class, find the probability that (i) 3 boys are selected, (ii) exactly 2 boys are selected, (iii) at least one boy is selected, (iv) exactly 2 girls are selected.

17.47. A pair of fair dice is tossed. Find the probability that the maximum of the two numbers is greater than 4.

17.48. Of 120 students, 60 are studying French, 50 are studying Spanish, and 20 are studying French and Spanish. If a student is chosen at random, find the probability that the student (i) is studying French or Spanish, (ii) is studying neither French nor Spanish.

17.49. Three boys and 3 girls sit in a row. Find the probability that (i) the 3 girls sit together, (ii) the boys and girls sit in alternate seats.

17.50. A die is tossed 50 times. The following table gives the six numbers and their frequency of occurrence:

Number	1	2	3	4	5	6
Frequency	7	9	8	7	9	10

Find the relative frequency of the event (i) a 4 appears, (ii) an odd number appears, (iii) a prime number appears.

Answers to Supplementary Problems

17.29. (i) $A \cup B^c$, (ii) $(A \cup B)^c$

17.30. (i) $(A \cap B^c \cap C^c) \cup (B \cap A^c \cap C^c) \cup (C \cap A^c \cap B^c)$ (iii) $(A \cup B \cup C)^c$
(ii) $(A \cap B) \cup (A \cap C) \cup (B \cap C)$ (iv) $(A \cup B) \cap C^c$

17.31. (i) S = {HH1, HH2, HH3, HH4, HH5, HH6, HT1, HT2, HT3, HT4, HT5, HT6, TH1, TH2, TH3, TH4, TH5, TH6, TT1, TT2, TT3, TT4, TT5, TT6}

(ii) A = {HH2, HH4, HH6}, B = {HH2, HT2, TH2, TT2}, C = {HT2, TH2, HT3, TH3, HT5, TH5}

(iii) (a) $A \cap B$ = {HH2}
(b) $B \setminus (A \cup C)$ = {TT2}
(c) $B \cup C$ = {HH2, HT2, TH2, TT2, HT3, TH3, HT5, TH5}

17.32. (i) no, (ii) no, (iii) yes, (iv) yes

17.33. (i) $\frac{5}{12}$, (ii) $\frac{1}{2}$, (iii) $\frac{1}{3}$, (iv) $\frac{1}{10}$

17.34. $P(A) = \frac{3}{8}$, $P(B) = \frac{3}{4}$, $P(A \cap B^c) = \frac{1}{8}$

17.35. $P(A \cap B) = \frac{1}{8}$, $P(A^c \cap B^c) = \frac{1}{4}$, $P(A^c \cup B^c) = \frac{7}{8}$, $P(B \cap A^c) = \frac{1}{4}$

17.36. $P(H) = \frac{3}{4}$, $P(T) = \frac{1}{4}$

17.37. $\frac{3}{5}$

17.38. (i) $\frac{2}{3}$, (ii) $\frac{4}{9}$, (iii) $\frac{1}{3}$, (iv) $\frac{2}{9}$

17.39. (i) $\frac{2}{3}$, (ii) $\frac{5}{16}$

17.40. $p = \frac{3}{5}$; the odds are 3 to 2.

17.41. (i) $\frac{1}{5}$, (ii) $\frac{3}{20}$, (iii) $\frac{11}{20}$

17.42. (i) $\frac{1}{5}$, (ii) $\frac{3}{10}$, (iii) $\frac{1}{10}$

17.43. (i) $\frac{1}{15}$, (ii) $\frac{7}{15}$, (iii) $\frac{8}{15}$

17.44. $\frac{3}{5}$

17.45. (i) $\frac{3}{10}$, (ii) $\frac{3}{10}$, (iii) $\frac{1}{15}$, (iv) $\frac{8}{15}$

17.46. (i) $\frac{3}{14}$, (ii) $\frac{27}{56}$, (iii) $\frac{27}{28}$, (iv) $\frac{15}{56}$

17.47. $\frac{5}{9}$

17.48. (i) $\frac{3}{4}$, (ii) $\frac{1}{4}$

17.49. (i) $\frac{1}{5}$, (ii) $\frac{1}{10}$

17.50. (i) $\frac{7}{50}$, (ii) $\frac{24}{50}$, (iii) $\frac{26}{50}$

Chapter 18

Conditional Probability. Independence

CONDITIONAL PROBABILITY

Let E be an arbitrary event in a sample space S with $P(E) > 0$. The probability that an event A occurs once E has occurred or, in other words, the conditional probability of A given E, written $P(A \mid E)$, is defined as follows:

$$P(A \mid E) = \frac{P(A \cap E)}{P(E)}$$

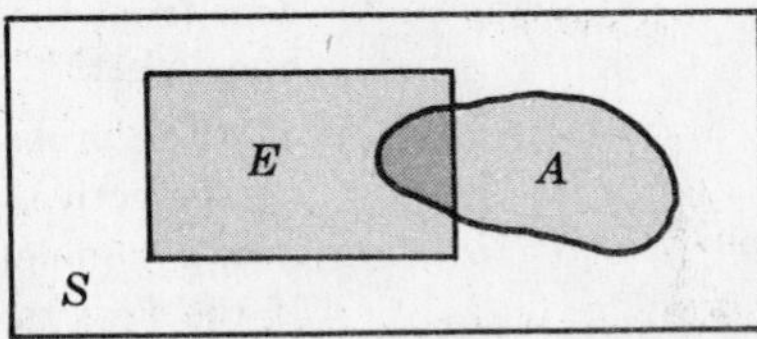

As seen in the adjoining Venn diagram, $P(A \mid E)$ in a certain sense measures the relative probability of A with respect to the reduced space E.

In particular, if S is an equiprobable space and we let $|A|$ denote the number of elements in the event A, then

$$P(A \cap E) = \frac{|A \cap E|}{|S|}, \quad P(E) = \frac{|E|}{|S|} \quad \text{and so} \quad P(A \mid E) = \frac{P(A \cap E)}{P(E)} = \frac{|A \cap E|}{|E|}$$

In other words,

Theorem 18.1: If S is an equiprobable space and A and E are events of S, then

$$P(A \mid E) = \frac{\text{number of elements in } A \cap E}{\text{number of elements in } E}$$

or

$$P(A \mid E) = \frac{\text{number of ways } A \text{ and } E \text{ can occur}}{\text{number of ways } E \text{ can occur}}$$

Example 1.1: Let a pair of fair dice be tossed. If the sum is 6, find the probability that one of the dice is a 2. In other words, if

$$E = \{\text{sum is } 6\} = \{(1,5), (2,4), (3,3), (4,2), (5,1)\}$$

and

$$A = \{\text{a 2 appears on at least one die}\}$$

find $P(A \mid E)$.

Now E consists of five elements and two of them, $(2,4)$ and $(4,2)$, belong to A: $A \cap E = \{(2,4), (4,2)\}$. Then $P(A \mid E) = \frac{2}{5}$.

On the other hand, since A consists of eleven elements,

$$A = \{(2,1), (2,2), (2,3), (2,4), (2,5), (2,6), (1,2), (3,2), (4,2), (5,2), (6,2)\}$$

and S consists of 36 elements, $P(A) = \frac{11}{36}$.

Example 1.2: A couple has two children; the sample space is $S = \{bb, bg, gb, gg\}$ with probability $\frac{1}{4}$ for each point. Find the probability p that both children are boys if (i) it is known that one of the children is a boy, (ii) it is known that the older child is a boy.

(i) Here the reduced space consists of three elements, $\{bb, bg, gb\}$; hence $p = \frac{1}{3}$.

(ii) Here the reduced space consists of only two elements, $\{bb, bg\}$; hence $p = \frac{1}{2}$.

MULTIPLICATION THEOREM FOR CONDITIONAL PROBABILITY

If we cross multiply the above equation defining conditional probability and use the fact that $A \cap E = E \cap A$, we obtain the following useful formula.

Theorem 18.2: $P(E \cap A) = P(E)\,P(A \mid E)$

This theorem can be extended by induction as follows:

Corollary 18.3:
$$P(A_1 \cap A_2 \cap \cdots \cap A_n) = P(A_1)\,P(A_2 \mid A_1)\,P(A_3 \mid A_1 \cap A_2) \\ P(A_4 \mid A_1 \cap A_2 \cap A_3) \cdots P(A_n \mid A_1 \cap A_2 \cap \cdots \cap A_{n-1})$$

We now apply the above theorem which is called, appropriately, the multiplication theorem.

Example 2.1: A lot contains 12 items of which 4 are defective. Three items are drawn at random from the lot one after the other. Find the probability p that all three are non-defective.

The probability that the first item is non-defective is $\frac{8}{12}$ since 8 of 12 items are non-defective. If the first item is non-defective, then the probability that the next item is non-defective is $\frac{7}{11}$ since only 7 of the remaining 11 items are non-defective. If the first two items are non-defective, then the probability that the last item is non-defective is $\frac{6}{10}$ since only 6 of the remaining 10 items are now non-defective. Thus by the multiplication theorem,

$$p = \frac{8}{12} \cdot \frac{7}{11} \cdot \frac{6}{10} = \frac{14}{55}$$

FINITE STOCHASTIC PROCESSES AND TREE DIAGRAMS

A (finite) sequence of experiments in which each experiment has a finite number of outcomes with given probabilities is called a *(finite) stochastic process*. A convenient way of describing such a process and computing the probability of any event is by a *tree diagram* as illustrated below; the multiplication theorem of the previous section is used to compute the probability that the result represented by any given path of the tree does occur.

Example 3.1: We are given three boxes as follows:

Box I has 10 light bulbs of which 4 are defective.

Box II has 6 light bulbs of which 1 is defective.

Box III has 8 light bulbs of which 3 are defective.

We select a box at random and then draw a bulb at random. What is the probability p that the bulb is defective?

Here we perform a sequence of two experiments:

(i) select one of the three boxes;

(ii) select a bulb which is either defective (D) or non-defective (N).

The following tree diagram describes this process and gives the probability of each branch of the tree:

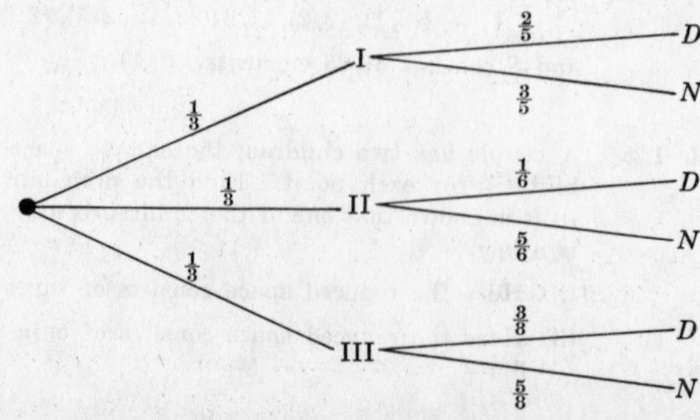

The probability that any particular path of the tree occurs is, by the multiplication theorem, the product of the probabilities of each branch of the path, e.g., the probability of selecting box I and then a defective bulb is $\frac{1}{3} \cdot \frac{2}{5} = \frac{2}{15}$.

Now since there are three mutually exclusive paths which lead to a defective bulb, the sum of the probabilities of these paths is the required probability:

$$p = \frac{1}{3} \cdot \frac{2}{5} + \frac{1}{3} \cdot \frac{1}{6} + \frac{1}{3} \cdot \frac{3}{8} = \frac{113}{360}$$

Example 3.2: A coin, weighted so that $P(\mathrm{H}) = \frac{2}{3}$ and $P(\mathrm{T}) = \frac{1}{3}$, is tossed. If heads appears, then a number is selected at random from the numbers 1 through 9; if tails appears, then a number is selected at random from the numbers 1 through 5. Find the probability p that an even number is selected.

The tree diagram with respective probabilities is

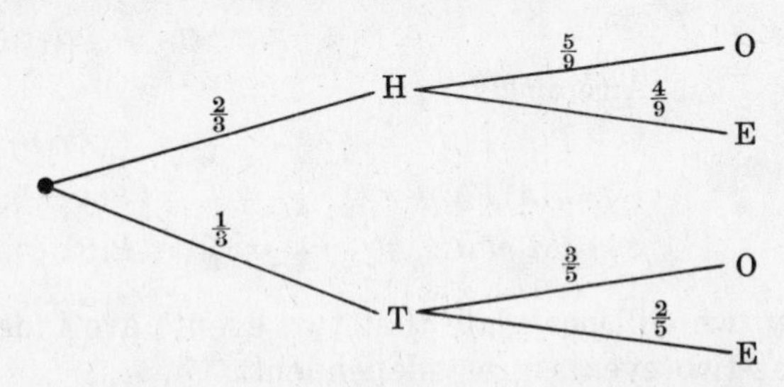

Note that the probability of selecting an even number from the numbers 1 through 9 is $\frac{4}{9}$ since there are 4 even numbers out of the 9 numbers, whereas the probability of selecting an even number from the numbers 1 through 5 is $\frac{2}{5}$ since there are 2 even numbers out of the 5 numbers. Two of the paths lead to an even number: HE and TE. Thus

$$p = P(\mathrm{E}) = \frac{2}{3} \cdot \frac{4}{9} + \frac{1}{3} \cdot \frac{2}{5} = \frac{58}{135}$$

Example 3.3: Consider the stochastic process of the preceding example. If an even number was selected, what is the probability that tails appeared on the coin? In other words, we seek the conditional probability of T given E: $P(\mathrm{T} \mid \mathrm{E})$.

Now tails and an even number can occur only on the bottom path which has probability $\frac{1}{3} \cdot \frac{2}{5} = \frac{2}{15}$; that is, $P(\mathrm{T} \cap \mathrm{E}) = \frac{2}{15}$. Furthermore, by the preceding example, $P(\mathrm{E}) = \frac{58}{135}$. Thus by the definition of conditional probability,

$$P(\mathrm{T} \mid \mathrm{E}) = \frac{P(\mathrm{T} \cap \mathrm{E})}{P(\mathrm{E})} = \frac{2/15}{58/135} = \frac{9}{29}$$

In other words, we divide the probability of the successful path by the probability of the reduced sample space.

INDEPENDENCE

An event B is said to be *independent* of an event A if the probability that B occurs is not influenced by whether A has or has not occurred. In other words, if the probability of B equals the conditional probability of B given A: $P(B) = P(B \mid A)$. Now substituting $P(B)$ for $P(B \mid A)$ in the multiplication theorem $P(A \cap B) = P(A)\,P(B \mid A)$, we obtain

$$P(A \cap B) = P(A)\,P(B)$$

We formally use the above equation as our definition of independence.

Definition: Events A and B are independent if $P(A \cap B) = P(A)\,P(B)$; otherwise they are dependent.

Example 4.1: Let a fair coin be tossed three times; we obtain the equiprobable space

$$S = \{\mathrm{HHH}, \mathrm{HHT}, \mathrm{HTH}, \mathrm{HTT}, \mathrm{THH}, \mathrm{THT}, \mathrm{TTH}, \mathrm{TTT}\}$$

Consider the events

$$A = \{\text{first toss is heads}\}, \qquad B = \{\text{second toss is heads}\}$$
$$C = \{\text{exactly two heads are tossed in a row}\}$$

Clearly A and B are independent events; this fact is verified below. On the other hand, the relationship between A and C or B and C is not obvious. We claim that A and C are independent, but that B and C are dependent. We have

$$P(A) = P(\{\text{HHH, HHT, HTH, HTT}\}) = \tfrac{4}{8} = \tfrac{1}{2}$$
$$P(B) = P(\{\text{HHH, HHT, THH, THT}\}) = \tfrac{4}{8} = \tfrac{1}{2}$$
$$P(C) = P(\{\text{HHT, THH}\}) = \tfrac{2}{8} = \tfrac{1}{4}$$

Then

$$P(A \cap B) = P(\{\text{HHH, HHT}\}) = \tfrac{1}{4}, \qquad P(A \cap C) = P(\{\text{HHT}\}) = \tfrac{1}{8},$$
$$P(B \cap C) = P(\{\text{HHT, THH}\}) = \tfrac{1}{4}$$

Accordingly,

$$P(A)\,P(B) = \tfrac{1}{2} \cdot \tfrac{1}{2} = \tfrac{1}{4} = P(A \cap B), \quad \text{and so } A \text{ and } B \text{ are independent;}$$
$$P(A)\,P(C) = \tfrac{1}{2} \cdot \tfrac{1}{4} = \tfrac{1}{8} = P(A \cap C), \quad \text{and so } A \text{ and } C \text{ are independent;}$$
$$P(B)\,P(C) = \tfrac{1}{2} \cdot \tfrac{1}{4} = \tfrac{1}{8} \neq P(B \cap C), \quad \text{and so } B \text{ and } C \text{ are dependent.}$$

Frequently, we will postulate that two events are independent, or the experiment itself will imply that two events are independent.

Example 4.2: The probability that A hits a target is $\frac{1}{4}$ and the probability that B hits it is $\frac{2}{5}$. What is the probability that the target will be hit if A and B each shoot at the target?

We are given that $P(A) = \frac{1}{2}$ and $P(B) = \frac{2}{5}$, and we seek $P(A \cup B)$. Furthermore, the probability that A or B hits the target is not influenced by what the other does; that is, the event that A hits the target is independent of the event that B hits the target: $P(A \cap B) = P(A)\,P(B)$. Thus

$$P(A \cup B) = P(A) + P(B) - P(A \cap B) = P(A) + P(B) - P(A)\,P(B)$$
$$= \tfrac{1}{4} + \tfrac{2}{5} - \tfrac{1}{4} \cdot \tfrac{2}{5} = \tfrac{11}{20}$$

Three events A, B and C are *independent* if:

(i) $P(A \cap B) = P(A)\,P(B)$, $P(A \cap C) = P(A)\,P(C)$ and $P(B \cap C) = P(B)\,P(C)$

i.e. if the events are pairwise independent, and

(ii) $P(A \cap B \cap C) = P(A)\,P(B)\,P(C)$

The next example shows that condition (ii) does not follow from condition (i); in other words, three events may be pairwise independent but not independent themselves.

Example 4.3: Let a pair of fair coins be tossed; here $S = \{\text{HH, HT, TH, TT}\}$ is an equiprobable space. Consider the events

$$A = \{\text{heads on the first coin}\} = \{\text{HH, HT}\}$$
$$B = \{\text{heads on the second coin}\} = \{\text{HH, TH}\}$$
$$C = \{\text{heads on exactly one coin}\} = \{\text{HT, TH}\}$$

Then $P(A) = P(B) = P(C) = \frac{2}{4} = \frac{1}{2}$ and

$$P(A \cap B) = P(\{\text{HH}\}) = \tfrac{1}{4}, \quad P(A \cap C) = P(\{\text{HT}\}) = \tfrac{1}{4}, \quad P(B \cap C) = (\{\text{TH}\}) = \tfrac{1}{4}$$

Thus condition (i) is satisfied, i.e., the events are pairwise independent. However, $A \cap B \cap C = \emptyset$ and so

$$P(A \cap B \cap C) = P(\emptyset) = 0 \neq P(A)\,P(B)\,P(C)$$

In other words, condition (ii) is not satisfied and so the three events are not independent.

Solved Problems

CONDITIONAL PROBABILITY IN EQUIPROBABLE SPACES

18.1. A pair of fair dice is thrown. Find the probability p that the sum is 10 or greater if (i) a 5 appears on the first die, (ii) a 5 appears on at least one of the dice.

(i) If a 5 appears on the first die, then the reduced sample space is

$$A = \{(5,1), (5,2), (5,3), (5,4), (5,5), (5,6)\}$$

The sum is 10 or greater on two of the six outcomes: (5, 5), (5, 6). Hence $p = \frac{2}{6} = \frac{1}{3}$.

(ii) If a 5 appears on at least one of the dice, then the reduced sample space has eleven elements:

$$B = \{(5,1), (5,2), (5,3), (5,4), (5,5), (5,6), (1,5), (2,5), (3,5), (4,5), (6,5)\}$$

The sum is 10 or greater on three of the eleven outcomes: (5, 5), (5, 6), (6, 5). Hence $p = \frac{3}{11}$.

18.2. Three fair coins are tossed. Find the probability p that they are all heads if (i) the first coin is heads, (ii) one of the coins is heads.

The sample space has eight elements: $S = \{\text{HHH, HHT, HTH, HTT, THH, THT, TTH, TTT}\}$.

(i) If the first coin is heads, the reduced sample space is $A = \{\text{HHH, HHT, HTH, HTT}\}$. Since the coins are all heads in 1 of 4 cases, $p = \frac{1}{4}$.

(ii) If one of the coins is heads, the reduced sample space is $B = \{\text{HHH, HHT, HTH, HTT, THH, THT, TTH}\}$. Since the coins are all heads in 1 of 7 cases, $p = \frac{1}{7}$.

18.3. A pair of fair dice is thrown. If the two numbers appearing are different, find the probability p that (i) the sum is six, (ii) an ace appears, (iii) the sum is 4 or less.

Of the 36 ways the pair of dice can be thrown, 6 will contain the same numbers: (1, 1), (2, 2), ..., (6, 6). Thus the reduced sample space will consist of $36 - 6 = 30$ elements.

(i) The sum six can appear in 4 ways: (1, 5), (2, 4), (4, 2), (5, 1). (We cannot include (3, 3) since the numbers are the same.) Hence $p = \frac{4}{30} = \frac{2}{15}$.

(ii) An ace can appear in 10 ways: (1, 2), (1, 3), ..., (1, 6) and (2, 1), (3, 1), ..., (6, 1). Hence $p = \frac{10}{30} = \frac{1}{3}$.

(iii) The sum of 4 or less can occur in 4 ways: (3, 1), (1, 3), (2, 1), (1, 2). Thus $p = \frac{4}{30} = \frac{2}{15}$.

18.4. Two digits are selected at random from the digits 1 through 9. If the sum is even, find the probability p that both numbers are odd.

The sum is even if both numbers are even or if both numbers are odd. There are 4 even numbers (2, 4, 6, 8); hence there are $\binom{4}{2} = 6$ ways to choose two even numbers. There are 5 odd numbers (1, 3, 5, 7, 9); hence there are $\binom{5}{2} = 10$ ways to choose two odd numbers. Thus there are $6 + 10 = 16$ ways to choose two numbers such that their sum is even; since 10 of these ways occur when both numbers are odd, $p = \frac{10}{16} = \frac{5}{8}$.

18.5. A man is dealt 4 spade cards from an ordinary deck of 52 cards. If he is given three more cards, find the probability p that at least one of the additional cards is also a spade.

Since he is dealt 4 spades, there are $52-4 = 48$ cards remaining of which $13-4 = 9$ are spades. There are $\binom{48}{3} = 17{,}296$ ways in which he can be dealt three more cards. Since there are $48-9 = 39$ cards which are not spades, there are $\binom{39}{3} = 9139$ ways he can be dealt three cards which are not spades. Thus the probability q that he is not dealt another spade is $q = \frac{9139}{17{,}296}$; hence $p = 1-q = \frac{8157}{17{,}296}$.

18.6. Four people, called North, South, East and West, are each dealt 13 cards from an ordinary deck of 52 cards.

(i) If South has no aces, find the probability p that his partner North has exactly two aces.

(ii) If North and South together have nine hearts, find the probability p that East and West each has two hearts.

(i) There are 39 cards, including 4 aces, divided among North, East and West. There are $\binom{39}{13}$ ways that North can be dealt 13 of the 39 cards. There are $\binom{4}{2}$ ways he can be dealt 2 of the four aces, and $\binom{35}{11}$ ways he can be dealt 11 cards from the $39-4 = 35$ cards which are not aces. Thus

$$p = \frac{\binom{4}{2}\binom{35}{11}}{\binom{39}{13}} = \frac{6 \cdot 12 \cdot 13 \cdot 25 \cdot 26}{36 \cdot 37 \cdot 38 \cdot 39} = \frac{650}{2109}$$

(ii) There are 26 cards, including 4 hearts, divided among East and West. There are $\binom{26}{13}$ ways that, say, East can be dealt 13 cards. (We need only analyze East's 13 cards since West must have the remaining cards.) There are $\binom{4}{2}$ ways East can be dealt 2 hearts from 4 hearts, and $\binom{22}{11}$ ways he can be dealt 11 non-hearts from the $26-4 = 22$ non-hearts. Thus

$$p = \frac{\binom{4}{2}\binom{22}{11}}{\binom{26}{13}} = \frac{6 \cdot 12 \cdot 13 \cdot 12 \cdot 13}{23 \cdot 24 \cdot 25 \cdot 26} = \frac{234}{575}$$

MULTIPLICATION THEOREM

18.7. A class has 12 boys and 4 girls. If three students are selected at random from the class, what is the probability p that they are all boys?

The probability that the first student selected is a boy is 12/16 since there are 12 boys out of 16 students. If the first student is a boy, then the probability that the second is a boy is 11/15 since there are 11 boys left out of 15 students. Finally, if the first two students selected were boys, then the probability that the third student is a boy is 10/14 since there are 10 boys left out of 14 students. Thus, by the multiplication theorem, the probability that all three are boys is

$$p = \frac{12}{16} \cdot \frac{11}{15} \cdot \frac{10}{14} = \frac{11}{28}$$

Another Method. There are $\binom{16}{3} = 560$ ways to select 3 students of the 16 students, and $\binom{12}{3} = 220$ ways to select 3 boys out of 12 boys; hence $p = \frac{220}{560} = \frac{11}{28}$.

A Third Method. If the students are selected one after the other, then there are $16 \cdot 15 \cdot 14$ ways to select three students, and $12 \cdot 11 \cdot 10$ ways to select three boys; hence $p = \frac{12 \cdot 11 \cdot 10}{16 \cdot 15 \cdot 14} = \frac{11}{28}$.

18.8. A man is dealt 5 cards one after the other from an ordinary deck of 52 cards. What is the probability p that they are all spades?

The probability that the first card is a spade is 13/52, the second is a spade is 12/51, the third is a spade is 11/50, the fourth is a spade is 10/49, and the last is a spade is 9/48. (We assumed in each case that the previous cards were spades.) Thus $p = \frac{13}{52} \cdot \frac{12}{51} \cdot \frac{11}{50} \cdot \frac{10}{49} \cdot \frac{9}{48} = \frac{33}{66{,}640}$.

18.9. An urn contains 7 red marbles and 3 white marbles. Three marbles are drawn from the urn one after the other. Find the probability p that the first two are red and the third is white.

The probability that the first marble is red is 7/10 since there are 7 red marbles out of 10 marbles. If the first marble is red, then the probability that the second marble is red is 6/9 since there are 6 red marbles remaining out of the 9 marbles. If the first two marbles are red, then the probability that the third marble is white is 3/8 since there are 3 white marbles out of the 8 marbles in the urn. Hence, by the multiplication theorem,

$$p = \frac{7}{10} \cdot \frac{6}{9} \cdot \frac{3}{8} = \frac{7}{40}$$

18.10. The students in a class are selected at random, one after the other, for an examination. Find the probability p that the boys and girls in the class alternate if (i) the class consists of 4 boys and 3 girls; (ii) the class consists of 3 boys and 3 girls.

(i) If the boys and girls are to alternate, then the first student examined must be a boy. The probability that the first is a boy is 4/7. If the first is a boy, then the probability that the second is a girl is 3/6 since there are 3 girls out of 6 students left. Continuing in this manner, we obtain the probability that the third is a boy is 3/5, the fourth is a girl is 2/4, the fifth is a boy is 2/3, the sixth is a girl is 1/2, and the last is a boy is 1/1. Thus

$$p = \frac{4}{7} \cdot \frac{3}{6} \cdot \frac{3}{5} \cdot \frac{2}{4} \cdot \frac{2}{3} \cdot \frac{1}{2} \cdot \frac{1}{1} = \frac{1}{35}$$

(ii) There are two mutually exclusive cases: the first pupil is a boy, and the first is a girl. If the first student is a boy, then by the multiplication theorem the probability p_1 that the students alternate is

$$p_1 = \frac{3}{6} \cdot \frac{3}{5} \cdot \frac{2}{4} \cdot \frac{2}{3} \cdot \frac{1}{2} \cdot \frac{1}{1} = \frac{1}{20}$$

If the first student is a girl, then by the multiplication theorem the probability p_2 that the students alternate is

$$p_2 = \frac{3}{6} \cdot \frac{3}{5} \cdot \frac{2}{4} \cdot \frac{2}{3} \cdot \frac{1}{2} \cdot \frac{1}{1} = \frac{1}{20}$$

Thus $p = p_1 + p_2 = \frac{1}{20} + \frac{1}{20} = \frac{1}{10}$.

CONDITIONAL PROBABILITY

18.11. In a certain college, 25% of the students failed mathematics, 15% of the students failed chemistry, and 10% of the students failed both mathematics and chemistry. A student is selected at random.

(i) If he failed chemistry, what is the probability that he failed mathematics?

(ii) If he failed mathematics, what is the probability that he failed chemistry?

(iii) What is the probability that he failed mathematics or chemistry?

Let M = {students who failed mathematics} and C = {students who failed chemistry}; then

$$P(M) = 25\% = .25, \quad P(C) = 15\% = .15, \quad P(M \cap C) = 10\% = .10$$

(i) The probability that a student failed mathematics, given that he has failed chemistry is

$$P(M \mid C) = \frac{P(M \cap C)}{P(C)} = \frac{.10}{.15} = \frac{2}{3}$$

(ii) The probability that a student failed chemistry, given that he has failed mathematics is

$$P(C \mid M) = \frac{P(C \cap M)}{P(M)} = \frac{.10}{.25} = \frac{2}{5}$$

(iii) $$P(M \cup C) = P(M) + P(C) - P(M \cap C) = .25 + .15 - .10 = .30 = \frac{3}{10}$$

18.12. Let A and B be events with $P(A) = \frac{1}{2}$, $P(B) = \frac{1}{3}$ and $P(A \cap B) = \frac{1}{4}$. Find (i) $P(A \mid B)$, (ii) $P(B \mid A)$, (iii) $P(A \cup B)$, (iv) $P(A^c \mid B^c)$, (v) $P(B^c \mid A^c)$.

(i) $P(A \mid B) = \dfrac{P(A \cap B)}{P(B)} = \dfrac{\frac{1}{4}}{\frac{1}{3}} = \dfrac{3}{4}$ (ii) $P(B \mid A) = \dfrac{P(B \cap A)}{P(A)} = \dfrac{\frac{1}{4}}{\frac{1}{2}} = \dfrac{1}{2}$

(iii) $P(A \cup B) = P(A) + P(B) - P(A \cap B) = \frac{1}{2} + \frac{1}{3} - \frac{1}{4} = \frac{7}{12}$

(iv) First compute $P(B^c)$ and $P(A^c \cap B^c)$. $P(B^c) = 1 - P(B) = 1 - \frac{1}{3} = \frac{2}{3}$. By De Morgan's Law, $(A \cup B)^c = A^c \cap B^c$; hence $P(A^c \cap B^c) = P((A \cup B)^c) = 1 - P(A \cup B) = 1 - \frac{7}{12} = \frac{5}{12}$.

Thus $P(A^c \mid B^c) = \dfrac{P(A^c \cap B^c)}{P(B^c)} = \dfrac{\frac{5}{12}}{\frac{2}{3}} = \dfrac{5}{8}$.

(v) $P(A^c) = 1 - P(A) = 1 - \frac{1}{2} = \frac{1}{2}$. Then $P(B^c \mid A^c) = \dfrac{P(B^c \cap A^c)}{P(A^c)} = \dfrac{\frac{5}{12}}{\frac{1}{2}} = \dfrac{5}{6}$.

18.13. Let A and B be events with $P(A) = \frac{3}{8}$, $P(B) = \frac{5}{8}$ and $P(A \cup B) = \frac{3}{4}$. Find $P(A \mid B)$ and $P(B \mid A)$.

First compute $P(A \cap B)$ using the formula $P(A \cup B) = P(A) + P(B) - P(A \cap B)$:

$$\tfrac{3}{4} = \tfrac{3}{8} + \tfrac{5}{8} - P(A \cap B) \quad \text{or} \quad P(A \cap B) = \tfrac{1}{4}$$

Then $P(A \mid B) = \dfrac{P(A \cap B)}{P(B)} = \dfrac{\frac{1}{4}}{\frac{5}{8}} = \dfrac{2}{5}$ and $P(B \mid A) = \dfrac{P(B \cap A)}{P(A)} = \dfrac{\frac{1}{4}}{\frac{3}{8}} = \dfrac{2}{3}$.

18.14. Let $S = \{a, b, c, d, e\}$ with $P(a) = P(b) = \frac{1}{10}$, $P(c) = \frac{1}{5}$ and $P(d) = P(e) = \frac{3}{10}$, and let $E = \{b, c, d\}$. Find: (i) $P(A \mid E)$ where $A = \{a, b, c\}$; (ii) $P(B \mid E)$ where $B = \{c, d, e\}$; (iii) $P(C \mid E)$ where $C = \{a, c, e\}$; (iv) $P(D \mid E)$ where $D = \{a, e\}$.

In each case use the definition of conditional probability: $P(* \mid E) = \dfrac{P(* \cap E)}{P(E)}$.

Here $P(E) = P(\{b, c, d\}) = P(b) + P(c) + P(d) = \frac{1}{10} + \frac{1}{5} + \frac{3}{10} = \frac{3}{5}$.

(i) $A \cap E = \{b, c\}$, $P(A \cap E) = P(b) + P(c) = \frac{1}{10} + \frac{1}{5} = \frac{3}{10}$, and $P(A \mid E) = \dfrac{P(A \cap E)}{P(E)} = \dfrac{\frac{3}{10}}{\frac{3}{5}} = \dfrac{1}{2}$.

(ii) $B \cap E = \{c, d\}$, $P(B \cap E) = P(c) + P(d) = \frac{1}{5} + \frac{3}{10} = \frac{1}{2}$, and $P(B \mid E) = \dfrac{P(B \cap E)}{P(E)} = \dfrac{\frac{1}{2}}{\frac{3}{5}} = \dfrac{5}{6}$.

(iii) $C \cap E = \{c\}$, $P(C \cap E) = P(c) = \frac{1}{5}$, and $P(C \mid E) = \dfrac{P(C \cap E)}{P(E)} = \dfrac{\frac{1}{5}}{\frac{3}{5}} = \dfrac{1}{3}$.

(iv) $D \cap E = \emptyset$, $P(D \cap E) = P(\emptyset) = 0$, and $P(D \mid E) = \dfrac{P(D \cap E)}{P(E)} = \dfrac{0}{\frac{3}{5}} = 0$.

18.15. Show that the conditional probability function $P(* \mid E)$ satisfies the three axioms of Theorem 17.1; that is:

[P_1] For any event A, $0 \leq P(A \mid E) \leq 1$.

[P_2] For the certain event S, $P(S \mid E) = 1$.

[P_3] If A and B are mutually exclusive, then $P(A \cup B \mid E) = P(A \mid E) + P(B \mid E)$.

Since $A \cap E \subset E$, $P(A \cap E) \leq P(E)$. Thus $P(A \mid E) = \dfrac{P(A \cap E)}{P(E)} \leq 1$ and is also non-negative; that is, $0 \leq P(A \mid E) \leq 1$.

Since $S \cap E = E$, $P(S \mid E) = \dfrac{P(S \cap E)}{P(E)} = \dfrac{P(E)}{P(E)} = 1$.

If $A \cap B = \emptyset$, then $A \cap B \cap E = (A \cap E) \cap (B \cap E) = \emptyset$, i.e. $A \cap E$ and $B \cap E$ are mutually exclusive events; hence $P((A \cap E) \cup (B \cap E)) = P(A \cap E) + P(B \cap E)$. But $(A \cap E) \cup (B \cap E) = (A \cup B) \cap E$. Thus $P((A \cup B) \cap E) = P(A \cap E) + P(B \cap E)$. Hence

$$P(A \cup B \mid E) = \frac{P((A \cup B) \cap E)}{P(E)} = \frac{P(A \cap E) + P(B \cap E)}{P(E)}$$
$$= \frac{P(A \cap E)}{P(E)} + \frac{P(B \cap E)}{P(E)} = P(A \mid E) + P(B \mid E)$$

Remark: Since $P(* \mid E)$ satisfies the above three axioms, it also satisfies any consequence of these axioms; for example,

$$P(A \cup B \mid E) = P(A \mid E) + P(B \mid E) - P(A \cap B \mid E)$$

18.16. Prove: Let $E_1, E_2, \ldots, E_n$ form a partition of a sample space S, and let A be any event. Then

$$P(A) = P(E_1)P(A \mid E_1) + P(E_2)P(A \mid E_2) + \cdots + P(E_n)P(A \mid E_n)$$

Since the E_i form a partition of S, the E_i are pairwise disjoint and $S = E_1 \cup E_2 \cup \cdots \cup E_n$. Hence

$$A = S \cap A = (E_1 \cup E_2 \cup \cdots \cup E_n) \cap A = (E_1 \cap A) \cup (E_2 \cap A) \cup \cdots \cup (E_n \cap A)$$

and the $E_i \cap A$ are also pairwise disjoint. Accordingly,

$$P(A) = P(E_1 \cap A) + P(E_2 \cap A) + \cdots + P(E_n \cap A)$$

By the multiplication theorem, $P(E_i \cap A) = P(E_i)P(A \mid E_i)$; hence

$$P(A) = P(E_1)P(A \mid E_1) + P(E_2)P(A \mid E_2) + \cdots + P(E_n)P(A \mid E_n)$$

18.17. Three machines A, B and C produce respectively 50%, 30% and 20% of the total number of items of a factory. The percentages of defective output of these machines are respectively 3%, 4% and 5%. If an item is selected at random, what is the probability p that the item is defective?

Let X be the event that an item is defective. Then by the preceding problem,

$$P(X) = P(A)P(X \mid A) + P(B)P(X \mid B) + P(C)P(X \mid C)$$
$$= 50\% \cdot 3\% + 30\% \cdot 4\% + 20\% \cdot 5\%$$
$$= \frac{50}{100} \cdot \frac{3}{100} + \frac{30}{100} \cdot \frac{4}{100} + \frac{20}{100} \cdot \frac{5}{100} = \frac{37}{1000} = 3.7\%$$

We can also consider this problem as a stochastic process having the adjoining tree diagram.

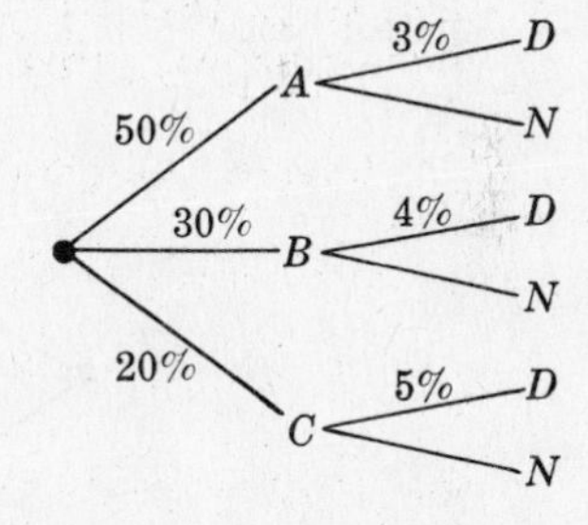

18.18. Prove Bayes' Theorem: Let $E_1, E_2, \ldots, E_n$ form a partition of a sample space S, and let A be any event. Then, for each i,

$$P(E_i \mid A) = \frac{P(E_i)P(A \mid E_i)}{P(E_1)P(A \mid E_1) + P(E_2)P(A \mid E_2) + \cdots + P(E_n)P(A \mid E_n)}$$

Using the definition of conditional probability, we obtain

$$P(E_i \mid A) = \frac{P(E_i \cap A)}{P(A)} = \frac{P(E_i)P(A \mid E_i)}{P(E_1)P(A \mid E_1) + P(E_2)P(A \mid E_2) + \cdots + P(E_n)P(A \mid E_n)}$$

Here we used the multiplication theorem, $P(E_i \cap A) = P(E_i)P(A \mid E_i)$, to obtain the numerator, and Problem 18.16 to obtain the denominator.

18.19. In a certain college, 4% of the men and 1% of the women are taller than 6 feet. Furthermore, 60% of the students are women. Now if a student is selected at random and is taller than 6 feet, what is the probability that the student is a woman?

Let $A = \{\text{students taller than 6 feet}\}$. We seek $P(W \mid A)$, the probability that a student is a woman given that the student is taller than 6 feet. By Bayes' theorem,

$$P(W \mid A) = \frac{P(W)P(A \mid W)}{P(W)P(A \mid W) + P(M)P(A \mid M)} = \frac{60\% \cdot 1\%}{40\% \cdot 4\% + 60\% \cdot 1\%} = \frac{3}{11}$$

18.20. Find $P(B \mid A)$ if (i) A is a subset of B, (ii) A and B are mutually exclusive.

(i) If A is a subset of B, then whenever A occurs B must occur; hence $P(B \mid A) = 1$. Alternately, if A is a subset of B then $A \cap B = A$; hence

$$P(B \mid A) = \frac{P(A \cap B)}{P(A)} = \frac{P(A)}{P(A)} = 1$$

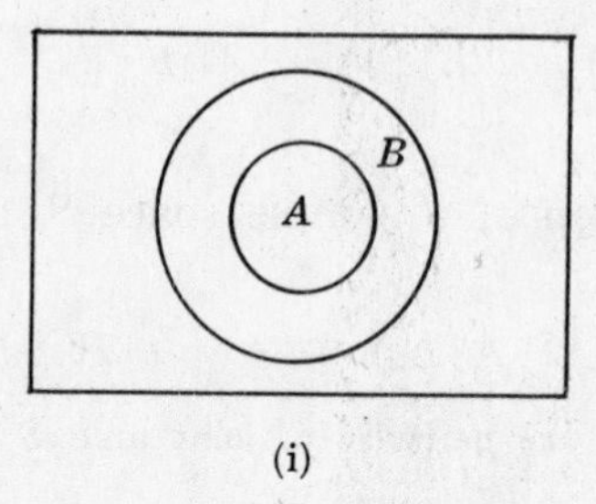

(i)

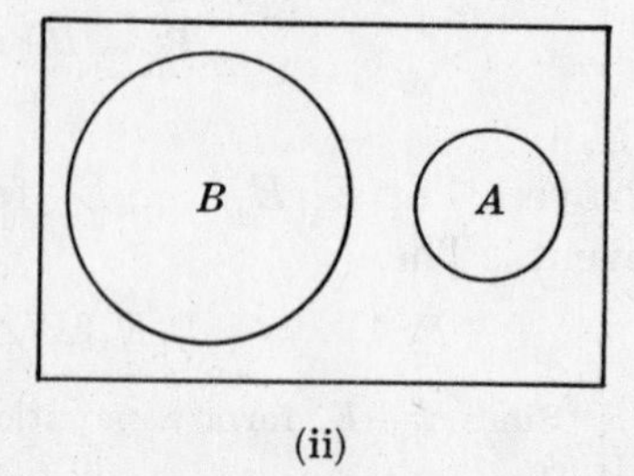

(ii)

(ii) If A and B are mutually exclusive, i.e. disjoint, then whenever A occurs B cannot occur; hence $P(B \mid A) = 0$. Alternately, if A and B are mutually exclusive then $A \cap B = \emptyset$; hence

$$P(B \mid A) = \frac{P(A \cap B)}{P(A)} = \frac{P(\emptyset)}{P(A)} = \frac{0}{P(A)} = 0$$

FINITE STOCHASTIC PROCESSES

18.21. A box contains three coins; one coin is fair, one coin is two-headed, and one coin is weighted so that the probability of heads appearing is $\frac{1}{3}$. A coin is selected at random and tossed. Find the probability p that heads appears.

Construct the tree diagram as shown in Figure (*a*) below. Note that I refers to the fair coin, II to the two-headed coin, and III to the weighted coin. Now heads appears along three of the paths; hence

$$p = \tfrac{1}{3}\cdot\tfrac{1}{2} + \tfrac{1}{3}\cdot 1 + \tfrac{1}{3}\cdot\tfrac{1}{3} = \tfrac{11}{18}$$

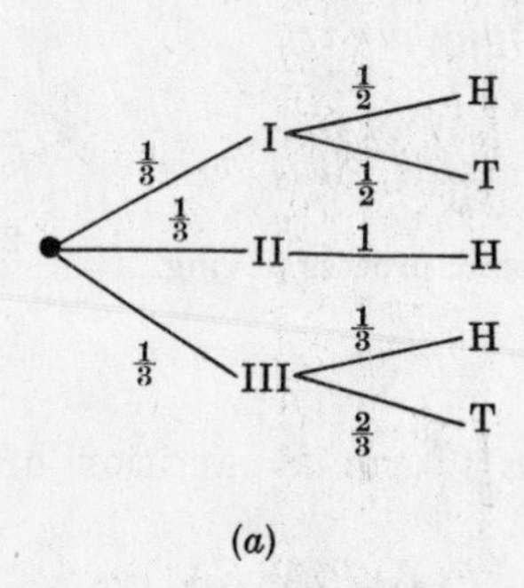

(*a*)

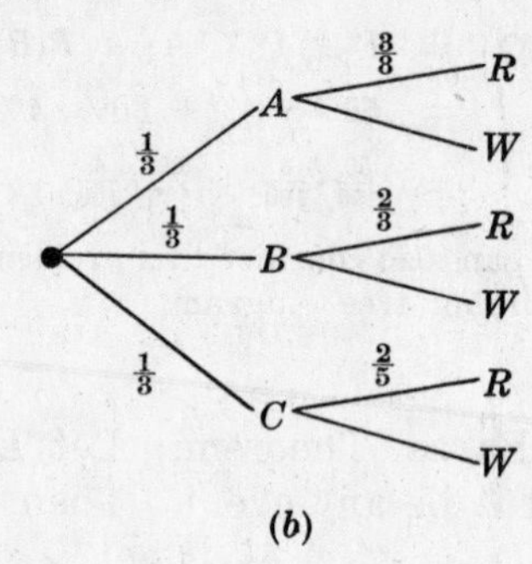

(*b*)

18.22. We are given three urns as follows:

Urn A contains 3 red and 5 white marbles.

Urn B contains 2 red and 1 white marble.

Urn C contains 2 red and 3 white marbles.

An urn is selected at random and a marble is drawn from the urn. If the marble is red, what is the probability that it came from urn A?

Construct the tree diagram as shown in Figure (*b*) above.

We seek the probability that A was selected, given that the marble is red; that is, $P(A \mid R)$. In order to find $P(A \mid R)$, it is necessary first to compute $P(A \cap R)$ and $P(R)$.

The probability that urn A is selected and a red marble drawn is $\frac{1}{3}\cdot\frac{3}{8} = \frac{1}{8}$; that is, $P(A \cap R) = \frac{1}{8}$. Since there are three paths leading to a red marble, $P(R) = \frac{1}{3}\cdot\frac{3}{8} + \frac{1}{3}\cdot\frac{2}{3} + \frac{1}{3}\cdot\frac{2}{5} = \frac{173}{360}$. Thus

$$P(A \mid R) = \frac{P(A \cap R)}{P(R)} = \frac{\frac{1}{3}}{\frac{173}{360}} = \frac{45}{173}$$

18.23. Box A contains nine cards numbered 1 through 9, and Box B contains five cards numbered 1 through 5. A box is chosen at random and a card drawn. If the number is even, find the probability that the card came from Box A.

The tree diagram of the process is shown in Figure (*a*) below.

We seek $P(A \mid E)$, the probability that A was selected, given that the number is even. The probability that Box A and an even number is drawn is $\frac{1}{2}\cdot\frac{4}{9} = \frac{2}{9}$; that is, $P(A \cap E) = \frac{2}{9}$. Since there are two paths which lead to an even number, $P(E) = \frac{1}{2}\cdot\frac{4}{9} + \frac{1}{2}\cdot\frac{2}{5} = \frac{19}{45}$. Thus

$$P(A \mid E) = \frac{P(A \cap E)}{P(E)} = \frac{\frac{2}{9}}{\frac{19}{45}} = \frac{10}{19}$$

(*a*) (*b*)

18.24. An urn contains 3 red marbles and 7 white marbles. A marble is drawn from the urn and a marble of the other color is then put into the urn. A second marble is drawn from the urn.

(i) Find the probability p that the second marble is red.

(ii) If both marbles were of the same color, what is the probability p that they were both white?

Construct the tree diagram as shown in Figure (*b*) above.

(i) Two paths of the tree lead to a red marble: $p = \frac{3}{10}\cdot\frac{2}{10} + \frac{7}{10}\cdot\frac{4}{10} = \frac{17}{50}$.

(ii) The probability that both marbles were white is $\frac{7}{10}\cdot\frac{6}{10} = \frac{21}{50}$. The probability that both marbles were of the same color, i.e. the probability of the reduced sample space, is $\frac{3}{10}\cdot\frac{2}{10} + \frac{7}{10}\cdot\frac{6}{10} = \frac{12}{25}$. Hence the conditional probability $p = \frac{21}{50} / \frac{12}{25} = \frac{7}{8}$.

18.25. We are given two urns as follows:

Urn A contains 3 red and 2 white marbles.

Urn B contains 2 red and 5 white marbles.

An urn is selected at random; a marble is drawn and put into the other urn; then a marble is drawn from the second urn. Find the probability p that both marbles drawn are of the same color.

Construct the following tree diagram:

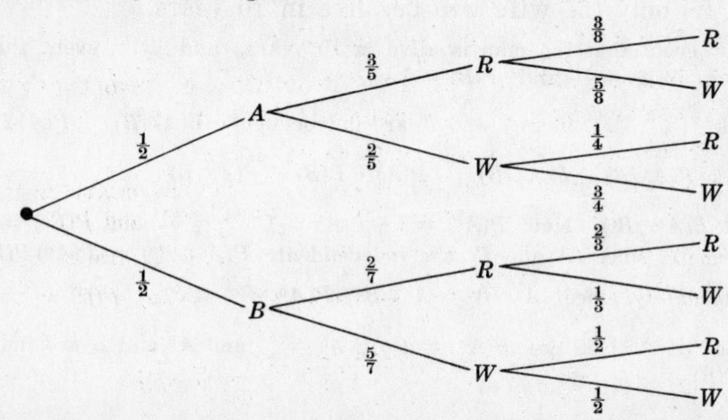

Note that if urn A is selected and a red marble drawn and put into urn B, then urn B has 3 red marbles and 5 white marbles.

Since there are four paths which lead to two marbles of the same color,

$$p = \frac{1}{2}\cdot\frac{3}{5}\cdot\frac{3}{8} + \frac{1}{2}\cdot\frac{2}{5}\cdot\frac{3}{4} + \frac{1}{2}\cdot\frac{2}{7}\cdot\frac{2}{3} + \frac{1}{2}\cdot\frac{5}{7}\cdot\frac{1}{2} = \frac{901}{1680}$$

INDEPENDENCE

18.26. Let A = event that a family has children of both sexes, and let B = event that a family has at most one boy. (i) Show that A and B are independent events if a family has three children. (ii) Show that A and B are dependent events if a family has two children.

(i) We have the equiprobable space $S = \{bbb, bbg, bgb, bgg, gbb, gbg, ggb, ggg\}$. Here

$$A = \{bbg, bgb, bgg, gbb, gbg, ggb\} \quad \text{and so} \quad P(A) = \tfrac{6}{8} = \tfrac{3}{4}$$

$$B = \{bgg, gbg, ggb, ggg\} \quad \text{and so} \quad P(B) = \tfrac{4}{8} = \tfrac{1}{2}$$

$$A \cap B = \{bgg, gbg, ggb\} \quad \text{and so} \quad P(A \cap B) = \tfrac{3}{8}$$

Since $P(A)\,P(B) = \frac{3}{4}\cdot\frac{1}{2} = \frac{3}{8} = P(A \cap B)$, A and B are independent.

(ii) We have the equiprobable space $S = \{bb, bg, gb, gg\}$. Here

$$A = \{bg, gb\} \quad \text{and so} \quad P(A) = \tfrac{1}{2}$$

$$B = \{bg, gb, gg\} \quad \text{and so} \quad P(B) = \tfrac{3}{4}$$

$$A \cap B = \{bg, gb\} \quad \text{and so} \quad P(A \cap B) = \tfrac{1}{2}$$

Since $P(A)\,P(B) \neq P(A \cap B)$, A and B are dependent.

18.27. Prove: If A and B are independent events, then A^c and B^c are independent events.

Let $P(A) = x$ and $P(B) = y$. Then $P(A^c) = 1 - x$ and $P(B^c) = 1 - y$. Since A and B are independent, $P(A \cap B) = P(A)\,P(B) = xy$. Furthermore,

$$P(A \cup B) = P(A) + P(B) - P(A \cap B) = x + y - xy$$

By De Morgan's theorem, $(A \cup B)^c = A^c \cap B^c$; hence

$$P(A^c \cap B^c) = P((A \cup B)^c) = 1 - P(A \cup B) = 1 - x - y + xy$$

On the other hand,

$$P(A^c)\,P(B^c) = (1 - x)(1 - y) = 1 - x - y + xy$$

Thus $P(A^c \cap B^c) = P(A^c)\,P(B^c)$, and so A^c and B^c are independent.

18.28. The probability that a man will live 10 more years is $\frac{1}{4}$, and the probability that his wife will live 10 more years is $\frac{1}{3}$. Find the probability that (i) both will be alive in 10 years, (ii) at least one will be alive in 10 years, (iii) neither will be alive in 10 years, (iv) only the wife will be alive in 10 years.

Let A = event that the man is alive in 10 years, and B = event that his wife is alive in 10 years; then $P(A) = \frac{1}{4}$ and $P(B) = \frac{1}{3}$.

(i) We seek $P(A \cap B)$. Since A and B are independent, $P(A \cap B) = P(A)\,P(B) = \frac{1}{4}\cdot\frac{1}{3} = \frac{1}{12}$.

(ii) We seek $P(A \cup B)$. $P(A \cup B) = P(A) + P(B) - P(A \cap B) = \frac{1}{4} + \frac{1}{3} - \frac{1}{12} = \frac{1}{2}$.

(iii) We seek $P(A^c \cap B^c)$. Now $P(A^c) = 1 - P(A) = 1 - \frac{1}{4} = \frac{3}{4}$ and $P(B^c) = 1 - P(B) = 1 - \frac{1}{3} = \frac{2}{3}$. Furthermore, since A^c and B^c are independent, $P(A^c \cap B^c) = P(A^c)\,P(B^c) = \frac{3}{4}\cdot\frac{2}{3} = \frac{1}{2}$.

Alternately, since $(A \cup B)^c = A^c \cap B^c$, $P(A^c \cap B^c) = P((A \cup B)^c) = 1 - P(A \cup B) = 1 - \frac{1}{2} = \frac{1}{2}$.

(iv) We seek $P(A^c \cap B)$. Since $P(A^c) = 1 - P(A) = \frac{3}{4}$ and A^c and B are independent, $P(A^c \cap B) = P(A^c)\,P(B) = \frac{1}{4}$.

18.29. Box A contains 8 items of which 3 are defective, and box B contains 5 items of which 2 are defective. An item is drawn at random from each box.

(i) What is the probability p that both items are non-defective?

(ii) What is the probability p that one item is defective and one not?

(iii) If one item is defective and one is not, what is the probability p that the defective item came from box A?

(i) The probability of choosing a non-defective item from A is $\frac{5}{8}$ and from B is $\frac{3}{5}$. Since the events are independent, $p = \frac{5}{8}\cdot\frac{3}{5} = \frac{3}{8}$.

(ii) **Method 1.** The probability of choosing two defective items is $\frac{3}{8}\cdot\frac{2}{5} = \frac{3}{20}$. From (i) the probability of choosing two non-defective items is $\frac{3}{8}$. Hence $p = 1-\frac{3}{8}-\frac{3}{20} = \frac{19}{40}$.

Method 2. The probability p_1 of choosing a defective item from A and a non-defective item from B is $\frac{3}{8}\cdot\frac{3}{5} = \frac{9}{40}$. The probability p_2 of choosing a non-defective item from A and a defective item from B is $\frac{5}{8}\cdot\frac{2}{5} = \frac{1}{4}$. Hence $p = p_1 + p_2 = \frac{9}{40}+\frac{1}{4} = \frac{19}{40}$.

(iii) Consider the events $X = \{\text{defective item from } A\}$ and $Y = \{\text{one item is defective and one non-defective}\}$. We seek $P(X \mid Y)$. By (ii), $P(X\cap Y) = p_1 = \frac{9}{40}$ and $P(Y) = \frac{19}{40}$. Hence

$$p = P(X \mid Y) = \frac{P(X\cap Y)}{P(Y)} = \frac{\frac{9}{40}}{\frac{19}{40}} = \frac{9}{19}$$

18.30. The probabilities that three men hit a target are respectively $\frac{1}{6}$, $\frac{1}{4}$ and $\frac{1}{3}$. Each shoots once at the target. (i) Find the probability p that exactly one of them hits the target. (ii) If only one hit the target, what is the probability that it was the first man?

Consider the events $A = \{\text{first man hits the target}\}$, $B = \{\text{second man hits the target}\}$, and $C = \{\text{third man hits the target}\}$; then $P(A) = \frac{1}{6}$, $P(B) = \frac{1}{4}$ and $P(C) = \frac{1}{3}$. The three events are independent, and $P(A^c) = \frac{5}{6}$, $P(B^c) = \frac{3}{4}$, $P(C^c) = \frac{2}{3}$.

(i) Let $E = \{\text{exactly one man hits the target}\}$. Then

$$E = (A\cap B^c\cap C^c) \cup (A^c\cap B\cap C^c) \cup (A^c\cap B^c\cap C)$$

In other words, if only one hit the target, then it was either only the first man, $A\cap B^c\cap C^c$, or only the second man, $A^c\cap B\cap C^c$, or only the third man, $A^c\cap B^c\cap C$. Since the three events are mutually exclusive,

$$\begin{aligned} p = P(E) &= P(A\cap B^c\cap C^c) + P(A^c\cap B\cap C^c) + P(A^c\cap B^c\cap C) \\ &= P(A)\,P(B^c)\,P(C^c) + P(A^c)\,P(B)\,P(C^c) + P(A^c)\,P(B^c)\,P(C) \\ &= \tfrac{1}{6}\cdot\tfrac{3}{4}\cdot\tfrac{2}{3} + \tfrac{5}{6}\cdot\tfrac{1}{4}\cdot\tfrac{2}{3} + \tfrac{5}{6}\cdot\tfrac{3}{4}\cdot\tfrac{1}{3} = \tfrac{1}{12} + \tfrac{5}{36} + \tfrac{5}{24} = \tfrac{31}{72} \end{aligned}$$

(ii) We seek $P(A \mid E)$, the probability that the first man hit the target given that only one man hit the target. Now $A\cap E = A\cap B^c\cap C^c$ is the event that only the first man hit the target. By (i), $P(A\cap E) = P(A\cap B^c\cap C^c) = \frac{1}{12}$ and $P(E) = \frac{31}{72}$; hence

$$P(A \mid E) = \frac{P(A\cap E)}{P(E)} = \frac{\frac{1}{12}}{\frac{31}{72}} = \frac{6}{31}$$

Supplementary Problems

CONDITIONAL PROBABILITY IN EQUIPROBABLE SPACES

18.31. A die is tossed. If the number is odd, what is the probability that it is prime?

18.32. A letter is chosen at random from the word "trained". If it is a consonant, what is the probability that it is the first letter of the word?

18.33. Three fair coins are tossed. If both heads and tails appear, determine the probability that exactly one head appears.

18.34. A pair of dice is tossed. If the numbers appearing are different, find the probability that the sum is even.

18.35. A man is dealt 5 red cards from an ordinary deck of 52 cards. What is the probability that they are all of the same suit, i.e. hearts or diamonds?

18.36. A man is dealt 3 spade cards from an ordinary deck of 52 cards. If he is given four more cards, determine the probability that at least two of the additional cards are also spades.

18.37. Two different digits are selected at random from the digits 1 through 9.

(i) If the sum is odd, what is the probability that 2 is one of the numbers selected?

(ii) If 2 is one of the digits selected, what is the probability that the sum is odd?

18.38. Four persons, called North, South, East and West, are each dealt 13 cards from an ordinary deck of 52 cards.

(i) If South has exactly one ace, what is the probability that his partner North has the other three aces?

(ii) If North and South together have 10 hearts, what is the probability that neither East nor West has the other 3 hearts?

18.39. A class has 10 boys and 5 girls. Three students are selected from the class at random, one after the other. Find the probability that (i) the first two are boys and the third is a girl, (ii) the first and third are boys and the second is a girl, (iii) the first and third are of the same sex, and the second is of the opposite sex.

18.40. In the preceding problem, if the first and third students selected are of the same sex and the second student is of the opposite sex, what is the probability that the second student is a girl?

CONDITIONAL PROBABILITY

18.41. In a certain town, 40% of the people have brown hair, 25% have brown eyes, and 15% have both brown hair and brown eyes. A person is selected at random from the town.

(i) If he has brown hair, what is the probability that he also has brown eyes?

(ii) If he has brown eyes, what is the probability that he does not have brown hair?

(iii) What is the probability that he has neither brown hair nor brown eyes?

18.42. Let A and B be events with $P(A) = \frac{1}{3}$, $P(B) = \frac{1}{4}$ and $P(A \cup B) = \frac{1}{2}$. Find (i) $P(A \mid B)$, (ii) $P(B \mid A)$, (iii) $P(A \cap B^c)$, (iv) $P(A \mid B^c)$.

18.43. Let $S = \{a, b, c, d, e, f\}$ with $P(a) = \frac{1}{16}$, $P(b) = \frac{1}{16}$, $P(c) = \frac{1}{8}$, $P(d) = \frac{3}{16}$, $P(e) = \frac{1}{4}$ and $P(f) = \frac{5}{16}$. Let $A = \{a, c, e\}$, $B = \{c, d, e, f\}$ and $C = \{b, c, f\}$. Find (i) $P(A \mid B)$, (ii) $P(B \mid C)$, (iii) $P(C \mid A^c)$, (iv) $P(A^c \mid C)$.

18.44. In a certain college, 25% of the boys and 10% of the girls are studying mathematics. The girls constitute 60% of the student body. If a student is selected at random and is studying mathematics, determine the probability that the student is a girl.

FINITE STOCHASTIC PROCESSES

18.45. We are given two urns as follows:

Urn A contains 5 red marbles, 3 white marbles and 8 blue marbles.

Urn B contains 3 red marbles and 5 white marbles.

A fair die is tossed; if 3 or 6 appears, a marble is chosen from B, otherwise a marble is chosen from A. Find the probability that (i) a red marble is chosen, (ii) a white marble is chosen, (iii) a blue marble is chosen.

18.46. Refer to the preceding problem. (i) If a red marble is chosen, what is the probability that it came from urn A? (ii) If a white marble is chosen, what is the probability that a 5 appeared on the die?

18.47. An urn contains 5 red marbles and 3 white marbles. A marble is selected at random from the urn, discarded, and two marbles of the other color are put into the urn. A second marble is then selected from the urn. Find the probability that (i) the second marble is red, (ii) both marbles are of the same color.

18.48. Refer to the preceding problem. (i) If the second marble is red, what is the probability that the first marble is red? (ii) If both marbles are of the same color, what is the probability that they are both white?

18.49. A box contains three coins, two of them fair and one two-headed. A coin is selected at random and tossed twice. If heads appears both times, what is the probability that the coin is two-headed?

18.50. We are given two urns as follows:

Urn A contains 5 red marbles and 3 white marbles.

Urn B contains 1 red marble and 2 white marbles.

A fair die is tossed; if a 3 or 6 appears, a marble is drawn from B and put into A and then a marble is drawn from A; otherwise, a marble is drawn from A and put into B and then a marble is drawn from B.

(i) What is the probability that both marbles are red?

(ii) What is the probability that both marbles are white?

18.51. Box A contains nine cards numbered 1 through 9, and box B contains five cards numbered 1 through 5. A box is chosen at random and a card drawn; if the card shows an even number, another card is drawn from the same box; if the card shows an odd number, a card is drawn from the other box.

(i) What is the probability that both cards show even numbers?

(ii) If both cards show even numbers, what is the probability that they came from box A?

(iii) What is the probability that both cards show odd numbers?

18.52. A box contains a fair coin and a two-headed coin. A coin is selected at random and tossed. If heads appears, the other coin is tossed; if tails appears, the same coin is tossed.

(i) Find the probability that heads appears on the second toss.

(ii) If heads appeared on the second toss, find the probability that it also appeared on the first toss.

18.53. A box contains three coins, two of them fair and one two-headed. A coin is selected at random and tossed. If heads appears the coin is tossed again; if tails appears, then another coin is selected from the two remaining coins and tossed.

(i) Find the probability that heads appears twice.

(ii) If the same coin is tossed twice, find the probability that it is the two-headed coin.

(iii) Find the probability that tails appears twice.

18.54. Urn A contains x red marbles and y white marbles, and urn B contains z red marbles and v white marbles.

(i) If an urn is selected at random and a marble drawn, what is the probability that the marble is red?

(ii) If a marble is drawn from urn A and put into urn B and then a marble is drawn from urn B, what is the probability that the second marble is red?

18.55. A box contains 5 radio tubes of which 2 are defective. The tubes are tested one after the other until the 2 defective tubes are discovered. What is the probability that the process stopped on the (i) second test, (ii) third test?

18.56. Refer to the preceding problem. If the process stopped on the third test, what is the probability that the first tube is non-defective?

INDEPENDENCE

18.57. Prove: If A and B are independent, then A and B^c are independent and A^c and B are independent.

18.58. Let A and B be events with $P(A) = \frac{1}{4}$, $P(A \cup B) = \frac{1}{3}$ and $P(B) = p$. (i) Find p if A and B are mutually exclusive. (ii) Find p if A and B are independent. (iii) Find p if A is a subset of B.

18.59. Urn A contains 5 red marbles and 3 white marbles, and urn B contains 2 red marbles and 6 white marbles.

(i) If a marble is drawn from each urn, what is the probability that they are both of the same color?

(ii) If two marbles are drawn from each urn, what is the probability that all four marbles are of the same color?

18.60. Let three fair coins be tossed. Let $A = \{\text{all heads or all tails}\}$, $B = \{\text{at least two heads}\}$ and $C = \{\text{at most two heads}\}$. Of the pairs (A, B), (A, C) and (B, C), which are independent and which are dependent?

18.61. The probability that A hits a target is $\frac{1}{4}$ and the probability that B hits a target is $\frac{1}{3}$.

(i) If each fires twice, what is the probability that the target will be hit at least once?

(ii) If each fires once and the target is hit only once, what is the probability that A hit the target?

(iii) If A can fire only twice, how many times must B fire so that there is at least a 90% probability that the target will be hit?

18.62. Let A and B be independent events with $P(A) = \frac{1}{2}$ and $P(A \cup B) = \frac{2}{3}$. Find (i) $P(B)$, (ii) $P(A \mid B)$, (iii) $P(B^c \mid A)$.

Answers to Supplementary Problems

18.31. $\frac{2}{3}$

18.32. $\frac{1}{4}$

18.33. $\frac{1}{2}$

18.34. $\frac{2}{5}$

18.35. $\dfrac{2\binom{13}{5}}{\binom{26}{5}} = \dfrac{9}{230}$

18.36. $1 - \dfrac{\binom{39}{4}}{\binom{49}{4}} - \dfrac{9\binom{39}{3}}{\binom{49}{4}}$

18.37. (i) $\frac{1}{4}$, (ii) $\frac{5}{8}$

18.38. (i) $\dfrac{\binom{36}{10}}{\binom{39}{13}} = \dfrac{22}{703}$ (ii) $\dfrac{2\binom{23}{10}}{\binom{26}{13}} = \dfrac{11}{50}$

18.39. (i) $\frac{10}{15}\cdot\frac{9}{14}\cdot\frac{5}{13} = \frac{15}{91}$

(ii) $\frac{10}{15}\cdot\frac{5}{14}\cdot\frac{9}{13} = \frac{15}{91}$

(iii) $\frac{15}{91} + \frac{20}{273} = \frac{5}{21}$

18.40. $\dfrac{\frac{15}{91}}{\frac{5}{21}} = \frac{9}{13}$

18.41. (i) $\frac{3}{8}$, (ii) $\frac{2}{5}$, (iii) $\frac{1}{2}$

18.42. (i) $\frac{1}{3}$, (ii) $\frac{1}{4}$, (iii) $\frac{1}{4}$, (iv) $\frac{1}{3}$

18.43. (i) $\frac{3}{7}$, (ii) $\frac{3}{4}$, (iii) $\frac{2}{3}$, (iv) $\frac{3}{4}$

18.44. $\frac{3}{8}$

18.45. (i) $\frac{1}{3}$

(ii) $\frac{1}{3}$

(iii) $\frac{1}{3}$

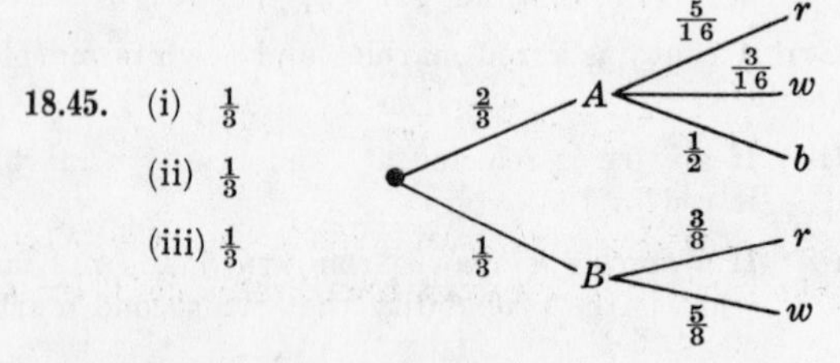

18.46. (i) $\frac{5}{8}$, (ii) $\frac{5}{32}$

18.47. (i) $\frac{41}{72}$, (ii) $\frac{13}{36}$

18.48. (i) $\frac{20}{41}$, (ii) $\frac{3}{13}$

18.49. $\frac{2}{3}$

18.50. (i) $\frac{5}{24} + \frac{2}{27} = \frac{61}{216}$, (ii) $\frac{3}{16} + \frac{8}{81} = \frac{371}{1296}$

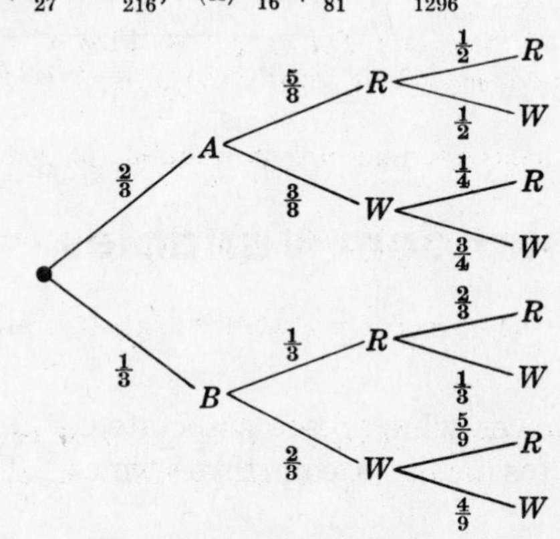

Tree diagram for Problem 18.50

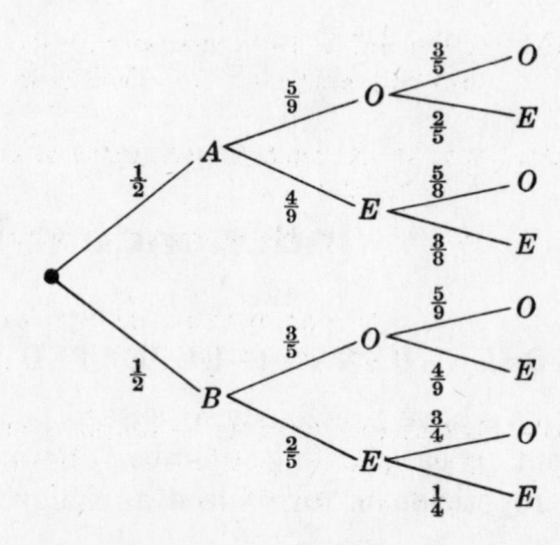

Tree diagram for Problem 18.51

18.51. (i) $\frac{1}{12} + \frac{1}{20} = \frac{2}{15}$, (ii) $\dfrac{\frac{1}{12}}{\frac{2}{15}} = \frac{5}{8}$, (iii) $\frac{1}{6} + \frac{1}{6} = \frac{1}{3}$

18.52. (i) $\frac{5}{8}$, (ii) $\frac{4}{5}$

18.53. (i) $\frac{1}{12} + \frac{1}{12} + \frac{1}{3} = \frac{1}{2}$, (ii) $\frac{2}{3}$, (iii) $\frac{1}{12}$

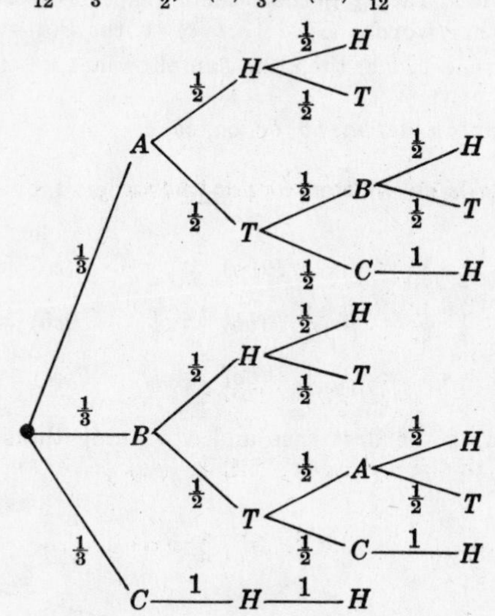

Tree diagram for Problem 18.53

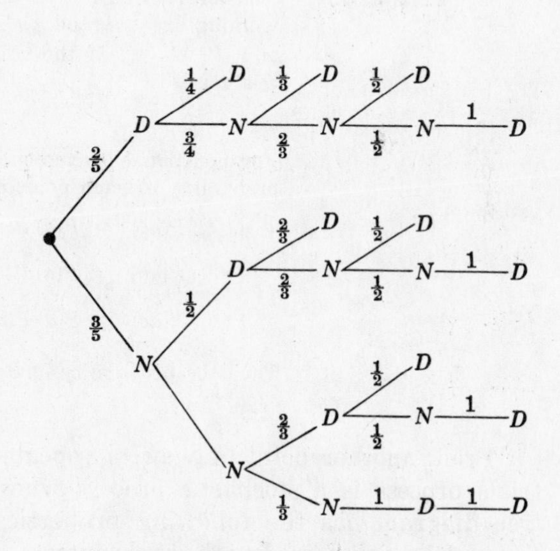

Tree diagram for Problem 18.55

18.54. (i) $\frac{1}{2}\left(\frac{x}{x+y} + \frac{z}{z+v}\right)$, (ii) $\frac{xz+x+yz}{(x+y)(z+v+1)}$

18.55. (i) $\frac{1}{10}$, (ii) $\frac{3}{10}$; we must include the case where the three non-defective tubes appear first, since the last two tubes must then be the defective ones.

18.56. $\frac{2}{3}$

18.58. (i) $\frac{1}{12}$, (ii) $\frac{1}{9}$, (iii) $\frac{1}{3}$

18.59. (i) $\frac{7}{16}$, (ii) $\frac{55}{784}$

18.60. Only A and B are independent.

18.61. (i) $\frac{3}{4}$, (ii) $\frac{2}{5}$, (iii) 5

18.62. (i) $\frac{1}{3}$, (ii) $\frac{1}{2}$, (iii) $\frac{2}{3}$

Chapter 19

Independent Trials, Random Variables

INDEPENDENT OR REPEATED TRIALS

We have previously discussed probability spaces which were associated with an experiment repeated a finite number of times, as the tossing of a coin three times. This concept of repetition is formalized as follows:

Definition: Let S^* be a finite probability space. By n *independent* or *repeated trials*, we mean the probability space S consisting of ordered n-tuples of elements of S^* with the probability of an n-tuple defined to be the product of the probabilities of its components:

$$P((s_1, s_2, \ldots, s_n)) \;=\; P(s_1)\,P(s_2)\cdots P(s_n)$$

Example 1.1: Whenever three horses a, b and c race together, their respective probabilities of winning are $\frac{1}{2}$, $\frac{1}{3}$ and $\frac{1}{6}$. In other words, $S^* = \{a, b, c\}$ with $P(a) = \frac{1}{2}$, $P(b) = \frac{1}{3}$ and $P(c) = \frac{1}{6}$. If the horses race twice, then the sample space of the 2 repeated trials is

$$S = \{aa, ab, ac, ba, bb, bc, ca, cb, cc\}$$

For notational convenience, we have written ac for the ordered pair (a, c). The probability of each point in S is

$$P(aa) = P(a)\,P(a) = \tfrac{1}{2}\cdot\tfrac{1}{2} = \tfrac{1}{4} \qquad P(ba) = \tfrac{1}{6} \qquad P(ca) = \tfrac{1}{12}$$

$$P(ab) = P(a)\,P(b) = \tfrac{1}{2}\cdot\tfrac{1}{3} = \tfrac{1}{6} \qquad P(bb) = \tfrac{1}{9} \qquad P(cb) = \tfrac{1}{18}$$

$$P(ac) = P(a)\,P(c) = \tfrac{1}{2}\cdot\tfrac{1}{6} = \tfrac{1}{12} \qquad P(bc) = \tfrac{1}{18} \qquad P(cc) = \tfrac{1}{36}$$

Thus the probability of c winning the first race and a winning the second race is $P(ca) = \frac{1}{12}$.

From another point of view, a repeated trials process is a stochastic process whose tree diagram has the following properties: (i) every branch point has the same outcomes; (ii) the probability is the same for each branch leading to the same outcome. For example, the tree diagram of the repeated trials process of the preceding experiment is as shown in the adjoining figure.

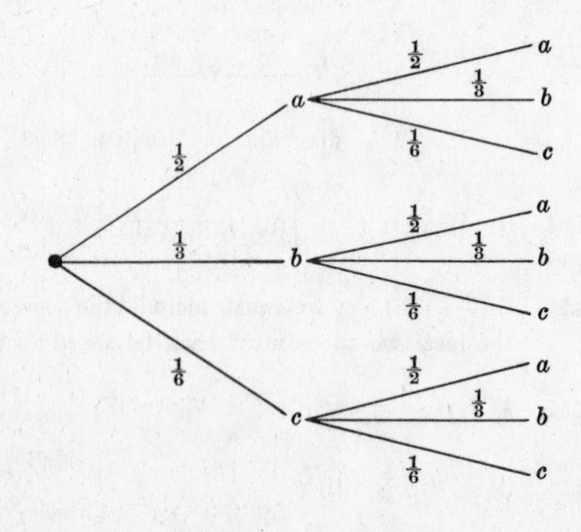

Observe that every branch point has the outcomes a, b and c, and each branch leading to outcome a has probability $\frac{1}{2}$, each branch leading to b has probability $\frac{1}{3}$, and each leading to c has probability $\frac{1}{6}$.

REPEATED TRIALS WITH TWO OUTCOMES

We now investigate repeated trials of an experiment with two outcomes; we call one of the outcomes *success* and the other outcome *failure*. Let p be the probability of success, and so $q = 1 - p$ is the probability of failure. Frequently we are interested in the number of successes and not in the order in which they occur. The following theorem applies.

Theorem 19.1: The probability of exactly k successes in n repeated trials is denoted and given by

$$f(n, k, p) \;=\; \binom{n}{k} p^k q^{n-k}$$

The probability of at least one success is $1 - q^n$.

Here $\binom{n}{k}$ is the binomial coefficient (see Chapter 13).

Example 2.1: A fair coin is tossed 6 times or, equivalently, six fair coins are tossed; call heads a success. Then $n = 6$ and $p = q = \frac{1}{2}$.

(i) The probability that exactly two heads occurs (i.e. $k = 2$) is

$$f(6, 2, \tfrac{1}{2}) \;=\; \binom{6}{2} (\tfrac{1}{2})^2 (\tfrac{1}{2})^4 \;=\; \tfrac{15}{64}$$

(ii) The probability of getting at least four heads (i.e. $k = 4, 5$ or 6) is

$$f(6,4,\tfrac{1}{2}) + f(6,5,\tfrac{1}{2}) + f(6,6,\tfrac{1}{2}) \;=\; \binom{6}{4}(\tfrac{1}{2})^4(\tfrac{1}{2})^2 + \binom{6}{5}(\tfrac{1}{2})^5(\tfrac{1}{2}) + \binom{6}{6}(\tfrac{1}{2})^6$$
$$=\; \tfrac{15}{64} + \tfrac{6}{64} + \tfrac{1}{64} \;=\; \tfrac{11}{32}$$

(iii) The probability of no heads (i.e. all failures) is $q^6 = (\frac{1}{2})^6 = \frac{1}{64}$, and so the probability of at least one heads is $1 - q^6 = 1 - \frac{1}{64} = \frac{63}{64}$.

Example 2.2: A fair die is tossed 7 times; call a toss a success if a 5 or a 6 appears. Then $n = 7$, $p = P(\{5, 6\}) = \frac{1}{3}$ and $q = 1 - p = \frac{2}{3}$.

(i) The probability that a 5 or a 6 occurs exactly 3 times (i.e. $k = 3$) is

$$f(7, 3, \tfrac{1}{3}) \;=\; \binom{7}{3} (\tfrac{1}{3})^3 (\tfrac{2}{3})^4 \;=\; \tfrac{560}{2187}$$

(ii) The probability that a 5 or a 6 never occurs (i.e. all failures) is $q^7 = (\frac{2}{3})^7 = \frac{128}{2187}$; hence the probability that a 5 or a 6 occurs at least once is $1 - q^7 = \frac{2059}{2187}$.

If we treat n and p as constants, then the function defined by $f(n, k, p)$ is called the *binomial distribution* since for $k = 0, 1, 2, \ldots, n$ it corresponds to the successive terms of the binomial expansion:

$$(q+p)^n \;=\; q^n + \binom{n}{1} q^{n-1} p + \binom{n}{2} q^{n-2} p^2 + \cdots + p^n$$
$$=\; f(n,0,p) + f(n,1,p) + f(n,2,p) + \cdots + f(n,n,p)$$

The use of the word distribution will be explained later in the chapter. This distribution is also called the Bernoulli distribution, and independent trials with two outcomes are called Bernoulli trials.

RANDOM VARIABLES

Let S be a sample space of some experiment. As noted previously, the outcomes of the experiment, i.e. the sample points of S, need not be numbers. However, we frequently wish to assign a specific number to each outcome of the experiment. Such an assignment is called a random variable. In other words:

Definition: A *random variable* X is a function from a sample space S into the real numbers.

We shall let R_X denote the range of a random variable X: $R_X = X(S)$. We shall refer to R_X as the *range space*.

Remark: If the sample space S is nondenumerable, then for theoretical reasons lying beyond the scope of this text certain functions defined on S are not random variables. However, the sample spaces here are finite, and every function defined on a finite sample space is a random variable.

Example 3.1: A pair of dice is tossed. The sample space S consists of the 36 ordered pairs of numbers between 1 and 6:

$$S = \{(1,1), (1,2), \ldots, (6,6)\}$$

Let X assign to each point in S the sum of the numbers; then X is a random variable with range space

$$R_X = \{2, 3, 4, 5, 6, 7, 8, 9, 10, 11, 12\}$$

Let Y assign to each point the maximum of the two numbers; then Y is a random variable with range space

$$R_Y = \{1, 2, 3, 4, 5, 6\}$$

Example 3.2: A sample of 3 items is selected from a box containing, say, 12 items of which 3 are defective. The sample space S consists of the $\binom{12}{3} = 220$ different samples of size 3. Let X denote the number of defective items in the sample; then X is a random variable with range space $R_X = \{0, 1, 2, 3\}$.

PROBABILITY SPACE AND DISTRIBUTION OF A RANDOM VARIABLE

Let $R_X = \{x_1, x_2, \ldots, x_t\}$ be the range space of a random variable X defined on a finite sample space S. Then X induces an assignment of probabilities on the range space R_X as follows:

$$p_i = P(x_i) = \text{sum of probabilities of points in } S \text{ whose image is } x_i$$

The function assigning p_i to x_i or, in other words, the set of ordered pairs $(x_1, p_1), \ldots, (x_t, p_t)$, usually given by a table

x_1	x_2	$\cdots$	x_t
p_1	p_2	$\cdots$	p_t

is called the *distribution* of the random variable X.

In the case that S is an equiprobable space, we can easily obtain the distribution of a random variable from the following result.

Theorem 19.2: Let S be an equiprobable space. If $R_X = \{x_1, \ldots, x_t\}$ is the range space of a random variable X defined on S, then

$$p_i = P(x_i) = \frac{\text{number of points in } S \text{ whose image is } x_i}{\text{number of points in } S}$$

Example 4.1: A pair of fair dice is tossed. We obtain the equiprobable space S consisting of the 36 ordered pairs of numbers between 1 and 6:

$$S = \{(1,1), (1,2), \ldots, (6,6)\}$$

Let X be the random variable which assigns to each point the sum of the two numbers; then X has range space

$$R_X = \{2, 3, 4, 5, 6, 7, 8, 9, 10, 11, 12\}$$

We use Theorem 19.2 to obtain the distribution of X. There is only one point $(1,1)$ whose image is 2; hence $P(2) = \frac{1}{36}$. There are two points in S, $(1,2)$ and $(2,1)$, having image 3; hence $P(3) = \frac{2}{36}$. There are three points in S, $(1,3)$, $(2,2)$ and $(3,1)$, having image 4; hence $P(4) = \frac{3}{36}$. Similarly, $P(5) = \frac{4}{36}$, $P(6) = \frac{5}{36}$, etc. The distribution of X consists of the points in R_X with their respective probabilities:

x_i	2	3	4	5	6	7	8	9	10	11	12
p_i	$\frac{1}{36}$	$\frac{2}{36}$	$\frac{3}{36}$	$\frac{4}{36}$	$\frac{5}{36}$	$\frac{6}{36}$	$\frac{5}{36}$	$\frac{4}{36}$	$\frac{3}{36}$	$\frac{2}{36}$	$\frac{1}{36}$

Example 4.2: A sample of 3 items is selected at random from a box containing 12 items of which 3 are defective. The sample space S consists of the $\binom{12}{3} = 220$ different equally likely samples of size 3. Let X be the random variable which assigns to each sample the number of defective items in the sample; then $R_X = \{0, 1, 2, 3\}$. Again we use Theorem 19.2 to obtain the distribution of X.

There are $\binom{9}{3} = 84$ samples of size 3 which contain no defective items; hence $P(0) = \frac{84}{220}$. There are $3 \cdot \binom{9}{2} = 108$ samples of size 3 containing 1 defective item; hence $P(1) = \frac{108}{220}$. There are $\binom{3}{2} \cdot 9 = 27$ samples of size 3 containing 2 defective items; hence $P(2) = \frac{27}{220}$. There is only one sample of size 3 containing the 3 defective items; hence $P(3) = \frac{1}{220}$. The distribution of X follows:

x_i	0	1	2	3
p_i	$\frac{84}{220}$	$\frac{108}{220}$	$\frac{27}{220}$	$\frac{1}{220}$

Example 4.3: A coin weighted so that $P(\text{H}) = \frac{2}{3}$ and $P(\text{T}) = \frac{1}{3}$ is tossed three times. The probabilities of the points in the sample space $S = \{\text{HHH, HHT, HTH, HTT, THH, THT, TTH, TTT}\}$ are as follows:

$$P(\text{HHH}) = \tfrac{2}{3}\cdot\tfrac{2}{3}\cdot\tfrac{2}{3} = \tfrac{8}{27} \qquad P(\text{THH}) = \tfrac{1}{3}\cdot\tfrac{2}{3}\cdot\tfrac{2}{3} = \tfrac{4}{27}$$
$$P(\text{HHT}) = \tfrac{2}{3}\cdot\tfrac{2}{3}\cdot\tfrac{1}{3} = \tfrac{4}{27} \qquad P(\text{THT}) = \tfrac{1}{3}\cdot\tfrac{2}{3}\cdot\tfrac{1}{3} = \tfrac{2}{27}$$
$$P(\text{HTH}) = \tfrac{2}{3}\cdot\tfrac{1}{3}\cdot\tfrac{2}{3} = \tfrac{4}{27} \qquad P(\text{TTH}) = \tfrac{1}{3}\cdot\tfrac{1}{3}\cdot\tfrac{2}{3} = \tfrac{2}{27}$$
$$P(\text{HTT}) = \tfrac{2}{3}\cdot\tfrac{1}{3}\cdot\tfrac{1}{3} = \tfrac{2}{27} \qquad P(\text{TTT}) = \tfrac{1}{3}\cdot\tfrac{1}{3}\cdot\tfrac{1}{3} = \tfrac{1}{27}$$

Let X be the random variable which assigns to each point in S the largest number of successive heads which occurs. Thus,

$$X(\text{TTT}) = 0$$
$$X(\text{HTH}) = 1, \quad X(\text{HTT}) = 1, \quad X(\text{THT}) = 1, \quad X(\text{TTH}) = 1$$
$$X(\text{HHT}) = 2, \quad X(\text{THH}) = 2$$
$$X(\text{HHH}) = 3$$

The range space of X is $R_X = \{0, 1, 2, 3\}$. We obtain $P(0)$, $P(1)$, $P(2)$ and $P(3)$ as follows:

$$P(0) = P(\text{TTT}) = \tfrac{1}{27}$$
$$P(1) = P(\text{HTH}) + P(\text{HTT}) + P(\text{THT}) + P(\text{TTH}) = \tfrac{4}{27} + \tfrac{2}{27} + \tfrac{2}{27} + \tfrac{2}{27} = \tfrac{10}{27}$$
$$P(2) = P(\text{HHT}) + P(\text{THH}) = \tfrac{4}{27} + \tfrac{4}{27} = \tfrac{8}{27}$$
$$P(3) = P(\text{HHH}) = \tfrac{8}{27}$$

The distribution of X consists of the points in the range space of X with their respective probabilities:

x_i	0	1	2	3
$P(x_i)$	$\frac{1}{27}$	$\frac{10}{27}$	$\frac{8}{27}$	$\frac{8}{27}$

Now consider n repeated trials of an experiment with two outcomes: that of success with probability p and that of failure with probability $q = 1-p$. To each sequence of n outcomes, let X_n assign the number of successes. Then X_n is a random variable with range space $R_{X_n} = \{0, 1, 2, \ldots, n\}$. Using Theorem 19.1, we obtain

$$P(0) = f(n, 0, p) = q^n$$
$$P(1) = f(n, 1, p) = \binom{n}{1} q^{n-1} p$$
$$P(2) = f(n, 2, p) = \binom{n}{2} q^{n-2} p^2$$
$$\cdots\cdots\cdots\cdots\cdots\cdots\cdots\cdots$$
$$P(n) = f(n, n, p) = p^n$$

In other words, the distribution of X_n is the binomial distribution as discussed previously:

x_i	0	1	2	$\cdots$	n
p_i	q^n	$\binom{n}{1} q^{n-1} p$	$\binom{n}{2} q^{n-2} p^2$	$\cdots$	p^n

EXPECTED VALUE

If the outcomes of an experiment are the numbers $x_1, x_2, \ldots, x_r$ which occur with respective probabilities $p_1, p_2, \ldots, p_r$, then the expected value is defined to be

$$E = x_1 p_1 + x_2 p_2 + \cdots + x_r p_r$$

In particular, if the x_i and p_i come from the distribution of a random variable X, then the expected value is called the expectation of the random variable and is denoted by $E(X)$.

Example 5.1: Consider the random variable X of Example 4.2. Its distribution gives the possible numbers of defective items in a sample of size 3 with their respective probabilities of occurrence. Then the expected number of defective items in a sample of size 3, i.e. the expectation of the random variable X, is

$$E(X) = 0 \cdot \tfrac{84}{220} + 1 \cdot \tfrac{108}{220} + 2 \cdot \tfrac{27}{220} + 3 \cdot \tfrac{1}{220} = \tfrac{3}{5}$$

Example 5.2: A fair coin is tossed 6 times. The possible numbers of heads which can occur with their respective probabilities are as follows:

x_i	0	1	2	3	4	5	6
p_i	$\frac{1}{64}$	$\frac{6}{64}$	$\frac{15}{64}$	$\frac{20}{64}$	$\frac{15}{64}$	$\frac{6}{64}$	$\frac{1}{64}$

Note that the above table is the binomial distribution with $n = 6$ and $p = \frac{1}{2}$. The expected value, i.e., the expected number of heads, is

$$E = 0 \cdot \tfrac{1}{64} + 1 \cdot \tfrac{6}{64} + 2 \cdot \tfrac{15}{64} + 3 \cdot \tfrac{20}{64} + 4 \cdot \tfrac{15}{64} + 5 \cdot \tfrac{6}{64} + 6 \cdot \tfrac{1}{64} = 3$$

Observe that $E = np$.

The result in the preceding example holds true in general; namely,

Theorem 19.3: For the binomial distribution

x_i	0	1	2	$\cdots$	n
p_i	q^n	$\binom{n}{1} q^{n-1} p$	$\binom{n}{2} q^{n-2} p^2$	$\cdots$	p^n

the expected value is $E = np$.

Example 5.3: The probability is 0.2 that an item produced by a factory is defective. If 3000 items are made by the factory, then the expected number of defective items is $E = np = (3000)(0.2) = 60$.

GAMBLING GAMES

In a gambling game, the expected value E is considered to be the value of the game to the player. The game is said to be *favorable* to the player if E is positive, and *unfavorable* if E is negative. If $E = 0$, the game is *fair*.

Example 6.1: A player tosses a fair die. If a prime number occurs he wins that number of dollars, but if a non-prime number occurs he loses that number of dollars. The possible outcomes of the game with their respective probabilities are as follows:

x_i	2	3	5	-1	-4	-6
p_i	$\frac{1}{6}$	$\frac{1}{6}$	$\frac{1}{6}$	$\frac{1}{6}$	$\frac{1}{6}$	$\frac{1}{6}$

The negative numbers -1, -4 and -6 correspond to the fact that the player loses if a non-prime number occurs. The expected value of the game is

$$E = 2\cdot\tfrac{1}{6} + 3\cdot\tfrac{1}{6} + 5\cdot\tfrac{1}{6} - 1\cdot\tfrac{1}{6} - 4\cdot\tfrac{1}{6} - 6\cdot\tfrac{1}{6} = -\tfrac{1}{6}$$

Thus the game is unfavorable to the player since the expected value is negative.

Example 6.2: A player tosses a fair coin. If a head occurs he receives 5 dollars, but if a tail occurs he receives only 1 dollar. His expected winnings $E = 5\cdot\frac{1}{2} + 1\cdot\frac{1}{2} = 3$.

Question: How much money should he pay to play the game so that the game is fair? We claim that the answer is 3 dollars. For if he pays 3 dollars to play the game, then each outcome is decreased by 3 dollars. Thus when a head occurs he wins $5 - 3 = 2$ dollars and when a tail occurs he wins $1 - 3 = -2$ dollars, i.e., he loses 2 dollars; hence the expected value is now $E = 2\cdot\frac{1}{2} + (-2)\cdot\frac{1}{2} = 0$.

The result in the preceding example holds true in general; namely,

Theorem 19.4: Let E be the expected value of an experiment. If a constant k is added to each outcome of the experiment, then the expected value E^* of the new experiment is $E + k$: $E^* = E + k$.

Example 6.3: A thousand tickets at \$1 each are sold in a lottery in which there is one prize of \$500, four prizes of \$100 each, and five prizes of \$10 each. The expected winnings of one ticket are

$$500\cdot\frac{1}{1000} + 100\cdot\frac{4}{1000} + 10\cdot\frac{5}{1000} = 0.95$$

However, since a ticket costs \$1, the expected value of the game is $0.95 - 1.00 = -\$0.05$, which is unfavorable to the player.

Solved Problems

REPEATED TRIALS WITH TWO OUTCOMES

19.1. Find (i) $f(5,2,\frac{1}{3})$, (ii) $f(6,3,\frac{1}{2})$, (iii) $f(4,3,\frac{1}{4})$. Here

$$f(n,k,p) = \binom{n}{k} p^k (1-p)^{n-k}$$

(i) $f(5,2,\frac{1}{3}) = \binom{5}{2}(\frac{1}{3})^2(\frac{2}{3})^3 = \frac{5\cdot 4}{1\cdot 2}(\frac{1}{3})^2(\frac{2}{3})^3 = \frac{80}{243}$

(ii) $f(6,3,\frac{1}{2}) = \binom{6}{3}(\frac{1}{2})^3(\frac{1}{2})^3 = \frac{6\cdot 5\cdot 4}{1\cdot 2\cdot 3}(\frac{1}{2})^3(\frac{1}{2})^3 = \frac{5}{16}$

(iii) $f(4,3,\frac{1}{4}) = \binom{4}{3}(\frac{1}{4})^3(\frac{3}{4}) = \binom{4}{1}(\frac{1}{4})^3(\frac{3}{4}) = \frac{4}{1}(\frac{1}{4})^3(\frac{3}{4}) = \frac{3}{64}$

19.2. A fair coin is tossed three times. Find the probability P that there will appear (i) three heads, (ii) two heads, (iii) one head, (iv) no heads.

Method 1. We obtain the following equiprobable space of eight elements:

$$S = \{\text{HHH, HHT, HTH, HTT, THH, THT, TTH, TTT}\}$$

(i) Three heads (HHH) occurs only once among the eight sample points; hence $P = \frac{1}{8}$.

(ii) Two heads occurs 3 times (HHT, HTH, and THH); hence $P = \frac{3}{8}$.

(iii) One head occurs 3 times (HTT, THT and TTH); hence $P = \frac{3}{8}$.

(iv) No heads, i.e. three tails (TTT), occurs only once; hence $P = \frac{1}{8}$.

Method 2. Use Theorem 19.1 with $n = 3$ and $p = q = \frac{1}{2}$.

(i) Here $k = 3$ and $P = f(3,3,\frac{1}{2}) = \binom{3}{3}(\frac{1}{2})^3(\frac{1}{2})^0 = 1\cdot\frac{1}{8}\cdot 1 = \frac{1}{8}$.

(ii) Here $k = 2$ and $P = f(3,2,\frac{1}{2}) = \binom{3}{2}(\frac{1}{2})^2(\frac{1}{2}) = 3\cdot\frac{1}{4}\cdot\frac{1}{2} = \frac{3}{8}$.

(iii) Here $k = 1$ and $P = f(3,1,\frac{1}{2}) = \binom{3}{1}(\frac{1}{2})^1(\frac{1}{2})^2 = 3\cdot\frac{1}{2}\cdot\frac{1}{4} = \frac{3}{8}$.

(iv) Here $k = 0$ and $P = f(3,0,\frac{1}{2}) = \binom{3}{0}(\frac{1}{2})^0(\frac{1}{2})^3 = 1\cdot 1\cdot\frac{1}{8} = \frac{1}{8}$.

19.3. Suppose that 20% of the items produced by a factory are defective. If 4 items are chosen at random, what is the probability P that (i) 2 are defective, (ii) 3 are defective, (iii) at least one is defective?

Use Theorem 19.1 with $n = 4$, $p = .2$ and $q = 1 - p = .8$.

(i) Here $k = 2$ and $P = f(4,2,.2) = \binom{4}{2}(.2)^2(.8)^2 = .1536$.

(ii) Here $k = 3$ and $P = f(4,3,.2) = \binom{4}{3}(.2)^3(.8) = .0256$.

(iii) $P = 1 - q^4 = 1 - (.8)^4 = 1 - .4096 = .5904$.

19.4. Team A has probability $\frac{2}{3}$ of winning whenever it plays. If A plays 4 games, find the probability that A wins (i) exactly 2 games, (ii) at least 1 game, (iii) more than half of the games.

Here $n = 4$, $p = \frac{2}{3}$ and $q = 1 - p = \frac{1}{3}$.

(i) We seek $f(n, k, p)$ when $k = 2$: $f(4, 2, \frac{2}{3}) = \binom{4}{2}(\frac{2}{3})^2(\frac{1}{3})^2 = \frac{8}{27}$.

(ii) Here $q^4 = (\frac{1}{3})^4 = \frac{1}{81}$ is the probability that A loses all four games. Then $1 - q^4 = \frac{80}{81}$ is the probability of winning at least one game.

(iii) A wins more than half the games if A wins 3 or 4 games. Hence the required probability is

$$f(4, 3, \tfrac{1}{3}) + f(4, 4, \tfrac{1}{3}) = \binom{4}{3}(\tfrac{2}{3})^3(\tfrac{1}{3}) + \binom{4}{4}(\tfrac{2}{3})^4 = \tfrac{32}{81} + \tfrac{16}{81} = \tfrac{16}{27}$$

19.5. A family has 6 children. Find the probability that there are (i) 3 boys and 3 girls, (ii) fewer boys than girls. Assume that the probability of the birth of either sex is $\frac{1}{2}$.

(i) The required probability is

$$f(6, 3, \tfrac{1}{2}) = \binom{6}{3}(\tfrac{1}{2})^3(\tfrac{1}{2})^3 = \tfrac{20}{64} = \tfrac{5}{16}$$

(ii) There are fewer boys than girls if there are 0, 1 or 2 boys. Hence the required probability is

$$f(6, 0, \tfrac{1}{2}) + f(6, 1, \tfrac{1}{2}) + f(6, 2, \tfrac{1}{2}) = (\tfrac{1}{2})^6 + \binom{6}{1}(\tfrac{1}{2})^5(\tfrac{1}{2}) + \binom{6}{2}(\tfrac{1}{2})^4(\tfrac{1}{2})^2 = \tfrac{11}{32}$$

19.6. The probability that a college student does not graduate is .3. Of 5 college students chosen at random, find the probability that (i) one will not graduate, (ii) three will not graduate, (iii) at least one will not graduate.

Here $n = 5$, $p = .3$ and $q = 1 - p = .7$.

(i) $f(5, 1, .3) = \binom{5}{1}(.3)(.7)^4 = .36015$

(ii) $f(5, 3, .3) = \binom{5}{3}(.3)^3(.7)^2 = .1323$

(iii) $1 - q^5 = 1 - (.7)^5 = 1 - .16807 = .83193$

19.7. Consider n repeated trials with two outcomes and $p = q = \frac{1}{2}$.

(i) Find the probability of exactly k successes.

(ii) Find the probability P of more successes than failures if n is (*a*) odd, (*b*) even.

(i) The probability of exactly k successes is

$$f(n, k, \tfrac{1}{2}) = \binom{n}{k}(\tfrac{1}{2})^k(\tfrac{1}{2})^{n-k} = \binom{n}{k}(\tfrac{1}{2})^n$$

(ii) Recall that the distribution of the number of successes appears in the expansion of the binomial $(q+p)^n = (\frac{1}{2} + \frac{1}{2})^n$:

$$1 = (\tfrac{1}{2} + \tfrac{1}{2})^n = (\tfrac{1}{2})^n + \binom{n}{1}(\tfrac{1}{2})^n + \binom{n}{2}(\tfrac{1}{2})^n + \cdots + \binom{n}{2}(\tfrac{1}{2})^n + \binom{n}{1}(\tfrac{1}{2})^n + (\tfrac{1}{2})^n$$

Since the binomial coefficients are symmetrical about the center of the expansion and since each term is $(\frac{1}{2})^n$ times a binomial coefficient, the above terms are also symmetrical about the center of the expansion.

(*a*) If n is odd, then there is an even number of terms in the above expansion and so there are more successes than failures in exactly the terms in the last half of the expansion. Thus $1 = P + P$ or $P = \frac{1}{2}$.

(*b*) If n is even, then there is an odd number of terms in the above expansion, with the middle term $M = \binom{n}{n/2}(\frac{1}{2})^n$. Now there are more successes than failures in the terms to the right of the middle term; hence by the symmetry, $1 = P + M + P$ or $P = \frac{1}{2}(1 - M)$.

19.8. The probability of team A winning any game is $\frac{1}{2}$. What is the probability P that A wins more than half of its games if A plays (i) 43 games, (ii) 8 games?

(i) By the preceding problem, $P = \frac{1}{2}$ since n is odd and $p = q = \frac{1}{2}$.

(ii) The middle term M of the expansion $(\frac{1}{2} + \frac{1}{2})^8$, i.e. the probability that A wins 4 games and loses 4 games, is $\binom{8}{4}(\frac{1}{2})^8 = \frac{35}{128}$. Then by the preceding problem, $P = \frac{1}{2}(1 - M) = \frac{1}{2}(1 - \frac{35}{128}) = \frac{93}{256}$.

19.9. How many dice must be thrown so that there is a better than even chance of obtaining a six?

The probability of not obtaining a six on n dice is $(\frac{5}{6})^n$. Hence we seek the smallest n for which $(\frac{5}{6})^n$ is *less* than $\frac{1}{2}$:

$$(\tfrac{5}{6})^1 = \tfrac{5}{6};\quad (\tfrac{5}{6})^2 = \tfrac{25}{36};\quad (\tfrac{5}{6})^3 = \tfrac{125}{216};\quad \text{but } (\tfrac{5}{6})^4 = \tfrac{625}{1296} < \tfrac{1}{2}$$

Thus 4 dice must be thrown.

19.10. The probability of a man hitting a target is $\frac{1}{4}$. (i) If he fires 7 times, what is the probability P of his hitting the target at least twice? (ii) How many times must he fire so that the probability of his hitting the target at least once is greater than $\frac{2}{3}$?

(i) We seek the sum of the probabilities for $k = 2, 3, 4, 5, 6$ and 7. It is simpler in this case to find the sum of the probabilities for $k = 0$ and 1, i.e. no hits or 1 hit, and then subtract it from 1.

$$f(7, 0, \tfrac{1}{4}) = (\tfrac{3}{4})^7 = \tfrac{2187}{16,384}, \qquad f(7, 1, \tfrac{1}{4}) = \binom{7}{1}(\tfrac{1}{4})(\tfrac{3}{4})^6 = \tfrac{5103}{16,384}$$

Then $P = 1 - \frac{2187}{16,384} - \frac{5103}{16,384} = \frac{4547}{8192}$.

(ii) The probability of not hitting the target is q^n. Thus we seek the smallest n for which q^n is less than $1 - \frac{2}{3} = \frac{1}{3}$, where $q = 1 - p = 1 - \frac{1}{4} = \frac{3}{4}$. Hence compute successive powers of q until $q^n < \frac{1}{3}$ is obtained:

$$(\tfrac{3}{4})^1 = \tfrac{3}{4} \nless \tfrac{1}{3};\quad (\tfrac{3}{4})^2 = \tfrac{9}{16} \nless \tfrac{1}{3};\quad (\tfrac{3}{4})^3 = \tfrac{27}{64} \nless \tfrac{1}{3};\quad \text{but } (\tfrac{3}{4})^4 = \tfrac{81}{256} < \tfrac{1}{3}$$

In other words, he must fire 4 times.

19.11. Prove Theorem 19.1: The probability of exactly k successes in n repeated trials is $f(n, k, p) = \binom{n}{k} p^k q^{n-k}$. The probability of at least one success is $1 - q^n$.

The sample space of the n repeated trials consists of all ordered n-tuples whose components are either s (success) or f (failure). The event A of k successes consists of all ordered n-tuples of which k components are s and the other $n - k$ components are f. The number of n-tuples in the event A is equal to the number of ways that k letters s can be distributed among the n components of an n-tuple; hence A consists of $\binom{n}{k}$ sample points. Since the probability of each point in A is $p^k q^{n-k}$, we have $P(A) = \binom{n}{k} p^k q^{n-k}$.

The event B of no successes consists of the single n-tuple $(f, f, \ldots, f)$, i.e. all failures, whose probability is q^n. Accordingly, the probability of at least one success is $1 - q^n$.

19.12. The mathematics department has 8 graduate assistants who are assigned to the same office. Each assistant is just as likely to study at home as in the office. How many desks must there be in the office so that each assistant has a desk at least 90% of the time?

If there are 8 desks, then 0% of the time an assistant will not have a desk at which to study.

Suppose there are 7 desks; then an assistant will not have a desk at which to study if all 8 assistants are in the office. The probability that 8 will study in the office is $(\frac{1}{2})^8 = \frac{1}{256} = 0.4\%$. Thus there will not be enough desks 0.4% of the time.

Suppose there are 6 desks; then an assistant will not have a desk if 8 or 7 assistants are in the office. The probability that 7 will study in the office is $f(8,7,\frac{1}{2}) = \frac{8}{256} = 3.2\%$. Thus there will not be enough desks $3.2\% + 0.4\% = 3.6\%$ of the time.

Suppose there are 5 desks; then an assistant will not have a desk if 8, 7 or 6 assistants are in the office. The probability that 6 will study in the office is $f(8,6,\frac{1}{2}) = \frac{28}{256} = 10.9\%$. Thus there will not be enough desks $10.9\% + 3.2\% + 0.4\% = 14.5\%$ of the time. In other words, there will be enough desks only 85.5% of the time.

Accordingly, 6 desks are required.

RANDOM VARIABLES

19.13. A pair of fair dice is thrown. Let X be the random variable which assigns to each point the maximum of the two numbers which occur: $X((a,b)) = \max(a,b)$. (i) Find the distribution of X. (ii) Find the expectation of X.

(i) The sample space S is the equiprobable space consisting of the 36 ordered pairs of numbers between 1 and 6: $S = \{(1,1), (1,2), \ldots, (6,6)\}$. Since X assigns to each point in S the maximum of the two numbers in the ordered pair, the range space of X consists of the numbers from 1 through 6: $R_X = \{1,2,3,4,5,6\}$.

There is only one point $(1,1)$ whose maximum is 1; hence $P(1) = \frac{1}{36}$. Each of three points in S, $(1,2)$, $(2,2)$ and $(2,1)$, has maximum 2; hence $P(2) = \frac{3}{36}$. Each of five points in S, $(1,3)$, $(2,3)$, $(3,3)$, $(3,2)$ and $(3,1)$, has 3; hence $P(3) = \frac{5}{36}$. Similarly, $P(4) = \frac{7}{36}$, $P(5) = \frac{9}{36}$ and $P(6) = \frac{11}{36}$.

The distribution of X consists of the points in R_X with their respective probabilities:

x_i	1	2	3	4	5	6
p_i	$\frac{1}{36}$	$\frac{3}{36}$	$\frac{5}{36}$	$\frac{7}{36}$	$\frac{9}{36}$	$\frac{11}{36}$

(ii) To obtain the expectation of X, i.e. the expected value, multiply each number x_i by its probability p_i and take the sum:

$$E = 1\cdot\tfrac{1}{36} + 2\cdot\tfrac{3}{36} + 3\cdot\tfrac{5}{36} + 4\cdot\tfrac{7}{36} + 5\cdot\tfrac{9}{36} + 6\cdot\tfrac{11}{36} = \tfrac{161}{36} = 4.5$$

19.14. Suppose a random variable X takes on the values -3, -1, 2 and 5 with respective probabilities $\dfrac{2x-3}{10}$, $\dfrac{x+1}{10}$, $\dfrac{x-1}{10}$ and $\dfrac{x-2}{10}$. Determine the distribution and expectation of X.

Set the sum of the probabilities equal to 1, and solve to find $x = 3$. Then put $x = 3$ into the above probabilities and obtain .3, .4, .2 and .1 respectively. The distribution of X is

x_i	-3	-1	2	5
p_i	.3	.4	.2	.1

The expectation is obtained by multiplying each value by its probability and taking the sum:

$$E = (-3)(.3) + (-1)(.4) + 2(.2) + 5(.1) = -.4$$

19.15. Four fair coins are tossed. Let X denote the number of heads occurring. Find the distribution and expectation of the random variable X.

Since this experiment consists of four independent trials, we obtain the binomial distribution with $n = 4$ and $p = \frac{1}{2}$:

$$P(0) = f(4, 0, \tfrac{1}{2}) = \tfrac{1}{16} \qquad P(3) = f(4, 3, \tfrac{1}{2}) = \tfrac{4}{16}$$
$$P(1) = f(4, 1, \tfrac{1}{2}) = \tfrac{4}{16} \qquad P(4) = f(4, 4, \tfrac{1}{2}) = \tfrac{1}{16}$$
$$P(2) = f(4, 2, \tfrac{1}{2}) = \tfrac{6}{16}$$

Thus the distribution is

x_i	0	1	2	3	4
p_i	$\frac{1}{16}$	$\frac{4}{16}$	$\frac{6}{16}$	$\frac{4}{16}$	$\frac{1}{16}$

and the expectation is $E = 0 \cdot \frac{1}{16} + 1 \cdot \frac{4}{16} + 2 \cdot \frac{6}{16} + 3 \cdot \frac{4}{16} + 4 \cdot \frac{1}{16} = 2$. This agrees with Theorem 19.3 which states that the expected value $E = np = 4 \cdot \frac{1}{2} = 2$.

19.16. A coin weighted so that $P(\text{H}) = \frac{3}{4}$ and $P(\text{T}) = \frac{1}{4}$ is tossed three times. Let X be the random variable which denotes the longest string of heads which occurs. Find the distribution and expectation of X.

The random variable X is defined on the sample space

$$S = \{\text{HHH, HHT, HTH, HTT, THH, THT, TTH, TTT}\}$$

The points in S have the following respective probabilities:

$$P(\text{HHH}) = \tfrac{3}{4} \cdot \tfrac{3}{4} \cdot \tfrac{3}{4} = \tfrac{27}{64} \qquad P(\text{THH}) = \tfrac{1}{4} \cdot \tfrac{3}{4} \cdot \tfrac{3}{4} = \tfrac{9}{64}$$
$$P(\text{HHT}) = \tfrac{3}{4} \cdot \tfrac{3}{4} \cdot \tfrac{1}{4} = \tfrac{9}{64} \qquad P(\text{THT}) = \tfrac{1}{4} \cdot \tfrac{3}{4} \cdot \tfrac{1}{4} = \tfrac{3}{64}$$
$$P(\text{HTH}) = \tfrac{3}{4} \cdot \tfrac{1}{4} \cdot \tfrac{3}{4} = \tfrac{9}{64} \qquad P(\text{TTH}) = \tfrac{1}{4} \cdot \tfrac{1}{4} \cdot \tfrac{3}{4} = \tfrac{3}{64}$$
$$P(\text{HTT}) = \tfrac{3}{4} \cdot \tfrac{1}{4} \cdot \tfrac{1}{4} = \tfrac{3}{64} \qquad P(\text{TTT}) = \tfrac{1}{4} \cdot \tfrac{1}{4} \cdot \tfrac{1}{4} = \tfrac{1}{64}$$

Since X denotes the longest string of heads,

$$X(\text{TTT}) = 0; \quad X(\text{HTT}) = 1, \ X(\text{HTH}) = 1, \ X(\text{THT}) = 1, \ X(\text{TTH}) = 1;$$
$$X(\text{HHT}) = 2, \ X(\text{THH}) = 2; \quad X(\text{HHH}) = 3$$

Thus the range space of X is $R_X = \{0, 1, 2, 3\}$. The probability of each number x_i in R_X is obtained by summing the probabilities of the points in S whose image is x_i:

$$P(0) = P(\text{TTT}) = \tfrac{1}{64} \qquad P(2) = P(\text{HHT}) + P(\text{THH}) = \tfrac{18}{64}$$
$$P(1) = P(\text{HTT}) + P(\text{HTH}) + P(\text{THT}) + P(\text{TTH}) = \tfrac{18}{64} \qquad P(3) = P(\text{HHH}) = \tfrac{27}{64}$$

The distribution of X consists of the numbers in the range space R_X with their respective probabilities:

x_i	0	1	2	3
p_i	$\frac{1}{64}$	$\frac{18}{64}$	$\frac{18}{64}$	$\frac{27}{64}$

The expectation of X is $E = 0 \cdot \frac{1}{64} + 1 \cdot \frac{18}{64} + 2 \cdot \frac{18}{64} + 3 \cdot \frac{27}{64} = \frac{135}{64} = 2.1$.

19.17. Prove Theorem 19.2: Let S be an equiprobable space. If $R_X = \{x_1, \ldots, x_t\}$ is the range space of a random variable X defined on S, then

$$p_i = P(x_i) = \frac{\text{number of points in } S \text{ whose image is } x_i}{\text{number of points in } S}$$

Let S have n points and let $s_1, s_2, \ldots, s_r$ be the points in S whose image is x_i. We wish to show that $P(x_i) = r/n$. By definition,

$$P(x_i) = \text{sum of the probabilities of the points in } S \text{ whose image is } x_i$$
$$= P(s_1) + P(s_2) + \cdots + P(s_r)$$

Since S is an equiprobable space, each of the n points in S has probability $1/n$. Hence

$$P(x_i) = \overbrace{\frac{1}{n} + \frac{1}{n} + \cdots + \frac{1}{n}}^{r \text{ times}} = \frac{r}{n}$$

19.18. Let X be a random variable defined on a sample space S and with distribution

x_1	x_2	$\cdots$	x_t
p_1	p_2	$\cdots$	p_t

Show that the following are equivalent ways of defining the expectation $E(X)$ of the random variable X:

$$E(X) = x_1p_1 + \cdots + x_tp_t = \sum_{s \in S} X(s)\,P(s)$$

Let $s_{i1}, s_{i2}, \ldots, s_{in_i}$ be the points in S with image x_i, that is, $X(s_{ij}) = x_i$. Then,

$$p_1 = P(x_1) = \text{sum of the probabilities of the points in } S \text{ whose image is } x_1$$
$$= P(s_{11}) + P(s_{12}) + \cdots + P(s_{1n_1}) = \sum_{j=1}^{n_1} P(s_{1j})$$
$$p_2 = P(x_2) = P(s_{21}) + P(s_{22}) + \cdots + P(s_{2n_2}) = \sum_{j=1}^{n_2} P(s_{2j})$$
$$\cdots\cdots\cdots\cdots\cdots\cdots\cdots\cdots\cdots\cdots\cdots\cdots$$
$$p_t = P(x_t) = P(s_{t1}) + P(s_{t2}) + \cdots + P(s_{tn_t}) = \sum_{j=1}^{n_t} P(s_{tj})$$

Since $X(s_{1j}) = x_1$,

$$x_1p_1 = x_1 \sum_{j=1}^{n_1} P(s_{1j}) = \sum_{j=1}^{n_1} x_1\,P(s_{1j}) = \sum_{j=1}^{n_1} X(s_{1j})\,P(s_{1j})$$

Similarly,

$$x_2p_2 = \sum_{j=1}^{n_2} X(s_{2j})\,P(s_{2j}), \quad \ldots, \quad x_tp_t = \sum_{j=1}^{n_t} X(s_{tj})\,P(s_{tj})$$

Since the s_{ij} range over all the points $s \in S$, we obtain

$$E(X) = x_1p_1 + \cdots + x_tp_t = \sum_{j=1}^{n_1} X(s_{1j})\,P(s_{1j}) + \cdots + \sum_{j=1}^{n_t} X(s_{tj})\,P(s_{tj})$$
$$= \sum_{i=1}^{t} \sum_{j=1}^{n_t} X(s_{ij})\,P(s_{ij}) = \sum_{s \in S} X(s)\,P(s)$$

19.19. Let X and Y be random variables defined on the same sample space S. Let $X+Y$ be the random variable on S defined by

$$(X+Y)(s) = X(s) + Y(s)$$

Prove that $E(X+Y) = E(X) + E(Y)$.

Using the result of the preceding problem, we obtain

$$E(X+Y) = \sum_{s \in S} (X+Y)(s)\,P(s) = \sum_{s \in S} (X(s) + Y(s))\,P(s)$$
$$= \sum_{s \in S} (X(s)\,P(s) + Y(s)\,P(s))$$
$$= \sum_{s \in S} X(s)\,P(s) + \sum_{s \in S} Y(s)\,P(s) = E(X) + E(Y)$$

EXPECTED VALUES

19.20. A player tosses two fair coins. He wins \$2 if 2 heads occur, and \$1 if 1 head occurs. On the other hand, he loses \$3 if no heads occur. Determine the expected value of the game and if it is favorable to the player.

Note that the probability that 2 heads occur is $\frac{1}{4}$, that 2 tails occur is $\frac{1}{4}$ and that 1 head occurs is $\frac{1}{2}$. Thus the probability of winning \$2 is $\frac{1}{4}$, of winning \$1 is $\frac{1}{2}$, and of losing \$3 is $\frac{1}{4}$. Accordingly, $E = 2\cdot\frac{1}{4} + 1\cdot\frac{1}{2} - 3\cdot\frac{1}{4} = \frac{1}{4} = .25$. In other words, the expected value of the game is 25¢ in favor of the player.

19.21. A player tosses two fair coins. He wins \$3 if 2 heads occur, and \$1 if 1 head occurs. If the game is to be fair, how much should he lose if no heads occur?

Let x denote the amount he should lose if no heads occur. We have that the probability of winning \$3 is $\frac{1}{4}$, of winning \$1 is $\frac{1}{2}$, and of losing \$$x$ is $\frac{1}{4}$. Since the game is to be fair, set $E = 0$:

$$E = 3\cdot\tfrac{1}{4} + 1\cdot\tfrac{1}{2} - x\cdot\tfrac{1}{4} = 0 \quad\text{or}\quad \tfrac{5}{4} - \tfrac{x}{4} = 0 \quad\text{or}\quad x = 5$$

In other words, the game is fair if he loses \$5 when no heads occur.

19.22. A player tosses two fair coins. He wins \$5 if 2 heads occur, \$2 if 1 head occurs and \$1 if no heads occur. (i) Find his expected winnings. (ii) How much should he pay to play the game if it is to be fair?

(i) The probability of winning \$5 is $\frac{1}{4}$, of winning \$2 is $\frac{1}{2}$, and of winning \$1 is $\frac{1}{4}$; hence $E = 5\cdot\frac{1}{4} + 2\cdot\frac{1}{2} + 1\cdot\frac{1}{4} = 2.50$, that is, the expected winnings are \$2.50.

(ii) If he pays \$2.50 to play the game, then the game is fair.

19.23. A box contains 10 items of which 2 are defective. If four items are selected from the box, what is the expected number of defective items in the sample of size 4?

There are $\binom{10}{4} = 210$ different samples of size 4 (without replacement). Since there are 8 non-defective items in the box, then $\binom{8}{4} = 70$ different samples contain no defective items. Hence the probability $P(0)$ of no defective items in the sample is $P(0) = \frac{70}{210} = \frac{1}{3}$.

$2\cdot\binom{8}{3} = 112$ different samples contain 1 defective item; hence $P(1) = \frac{112}{210} = \frac{8}{15}$.

$\binom{8}{2} = 28$ different samples contain the 2 defective items; hence $P(2) = \frac{28}{210} = \frac{2}{15}$.

Thus the expected number of defective items is $E = 0\cdot\frac{1}{3} + 1\cdot\frac{8}{15} + 2\cdot\frac{2}{15} = \frac{4}{5}$.

19.24. A fair die is tossed 300 times. Compute the expected number of sixes.

This is an independent trials experiment with $n = 300$ and $p = \frac{1}{6}$.

Hence by Theorem 19.3, $E = np = 300\cdot\frac{1}{6} = 50$.

19.25. Determine the expected number of boys in a family with 4 children, assuming the sex distribution to be equally probable. What is the probability that the expected number of boys does occur?

This is an independent (repeated) trials experiment with $n = 4$ and $p = \frac{1}{2}$; hence by Theorem 19.3, $E = np = 4\cdot\frac{1}{2} = 2$. In other words, the family can expect to have 2 boys. The probability that the family has 2 boys is $f(4, 2, \frac{1}{2}) = \frac{3}{8}$.

19.26. Prove Theorem 19.3: For the binomial distribution

x_i	0	1	$\cdots$	k	$\cdots$	n
p_i	q^n	$\binom{n}{1} q^{n-1} p$	$\cdots$	$\binom{n}{k} q^{n-k} p^k$	$\cdots$	p^n

the expected value is $E = np$.

$$E = \sum_{k=0}^{n} k \binom{n}{k} q^{n-k} p^k = \sum_{k=1}^{n} k \frac{n!}{k!\,(n-k)!} q^{n-k} p^k$$

$$= np \sum_{k=1}^{n} \frac{(n-1)!}{(k-1)!\,(n-k)!} q^{n-k} p^{k-1} = np\,(q+p)^{n-1} = np$$

since $q + p = 1$. (In the above summation, we dropped the term when $k = 0$ since it is $0 \cdot q^n = 0$.)

19.27. A fair coin is tossed until a head or 4 tails occurs. Find the expected number of tosses of the coin.

Only one toss occurs if heads occurs the first time, i.e. the event H. Two tosses occur if the first is tails and the second is heads, i.e. the event TH. Three tosses occur if the first two are tails and the third is heads, i.e. the event TTH. Four tosses occur if either TTTH or TTTT occurs. Hence $P(1) = P(\text{H}) = \frac{1}{2}$, $P(2) = P(\text{TH}) = \frac{1}{4}$, $P(3) = P(\text{TTH}) = \frac{1}{8}$, and

$$P(4) = P(\text{TTTH}) + P(\text{TTTT}) = \tfrac{1}{16} + \tfrac{1}{16} = \tfrac{1}{8}$$

Thus $E = 1 \cdot \frac{1}{2} + 2 \cdot \frac{1}{4} + 3 \cdot \frac{1}{8} + 4 \cdot \frac{1}{8} = \frac{15}{8}$.

19.28. A box contains 8 light bulbs of which 3 are defective. A bulb is selected from the box and tested until a non-defective bulb is chosen. Find the expected number of bulbs to be chosen.

The probability of choosing a non-defective bulb the first time is $P(1) = \frac{5}{8}$. The probability of first choosing a defective bulb and then a non-defective one is $P(2) = \frac{3}{8} \cdot \frac{5}{7} = \frac{15}{56}$. The probability of first selecting two defective bulbs and then a non-defective one is $P(3) = \frac{3}{8} \cdot \frac{2}{7} \cdot \frac{5}{6} = \frac{5}{56}$. Lastly, the probability of first selecting the three defective bulbs and then a non-defective one is $P(4) = \frac{3}{8} \cdot \frac{2}{7} \cdot \frac{1}{6} \cdot \frac{5}{5} = \frac{1}{56}$. Hence

$$E = 1 \cdot \tfrac{5}{8} + 2 \cdot \tfrac{15}{56} + 3 \cdot \tfrac{5}{56} + 4 \cdot \tfrac{1}{56} = \tfrac{84}{56}$$

19.29. Prove Theorem 19.4: Let E be the expected value of an experiment. If a constant k is added to each outcome of the experiment, then the expected value E^* of the new experiment is $E + k$.

Let $x_1, x_2, \ldots, x_t$ be the outcomes of the original experiment with respective probabilities $p_1, p_2, \ldots, p_t$. Note $p_1 + p_2 + \cdots + p_t = 1$. Thus

$$\begin{aligned} E^* &= (x_1 + k)p_1 + (x_2 + k)p_2 + \cdots + (x_t + k)p_t \\ &= x_1p_1 + kp_1 + x_2p_2 + kp_2 + \cdots + x_tp_t + kp_t \\ &= x_1p_1 + x_2p_2 + \cdots + x_tp_t + k(p_1 + p_2 + \cdots + p_t) \\ &= E + k \end{aligned}$$

Supplementary Problems

REPEATED TRIALS WITH TWO OUTCOMES

19.30. Find (i) $f(5, 1, \frac{1}{3})$, (ii) $f(7, 2, \frac{1}{2})$, (iii) $f(4, 2, \frac{1}{4})$.

19.31. A card is drawn and replaced three times from an ordinary deck of 52 cards. Find the probability that (i) two hearts are drawn, (ii) three hearts are drawn, (iii) at least one heart is drawn.

19.32. A baseball player's batting average is .300. If he comes to bat 4 times, what is the probability that he will get (i) two hits, (ii) at least one hit?

19.33. A box contains 3 red marbles and 2 white marbles. A marble is drawn and replaced three times from the box. Find the probability that (i) 1 red marble was drawn, (ii) 2 red marbles were drawn, (iii) at least one red marble was drawn.

19.34. Team A has probability $\frac{2}{5}$ of winning whenever it plays. If A plays 4 games, find the probability that A wins (i) 2 games, (ii) at least 1 game, (iii) more than half of the games.

19.35. A card is drawn and replaced in an ordinary deck of 52 cards. How many times must a card be drawn so that (i) there is at least an even chance of drawing a heart, (ii) the probability of drawing a heart is greater than $\frac{3}{4}$?

19.36. The probability of a man hitting a target is $\frac{1}{3}$. (i) If he fires 5 times, what is the probability of hitting the target at least twice? (ii) How many times must he fire so that the probability of hitting the target at least once is more than 90%.

19.37. Ten typists are assigned to the same office. Each typist is as likely to work at home as in the office. How many typewriters must there be in the office so that each typist has a typewriter at least 90% of the time?

RANDOM VARIABLES

19.38. A pair of fair dice is thrown. Let X be the random variable which denotes the minimum of the two numbers which occur. Find the distribution and expectation of X.

19.39. Five fair coins are tossed. Let X be the random variable which denotes the number of heads occurring. Find the distribution and expectation of X.

19.40. A fair coin is tossed four times. Let X be the random variable which denotes the longest string of heads. Find the distribution and expectation of X.

19.41. Solve the preceding problem in the case that the coin is weighted so that $P(\text{H}) = \frac{2}{3}$ and $P(\text{T}) = \frac{1}{3}$.

19.42. Two cards are selected from a box which contains five cards numbered 1, 1, 2, 2 and 3. (i) Let X be the random variable which denotes the sum of the two numbers. Find the distribution and expectation of X. (ii) Let Y be the random variable which denotes the maximum of the two numbers. Find the distribution and expectation of Y.

19.43. Refer to the preceding problem. Let $X + Y$ be the random variable defined by $(X+Y)(s) = X(s) + Y(s)$, i.e. which denotes the sum of the two numbers added to the maximum number. For example, $(X+Y)(2,3) = 5 + 3 = 8$. Find the distribution and expectation of $X + Y$. Verify that $E(X+Y) = E(X) + E(Y)$.

19.44. A man stands at the origin. He tosses a coin. If a head occurs, he moves one unit to the right; if a tail occurs he moves one unit to the left. (i) If X is the random variable which denotes his distance from the origin after 4 tosses of the coin, find the distribution and expectation of X. (ii) If Y is the random variable which denotes his position after 4 tosses of the coin, find the distribution and expectation of Y.

19.45. Solve (i) and (ii) of the preceding problem for the case of 5 tosses of the coin.

EXPECTED VALUE

19.46. A player tosses three fair coins. He wins \$5 if 3 heads occur, \$3 if 2 heads occur, and \$1 if only 1 head occurs. On the other hand, he loses \$15 if 3 tails occur. Find the value of the game to the player.

19.47. A player tosses three fair coins. He wins \$8 if 3 heads occur, \$3 if 2 heads occur, and \$1 if only 1 head occurs. If the game is to be fair, how much should he lose if no heads occur?

19.48. A player tosses three fair coins. He wins \$8 if 3 heads occur, \$3 if 2 heads occur, \$1 if 1 head occurs, but nothing if no heads occur. If the game is to be fair, how much should he pay to play the game?

19.49. A box contains 8 items of which 2 are defective. A man selects 3 items from the box. Find the expected number of defective items he has drawn.

19.50. Of the bolts produced by a factory, 2% are defective. In a lot of 8000 bolts from the factory, how many are expected to be defective?

19.51. Find the expected number of girls in a family with 6 children, assuming sex distribution to be equally probable. What is the probability that the expected number of girls does occur?

19.52. A lottery with 500 tickets gives one prize of \$100, 3 prizes of \$50 each, and 5 prizes of \$25 each. (i) Find the expected winnings of a ticket. (ii) If a ticket costs \$1, what is the expected value of the game?

19.53. A fair coin is tossed until a head or five tails occur. Find the expected number of tosses of the coin.

19.54. A coin weighted so that $P(\text{H}) = \frac{1}{3}$ and $P(\text{T}) = \frac{2}{3}$ is tossed until a head or five tails occurs. Find the expected number of tosses of the coin.

19.55. The probability of team A winning any game is $\frac{1}{2}$. A plays B in a tournament. The first team to win 2 games in a row or a total of three games wins the tournament. Find the expected number of games in the tournament.

19.56. A box contains 10 transistors of which 2 are defective. A transistor is selected from the box and tested until a non-defective one is chosen. Find the expected number of transistors to be chosen.

19.57. Solve the preceding problem in the case that 3 of the 10 items are defective.

Answers to Supplementary Problems

19.30. (i) $\frac{80}{243}$, (ii) $\frac{21}{128}$, (iii) $\frac{27}{128}$

19.31. (i) $\frac{9}{64}$, (ii) $\frac{1}{64}$, (iii) $\frac{37}{64}$

19.32. (i) 0.2646, (ii) 0.7599

19.33. (i) $\frac{36}{125}$, (ii) $\frac{54}{125}$, (iii) $\frac{117}{125}$

19.34. (i) $\frac{216}{625}$, (ii) $\frac{544}{625}$, (iii) $\frac{112}{625}$

19.35. (i) 3, (ii) 5

19.36. (i) $\frac{131}{243}$, (ii) 6

19.37. 7

19.38.

x_i	1	2	3	4	5	6
p_i	$\frac{11}{36}$	$\frac{9}{36}$	$\frac{7}{36}$	$\frac{5}{36}$	$\frac{3}{36}$	$\frac{1}{36}$

$E(X) = \frac{91}{36}$

19.39.

x_i	0	1	2	3	4	5
p_i	$\frac{1}{32}$	$\frac{5}{32}$	$\frac{10}{32}$	$\frac{10}{32}$	$\frac{5}{32}$	$\frac{1}{32}$

$E(X) = 2.5$

19.40.

x_i	0	1	2	3	4
p_i	$\frac{1}{16}$	$\frac{7}{16}$	$\frac{5}{16}$	$\frac{2}{16}$	$\frac{1}{16}$

$E(X) = \frac{27}{16}$

19.41.

x_i	0	1	2	3	4
p_i	$\frac{1}{81}$	$\frac{20}{81}$	$\frac{28}{81}$	$\frac{16}{81}$	$\frac{16}{81}$

$E(X) = \frac{188}{81}$

19.42. (i)

x_i	2	3	4	5
p_i	.1	.4	.3	.2

$E(X) = 3.6$

(ii)

x_i	1	2	3
p_i	.1	.5	.4

$E(Y) = 2.3$

19.43.

x_i	3	5	6	7	8
p_i	.1	.4	.1	.2	.2

$E(X+Y) = 5.9$

19.44. (i)

x_i	0	2	4
p_i	$\frac{3}{8}$	$\frac{4}{8}$	$\frac{1}{8}$

$E(X) = 1.5$

(ii)

x_i	−4	−2	0	2	4
p_i	$\frac{1}{16}$	$\frac{4}{16}$	$\frac{6}{16}$	$\frac{4}{16}$	$\frac{1}{16}$

$E(Y) = 0$

19.45. (i)

x_i	1	3	5
p_i	$\frac{10}{16}$	$\frac{5}{16}$	$\frac{1}{16}$

$E(X) = \frac{15}{8}$

(ii)

x_i	−5	−3	−1	1	3	5
p_i	$\frac{1}{32}$	$\frac{5}{32}$	$\frac{10}{32}$	$\frac{10}{32}$	$\frac{5}{32}$	$\frac{1}{32}$

$E(Y) = 0$

19.46. 25¢

19.47. $20

19.48. $2.50

19.49. $\frac{23}{28}$

19.50. 160

19.51. 3, $\frac{5}{16}$

19.52. (i) 75¢, (ii) −25¢

19.53. $\frac{31}{16}$

19.54. $\frac{211}{81}$

19.55. $\frac{23}{8}$

19.56. $\frac{11}{9}$

19.57. $\frac{11}{8}$

Chapter 20

Markov Chains

PROBABILITY VECTORS

A row vector $u = (u_1, u_2, \ldots, u_n)$ is called a *probability vector* if its components are non-negative and their sum is 1.

Example 1.1: Consider the following vectors:

$$u = (\tfrac{3}{4}, 0, -\tfrac{1}{4}, \tfrac{1}{2}), \quad v = (\tfrac{3}{4}, \tfrac{1}{2}, 0, \tfrac{1}{4}) \quad \text{and} \quad w = (\tfrac{1}{4}, \tfrac{1}{4}, 0, \tfrac{1}{2})$$

Then:

u is not a probability vector since its third component is negative;

v is not a probability vector since the sum of its components is greater than 1;

w is a probability vector.

Example 1.2: The non-zero vector $v = (2, 3, 5, 0, 1)$ is not a probability vector since the sum of its components is $2+3+5+0+1 = 11$. However, since the components of v are non-negative, v has a unique scalar multiple λv which is a probability vector; it can be obtained from v by dividing each component of v by the sum of the components of v: $\frac{1}{11}v = (\frac{2}{11}, \frac{3}{11}, \frac{5}{11}, 0, \frac{1}{11})$.

Remark: Since the sum of the components of a probability vector is one, an arbitrary probability vector with n components can be represented in terms of $n-1$ unknowns as follows:

$$(x_1, x_2, \ldots, x_{n-1}, 1 - x_1 - \cdots - x_{n-1})$$

In particular, arbitrary probability vectors with 2 and 3 components can be represented, respectively, in the form

$$(x, 1-x) \quad \text{and} \quad (x, y, 1-x-y)$$

STOCHASTIC AND REGULAR STOCHASTIC MATRICES

A square matrix $P = (p_{ij})$ is called a *stochastic matrix* if each of its rows is a probability vector, i.e. if each entry of P is non-negative and the sum of the entries in every row is 1.

Example 2.1: Consider the following matrices:

$$\begin{pmatrix} \frac{1}{3} & 0 & \frac{2}{3} \\ \frac{3}{4} & \frac{1}{2} & -\frac{1}{4} \\ \frac{1}{3} & \frac{1}{3} & \frac{1}{3} \end{pmatrix} \qquad \begin{pmatrix} \frac{1}{4} & \frac{3}{4} \\ \frac{1}{3} & \frac{1}{3} \end{pmatrix} \qquad \begin{pmatrix} 0 & 1 & 0 \\ \frac{1}{2} & \frac{1}{6} & \frac{1}{3} \\ \frac{1}{3} & \frac{2}{3} & 0 \end{pmatrix}$$

(i) (ii) (iii)

(i) is not a stochastic matrix since the entry in the second row and third column is negative;

(ii) is not a stochastic matrix since the sum of the entries in the second row is not 1;

(iii) is a stochastic matrix since each row is a probability vector.

We shall prove (Problem 20.7)

Theorem 20.1: If A and B are stochastic matrices, then the product AB is a stochastic matrix. Therefore, in particular, all powers A^n are stochastic matrices.

We now define an important class of stochastic matrices whose properties shall be investigated subsequently.

Definition: A stochastic matrix P is said to be regular if all the entries of some power P^m are positive.

Example 2.2: The stochastic matrix $A = \begin{pmatrix} 0 & 1 \\ \frac{1}{2} & \frac{1}{2} \end{pmatrix}$ is regular since

$$A^2 = \begin{pmatrix} 0 & 1 \\ \frac{1}{2} & \frac{1}{2} \end{pmatrix}\begin{pmatrix} 0 & 1 \\ \frac{1}{2} & \frac{1}{2} \end{pmatrix} = \begin{pmatrix} \frac{1}{2} & \frac{1}{2} \\ \frac{1}{4} & \frac{3}{4} \end{pmatrix}$$

is positive in every entry.

Example 2.3: Consider the stochastic matrix $A = \begin{pmatrix} 1 & 0 \\ \frac{1}{2} & \frac{1}{2} \end{pmatrix}$. Here

$$A^2 = \begin{pmatrix} 1 & 0 \\ \frac{3}{4} & \frac{1}{4} \end{pmatrix}, \quad A^3 = \begin{pmatrix} 1 & 0 \\ \frac{7}{8} & \frac{1}{8} \end{pmatrix}, \quad A^4 = \begin{pmatrix} 1 & 0 \\ \frac{15}{16} & \frac{1}{16} \end{pmatrix}$$

In fact every power A^m will have 1 and 0 in the first row; hence A is not regular.

FIXED POINTS OF SQUARE MATRICES

A non-zero row vector $u = (u_1, u_2, \ldots, u_n)$ is called a *fixed point* of a square matrix A if u is left fixed, i.e. is not changed, when multiplied by A:

$$uA = u$$

We do not include the zero vector 0 as a fixed point of a matrix since it is always left fixed by every matrix A: $0A = 0$.

Example 3.1: Consider the matrix $A = \begin{pmatrix} 2 & 1 \\ 2 & 3 \end{pmatrix}$. Then the vector $u = (2, -1)$ is a fixed point of A. For,

$$uA = (2, -1)\begin{pmatrix} 2 & 1 \\ 2 & 3 \end{pmatrix} = (2\cdot2 - 1\cdot2,\ 2\cdot1 - 1\cdot3) = (2, -1) = u$$

Example 3.2: Suppose u is a fixed point of a matrix A: $uA = u$. We claim that every scalar multiple of u, say λu, is also a fixed point of A. For,

$$(\lambda u)A = \lambda(uA) = \lambda u$$

Thus, in particular, the vector $2u = (4, -2)$ is a fixed point of the matrix A of the example above:

$$(4, -2)\begin{pmatrix} 2 & 1 \\ 2 & 3 \end{pmatrix} = (4\cdot2 - 2\cdot2,\ 4\cdot1 - 2\cdot3) = (4, -2)$$

We state the result of the preceding example as a theorem.

Theorem 20.2: If u is a fixed vector of a matrix A, then for any real number $\lambda \neq 0$ the scalar multiple λu is also a fixed vector of A.

FIXED POINTS AND REGULAR STOCHASTIC MATRICES

The main relationship between regular stochastic matrices and fixed points is contained in the next theorem whose proof lies beyond the scope of this text.

Theorem 20.3: Let P be a regular stochastic matrix. Then:

(i) P has a unique fixed probability vector t, and the components of t are all positive;

(ii) the sequence $P, P^2, P^3, \ldots$ of powers of P approaches the matrix T whose rows are each the fixed point t;

(iii) if p is any probability vector, then the sequence of vectors $pP, pP^2, pP^3, \ldots$ approaches the fixed point t.

Note: P^n approaches T means that each entry of P^n approaches the corresponding entry of T, and pP^n approaches t means that each component of pP^n approaches the corresponding component of t.

Example 4.1: Consider the regular stochastic matrix $P = \begin{pmatrix} 0 & 1 \\ \frac{1}{2} & \frac{1}{2} \end{pmatrix}$. We seek a probability vector $t = (x, 1-x)$ such that $tP = t$:

$$(x, 1-x)\begin{pmatrix} 0 & 1 \\ \frac{1}{2} & \frac{1}{2} \end{pmatrix} = (x, 1-x)$$

Multiplying the left side of the above matrix equation, we obtain

$$(\tfrac{1}{2} - \tfrac{1}{2}x,\ \tfrac{1}{2} + \tfrac{1}{2}x) = (x,\ 1-x) \quad \text{or} \quad \begin{cases} \frac{1}{2} - \frac{1}{2}x = x \\ \frac{1}{2} + \frac{1}{2}x = 1 - x \end{cases} \quad \text{or} \quad x = \tfrac{1}{3}$$

Thus $t = (\frac{1}{3}, 1 - \frac{1}{3}) = (\frac{1}{3}, \frac{2}{3})$ is the unique fixed probability vector of P. By Theorem 20.3, the sequence $P, P^2, P^3, \ldots$ approaches the matrix T whose rows are each the vector t:

$$T = \begin{pmatrix} \frac{1}{3} & \frac{2}{3} \\ \frac{1}{3} & \frac{2}{3} \end{pmatrix} = \begin{pmatrix} .33 & .67 \\ .33 & .67 \end{pmatrix}$$

We exhibit some of the powers of P to indicate the above result:

$$P^2 = \begin{pmatrix} \frac{1}{2} & \frac{1}{2} \\ \frac{1}{4} & \frac{3}{4} \end{pmatrix} = \begin{pmatrix} .50 & .50 \\ .25 & .75 \end{pmatrix}; \quad P^3 = \begin{pmatrix} \frac{1}{4} & \frac{3}{4} \\ \frac{3}{8} & \frac{5}{8} \end{pmatrix} = \begin{pmatrix} .25 & .75 \\ .37 & .63 \end{pmatrix}$$

$$P^4 = \begin{pmatrix} \frac{3}{8} & \frac{5}{8} \\ \frac{5}{16} & \frac{11}{16} \end{pmatrix} = \begin{pmatrix} .37 & .63 \\ .31 & .69 \end{pmatrix}; \quad P^5 = \begin{pmatrix} \frac{5}{16} & \frac{11}{16} \\ \frac{11}{32} & \frac{21}{32} \end{pmatrix} = \begin{pmatrix} .31 & .69 \\ .34 & .66 \end{pmatrix}$$

Example 4.2: Find the unique fixed probability vector of the regular stochastic matrix

$$P = \begin{pmatrix} 0 & 1 & 0 \\ 0 & 0 & 1 \\ \frac{1}{2} & \frac{1}{2} & 0 \end{pmatrix}$$

Method 1. We seek a probability vector $t = (x, y, 1-x-y)$ such that $tP = t$:

$$(x, y, 1-x-y)\begin{pmatrix} 0 & 1 & 0 \\ 0 & 0 & 1 \\ \frac{1}{2} & \frac{1}{2} & 0 \end{pmatrix} = (x, y, 1-x-y)$$

Multiplying the left side of the above matrix equation and then setting corresponding components equal to each other, we obtain the system

$$\begin{array}{rcl} \tfrac{1}{2} - \tfrac{1}{2}x - \tfrac{1}{2}y &=& x \\ x + \tfrac{1}{2} - \tfrac{1}{2}x - \tfrac{1}{2}y &=& y \\ y &=& 1 - x - y \end{array} \quad \text{or} \quad \begin{array}{rcl} 3x + y &=& 1 \\ x - 3y &=& -1 \\ x + 2y &=& 1 \end{array} \quad \text{or} \quad \begin{array}{rcl} x &=& \tfrac{1}{5} \\ y &=& \tfrac{2}{5} \end{array}$$

Thus $t = (\frac{1}{5}, \frac{2}{5}, \frac{2}{5})$ is the unique fixed probability vector of P.

Method 2. We first seek any fixed vector $u = (x, y, z)$ of the matrix P:

$$(x, y, z)\begin{pmatrix} 0 & 1 & 0 \\ 0 & 0 & 1 \\ \frac{1}{2} & \frac{1}{2} & 0 \end{pmatrix} = (x, y, z) \quad \text{or} \quad \begin{cases} \frac{1}{2}z = x \\ x + \frac{1}{2}z = y \\ y = z \end{cases}$$

We know that the system has a non-zero solution; hence we can arbitrarily assign a value to one of the unknowns. Set $z = 2$. Then by the first equation $x = 1$, and by the third equation $y = 2$. Thus $u = (1, 2, 2)$ is a fixed point of P. But every multiple of u is a fixed point of P; hence multiply u by $\frac{1}{5}$ to obtain the required fixed probability vector $t = \frac{1}{5}u = (\frac{1}{5}, \frac{2}{5}, \frac{2}{5})$.

MARKOV CHAINS

We now consider a sequence of trials whose outcomes, say, $X_1, X_2, \ldots$, satisfy the following two properties:

(i) Each outcome belongs to a finite set of outcomes $\{a_1, a_2, \ldots, a_m\}$ called the *state space* of the system; if the outcome on the nth trial is a_i, then we say that the system is in state a_i at time n or at the nth step.

(ii) The outcome of any trial depends at most upon the outcome of the immediately preceding trial and not upon any other previous outcome; with each pair of states (a_i, a_j) there is given the probability p_{ij} that a_j occurs immediately after a_i occurs.

Such a stochastic process is called a (finite) *Markov chain*. The numbers p_{ij}, called the *transition probabilities*, can be arranged in a matrix

$$P = \begin{pmatrix} p_{11} & p_{12} & \cdots & p_{1m} \\ p_{21} & p_{22} & \cdots & p_{2m} \\ \cdots & \cdots & \cdots & \cdots \\ p_{m1} & p_{m2} & \cdots & p_{mm} \end{pmatrix}$$

called the *transition matrix*.

Thus with each state a_i there corresponds the ith row $(p_{i1}, p_{i2}, \ldots, p_{im})$ of the transition matrix P; if the system is in state a_i, then this row vector represents the probabilities of all the possible outcomes of the next trial and so it is a probability vector. Accordingly,

Theorem 20.4: The transition matrix P of a Markov chain is a stochastic matrix.

Example 5.1: A man either drives his car or takes a train to work each day. Suppose he never takes the train two days in a row; but if he drives to work, then the next day he is just as likely to drive again as he is to take the train.

The state space of the system is $\{t\,(\text{train}), d\,(\text{drive})\}$. This stochastic process is a Markov chain since the outcome on any day depends only on what happened the preceding day. The transition matrix of the Markov chain is

$$\begin{array}{c} \\ t \\ d \end{array}\begin{array}{c} \begin{array}{cc} t & d \end{array} \\ \begin{pmatrix} 0 & 1 \\ \frac{1}{2} & \frac{1}{2} \end{pmatrix} \end{array}$$

The first row of the matrix corresponds to the fact that he never takes the train two days in a row and so he definitely will drive the day after he takes the train. The second row of the matrix corresponds to the fact that the day after he drives he will drive or take the train with equal probability.

Example 5.2: Three boys A, B and C are throwing a ball to each other. A always throws the ball to B and B always throws the ball to C; but C is just as likely to throw the ball to B as to A. Let X_n denote the nth person to be thrown the ball. The state space of the system is $\{A, B, C\}$. This is a Markov chain since the person throwing the ball is not influenced by those who previously had the ball. The transition matrix of the Markov chain is

$$\begin{array}{c} \\ A \\ B \\ C \end{array} \begin{array}{c} \begin{array}{ccc} A & B & C \end{array} \\ \begin{pmatrix} 0 & 1 & 0 \\ 0 & 0 & 1 \\ \frac{1}{2} & \frac{1}{2} & 0 \end{pmatrix} \end{array}$$

The first row of the matrix corresponds to the fact that A always throws the ball to B. The second row corresponds to the fact that B always throws the ball to C. The last row corresponds to the fact that C throws the ball to A or B with equal probability (and does not throw it to himself).

Example 5.3: A school contains 200 boys and 150 girls. One student is selected after another to take an eye examination. Let X_n denote the sex of the nth student who takes the examination. The state space of the stochastic process is $\{m \text{ (male)}, f \text{ (female)}\}$. However, this process is not a Markov chain since, for example, the probability that the third person is a girl depends not only on the outcome of the second trial but on both the first and second trials.

Example 5.4: (Random walk with reflecting barriers.) A man is at an integral point on the x-axis between the origin O and, say, the point 5. He takes a unit step to the right with probability p or to the left with probability $q = 1 - p$, unless he is at the origin where he takes a step to the right to 1 or at the point 5 where he takes a step to the left to 4. Let X_n denote his position after n steps. This is a Markov chain with state space $\{a_0, a_1, a_2, a_3, a_4, a_5\}$ where a_i means that the man is at the point i. The transition matrix is

$$P \quad = \quad \begin{array}{c} \\ a_0 \\ a_1 \\ a_2 \\ a_3 \\ a_4 \\ a_5 \end{array} \begin{array}{c} \begin{array}{cccccc} a_0 & a_1 & a_2 & a_3 & a_4 & a_5 \end{array} \\ \begin{pmatrix} 0 & 1 & 0 & 0 & 0 & 0 \\ q & 0 & p & 0 & 0 & 0 \\ 0 & q & 0 & p & 0 & 0 \\ 0 & 0 & q & 0 & p & 0 \\ 0 & 0 & 0 & q & 0 & p \\ 0 & 0 & 0 & 0 & 1 & 0 \end{pmatrix} \end{array}$$

Each row of the matrix, except the first and last, corresponds to the fact that the man moves from state a_i to state a_{i+1} with probability p or back to state a_{i-1} with probability $q = 1 - p$. The first row corresponds to the fact that the man must move from state a_0 to state a_1, and the last row that the man must move from state a_5 to state a_4.

HIGHER TRANSITION PROBABILITIES

The entry p_{ij} in the transition matrix P of a Markov chain is the probability that the system changes from the state a_i to the state a_j in one step: $a_i \to a_j$. Question: What is the probability, denoted by $p_{ij}^{(n)}$, that the system changes from the state a_i to the state a_j in exactly n steps:

$$a_i \to a_{k_1} \to a_{k_2} \to \cdots \to a_{k_{n-1}} \to a_j$$

The next theorem answers this question; here the $p_{ij}^{(n)}$ are arranged in a matrix $P^{(n)}$ called the *n-step transition matrix*:

Theorem 20.5: Let P be the transition matrix of a Markov chain process. Then the n-step transition matrix is equal to the nth power of P: $P^{(n)} = P^n$.

Now suppose that, at some arbitrary time, the probability that the system is in state a_i is p_i; we denote these probabilities by the probability vector $p = (p_1, p_2, \ldots, p_m)$ which is called the *probability distribution* of the system at that time. In particular, we shall let

$$p^{(0)} = (p_1^{(0)}, p_2^{(0)}, \ldots, p_m^{(0)})$$

denote the *initial probability distribution*, i.e. the distribution when the process begins, and we shall let

$$p^{(n)} = (p_1^{(n)}, p_2^{(n)}, \ldots, p_m^{(n)})$$

denote the *nth step probability distribution*, i.e. the distribution after the first n steps. The following theorem applies.

Theorem 20.6: Let P be the transition matrix of a Markov chain process. If $p = (p_i)$ is the probability distribution of the system at some arbitrary time, then pP is the probability distribution of the system one step later and pP^n is the probability distribution of the system n steps later. In particular,

$$p^{(1)} = p^{(0)}P,\ p^{(2)} = p^{(1)}P,\ p^{(3)} = p^{(2)}P,\ \ldots,\ p^{(n)} = p^{(0)}P^n$$

Example 6.1: Consider the Markov chain of Example 5.1 whose transition matrix is

$$P = \begin{matrix} & t & d \\ t & 0 & 1 \\ d & \frac{1}{2} & \frac{1}{2} \end{matrix}$$

Here t is the state of taking a train to work and d of driving to work. By Example 4.1,

$$P^4 = \begin{pmatrix} \frac{3}{8} & \frac{5}{8} \\ \frac{5}{16} & \frac{11}{16} \end{pmatrix}$$

Thus the probability that the system changes from, say, state t to state d in exactly 4 steps is $\frac{5}{8}$, i.e. $p_{td}^{(4)} = \frac{5}{8}$. Similarly, $p_{tt}^{(4)} = \frac{3}{8}$, $p_{dt}^{(4)} = \frac{5}{16}$ and $p_{dd}^{(4)} = \frac{11}{16}$.

Now suppose that on the first day of work, the man tossed a fair die and drove to work if and only if a 6 appeared. In other words, $p^{(0)} = (\frac{5}{6}, \frac{1}{6})$ is the initial probability distribution. Then

$$p^{(4)} = p^{(0)}P^4 = (\tfrac{5}{6}, \tfrac{1}{6})\begin{pmatrix} \frac{3}{8} & \frac{5}{8} \\ \frac{5}{16} & \frac{11}{16} \end{pmatrix} = (\tfrac{35}{96}, \tfrac{61}{96})$$

is the probability distribution after 4 days, i.e. $p_t^{(4)} = \frac{35}{96}$ and $p_d^{(4)} = \frac{61}{96}$.

Example 6.2: Consider the Markov chain of Example 5.2 whose transition matrix is

$$P = \begin{matrix} & A & B & C \\ A & 0 & 1 & 0 \\ B & 0 & 0 & 1 \\ C & \frac{1}{2} & \frac{1}{2} & 0 \end{matrix}$$

Suppose C was the first person with the ball, i.e. suppose $p^{(0)} = (0, 0, 1)$ is the initial probability distribution. Then

$$p^{(1)} = p^{(0)}P = (0, 0, 1)\begin{pmatrix} 0 & 1 & 0 \\ 0 & 0 & 1 \\ \frac{1}{2} & \frac{1}{2} & 0 \end{pmatrix} = (\tfrac{1}{2}, \tfrac{1}{2}, 0)$$

$$p^{(2)} = p^{(1)}P = (\tfrac{1}{2}, \tfrac{1}{2}, 0)\begin{pmatrix} 0 & 1 & 0 \\ 0 & 0 & 1 \\ \tfrac{1}{2} & \tfrac{1}{2} & 0 \end{pmatrix} = (0, \tfrac{1}{2}, \tfrac{1}{2})$$

$$p^{(3)} = p^{(2)}P = (0, \tfrac{1}{2}, \tfrac{1}{2})\begin{pmatrix} 0 & 1 & 0 \\ 0 & 0 & 1 \\ \tfrac{1}{2} & \tfrac{1}{2} & 0 \end{pmatrix} = (\tfrac{1}{4}, \tfrac{1}{4}, \tfrac{1}{2})$$

Thus, after three throws, the probability that A has the ball is $\frac{1}{4}$, that B has the ball is $\frac{1}{4}$ and that C has the ball is $\frac{1}{2}$: $p_A^{(3)} = \frac{1}{4}$, $p_B^{(3)} = \frac{1}{4}$ and $p_C^{(3)} = \frac{1}{2}$.

Example 6.3: Consider the random walk problem of Example 5.4. Suppose the man began at the point 2; find the probability distribution after 3 steps and after 4 steps, i.e. $p^{(3)}$ and $p^{(4)}$.

Now $p^{(0)} = (0,0,1,0,0,0)$ is the initial probability distribution. Then

$$\begin{aligned} p^{(1)} &= p^{(0)}P = (0, q, 0, p, 0, 0) \\ p^{(2)} &= p^{(1)}P = (q^2, 0, 2pq, 0, p^2, 0) \\ p^{(3)} &= p^{(2)}P = (0, q^2 + 2p^2q, 0, 3p^2q, 0, p^3) \\ p^{(4)} &= p^{(3)}P = (q^3 + 2p^2q^2, 0, pq^2 + 2p^3q + 3p^2q^2, 0, 3p^3q + p^3, 0) \end{aligned}$$

Thus after 4 steps he is at, say, the origin with probability $q^3 + 2p^2q^2$.

STATIONARY DISTRIBUTION OF REGULAR MARKOV CHAINS

Suppose that a Markov chain is regular, i.e. that its transition matrix P is regular. By Theorem 20.3, the sequence of n-step transition matrices P^n approaches the matrix T whose rows are each the unique fixed probability vector t of P; hence the probability $p_{ij}^{(n)}$ that a_j occurs for sufficiently large n is independent of the original state a_i and it approaches the component t_j of t. In other words,

Theorem 20.7: Let the transition matrix P of a Markov chain be regular. Then, in the long run, the probability that any state a_j occurs is approximately equal to the component t_j of the unique fixed probability vector t of P.

Thus we see that the effect of the initial state or the initial probability distribution of the process wears off as the number of steps of the process increase. Furthermore, every sequence of probability distributions approaches the fixed probability vector t of P, called the *stationary distribution* of the Markov chain.

Example 7.1: Consider the Markov chain process of Example 5.1 whose transition matrix is

$$P = \begin{matrix} & t & d \\ t & 0 & 1 \\ d & \tfrac{1}{2} & \tfrac{1}{2} \end{matrix}$$

By Example 4.1, the unique fixed probability vector of the above matrix is $(\frac{1}{3}, \frac{2}{3})$. Thus, in the long run, the man will take the train to work $\frac{1}{3}$ of the time, and drive to work the other $\frac{2}{3}$ of the time.

Example 7.2: Consider the Markov chain process of Example 5.2 whose transition matrix is

$$P = \begin{matrix} & A & B & C \\ A & 0 & 1 & 0 \\ B & 0 & 0 & 1 \\ C & \tfrac{1}{2} & \tfrac{1}{2} & 0 \end{matrix}$$

By Example 4.2, the unique fixed probability vector of the above matrix is $(\frac{1}{5}, \frac{2}{5}, \frac{2}{5})$. Thus, in the long run, A will be thrown the ball 20% of the time, and B and C 40% of the time.

ABSORBING STATES

A state a_i of a Markov chain is called *absorbing* if the system remains in the state a_i once it enters there. Thus a state a_i is absorbing if and only if the ith row of the transition matrix P has a 1 on the main diagonal and zeros everywhere else.

Example 8.1: Suppose the following matrix is the transition matrix of a Markov chain:

$$P = \begin{array}{c} \\ a_1 \\ a_2 \\ a_3 \\ a_4 \\ a_5 \end{array} \begin{array}{c} \begin{array}{ccccc} a_1 & a_2 & a_3 & a_4 & a_5 \end{array} \\ \begin{pmatrix} \frac{1}{4} & 0 & \frac{1}{4} & \frac{1}{4} & \frac{1}{4} \\ 0 & 1 & 0 & 0 & 0 \\ \frac{1}{2} & 0 & \frac{1}{4} & \frac{1}{4} & 0 \\ 0 & 1 & 0 & 0 & 0 \\ 0 & 0 & 0 & 0 & 1 \end{pmatrix} \end{array}$$

The states a_2 and a_5 are each absorbing, since each of the second and fifth rows has a 1 on the main diagonal.

Example 8.2: (Random walk with absorbing barriers.) Consider the random walk problem of Example 5.4, except now we assume that the man remains at either endpoint whenever he reaches there. This is also a Markov chain and the transition matrix is given by

$$P = \begin{array}{c} \\ a_0 \\ a_1 \\ a_2 \\ a_3 \\ a_4 \\ a_5 \end{array} \begin{array}{c} \begin{array}{cccccc} a_0 & a_1 & a_2 & a_3 & a_4 & a_5 \end{array} \\ \begin{pmatrix} 1 & 0 & 0 & 0 & 0 & 0 \\ q & 0 & p & 0 & 0 & 0 \\ 0 & q & 0 & p & 0 & 0 \\ 0 & 0 & q & 0 & p & 0 \\ 0 & 0 & 0 & q & 0 & p \\ 0 & 0 & 0 & 0 & 0 & 1 \end{pmatrix} \end{array}$$

We call this process a random walk with absorbing barriers, since the a_0 and a_5 are absorbing states. In this case, $p_0^{(n)}$ denotes the probability that the man reaches the state a_0 on or before the nth step. Similarly, $p_5^{(n)}$ denotes the probability that he reaches the a_5 on or before the nth step.

Example 8.3: A player has, say, x dollars. He bets one dollar at a time and wins with probability p and loses with probability $q = 1 - p$. The game ends when he loses all his money, i.e. has 0 dollars, or when he wins $N - x$ dollars, i.e. has N dollars. This game is identical to the random walk of the preceding example except that here the absorbing barriers are at 0 and N.

Example 8.4: A man tosses a fair coin until 3 heads occur in a row. Let $X_n = k$ if, at the nth trial, the last tail occurred at the $(n-k)$-th trial, i.e. X_n denotes the longest string of heads ending at the nth trial. This is a Markov chain process with state space $\{a_0, a_1, a_2, a_3\}$, where a_i means the string of heads has length i. The transition matrix is

$$\begin{array}{c} \\ a_0 \\ a_1 \\ a_2 \\ a_3 \end{array} \begin{array}{c} \begin{array}{cccc} a_0 & a_1 & a_2 & a_3 \end{array} \\ \begin{pmatrix} \frac{1}{2} & \frac{1}{2} & 0 & 0 \\ \frac{1}{2} & 0 & \frac{1}{2} & 0 \\ \frac{1}{2} & 0 & 0 & \frac{1}{2} \\ 0 & 0 & 0 & 1 \end{pmatrix} \end{array}$$

Each row, except the last, corresponds to the fact that a string of heads is either broken if a tail occurs or is extended by one if a head occurs. The last line corresponds to the fact that the game ends if three heads are tossed in a row. Note that a_3 is an absorbing state.

Let a_i be an absorbing state of a Markov chain with transition matrix P. Then, for $i \neq j$, the n-step transition probability $p_{ij}^{(n)} = 0$ for every n. Accordingly, every power of P has a zero entry and so P is not regular. Thus:

Theorem 20.8: If a stochastic matrix P has a 1 on the main diagonal, then P is not regular (unless P is a 1×1 matrix).

Solved Problems

PROBABILITY VECTORS AND STOCHASTIC MATRICES

20.1. Which vectors are probability vectors?
(i) $u = (\frac{1}{3}, 0, -\frac{1}{6}, \frac{1}{2}, \frac{1}{3})$, (ii) $v = (\frac{1}{3}, 0, \frac{1}{6}, \frac{1}{2}, \frac{1}{3})$, (iii) $w = (\frac{1}{3}, 0, 0, \frac{1}{6}, \frac{1}{2})$.

A vector is a probability vector if its components are non-negative and their sum is 1.

(i) u is not a probability vector since its third component is negative.

(ii) v is not a probability vector since the sum of the components is greater than 1.

(iii) w is a probability vector since its components are non-negative and their sum is 1.

20.2. Multiply each vector by the appropriate scalar to form a probability vector:
(i) $(2, 1, 0, 2, 3)$, (ii) $(4, 0, 1, 2, 0, 5)$, (iii) $(3, 0, -2, 1)$, (iv) $(0, 0, 0, 0, 0)$.

(i) The sum of the components is $2+1+0+3+2 = 8$; hence multiply the vector, i.e. each component, by $\frac{1}{8}$ to obtain the probability vector $(\frac{1}{4}, \frac{1}{8}, 0, \frac{1}{4}, \frac{3}{8})$.

(ii) The sum of the components is $4+0+1+2+0+5 = 12$; hence multiply the vector, i.e. each component, by $\frac{1}{12}$ to obtain the probability vector $(\frac{1}{3}, 0, \frac{1}{12}, \frac{1}{6}, 0, \frac{5}{12})$.

(iii) The first component is positive and the third is negative; hence it is impossible to multiply the vector by a scalar to form a vector with non-negative components. Thus no scalar multiple of the vector is a probability vector.

(iv) Every scalar multiple of the zero vector is the zero vector whose components add to 0. Thus no multiple of the zero vector is a probability vector.

20.3. Find a multiple of each vector which is a probability vector:
(i) $(\frac{1}{2}, \frac{2}{3}, 0, 2, \frac{5}{6})$, (ii) $(0, \frac{2}{3}, 1, \frac{3}{5}, \frac{5}{6})$.

In each case, first multiply each vector by a scalar so that the fractions are eliminated.

(i) First multiply the vector by 6 to obtain $(3, 4, 0, 12, 5)$. Then multiply by $1/(3+4+0+12+5) = \frac{1}{24}$ to obtain $(\frac{1}{8}, \frac{1}{6}, 0, \frac{1}{2}, \frac{5}{24})$ which is a probability vector.

(ii) First multiply the vector by 30 to obtain $(0, 20, 30, 18, 25)$. Then multiply by $1/(0+20+30+18+25) = \frac{1}{93}$ to obtain $(0, \frac{20}{93}, \frac{30}{93}, \frac{18}{93}, \frac{25}{93})$ which is a probability vector.

20.4. Which of the following matrices are stochastic matrices?

$$\text{(i) } A = \begin{pmatrix} \frac{1}{3} & \frac{1}{3} & \frac{1}{3} \\ \frac{1}{2} & 0 & \frac{1}{2} \end{pmatrix} \quad \text{(ii) } B = \begin{pmatrix} \frac{15}{16} & \frac{1}{16} \\ \frac{2}{3} & \frac{2}{3} \end{pmatrix} \quad \text{(iii) } C = \begin{pmatrix} 1 & 0 \\ \frac{1}{2} & \frac{1}{2} \end{pmatrix} \quad \text{(iv) } D = \begin{pmatrix} \frac{3}{2} & -\frac{1}{2} \\ \frac{1}{4} & \frac{3}{4} \end{pmatrix}$$

(i) A is not a stochastic matrix since it is not a square matrix.

(ii) B is not a stochastic matrix since the sum of the components in the last row is greater than 1.

(iii) C is a stochastic matrix.

(iv) D is not a stochastic matrix since the entry in the first row, second column is negative.

20.5. Let $A = \begin{pmatrix} a_1 & b_1 & c_1 \\ a_2 & b_2 & c_2 \\ a_3 & b_3 & c_3 \end{pmatrix}$ be a stochastic matrix and let $u = (u_1, u_2, u_3)$ be a probability vector. Show that uA is also a probability vector.

$$uA = (u_1, u_2, u_3)\begin{pmatrix} a_1 & b_1 & c_1 \\ a_2 & b_2 & c_2 \\ a_3 & b_3 & c_3 \end{pmatrix} = (u_1a_1 + u_2a_2 + u_3a_3,\ u_1b_1 + u_2b_2 + u_3b_3,\ u_1c_1 + u_2c_2 + u_3c_3)$$

Since the u_i, a_i, b_i and c_i are non-negative and since the products and sums of non-negative numbers are non-negative, the components of uA are non-negative as required. Thus we only need to show that the sum of the components of uA is 1. Here we use the fact that $u_1 + u_2 + u_3$, $a_1 + b_1 + c_1$, $a_2 + b_2 + c_2$ and $a_3 + b_3 + c_3$ are each 1:

$$\begin{aligned} & u_1a_1 + u_2a_2 + u_3a_3 + u_1b_1 + u_2b_2 + u_3b_3 + u_1c_1 + u_2c_2 + u_3c_3 \\ &= u_1(a_1 + b_1 + c_1) + u_2(a_2 + b_2 + c_2) + u_3(a_3 + b_3 + c_3) \\ &= u_1 \cdot 1 + u_2 \cdot 1 + u_3 \cdot 1 = u_1 + u_2 + u_3 = 1 \end{aligned}$$

20.6. Prove: If $A = (a_{ij})$ is a stochastic matrix of order n and $u = (u_1, u_2, \ldots, u_n)$ is a probability vector, then uA is also a probability vector.

The proof is similar to that of the preceding problem for the case $n = 3$:

$$\begin{aligned} uA &= (u_1, u_2, \ldots, u_n)\begin{pmatrix} a_{11} & a_{12} & \cdots & a_{1n} \\ a_{21} & a_{22} & \cdots & a_{2n} \\ \cdots & \cdots & \cdots & \cdots \\ a_{n1} & a_{n2} & \cdots & a_{nn} \end{pmatrix} \\ &= (u_1a_{11} + u_2a_{21} + \cdots + u_na_{n1},\ u_1a_{12} + u_2a_{22} + \cdots + u_na_{n2},\ \ldots,\ u_1a_{1n} + u_2a_{2n} + \cdots + u_na_{nn}) \end{aligned}$$

Since the u_i and a_{ij} are non-negative, the components of uA are also non-negative. Thus we only need to show that the sum of the components of uA is 1:

$$\begin{aligned} & u_1a_{11} + u_2a_{21} + \cdots + u_na_{n1} + u_1a_{12} + u_2a_{22} + \cdots + u_na_{n2} + \cdots + u_1a_{1n} + u_2a_{2n} + \cdots + u_na_{nn} \\ &= u_1(a_{11} + a_{12} + \cdots + a_{1n}) + u_2(a_{21} + a_{22} + \cdots + a_{2n}) + \cdots + u_n(a_{n1} + a_{n2} + \cdots + a_{nn}) \\ &= u_1 \cdot 1 + u_2 \cdot 1 + \cdots + u_n \cdot 1 = u_1 + u_2 + \cdots + u_n = 1 \end{aligned}$$

20.7. Prove Theorem 20.1: If A and B are stochastic matrices, then the product AB is a stochastic matrix. Therefore, in particular, all powers A^n are stochastic matrices.

The ith-row s_i of the product matrix AB is obtained by multiplying the ith-row r_i of A by the matrix B: $s_i = r_iB$. Since each r_i is a probability vector and B is a stochastic matrix, by the preceding problem, s_i is also a probability vector. Hence AB is a stochastic matrix.

20.8. Prove: Let $p = (p_1, p_2, \ldots, p_m)$ be a probability vector, and let T be a matrix whose rows are each the same vector $t = (t_1, t_2, \ldots, t_m)$. Then $pT = t$.

Using the fact that $p_1 + p_2 + \cdots + p_m = 1$, we have

$$\begin{aligned} pT &= (p_1, p_2, \ldots, p_m)\begin{pmatrix} t_1 & t_2 & \cdots & t_m \\ t_1 & t_2 & \cdots & t_m \\ \cdots & \cdots & \cdots & \cdots \\ t_1 & t_2 & \cdots & t_m \end{pmatrix} \\ &= (p_1t_1 + p_2t_1 + \cdots + p_mt_1,\ p_1t_2 + p_2t_2 + \cdots + p_mt_2,\ \ldots,\ p_1t_m + p_2t_m + \cdots + p_mt_m) \\ &= ((p_1 + p_2 + \cdots + p_m)t_1,\ (p_1 + p_2 + \cdots + p_m)t_2,\ \ldots,\ (p_1 + p_2 + \cdots + p_m)t_m) \\ &= (1 \cdot t_1,\ 1 \cdot t_2,\ \ldots,\ 1 \cdot t_m) = (t_1, t_2, \ldots, t_m) = t \end{aligned}$$

REGULAR STOCHASTIC MATRICES AND FIXED PROBABILITY VECTORS

20.9. Find the unique fixed probability vector of the regular stochastic matrix $A = \begin{pmatrix} \frac{3}{4} & \frac{1}{4} \\ \frac{1}{2} & \frac{1}{2} \end{pmatrix}$. What matrix does A^n approach?

We seek a probability vector $t = (x, 1-x)$ such that $tA = t$:

$$(x, 1-x)\begin{pmatrix} \frac{3}{4} & \frac{1}{4} \\ \frac{1}{2} & \frac{1}{2} \end{pmatrix} = (x, 1-x)$$

Multiply the left side of the above matrix equation and then set corresponding components equal to each other to obtain the two equations

$$\tfrac{3}{4}x + \tfrac{1}{2} - \tfrac{1}{2}x = x, \quad \tfrac{1}{4}x + \tfrac{1}{2} - \tfrac{1}{2}x = 1 - x$$

Solve either equation to obtain $x = \frac{2}{3}$. Thus $t = (\frac{2}{3}, \frac{1}{3})$ is the required probability vector.

Check the answer by computing the product tA:

$$(\tfrac{2}{3}, \tfrac{1}{3})\begin{pmatrix} \frac{3}{4} & \frac{1}{4} \\ \frac{1}{2} & \frac{1}{2} \end{pmatrix} = (\tfrac{1}{2}+\tfrac{1}{6}, \tfrac{1}{6}+\tfrac{1}{6}) = (\tfrac{2}{3}, \tfrac{1}{3})$$

The answer checks since $tA = t$.

The matrix A^n approaches the matrix T whose rows are each the fixed point t: $T = \begin{pmatrix} \frac{2}{3} & \frac{1}{3} \\ \frac{2}{3} & \frac{1}{3} \end{pmatrix}$.

20.10. (i) Show that the vector $u = (b, a)$ is a fixed point of the general 2×2 stochastic matrix $P = \begin{pmatrix} 1-a & a \\ b & 1-b \end{pmatrix}$.

(ii) Use the result of (i) to find the unique fixed probability vector of each of the following matrices:

$$A = \begin{pmatrix} \frac{1}{3} & \frac{2}{3} \\ 1 & 0 \end{pmatrix} \qquad B = \begin{pmatrix} \frac{1}{2} & \frac{1}{2} \\ \frac{2}{3} & \frac{1}{3} \end{pmatrix} \qquad C = \begin{pmatrix} .7 & .3 \\ .8 & .2 \end{pmatrix}$$

(i) $uP = (b, a)\begin{pmatrix} 1-a & a \\ b & 1-b \end{pmatrix} = (b - ab + ab, \; ab + a - ab) = (b, a) = u.$

(ii) By (i), $u = (1, \frac{2}{3})$ is a fixed point of A. Multiply u by 3 to obtain the fixed point $(3, 2)$ of A which has no fractions. Then multiply $(3, 2)$ by $1/(3+2) = \frac{1}{5}$ to obtain the required unique fixed probability vector $(\frac{3}{5}, \frac{2}{5})$.

By (i), $u = (\frac{2}{3}, \frac{1}{2})$ is a fixed point of B. Multiply u by 6 to obtain the fixed point $(4, 3)$, and then multiply by $1/(4+3) = \frac{1}{7}$ to obtain the required unique fixed probability vector $(\frac{4}{7}, \frac{3}{7})$.

By (i), $u = (.8, .3)$ is a fixed point of C. Hence $(8, 3)$ and the probability vector $(\frac{8}{11}, \frac{3}{11})$ are also fixed points of C.

20.11. Find the unique fixed probability vector of the regular stochastic matrix

$$P = \begin{pmatrix} \frac{1}{2} & \frac{1}{4} & \frac{1}{4} \\ \frac{1}{2} & 0 & \frac{1}{2} \\ 0 & 1 & 0 \end{pmatrix}$$

Method 1. We seek a probability vector $t = (x, y, 1-x-y)$ such that $tP = t$:

$$(x, y, 1-x-y)\begin{pmatrix} \frac{1}{2} & \frac{1}{4} & \frac{1}{4} \\ \frac{1}{2} & 0 & \frac{1}{2} \\ 0 & 1 & 0 \end{pmatrix} = (x, y, 1-x-y)$$

Multiply the left side of the above matrix equation and then set corresponding components equal to each other to obtain the system of three equations

$$\begin{cases} \frac{1}{2}x + \frac{1}{2}y = x \\ \frac{1}{4}x + 1 - x - y = y \\ \frac{1}{4}x + \frac{1}{2}y = 1 - x - y \end{cases} \quad \text{or} \quad \begin{cases} x - y = 0 \\ 3x + 8y = 4 \\ 5x + 6y = 4 \end{cases}$$

Choose any two of the equations and solve for x and y to obtain $x = \frac{4}{11}$ and $y = \frac{4}{11}$. Check the solution by substituting for x and y into the third equation. Since $1 - x - y = \frac{3}{11}$, the required fixed probability vector is $t = (\frac{4}{11}, \frac{4}{11}, \frac{3}{11})$.

Method 2. We seek any fixed vector $u = (x, y, z)$ of the matrix P:

$$(x, y, z)\begin{pmatrix} \frac{1}{2} & \frac{1}{4} & \frac{1}{4} \\ \frac{1}{2} & 0 & \frac{1}{2} \\ 0 & 1 & 0 \end{pmatrix} = (x, y, z)$$

Multiply the left side of the above matrix equation and set corresponding components equal to each other to obtain the system of three equations

$$\begin{cases} \frac{1}{2}x + \frac{1}{2}y = x \\ \frac{1}{4}x + z = y \\ \frac{1}{4}x + \frac{1}{2}y = z \end{cases} \quad \text{or} \quad \begin{cases} x - y = 0 \\ x - 4y + 4z = 0 \\ x + 2y - 4z = 0 \end{cases}$$

We know that the system has a non-zero solution; hence we can arbitrarily assign a value to one of the unknowns. Set $y = 4$. Then by the first equation $x = 4$, and by the third equation $z = 3$. Thus $u = (4, 4, 3)$ is a fixed point of P. Multiply u by $1/(4+4+3) = \frac{1}{11}$ to obtain $t = \frac{1}{11}u = (\frac{4}{11}, \frac{4}{11}, \frac{3}{11})$ which is a probability vector and is also a fixed point of P.

20.12. Find the unique fixed probability vector of the regular stochastic matrix

$$P = \begin{pmatrix} 0 & 1 & 0 \\ \frac{1}{6} & \frac{1}{2} & \frac{1}{3} \\ 0 & \frac{2}{3} & \frac{1}{3} \end{pmatrix}$$

What matrix does P^n approach?

We first seek any fixed vector $u = (x, y, z)$ of the matrix P:

$$(x, y, z)\begin{pmatrix} 0 & 1 & 0 \\ \frac{1}{6} & \frac{1}{2} & \frac{1}{3} \\ 0 & \frac{2}{3} & \frac{1}{3} \end{pmatrix} = (x, y, z)$$

Multiply the left side of the above matrix equation and set corresponding components equal to each other to obtain the system of three equations

$$\begin{cases} \frac{1}{6}y = x \\ x + \frac{1}{2}y + \frac{2}{3}z = y \\ \frac{1}{3}y + \frac{1}{3}z = z \end{cases} \quad \text{or} \quad \begin{cases} y = 6x \\ 6x + 3y + 4z = 6y \\ y + z = 3z \end{cases} \quad \text{or} \quad \begin{cases} y = 6x \\ 6x + 4z = 3y \\ y = 2z \end{cases}$$

We know that the system has a non-zero solution; hence we can arbitrarily assign a value to one of the unknowns. Set $x = 1$. Then by the first equation $y = 6$, and by the last equation $z = 3$. Thus $u = (1, 6, 3)$ is a fixed point of P. Since $1 + 6 + 3 = 10$, the vector $t = (\frac{1}{10}, \frac{6}{10}, \frac{3}{10})$ is the required unique fixed probability vector of P.

P^n approaches the matrix T whose rows are each the fixed point t: $T = \begin{pmatrix} \frac{1}{10} & \frac{6}{10} & \frac{3}{10} \\ \frac{1}{10} & \frac{6}{10} & \frac{3}{10} \\ \frac{1}{10} & \frac{6}{10} & \frac{3}{10} \end{pmatrix}$.

20.13. If $t = (\frac{1}{4}, 0, \frac{1}{2}, \frac{1}{4}, 0)$ is a fixed point of a stochastic matrix P, why is P not regular?

If P is regular then, by Theorem 20.3, P has a unique fixed probability vector, and the components of the vector are positive. Since the components of the given fixed probability vector are not all positive, P cannot be regular.

20.14. Which of the following stochastic matrices are regular?

$$\text{(i) } A = \begin{pmatrix} \frac{1}{2} & \frac{1}{2} \\ 0 & 1 \end{pmatrix} \quad \text{(ii) } B = \begin{pmatrix} 0 & 1 \\ 1 & 0 \end{pmatrix} \quad \text{(iii) } C = \begin{pmatrix} \frac{1}{2} & \frac{1}{4} & \frac{1}{4} \\ 0 & 1 & 0 \\ \frac{1}{2} & \frac{1}{2} & 0 \end{pmatrix} \quad \text{(iv) } D = \begin{pmatrix} 0 & 0 & 1 \\ \frac{1}{2} & \frac{1}{4} & \frac{1}{4} \\ 0 & 1 & 0 \end{pmatrix}$$

Recall that a stochastic matrix is regular if a power of the matrix has only positive entries.

(i) A is not regular since there is a 1 on the main diagonal (in the second row).

(ii) $$B^2 = \begin{pmatrix} 0 & 1 \\ 1 & 0 \end{pmatrix}\begin{pmatrix} 0 & 1 \\ 1 & 0 \end{pmatrix} = \begin{pmatrix} 1 & 0 \\ 0 & 1 \end{pmatrix} = \text{the identity matrix } I$$

$$B^3 = \begin{pmatrix} 1 & 0 \\ 0 & 1 \end{pmatrix}\begin{pmatrix} 0 & 1 \\ 1 & 0 \end{pmatrix} = \begin{pmatrix} 0 & 1 \\ 1 & 0 \end{pmatrix} = B$$

Thus every even power of B is the identity matrix I and every odd power of B is the matrix B. Accordingly every power of B has zero entries, and so B is not regular.

(iii) C is not regular since it has a 1 on the main diagonal.

(iv) $$D^2 = \begin{pmatrix} 0 & 1 & 0 \\ \frac{1}{8} & \frac{5}{16} & \frac{5}{16} \\ \frac{1}{2} & \frac{1}{4} & \frac{1}{4} \end{pmatrix} \quad \text{and} \quad D^3 = \begin{pmatrix} \frac{1}{2} & \frac{1}{4} & \frac{1}{4} \\ \frac{5}{32} & \frac{45}{64} & \frac{9}{64} \\ \frac{1}{8} & \frac{5}{16} & \frac{5}{16} \end{pmatrix}$$

Since all the entries of D^3 are positive, D is regular.

MARKOV CHAINS

20.15. A student's study habits are as follows. If he studies one night, he is 70% sure not to study the next night. On the other hand, the probability that he does not study two nights in succession is .6. In the long run, how often does he study?

The states of the system are S (studying) and T (not studying). The transition matrix is

$$P = \begin{array}{c} \\ S \\ T \end{array} \begin{array}{c} \begin{array}{cc} S & T \end{array} \\ \begin{pmatrix} .3 & .7 \\ .4 & .6 \end{pmatrix} \end{array}$$

To discover what happens in the long run, we must find the unique fixed probability vector t of P. By Problem 20.10, $u = (.4, .7)$ is a fixed point of P and so $t = (\frac{4}{11}, \frac{7}{11})$ is the required probability vector. Thus in the long run the student studies $\frac{4}{11}$ of the time.

20.16. A psychologist makes the following assumptions concerning the behavior of mice subjected to a particular feeding schedule. For any particular trial 80% of the mice that went right on the previous experiment will go right on this trial, and 60% of those mice that went left on the previous experiment will go right on this trial. If 50% went right on the first trial, what would he predict for (i) the second trial, (ii) the third trial, (iii) the thousandth trial?

The states of the system are R (right) and L (left). The transition matrix is

$$P = \begin{matrix} & R & L \\ R & .8 & .2 \\ L & .6 & .4 \end{matrix}$$

The probability distribution for the first trial is $p = (.5, .5)$. To compute the probability distribution for the next step, i.e. the second trial, multiply p by the transition matrix P:

$$(.5, .5)\begin{pmatrix} .8 & .2 \\ .6 & .4 \end{pmatrix} = (.7, .3)$$

Thus on the second trial he predicts that 70% of the mice will go right and 30% will go left. To compute the probability distribution for the third trial, multiply that of the second trial by P:

$$(.7, .3)\begin{pmatrix} .8 & .2 \\ .6 & .4 \end{pmatrix} = (.74, .26)$$

Thus on the third trial he predicts that 74% of the mice will go right and 26% will go left.

We assume that the probability distribution for the thousandth trial is essentially the stationary probability distribution of the Markov chain, i.e. the unique fixed probability vector t of the transition matrix P. By Problem 20.10, $u = (.6, .2)$ is a fixed point of P and so $t = (\frac{3}{4}, \frac{1}{4}) = (.75, .25)$. Thus he predicts that, on the thousandth trial, 75% of the mice will go to the right and 25% will go to the left.

20.17. Given the transition matrix $P = \begin{pmatrix} 1 & 0 \\ \frac{1}{2} & \frac{1}{2} \end{pmatrix}$ with initial probability distribution $p^{(0)} = (\frac{1}{3}, \frac{2}{3})$. Define and find: (i) $p_{21}^{(3)}$, (ii) $p^{(3)}$, (iii) $p_2^{(3)}$.

(i) $p_{21}^{(3)}$ is the probability of moving from state a_2 to state a_1 in 3 steps. It can be obtained from the 3-step transition matrix P^3; hence first compute P^3:

$$P^2 = \begin{pmatrix} 1 & 0 \\ \frac{3}{4} & \frac{1}{4} \end{pmatrix}, \quad P^3 = \begin{pmatrix} 1 & 0 \\ \frac{7}{8} & \frac{1}{8} \end{pmatrix}$$

Then $p_{21}^{(3)}$ is the entry in the second row first column of P^3: $p_{21}^{(3)} = \frac{7}{8}$.

(ii) $p^{(3)}$ is the probability distribution of the system after three steps. It can be obtained by successively computing $p^{(1)}$, $p^{(2)}$ and then $p^{(3)}$:

$$p^{(1)} = p^{(0)}P = (\tfrac{1}{3}, \tfrac{2}{3})\begin{pmatrix} 1 & 0 \\ \frac{1}{2} & \frac{1}{2} \end{pmatrix} = (\tfrac{2}{3}, \tfrac{1}{3})$$

$$p^{(2)} = p^{(1)}P = (\tfrac{2}{3}, \tfrac{1}{3})\begin{pmatrix} 1 & 0 \\ \frac{1}{2} & \frac{1}{2} \end{pmatrix} = (\tfrac{5}{6}, \tfrac{1}{6})$$

$$p^{(3)} = p^{(2)}P = (\tfrac{5}{6}, \tfrac{1}{6})\begin{pmatrix} 1 & 0 \\ \frac{1}{2} & \frac{1}{2} \end{pmatrix} = (\tfrac{11}{12}, \tfrac{1}{12})$$

However, since the 3-step transition matrix P^3 has already been computed in (i), $p^{(3)}$ can also be obtained as follows:

$$p^{(3)} = p^{(0)}P^3 = (\tfrac{1}{3}, \tfrac{2}{3})\begin{pmatrix} 1 & 0 \\ \frac{7}{8} & \frac{1}{8} \end{pmatrix} = (\tfrac{11}{12}, \tfrac{1}{12})$$

(iii) $p_2^{(3)}$ is the probability that the process is in the state a_2 after 3 steps; it is the second component of the 3-step probability distribution $p^{(3)}$: $p_2^{(3)} = \frac{1}{12}$.

20.18. Given the transition matrix $P = \begin{pmatrix} 0 & \frac{1}{2} & \frac{1}{2} \\ \frac{1}{2} & \frac{1}{2} & 0 \\ 0 & 1 & 0 \end{pmatrix}$ and the initial probability distribution $p^{(0)} = (\frac{2}{3}, 0, \frac{1}{3})$. Find: (i) $p_{32}^{(2)}$ and $p_{13}^{(2)}$, (ii) $p^{(4)}$ and $p_3^{(4)}$, (iii) the vector that $p^{(0)}P^n$ approaches, (iv) the matrix that P^n approaches.

(i) First compute the 2-step transition matrix P^2:

$$P^2 = \begin{pmatrix} 0 & \frac{1}{2} & \frac{1}{2} \\ \frac{1}{2} & \frac{1}{2} & 0 \\ 0 & 1 & 0 \end{pmatrix}\begin{pmatrix} 0 & \frac{1}{2} & \frac{1}{2} \\ \frac{1}{2} & \frac{1}{2} & 0 \\ 0 & 1 & 0 \end{pmatrix} = \begin{pmatrix} \frac{1}{4} & \frac{3}{4} & 0 \\ \frac{1}{4} & \frac{1}{2} & \frac{1}{4} \\ \frac{1}{2} & \frac{1}{2} & 0 \end{pmatrix}$$

Then $p_{32}^{(2)} = \frac{1}{2}$ and $p_{13}^{(2)} = 0$, since these numbers refer to the entries in P^2.

(ii) To compute $p^{(4)}$, use the 2-step transition matrix P^2 and the initial probability distribution $p^{(0)}$:

$$p^{(2)} = p^{(0)}P^2 = (\tfrac{1}{3}, \tfrac{2}{3}, 0) \qquad \text{and} \qquad p^{(4)} = p^{(2)}P^2 = (\tfrac{1}{4}, \tfrac{7}{12}, \tfrac{1}{6})$$

Since $p_3^{(4)}$ is the third component of $p^{(4)}$, $p_3^{(4)} = \frac{1}{6}$.

(iii) By Theorem 20.3, $p^{(0)}P^n$ approaches the unique fixed probability vector t of P. To obtain t, first find any fixed vector $u = (x, y, z)$:

$$(x, y, z)\begin{pmatrix} 0 & \frac{1}{2} & \frac{1}{2} \\ \frac{1}{2} & \frac{1}{2} & 0 \\ 0 & 1 & 0 \end{pmatrix} = (x, y, z) \qquad \text{or} \qquad \begin{cases} \frac{1}{2}y = x \\ \frac{1}{2}x + \frac{1}{2}y + z = y \\ \frac{1}{2}x = z \end{cases}$$

Find any non-zero solution of the above system of equations. Set $z = 1$; then by the third equation $x = 2$, and by the first equation $y = 4$. Thus $u = (2, 4, 1)$ is a fixed point of P and so $t = (\frac{2}{7}, \frac{4}{7}, \frac{1}{7})$. In other words, $p^{(0)}P^n$ approaches $(\frac{2}{7}, \frac{4}{7}, \frac{1}{7})$.

(iv) P^n approaches the matrix T whose rows are each the fixed probability vector of P; hence

$$P^n \text{ approaches } \begin{pmatrix} \frac{2}{7} & \frac{4}{7} & \frac{1}{7} \\ \frac{2}{7} & \frac{4}{7} & \frac{1}{7} \\ \frac{2}{7} & \frac{4}{7} & \frac{1}{7} \end{pmatrix}.$$

20.19. A salesman's territory consists of three cities, A, B and C. He never sells in the same city on successive days. If he sells in city A, then the next day he sells in city B. However, if he sells in either B or C, then the next day he is twice as likely to sell in city A as in the other city. In the long run, how often does he sell in each of the cities?

The transition matrix of the problem is as follows:

$$P = \begin{matrix} & \begin{matrix} A & B & C \end{matrix} \\ \begin{matrix} A \\ B \\ C \end{matrix} & \begin{pmatrix} 0 & 1 & 0 \\ \frac{2}{3} & 0 & \frac{1}{3} \\ \frac{2}{3} & \frac{1}{3} & 0 \end{pmatrix} \end{matrix}$$

We seek the unique fixed probability vector t of the matrix P. First find any fixed vector $u = (x, y, z)$:

$$(x, y, z)\begin{pmatrix} 0 & 1 & 0 \\ \frac{2}{3} & 0 & \frac{1}{3} \\ \frac{2}{3} & \frac{1}{3} & 0 \end{pmatrix} = (x, y, z) \qquad \text{or} \qquad \begin{cases} \frac{2}{3}y + \frac{2}{3}z = x \\ x + \frac{1}{3}z = y \\ \frac{1}{3}y = z \end{cases}$$

Set, say, $z = 1$. Then by the third equation $y = 3$, and by the first equation $x = \frac{8}{3}$. Thus $u = (\frac{8}{3}, 3, 1)$. Also $3u = (8, 9, 3)$ is a fixed vector of P. Multiply $3u$ by $1/(8+9+3) = \frac{1}{20}$ to obtain the required fixed probability vector $t = (\frac{2}{5}, \frac{9}{20}, \frac{3}{20}) = (.40, .45, .15)$. Thus in the long run he sells 40% of the time in city A, 45% of the time in B and 15% of the time in C.

20.20. There are 2 white marbles in urn A and 3 red marbles in urn B. At each step of the process a marble is selected from each urn and the two marbles selected are interchanged. Let the state a_i of the system be the number i of red marbles in urn A. (i) Find the transition matrix P. (ii) What is the probability that there are 2 red marbles in urn A after 3 steps? (iii) In the long run, what is the probability that there are 2 red marbles in urn A?

(i) There are three states, a_0, a_1 and a_2 described by the following diagrams:

A	B	A	B	A	B
2 W	3 R	1 W, 1 R	1 W, 2 R	2 R	2 W, 1 R
a_0		a_1		a_2	

If the system is in state a_0, then a white marble must be selected from urn A and a red marble from urn B, so the system must move to state a_1. Accordingly, the first row of the transition matrix is $(0, 1, 0)$.

Suppose the system is in state a_1. It can move to state a_0 if and only if a red marble is selected from urn A and a white marble from urn B; the probability of that happening is $\frac{1}{2}\cdot\frac{1}{3} = \frac{1}{6}$. Thus $p_{10} = \frac{1}{6}$. The system can move from state a_1 to a_2 if and only if a white marble is selected from urn A and a red marble from urn B; the probability of that happening is $\frac{1}{2}\cdot\frac{2}{3} = \frac{1}{3}$. Thus $p_{12} = \frac{1}{3}$. Accordingly, the probability that the system remains in state a_1 is $p_{11} = 1 - \frac{1}{6} - \frac{1}{3} = \frac{1}{2}$. Thus the second row of the transition matrix is $(\frac{1}{6}, \frac{1}{2}, \frac{1}{3})$. (Note that p_{11} can also be obtained from the fact that the system remains in the state a_1 if either a white marble is drawn from each urn, probability $\frac{1}{2}\cdot\frac{1}{3} = \frac{1}{6}$, or a red marble is drawn from each urn, probability $\frac{1}{2}\cdot\frac{2}{3} = \frac{1}{3}$; thus $p_{11} = \frac{1}{6} + \frac{1}{3} = \frac{1}{2}$.)

Now suppose the system is in state a_2. A red marble must be drawn from urn A. If a red marble is selected from urn B, probability $\frac{1}{3}$, then the system remains in state a_2; and if a white marble is selected from urn B, probability $\frac{2}{3}$, then the system moves to state a_1. Note that the system can never move from state a_2 to the state a_0. Thus the third row of the transition matrix is $(0, \frac{2}{3}, \frac{1}{3})$. That is,

$$P = \begin{array}{c} \\ a_0 \\ a_1 \\ a_2 \end{array} \begin{array}{c} \begin{array}{ccc} a_0 & a_1 & a_2 \end{array} \\ \begin{pmatrix} 0 & 1 & 0 \\ \frac{1}{6} & \frac{1}{2} & \frac{1}{3} \\ 0 & \frac{2}{3} & \frac{1}{3} \end{pmatrix} \end{array}$$

(ii) The system began in state a_0, i.e. $p^{(0)} = (1, 0, 0)$. Thus:

$$p^{(1)} = p^{(0)}P = (0, 1, 0), \quad p^{(2)} = p^{(1)}P = (\tfrac{1}{6}, \tfrac{1}{2}, \tfrac{1}{3}), \quad p^{(3)} = p^{(2)}P = (\tfrac{1}{12}, \tfrac{23}{36}, \tfrac{5}{18})$$

Accordingly, the probability that there are 2 red marbles in urn A after 3 steps is $\frac{5}{18}$.

(iii) We seek the unique fixed probability vector t of the transition matrix P. First find any fixed vector $u = (x, y, z)$:

$$(x, y, z)\begin{pmatrix} 0 & 1 & 0 \\ \frac{1}{6} & \frac{1}{2} & \frac{1}{3} \\ 0 & \frac{2}{3} & \frac{1}{3} \end{pmatrix} = (x, y, z) \qquad \text{or} \qquad \begin{cases} \frac{1}{6}y = x \\ x + \frac{1}{2}y + \frac{2}{3}z = y \\ \frac{1}{3}y + \frac{1}{3}z = z \end{cases}$$

Set, say, $x = 1$. Then by the first equation $y = 6$, and by the third equation $z = 3$. Hence $u = (1, 6, 3)$. Multiply u by $1/(1+6+3) = \frac{1}{10}$ to obtain the required unique fixed probability vector $t = (.1, .6, .3)$. Thus, in the long run, 30% of the time there will be 2 red marbles in urn A.

Note that the long run probability distribution is the same as if the five marbles were placed in an urn and 2 were selected at random to put into urn A.

20.21. A player has \$2. He bets \$1 at a time and wins with probability $\frac{1}{2}$. He stops playing if he loses the \$2 or wins \$4. (i) What is the probability that he has lost his money at the end of, at most, 5 plays? (ii) What is the probability that the game lasts more than 7 plays?

This is a random walk with absorbing barriers at 0 and 6 (see Examples 8.2 and 8.3). The transition matrix is

$$P \quad = \quad \begin{array}{c} \\ a_0 \\ a_1 \\ a_2 \\ a_3 \\ a_4 \\ a_5 \\ a_6 \end{array} \begin{array}{c} \begin{array}{ccccccc} a_0 & a_1 & a_2 & a_3 & a_4 & a_5 & a_6 \end{array} \\ \begin{pmatrix} 1 & 0 & 0 & 0 & 0 & 0 & 0 \\ \frac{1}{2} & 0 & \frac{1}{2} & 0 & 0 & 0 & 0 \\ 0 & \frac{1}{2} & 0 & \frac{1}{2} & 0 & 0 & 0 \\ 0 & 0 & \frac{1}{2} & 0 & \frac{1}{2} & 0 & 0 \\ 0 & 0 & 0 & \frac{1}{2} & 0 & \frac{1}{2} & 0 \\ 0 & 0 & 0 & 0 & \frac{1}{2} & 0 & \frac{1}{2} \\ 0 & 0 & 0 & 0 & 0 & 0 & 1 \end{pmatrix} \end{array}$$

with initial probability distribution $p^{(0)} = (0,0,1,0,0,0,0)$ since he began with \$2.

(i) We seek $p_0^{(5)}$, the probability that the system is in state a_0 after five steps. Compute the 5th step probability distribution $p^{(5)}$:

$$p^{(1)} = p^{(0)}P = (0, \tfrac{1}{2}, 0, \tfrac{1}{2}, 0, 0, 0) \qquad p^{(4)} = p^{(3)}P = (\tfrac{3}{8}, 0, \tfrac{5}{16}, 0, \tfrac{1}{4}, 0, \tfrac{1}{16})$$

$$p^{(2)} = p^{(1)}P = (\tfrac{1}{4}, 0, \tfrac{1}{2}, 0, \tfrac{1}{4}, 0, 0) \qquad p^{(5)} = p^{(4)}P = (\tfrac{3}{8}, \tfrac{5}{32}, 0, \tfrac{9}{32}, 0, \tfrac{1}{8}, \tfrac{1}{16})$$

$$p^{(3)} = p^{(2)}P = (\tfrac{1}{4}, \tfrac{1}{4}, 0, \tfrac{3}{8}, 0, \tfrac{1}{8}, 0)$$

Thus $p_0^{(5)}$, the probability that he has no money after 5 plays, is $\frac{3}{8}$.

(ii) Compute $p^{(7)}$: $p^{(6)} = p^{(5)}P = (\frac{29}{64}, 0, \frac{7}{32}, 0, \frac{13}{64}, 0, \frac{1}{8})$, $p^{(7)} = p^{(6)}P = (\frac{29}{64}, \frac{7}{64}, 0, \frac{27}{128}, 0, \frac{13}{128}, \frac{1}{8})$

The probability that the game lasts more than 7 plays, i.e. that the system is not in state a_0 or a_6 after 7 steps, is $\frac{7}{64} + \frac{27}{128} + \frac{13}{128} = \frac{27}{64}$.

20.22. Consider repeated tosses of a fair die. Let X_n be the maximum of the numbers occurring in the first n trials.

(i) Find the transition matrix P of the Markov chain. Is the matrix regular?
(ii) Find $p^{(1)}$, the probability distribution after the first toss.
(iii) Find $p^{(2)}$ and $p^{(3)}$.

(i) The state space of the Markov chain is $\{1,2,3,4,5,6\}$. The transition matrix is

$$P \quad = \quad \begin{array}{c} \\ 1 \\ 2 \\ 3 \\ 4 \\ 5 \\ 6 \end{array} \begin{array}{c} \begin{array}{cccccc} 1 & 2 & 3 & 4 & 5 & 6 \end{array} \\ \begin{pmatrix} \frac{1}{6} & \frac{1}{6} & \frac{1}{6} & \frac{1}{6} & \frac{1}{6} & \frac{1}{6} \\ 0 & \frac{2}{6} & \frac{1}{6} & \frac{1}{6} & \frac{1}{6} & \frac{1}{6} \\ 0 & 0 & \frac{3}{6} & \frac{1}{6} & \frac{1}{6} & \frac{1}{6} \\ 0 & 0 & 0 & \frac{4}{6} & \frac{1}{6} & \frac{1}{6} \\ 0 & 0 & 0 & 0 & \frac{5}{6} & \frac{1}{6} \\ 0 & 0 & 0 & 0 & 0 & 1 \end{pmatrix} \end{array}$$

We obtain, for example, the third row of the matrix as follows. Suppose the system is in state 3, i.e. the maximum of the numbers occurring on the first n trials is 3. Then the system remains in state 3 if a 1, 2, or 3 occurs on the $(n+1)$-st trial; hence $p_{33} = \frac{3}{6}$. On the other hand, the system moves to state 4, 5 or 6, respectively, if a 4, 5 or 6 occurs on the $(n+1)$-st trial; hence $p_{34} = p_{35} = p_{36} = \frac{1}{6}$. The system can never move to state 1 or 2 since a 3 has occurred on one of the trials; hence $p_{31} = p_{32} = 0$. Thus the third row of the transition matrix is $(0, 0, \frac{3}{6}, \frac{1}{6}, \frac{1}{6}, \frac{1}{6})$. The other rows are obtained similarly.

The matrix is not regular since state 6 is absorbing, i.e. there is a 1 on the main diagonal in row 6.

(ii) On the first toss of the die, the state of the system X_1 is the number occurring; hence $p^{(1)} = (\frac{1}{6}, \frac{1}{6}, \frac{1}{6}, \frac{1}{6}, \frac{1}{6}, \frac{1}{6})$.

(iii) $p^{(2)} = p^{(1)}P = (\frac{1}{36}, \frac{3}{36}, \frac{5}{36}, \frac{7}{36}, \frac{9}{36}, \frac{11}{36})$. $p^{(3)} = p^{(2)}P = (\frac{1}{216}, \frac{7}{216}, \frac{19}{216}, \frac{37}{216}, \frac{61}{216}, \frac{91}{216})$.

20.23. Two boys b_1 and b_2 and two girls g_1 and g_2 are throwing a ball from one to the other. Each boy throws the ball to the other boy with probability $\frac{1}{2}$ and to each girl with probability $\frac{1}{4}$. On the other hand, each girl throws the ball to each boy with probability $\frac{1}{2}$ and never to the other girl. In the long run, how often does each receive the ball?

This is a Markov chain with state space $\{b_1, b_2, g_1, g_2\}$ and transition matrix

$$P = \begin{matrix} & b_1 & b_2 & g_1 & g_2 \\ b_1 & 0 & \frac{1}{2} & \frac{1}{4} & \frac{1}{4} \\ b_2 & \frac{1}{2} & 0 & \frac{1}{4} & \frac{1}{4} \\ g_1 & \frac{1}{2} & \frac{1}{2} & 0 & 0 \\ g_2 & \frac{1}{2} & \frac{1}{2} & 0 & 0 \end{matrix}$$

We seek a fixed vector $u = (x, y, z, w)$ of P: $(x, y, z, w)P = (x, y, z, w)$. Set the corresponding components of uP equal to u to obtain the system

$$\begin{aligned} \tfrac{1}{2}y + \tfrac{1}{2}z + \tfrac{1}{2}w &= x \\ \tfrac{1}{2}x + \tfrac{1}{2}z + \tfrac{1}{2}w &= y \\ \tfrac{1}{4}x + \tfrac{1}{4}y &= z \\ \tfrac{1}{4}x + \tfrac{1}{4}y &= w \end{aligned}$$

We seek any non-zero solution. Set, say, $z = 1$; then $w = 1$, $x = 2$ and $y = 2$. Thus $u = (2, 2, 1, 1)$ and so the unique fixed probability of P is $t = (\frac{1}{3}, \frac{1}{3}, \frac{1}{6}, \frac{1}{6})$. Thus, in the long run, each boy receives the ball $\frac{1}{3}$ of the time and each girl $\frac{1}{6}$ of the time.

20.24. Prove Theorem 20.6: Let $P = (p_{ij})$ be the transition matrix of a Markov chain. If $p = (p_i)$ is the probability distribution of the system at some arbitrary time k, then pP is the probability distribution of the system one step later, i.e. at time $k+1$; hence pP^n is the probability distribution of the system n steps later, i.e. at time $k+n$. In particular, $p^{(1)} = p^{(0)}P$, $p^{(2)} = p^{(1)}P$, ... and also $p^{(n)} = p^{(0)}P^n$.

Suppose the state space is $\{a_1, a_2, \ldots, a_m\}$. The probability that the system is in state a_j at time k and then in state a_i at time $k+1$ is the product $p_j p_{ji}$. Thus the probability that the system is in state a_i at time $k+1$ is the sum

$$p_1 p_{1i} + p_2 p_{2i} + \cdots + p_m p_{mi} = \sum_{j=1}^{m} p_j p_{ji}$$

Thus the probabiilty distribution at time $k+1$ is

$$p^* = \left(\sum_{j=1}^{m} p_j p_{j1}, \sum_{j=1}^{m} p_j p_{j2}, \ldots, \sum_{j=1}^{m} p_j p_{jm} \right)$$

However, this vector is precisely the product of the vector $p = (p_i)$ by the matrix $P = (p_{ij})$: $p^* = pP$.

20.25. Prove Theorem 20.5: Let P be the transition matrix of a Markov chain. Then the n-step transition matrix is equal to the nth power of P: $P^{(n)} = P^n$.

Suppose the system is in state a_i at, say, time k. We seek the probability $p_{ij}^{(n)}$ that the system is in state a_j at time $k+n$. Now the probability distribution of the system at time k, since the system is in state a_i, is the vector $e_i = (0, \ldots, 0, 1, 0, \ldots, 0)$ which has a 1 at the ith position and zeros everywhere else. By the preceding problem, the probability distribution at time $k+n$ is the product $e_i P^n$. But $e_i P^n$ is the ith row of the matrix P^n. Thus $P_{ij}^{(n)}$ is the jth component of the ith row of P^n, and so $P^{(n)} = P^n$.

MISCELLANEOUS PROBLEMS

20.26. The transition probabilities of a Markov chain can be represented by a diagram, called a *transition diagram*, where a positive probability p_{ij} is denoted by an arrow from the state a_i to the state a_j. Find the transition matrix of each of the following transition diagrams:

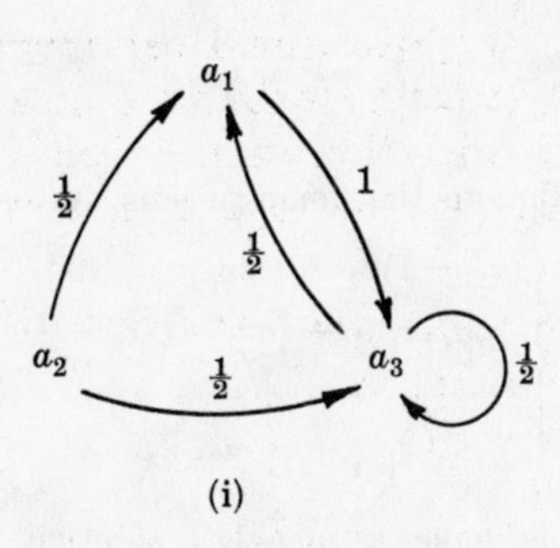

(i)

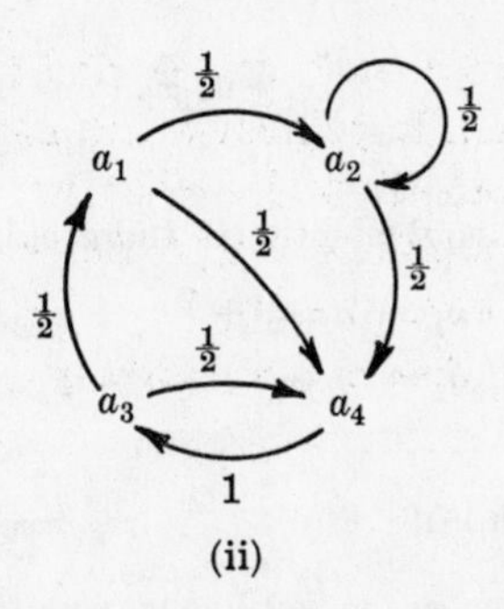

(ii)

(i) Note first that the state space is $\{a_1, a_2, a_3\}$ and so the transition matrix is of the form

$$P = \begin{matrix} & a_1 \quad a_2 \quad a_3 \\ \begin{matrix} a_1 \\ a_2 \\ a_3 \end{matrix} & \begin{pmatrix} & & \\ & & \\ & & \end{pmatrix} \end{matrix}$$

The ith row of the matrix is obtained by finding those arrows which emanate from a_i in the diagram; the number attached to the arrow from a_i to a_j is the jth component of the ith row. Thus the transition matrix is

$$P = \begin{matrix} & a_1 \quad a_2 \quad a_3 \\ \begin{matrix} a_1 \\ a_2 \\ a_3 \end{matrix} & \begin{pmatrix} 0 & 0 & 1 \\ \frac{1}{2} & 0 & \frac{1}{2} \\ \frac{1}{2} & 0 & \frac{1}{2} \end{pmatrix} \end{matrix}$$

(ii) The state space is $\{a_1, a_2, a_3, a_4\}$. The transition matrix is

$$P = \begin{matrix} & a_1 \quad a_2 \quad a_3 \quad a_4 \\ \begin{matrix} a_1 \\ a_2 \\ a_3 \\ a_4 \end{matrix} & \begin{pmatrix} 0 & \frac{1}{2} & 0 & \frac{1}{2} \\ 0 & \frac{1}{2} & 0 & \frac{1}{2} \\ \frac{1}{2} & 0 & 0 & \frac{1}{2} \\ 0 & 0 & 1 & 0 \end{pmatrix} \end{matrix}$$

20.27. Suppose the transition matrix of a Markov chain is as follows:

$$P = \begin{matrix} & a_1 \quad a_2 \quad a_3 \quad a_4 \\ \begin{matrix} a_1 \\ a_2 \\ a_3 \\ a_4 \end{matrix} & \begin{pmatrix} \frac{1}{2} & \frac{1}{2} & 0 & 0 \\ \frac{1}{2} & \frac{1}{2} & 0 & 0 \\ \frac{1}{4} & \frac{1}{4} & \frac{1}{4} & \frac{1}{4} \\ \frac{1}{4} & \frac{1}{4} & \frac{1}{4} & \frac{1}{4} \end{pmatrix} \end{matrix}$$

Is the Markov chain regular?

Note that once the system enters the state a_1 or the state a_2, then it can never move to state a_3 or state a_4, i.e. the system remains in the state subspace $\{a_1, a_2\}$. Thus, in particular, $p_{13}^{(n)} = 0$ for every n and so every power P^n will contain a zero entry. Hence P is not regular.

20.28. Prove: An $m \times m$ matrix $A = (a_{ij})$ has a non-zero fixed point if and only if the determinant of the matrix $A - I$ is zero. Here I is the identity matrix.

The matrix $A = (a_{ij})$ has a non-zero fixed point if and only if there exists a non-zero vector $u = (x_1, x_2, \ldots, x_m)$ such that

$$(x_1, x_2, \ldots, x_m) \begin{pmatrix} a_{11} & a_{12} & \cdots & a_{1m} \\ a_{21} & a_{22} & \cdots & a_{2m} \\ \cdots & \cdots & \cdots & \cdots \\ a_{m1} & a_{m2} & \cdots & a_{mm} \end{pmatrix} = (x_1, x_2, \ldots, x_m)$$

or
$$(a_{11}x_1 + a_{21}x_2 + \cdots + a_{m1}x_m,\ a_{12}x_1 + a_{22}x_2 + \cdots + a_{m2}x_m,\ \ldots,\ a_{1m}x_1 + a_{2m}x_2 + \cdots + a_{mm}x_m) = (x_1, x_2, \ldots, x_m)$$

or, equivalently, if there exists a non-zero solution to the homogeneous system

$$\begin{cases} a_{11}x_1 + a_{21}x_2 + \cdots + a_{m1}x_m = x_1 \\ a_{12}x_1 + a_{22}x_2 + \cdots + a_{m2}x_m = x_2 \\ \cdots\cdots\cdots\cdots\cdots\cdots \\ a_{1m}x_1 + a_{2m}x_2 + \cdots + a_{mm}x_m = x_m \end{cases} \quad \text{or} \quad \begin{cases} (a_{11}-1)x_1 + a_{21}x_2 + \cdots + a_{m1}x_m = 0 \\ a_{12}x_1 + (a_{22}-1)x_2 + \cdots + a_{m2}x_m = 0 \\ \cdots\cdots\cdots\cdots\cdots\cdots \\ a_{1m}x_1 + a_{2m}x_2 + \cdots + (a_{mm}-1)x_m = 0 \end{cases}$$

Now m homogeneous equations in m unknowns have a non-zero solution if and only if the determinant of the matrix of coefficients is zero. But the matrix of coefficients of the above homogeneous system of equations is the transpose of the matrix

$$A - I = \begin{pmatrix} a_{11} & a_{12} & \cdots & a_{1m} \\ a_{21} & a_{22} & \cdots & a_{2m} \\ \cdots & \cdots & \cdots & \cdots \\ a_{m1} & a_{m2} & \cdots & a_{mm} \end{pmatrix} - \begin{pmatrix} 1 & 0 & \cdots & 0 \\ 0 & 1 & \cdots & 0 \\ \cdots & \cdots & \cdots & \cdots \\ 0 & 0 & \cdots & 1 \end{pmatrix} = \begin{pmatrix} a_{11}-1 & a_{12} & \cdots & a_{1m} \\ a_{21} & a_{22}-1 & \cdots & a_{2m} \\ \cdots & \cdots & \cdots & \cdots \\ a_{m1} & a_{m2} & \cdots & a_{mm}-1 \end{pmatrix}$$

Since the determinant of a matrix is equal to the determinant of its transpose, A has a non-zero fixed point if and only if $\det(A - I) = 0$.

20.29. Prove: Every stochastic matrix $P = (p_{ij})$ has a non-zero fixed point.

Since the rows of the stochastic matrix P are probability vectors, the sum of its column vectors is

$$\begin{pmatrix} p_{11} \\ p_{21} \\ \cdots \\ p_{m1} \end{pmatrix} + \begin{pmatrix} p_{12} \\ p_{22} \\ \cdots \\ p_{m2} \end{pmatrix} + \cdots + \begin{pmatrix} p_{1m} \\ p_{2m} \\ \cdots \\ p_{mm} \end{pmatrix} = \begin{pmatrix} 1 \\ 1 \\ \cdots \\ 1 \end{pmatrix}$$

Accordingly, the sum of the columns of the matrix $P - I$ is

$$\begin{pmatrix} p_{11}-1 \\ p_{21} \\ \cdots \\ p_{m1} \end{pmatrix} + \begin{pmatrix} p_{12} \\ p_{22}-1 \\ \cdots \\ p_{m2} \end{pmatrix} + \cdots + \begin{pmatrix} p_{1m} \\ p_{2m} \\ \cdots \\ p_{mm}-1 \end{pmatrix} = \begin{pmatrix} 0 \\ 0 \\ \cdots \\ 0 \end{pmatrix}$$

Thus the columns of the matrix $P - I$ are dependent (see Page 114). But the determinant of a matrix is zero iff its columns are dependent; hence $\det(P - I) = 0$ and so, by the preceding problem, $P - I$ has a non-zero fixed point.

20.30. Suppose m points on a circle are numbered respectively $1, 2, \ldots, m$ in a counterclockwise direction. A particle performs a "random walk" on the circle; it moves one step counterclockwise with probability p or one step clockwise with probability $q = 1 - p$. Find the transition matrix of this Markov chain.

The state space is $\{1, 2, \ldots, m\}$. The diagram to the right below can be used to obtain the transition matrix which appears to the left below.

$$P = \begin{array}{c} \\ 1 \\ 2 \\ 3 \\ \vdots \\ m-1 \\ m \end{array} \begin{array}{c} \begin{array}{cccccccc} 1 & 2 & 3 & 4 & \cdots & m-2 & m-1 & m \end{array} \\ \begin{pmatrix} 0 & p & 0 & 0 & \cdots & 0 & 0 & q \\ q & 0 & p & 0 & \cdots & 0 & 0 & 0 \\ 0 & q & 0 & p & \cdots & 0 & 0 & 0 \\ \cdots & \cdots & \cdots & \cdots & \cdots & \cdots & \cdots & \cdots \\ 0 & 0 & 0 & 0 & \cdots & q & 0 & p \\ p & 0 & 0 & 0 & \cdots & 0 & q & 0 \end{pmatrix} \end{array}$$

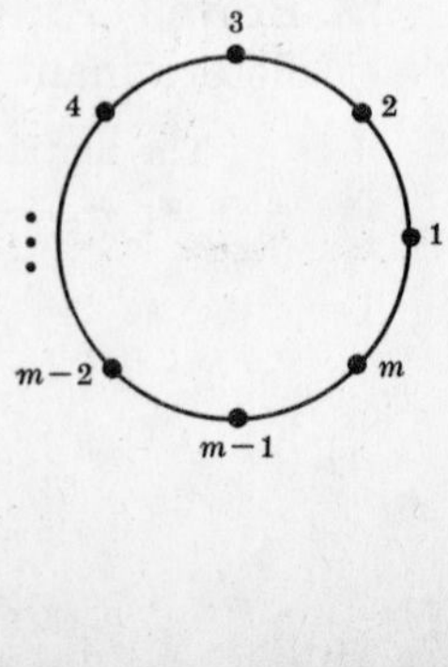

Supplementary Problems

PROBABILITY VECTORS AND STOCHASTIC MATRICES

20.31. Which vectors are probability vectors?

(i) $(\frac{1}{4}, \frac{1}{2}, -\frac{1}{4}, \frac{1}{2})$ (ii) $(\frac{1}{2}, 0, \frac{1}{3}, \frac{1}{6}, \frac{1}{6})$ (iii) $(\frac{1}{12}, \frac{1}{2}, \frac{1}{6}, 0, \frac{1}{4})$.

20.32. Find a scalar multiple of each vector which is a probability vector:

(i) $(3, 0, 2, 5, 3)$ (ii) $(2, \frac{1}{2}, 0, \frac{1}{4}, \frac{3}{4}, 0, 1)$ (iii) $(\frac{1}{3}, 2, \frac{1}{2}, 0, \frac{1}{4}, \frac{2}{3})$.

20.33. Which matrices are stochastic?

(i) $\begin{pmatrix} 0 & 1 & 0 \\ \frac{1}{2} & \frac{1}{4} & \frac{1}{4} \end{pmatrix}$ (ii) $\begin{pmatrix} 1 & 0 \\ 0 & 1 \end{pmatrix}$ (iii) $\begin{pmatrix} 0 & 1 \\ \frac{1}{2} & \frac{1}{4} \end{pmatrix}$ (iv) $\begin{pmatrix} \frac{1}{2} & \frac{1}{2} \\ \frac{1}{2} & \frac{1}{2} \end{pmatrix}$ (v) $\begin{pmatrix} 0 & 1 \\ -\frac{1}{2} & \frac{3}{2} \end{pmatrix}$

REGULAR STOCHASTIC MATRICES AND FIXED PROBABILITY VECTORS

20.34. Find the unique fixed probability vector of each matrix:

(i) $\begin{pmatrix} \frac{2}{3} & \frac{1}{3} \\ \frac{2}{5} & \frac{3}{5} \end{pmatrix}$ (ii) $\begin{pmatrix} \frac{1}{4} & \frac{3}{4} \\ \frac{5}{6} & \frac{1}{6} \end{pmatrix}$ (iii) $\begin{pmatrix} .2 & .8 \\ .5 & .5 \end{pmatrix}$ (iv) $\begin{pmatrix} .7 & .3 \\ .6 & .4 \end{pmatrix}$

20.35. (i) Find the unique fixed probability vector t of $P = \begin{pmatrix} 0 & \frac{3}{4} & \frac{1}{4} \\ \frac{1}{2} & \frac{1}{2} & 0 \\ 0 & 1 & 0 \end{pmatrix}$.

(ii) What matrix does P^n approach? (iii) What vector does $(\frac{1}{4}, \frac{1}{4}, \frac{1}{2})P^n$ approach?

20.36. Find the unique fixed probability vector t of each matrix:

(i) $A = \begin{pmatrix} 0 & \frac{1}{2} & \frac{1}{2} \\ \frac{1}{3} & \frac{2}{3} & 0 \\ 0 & 1 & 0 \end{pmatrix}$ (ii) $B = \begin{pmatrix} 0 & 1 & 0 \\ \frac{1}{2} & 0 & \frac{1}{2} \\ \frac{1}{2} & \frac{1}{4} & \frac{1}{4} \end{pmatrix}$

20.37. (i) Find the unique fixed probability vector t of $P = \begin{pmatrix} 0 & \frac{1}{2} & \frac{1}{2} & 0 \\ \frac{1}{2} & \frac{1}{4} & 0 & \frac{1}{4} \\ 0 & 0 & 0 & 1 \\ 0 & \frac{1}{2} & 0 & \frac{1}{2} \end{pmatrix}$.

(ii) What matrix does P^n approach?

(iii) What vector does $(\frac{1}{4}, 0, \frac{1}{2}, \frac{1}{4})P^n$ approach?

(iv) What vector does $(\frac{1}{2}, 0, 0, \frac{1}{2})P^n$ approach?

20.38. (i) Given that $t = (\frac{1}{2}, 0, \frac{1}{4}, \frac{1}{4})$ is a fixed point of a stochastic matrix P, is P regular?

(ii) Given that $t = (\frac{1}{4}, \frac{1}{4}, \frac{1}{4}, \frac{1}{4})$ is a fixed point of a stochastic matrix P, is P regular?

20.39. Which of the stochastic matrices are regular?

(i) $\begin{pmatrix} \frac{1}{2} & \frac{1}{4} & \frac{1}{4} \\ 0 & 1 & 0 \\ \frac{1}{2} & 0 & \frac{1}{2} \end{pmatrix}$ (ii) $\begin{pmatrix} \frac{1}{2} & \frac{1}{2} & 0 \\ \frac{1}{2} & \frac{1}{2} & 0 \\ \frac{1}{4} & \frac{1}{4} & \frac{1}{2} \end{pmatrix}$ (iii) $\begin{pmatrix} 0 & 0 & 1 \\ \frac{1}{2} & 0 & \frac{1}{2} \\ 0 & 1 & 0 \end{pmatrix}$

20.40. Show that $(cf + ce + de,\ af + bf + ae,\ ad + bd + bc)$ is a fixed point of the matrix

$$P = \begin{pmatrix} 1-a-b & a & b \\ c & 1-c-d & d \\ e & f & 1-e-f \end{pmatrix}$$

MARKOV CHAINS

20.41. A man's smoking habits are as follows. If he smokes filter cigarettes one week, he switches to non-filter cigarettes the next week with probability .2. On the other hand, the probability that he smokes non-filter cigarettes two weeks in succession is .7. In the long run, how often does he smoke filter cigarettes?

20.42. A gambler's luck follows a pattern. If he wins a game, the probability of winning the next game is .6. However, if he loses a game, the probability of losing the next game is .7. There is an even chance that the gambler wins the first game.

(i) What is the probability that he wins the second game?

(ii) What is the probability that he wins the third game?

(iii) In the long run, how often will he win?

20.43. For a Markov chain, the transition matrix is $P = \begin{pmatrix} \frac{1}{2} & \frac{1}{2} \\ \frac{3}{4} & \frac{1}{4} \end{pmatrix}$ with initial probability distribution $p^{(0)} = (\frac{1}{4}, \frac{3}{4})$. Find: (i) $p_{21}^{(2)}$; (ii) $p_{12}^{(2)}$; (iii) $p^{(2)}$; (iv) $p_1^{(2)}$; (v) the vector $p^{(0)}P^n$ approaches; (vi) the matrix P^n approaches.

20.44. For a Markov chain, the transition matrix is $P = \begin{pmatrix} \frac{1}{2} & 0 & \frac{1}{2} \\ 1 & 0 & 0 \\ \frac{1}{4} & \frac{1}{2} & \frac{1}{4} \end{pmatrix}$ and the initial probability distribution is $p^{(0)} = (\frac{1}{2}, \frac{1}{2}, 0)$. Find (i) $p_{13}^{(2)}$, (ii) $p_{23}^{(2)}$, (iii) $p^{(2)}$, (iv) $p_1^{(2)}$.

20.45. Each year a man trades his car for a new car. If he has a Buick, he trades it for a Plymouth. If he has a Plymouth, he trades it for a Ford. However, if he has a Ford, he is just as likely to trade it for a new Ford as to trade it for a Buick or a Plymouth. In 1955 he bought his first car which was a Ford.

(i) Find the probability that he has a (*a*) 1957 Ford, (*b*) 1957 Buick, (*c*) 1958 Plymouth, (*d*) 1958 Ford.

(ii) In the long run, how often will he have a Ford?

20.46. There are 2 white marbles in urn A and 4 red marbles in urn B. At each step of the process a marble is selected from each urn, and the two marbles selected are interchanged. Let X_n be the number of red marbles in urn A after n interchanges. (i) Find the transition matrix P. (ii) What is the probability that there are 2 red marbles in urn A after 3 steps? (iii) In the long run, what is the probability that there are 2 red marbles in urn A?

20.47. Solve the preceding problem in the case that there are 3 white marbles in urn A and 3 red marbles in urn B.

20.48. A fair coin is tossed until 3 heads occur in a row. Let X_n be the length of the sequence of heads ending at the nth trial. (See Example 8.4.) What is the probability that there are at least 8 tosses of the coin?

20.49. A player has 3 dollars. At each play of a game, he loses one dollar with probability $\frac{3}{4}$ but wins two dollars with probability $\frac{1}{4}$. He stops playing if he has lost his 3 dollars or he has won at least 3 dollars.

(i) Find the transition matrix of the Markov chain.

(ii) What is the probability that there are at least 4 plays to the game?

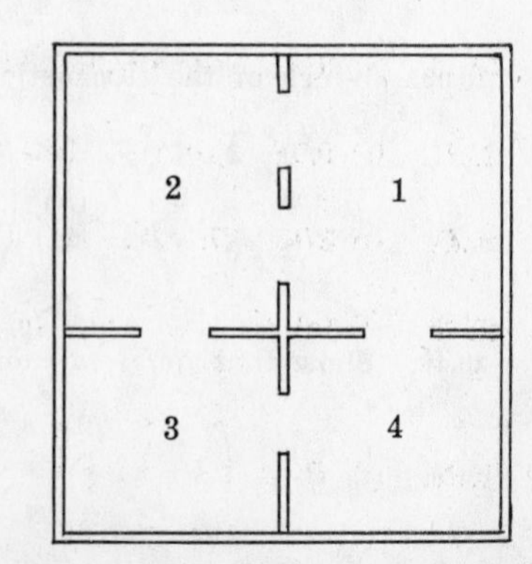

20.50. The diagram on the right shows four compartments with doors leading from one to another. A mouse in any compartment is equally likely to pass through each of the doors of the compartment. Find the transition matrix of the Markov chain.

20.51. Find the transition matrix corresponding to each transition diagram:

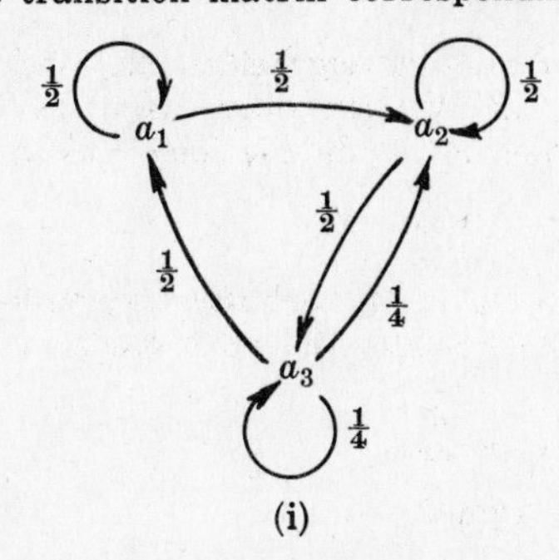

(i)

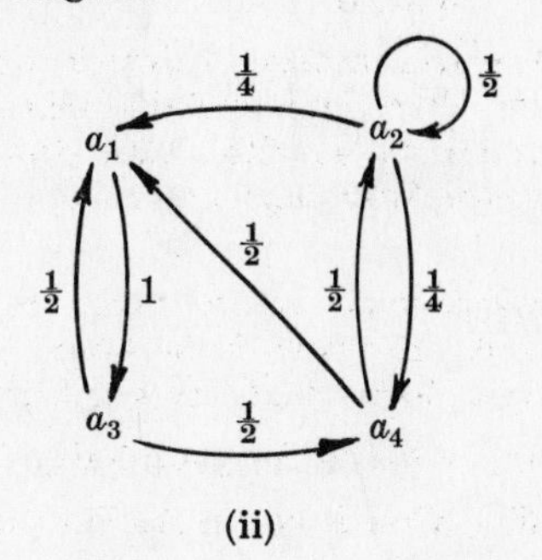

(ii)

20.52. Draw a transition diagram for each transition matrix:

(i) $P = \begin{array}{c} \\ a_1 \\ a_2 \end{array}\begin{array}{c} \begin{array}{cc} a_1 & a_2 \end{array} \\ \begin{pmatrix} \frac{1}{2} & \frac{1}{2} \\ \frac{1}{3} & \frac{2}{3} \end{pmatrix} \end{array}$ (ii) $P = \begin{array}{c} \\ a_1 \\ a_2 \\ a_3 \end{array}\begin{array}{c} \begin{array}{ccc} a_1 & a_2 & a_3 \end{array} \\ \begin{pmatrix} 0 & \frac{1}{2} & \frac{1}{2} \\ \frac{1}{4} & \frac{1}{4} & \frac{1}{2} \\ 0 & \frac{1}{2} & \frac{1}{2} \end{pmatrix} \end{array}$

Answers to Supplementary Problems

20.31. Only (iii).

20.32. (i) (3/13, 0, 2/13, 5/13, 3/13)
(ii) (8/18, 2/18, 0, 1/18, 3/18, 0, 4/18)
(iii) (4/45, 24/45, 6/45, 0, 3/45, 8/45)

20.33. Only (ii) and (iv).

20.34. (i) (6/11, 5/11), (ii) (10/19, 9/19), (iii) (5/13, 8/13), (iv) $(\frac{2}{3}, \frac{1}{3})$

20.35. (i) $t = (4/13, 8/13, 1/13)$, (iii) $t = (4/13, 8/13, 1/13)$

20.36. (i) $t = (2/9, 6/9, 1/9)$, (ii) $t = (5/15, 6/15, 4/15)$

20.37. (i) $t = (2/11, 4/11, 1/11, 4/11)$, (iii) t, (iv) t

20.39. Only (iii).

20.41. 60% of the time.

20.42. (i) 9/20, (ii) 87/200, (iii) 3/7 of the time.

20.43. (i) 9/16, (ii) 3/8, (iii) (37/64, 27/64), (iv) 27/64, (v) (.6, .4), (vi) $\begin{pmatrix} .6 & .4 \\ .6 & .4 \end{pmatrix}$

20.44. (i) 3/8, (ii) 1/2, (iii) (7/16, 2/16, 7/16), (iv) 7/16

20.45. (i) (*a*) 4/9, (*b*) 1/9, (*c*) 7/27, (*d*) 16/27. (ii) 50% of the time

20.46. (i) $P = \begin{pmatrix} 0 & 1 & 0 \\ \frac{1}{8} & \frac{1}{2} & \frac{3}{8} \\ 0 & \frac{1}{2} & \frac{1}{2} \end{pmatrix}$ (ii) 3/8 (iii) 2/5

20.47. (i) $P = \begin{pmatrix} 0 & 1 & 0 & 0 \\ \frac{1}{9} & \frac{4}{9} & \frac{4}{9} & 0 \\ 0 & \frac{4}{9} & \frac{4}{9} & \frac{1}{9} \\ 0 & 0 & 1 & 0 \end{pmatrix}$ (ii) 32/81 (iii) 9/20

20.48. 47/128

20.49. (i) $P = \begin{pmatrix} 1 & 0 & 0 & 0 & 0 & 0 & 0 \\ \frac{3}{4} & 0 & 0 & \frac{1}{4} & 0 & 0 & 0 \\ 0 & \frac{3}{4} & 0 & 0 & \frac{1}{4} & 0 & 0 \\ 0 & 0 & \frac{3}{4} & 0 & 0 & \frac{1}{4} & 0 \\ 0 & 0 & 0 & \frac{3}{4} & 0 & 0 & \frac{1}{4} \\ 0 & 0 & 0 & 0 & \frac{3}{4} & 0 & \frac{1}{4} \\ 0 & 0 & 0 & 0 & 0 & 0 & 1 \end{pmatrix}$ (ii) 27/64

20.50. $P = \begin{pmatrix} 0 & \frac{2}{3} & 0 & \frac{1}{3} \\ \frac{2}{3} & 0 & \frac{1}{3} & 0 \\ 0 & \frac{1}{2} & 0 & \frac{1}{2} \\ \frac{1}{2} & 0 & \frac{1}{2} & 0 \end{pmatrix}$

20.51. (i) $\begin{pmatrix} \frac{1}{2} & \frac{1}{2} & 0 \\ 0 & \frac{1}{2} & \frac{1}{2} \\ \frac{1}{2} & \frac{1}{4} & \frac{1}{4} \end{pmatrix}$ (ii) $\begin{pmatrix} 0 & 0 & 1 & 0 \\ \frac{1}{4} & \frac{1}{2} & 0 & \frac{1}{4} \\ \frac{1}{2} & 0 & 0 & \frac{1}{2} \\ \frac{1}{2} & \frac{1}{2} & 0 & 0 \end{pmatrix}$

20.52. (i)

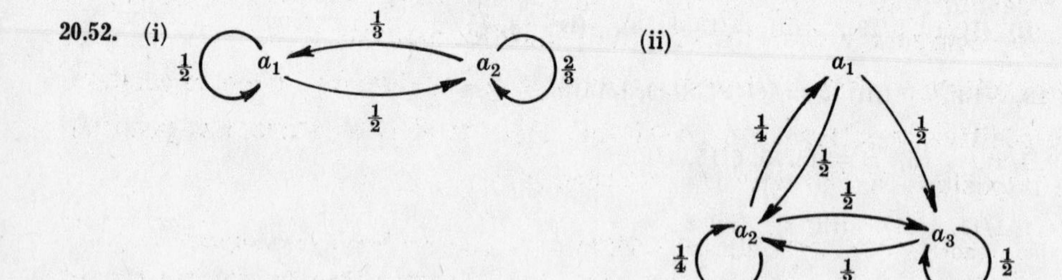

Chapter 21

Inequalities

THE REAL LINE

The set of real numbers, denoted by **R**, plays a dominant role in mathematics. We assume the reader is familiar with the geometric representation of **R** by means of the points on a straight line. As shown below, a point, called the *origin*, is chosen to represent 0 and another point, usually to the right of 0, to represent 1. Then there is a natural way to pair off the points on the line and the real numbers, i.e. each point will represent a unique real number and each real number will be represented by a unique point. For this reason we refer to **R** as the *real line* and use the words point and number interchangeably.

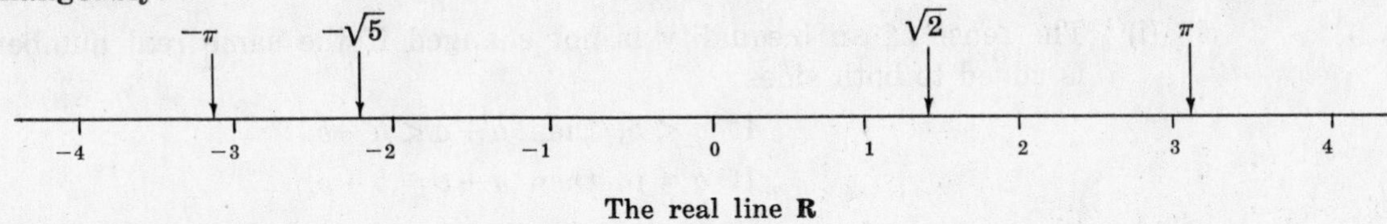

The real line **R**

POSITIVE NUMBERS

Those numbers to the right of 0 on the real line **R**, i.e. on the same side as 1, are the *positive numbers*; those numbers to the left of 0 are the *negative numbers*. The set of positive numbers can be completely described by the following axioms:

[P_1] If $a \in \mathbf{R}$, then exactly one of the following is true: a is positive; $a = 0$; $-a$ is positive.

[P_2] If $a, b \in \mathbf{R}$ are positive, then their sum $a + b$ and their product $a \cdot b$ are also positive.

It follows that a is positive if and only if $-a$ is negative.

Example 1.1: We show, using only [P_1] and [P_2], that the real number 1 is positive.

By [P_1], either 1 or -1 is positive. If -1 is positive then, by [P_2], the product $(-1)(-1) = 1$ is positive. But this contradicts [P_1] which states that 1 and -1 cannot both be positive. Hence the assumption that -1 is positive is false and so 1 is positive.

Example 1.2: The real number -2 is negative. For, by the preceding example, 1 is positive and so, by [P_2], the sum $1 + 1 = 2$ is positive; hence -2 is not positive, i.e. -2 is negative.

ORDER

An order relation in **R** is defined using the concept of positiveness.

Definition: The real number a is *less than* the real number b, written $a < b$, if the difference $b - a$ is positive.

The following notation is also used:

$a > b$, read a is greater than b, means $b < a$

$a \leq b$, read a is less than or equal to b, means $a < b$ or $a = b$

$a \geq b$, read a is greater than or equal to b, means $b \leq a$

Geometrically speaking,

$a < b$ means a is to the left of b on the real line **R**

$a > b$ means a is to the right of b on the real line **R**

Example 2.1: $2 < 5$, $-6 \leq -3$, $4 \leq 4$, $5 > -8$

Example 2.2: A real number x is positive iff $x > 0$, and x is negative iff $x < 0$. For $x \neq 0$, x^2 is always greater than 0: $x^2 > 0$.

Example 2.3: The notation $2 < x < 5$ means $2 < x$ and also $x < 5$; hence x will lie between 2 and 5 on the real line.

We refer to the relations $<$, $>$, $\leq$ and $\geq$ as inequalities in order to distinguish them from the equality relation $=$. We also shall refer to $<$ and $>$ as strict inequalities. We now state basic properties about inequalities which shall be used throughout.

Theorem 21.1: Let a, b and c be real numbers.

(i) The sense of an inequality is not changed if the same real number is added to both sides:

If $a < b$, then $a + c < b + c$.

If $a \leq b$, then $a + c \leq b + c$.

(ii) The sense of an inequality is not changed if both sides are multiplied by the same positive real number:

If $a < b$ and $c > 0$, then $ac < bc$.

If $a \leq b$ and $c > 0$, then $ac \leq bc$.

(iii) The sense of an inequality is reversed if both sides are multiplied by the same *negative* number:

If $a < b$ and $c < 0$, then $ac > bc$.

If $a \leq b$ and $c < 0$, then $ac \geq bc$.

(FINITE) INTERVALS

Let a and b be real numbers such that $a < b$. Then the set of all real numbers x satisfying

$a < x < b$ is called the open interval from a to b

$a \leq x \leq b$ is called the closed interval from a to b

$a < x \leq b$ is called the open-closed interval from a to b

$a \leq x < b$ is called the closed-open interval from a to b

The points a and b are called the endpoints of the interval. Observe that a closed interval contains both its endpoints, an open interval contains neither endpoint, and an open-closed and a closed-open interval contains exactly one of its endpoints:

open interval: $a < x < b$ closed interval: $a \leq x \leq b$

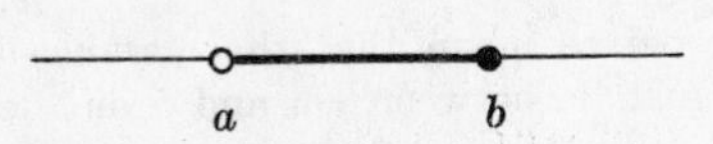

open-closed interval: $a < x \leqq b$

closed-open interval: $a \leqq x < b$

The open-closed and closed-open intervals are also referred to as being half-open (or: half-closed).

INFINITE INTERVALS

Let a be any real number. Then the set of all real numbers x satisfying $x < a$, $x \leqq a$, $x > a$ or $x \geqq a$ is called an *infinite interval.*

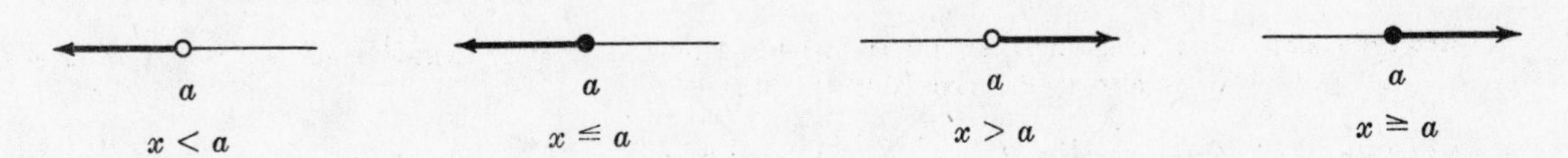

$x < a$ $\qquad$ $x \leqq a$ $\qquad$ $x > a$ $\qquad$ $x \geqq a$

LINEAR INEQUALITIES IN ONE UNKNOWN

Every linear inequality in one unknown x can be reduced to the form

$$ax < b, \quad ax \leqq b, \quad ax > b \quad \text{or} \quad ax \geqq b$$

If $a \neq 0$, then both sides of the inequality can be multiplied by $1/a$ with the sense of the inequality reversed if a is negative. Thus the inequality can be further reduced to the form

$$x < c, \quad x \leqq c, \quad x > c \quad \text{or} \quad x \geqq c$$

whose solution set is an infinite interval.

Example 3.1: Consider the inequality $5x + 7 \leqq 2x + 1$. Adding $-2x - 7$ to both sides (or: transposing), we obtain

$$5x - 2x \leqq 1 - 7 \quad \text{or} \quad 3x \leqq -6$$

Multiplying both sides by $\frac{1}{3}$ (or: dividing both sides by 3) we finally obtain

$$x \leqq -2$$

Example 3.2: Consider the inequality $2x + 3 < 4x + 9$. Transposing, we obtain

$$2x - 4x < 9 - 3 \quad \text{or} \quad -2x < 6$$

Multiplying both sides of the inequality by $-\frac{1}{2}$ (or: dividing both sides by -2) and reversing the inequality since $-\frac{1}{2}$ is negative, we finally obtain

$$x > -3$$

ABSOLUTE VALUE

The absolute value of a real number x, written $|x|$, is defined by

$$|x| = \begin{cases} x & \text{if } x \geqq 0 \\ -x & \text{if } x < 0 \end{cases}$$

that is, if x is non-negative then $|x| = x$, and if x is negative then $|x| = -x$. Thus the absolute value of every real number is non-negative: $|x| \geqq 0$ for every $x \in \mathbf{R}$.

Geometrically speaking, the absolute value of x is the distance between the point x on the real line and the origin, i.e. the point 0. Furthermore, the distance between any two points $a, b \in \mathbf{R}$ is $|a - b| = |b - a|$.

Example 4.1: $|-2| = 2$, $|7| = 7$, $|-\pi| = \pi$, $|-\sqrt{2}| = \sqrt{2}$

Example 4.2: $|3 - 8| = |-5| = 5$ and $|8 - 3| = |5| = 5$

Example 4.3: The statement $|x| < 5$ can be interpreted to mean that the distance between x and the origin is less than 5; hence x must lie between -5 and 5 on the real line. In other words,

$$|x| < 5 \quad \text{and} \quad -5 < x < 5$$

−5 0 5

have identical meaning and, similarly,

$$|x| \leq 5 \quad \text{and} \quad -5 \leq x \leq 5$$

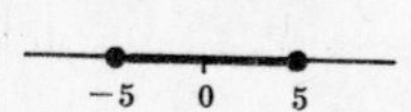

have identical meaning.

The graph of the function $f(x) = |x|$, i.e. the absolute value function, lies entirely in the upper half plane since $f(x) \geq 0$ for every $x \in \mathbf{R}$:

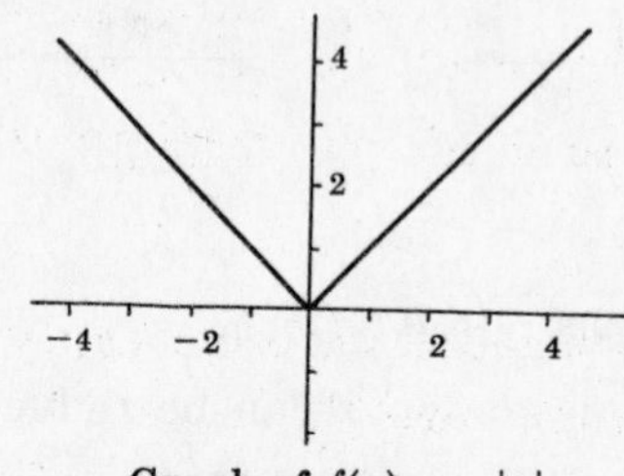

Graph of $f(x) = |x|$

The central facts about the absolute value function are the following:

Theorem 21.2: Let a and b be any real numbers. Then:

(i) $|a| \geq 0$, and $|a| = 0$ iff $a = 0$

(ii) $-|a| \leq a \leq |a|$

(iii) $|ab| = |a| \cdot |b|$

(iv) $|a+b| \leq |a| + |b|$

(v) $|a+b| \geq |a| - |b|$

Solved Problems

INEQUALITIES

21.1. Write each statement in notational form:

(i) a is less than b.

(ii) a is not greater than b.

(iii) a is less than or equal to b.

(iv) a is greater than b.

(v) a is not less than b.

(vi) a is not greater than or equal to b.

A vertical or slant line through a symbol designates the opposite meaning of the symbol.

(i) $a < b$, (ii) $a \ngtr b$, (iii) $a \leq b$, (iv) $a > b$, (v) $a \nless b$, (vi) $a \ngeq b$

21.2. Rewrite the following geometric relationships between the given real numbers using the inequality notation:

(i) y lies to the right of 8.

(ii) z lies to the left of -3.

(iii) x lies between -3 and 7.

(iv) w lies between 5 and 1.

Recall that $a < b$ means a lies to the left of b on the real line:

(i) $y > 8$ or $8 < y$.

(ii) $z < -3$.

(iii) $-3 < x$ and $x < 7$ or, more concisely, $-3 < x < 7$.

(iv) $1 < w < 5$.

21.3. Describe and diagram each of the following intervals:
(i) $2 < x < 4$, (ii) $-1 \leqq x \leqq 2$, (iii) $-3 < x \leqq 1$, (iv) $-4 \leqq x < -1$, (v) $x > -1$, (vi) $x \leqq 2$.

(i) All numbers greater than 2 and less than 4:

(ii) All numbers greater than or equal to -1 and less than or equal to 2:

(iii) All numbers greater than -3 and less than or equal to 1:

(iv) All numbers greater than or equal to -4 and less than -1:

(v) All numbers greater than -1:

(vi) All numbers less than or equal to 2:

21.4. Solve and diagram the solution set of each inequality:
(i) $4x - 3 \leqq 2x + 3$, (ii) $3x + 2 < 6x - 7$.

(i) Transpose to obtain: $4x - 2x \leqq 3 + 3$
or: $2x \leqq 6$
Divide by 2: $x \leqq 3$

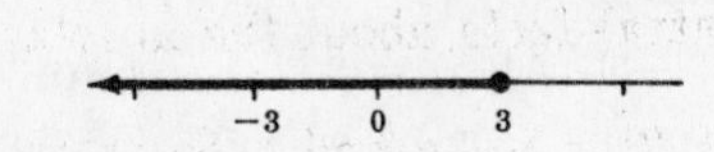

(ii) Transpose to obtain: $3x - 6x < -7 - 2$
or: $-3x < -9$
Divide by -3: $x > 3$

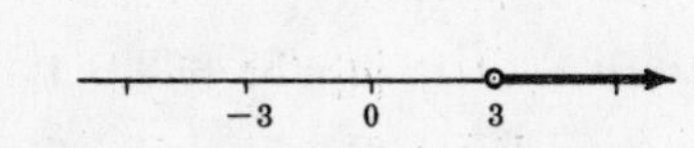

Note that the inequality is reversed since -3 is negative.

21.5. Solve each inequality and diagram its solution set:
(i) $3x - 1 \geqq 4x + 2$, (ii) $x - 3 > 1 + 3x$, (iii) $2x - 3 \leqq 5x - 9$.

(i) Transpose to obtain: $3x - 4x \geqq 2 + 1$
or: $-x \geqq 3$
Divide by -1: $x \leqq -3$

(ii) Transpose to obtain: $x - 3x > 1 + 3$
or: $-2x > 4$
Divide by -2: $x < -2$

(iii) Transpose to obtain: $2x - 5x \leqq -9 + 3$
or: $-3x \leqq -6$
Divide by -3: $x \geqq 2$

In each case the inequality was reversed since division was by a negative number.

21.6. Solve: $\frac{1}{2}x + \dfrac{x-2}{3} < 2x - \frac{1}{12}$

Multiply both sides by 12: $6x + 4(x-2) < 24x - 1$
or: $6x + 4x - 8 < 24x - 1$
Transpose: $6x + 4x - 24x < 8 - 1$
or: $-14x < 7$
Divide by -14: $x > -\frac{1}{2}$

21.7. Solve each inequality, i.e. rewrite the inequality so that x is alone between the inequality signs:

(i) $3 \leq x-4 \leq 8$, (ii) $-1 \leq x+3 \leq 2$, (iii) $-9 \leq 3x \leq 12$, (iv) $-6 \leq -2x \leq 4$.

(i) Add 4 to each side to obtain $7 \leq x \leq 12$.

(ii) Add -3 to each side to obtain $-4 \leq x \leq -1$.

(iii) Divide each side by 3 (or: multiply by $\frac{1}{3}$) to obtain $-3 \leq x \leq 4$.

(iv) Divide each side by -2 (or: multiply by $-\frac{1}{2}$) and reverse the inequalities to obtain $-2 \leq x \leq 3$.

21.8. Solve each inequality: (i) $3 < 2x-5 < 7$, (ii) $-7 \leq -2x+3 \leq 5$.

(i) Add 5 to each side to obtain: $8 < 2x < 12$

Divide each side by 2: $4 < x < 6$

(ii) Add -3 to each side to obtain: $-10 \leq -2x \leq 2$.

Divide each side by -2 and reverse inequalities: $-1 \leq x \leq 5$.

ABSOLUTE VALUE

21.9. Evaluate: (i) $|3-5|$, (ii) $|-3+5|$, (iii) $|-3-5|$, (iv) $|3-7|-|-5|$.

(i) $|3-5| = |-2| = 2$

(ii) $|-3+5| = |2| = 2$

(iii) $|-3-5| = |-8| = 8$

(iv) $|3-7|-|-5| = |-4|-|-5| = 4-5 = -1$

21.10. Evaluate: (i) $|2-8|+|3-1|$ (ii) $|2-5|-|4-7|$ (iii) $4+|-1-5|-|-8|$ (iv) $|3-6|-|-2+4|-|-2-3|$

(i) $|2-8|+|3-1| = |-6|+|2| = 6+2 = 8$

(ii) $|2-5|-|4-7| = |-3|-|-3| = 3-3 = 0$

(iii) $4+|-1-5|-|-8| = 4+|-6|-|-8| = 4+6-8 = 2$

(iv) $|3-6|-|-2+4|-|-2-3| = 3-2-5 = -4$

21.11. Rewrite without the absolute value sign:

(i) $|x| \leq 3$, (ii) $|x-2| < 5$, (iii) $|2x-3| \leq 7$.

(i) $-3 \leq x \leq 3$

(ii) $-5 < x-2 < 5$ or $-3 < x < 7$

(iii) $-7 \leq 2x-3 \leq 7$ or $-4 \leq 2x \leq 10$ or $-2 \leq x \leq 5$

21.12. Rewrite using an absolute value sign: (i) $-2 \leq x \leq 6$, (ii) $4 < x < 10$.

First rewrite the inequality so that a number and its negative appear at the ends of the inequality.

(i) Add -2 to each side to obtain $-4 \leq x-2 \leq 4$ which is equivalent to $|x-2| \leq 4$.

(ii) Add -7 to each side to obtain $-3 < x-7 < 3$ which is equivalent to $|x-7| < 3$.

PROOFS

21.13. Prove: If $a < b$ and $b < c$, then $a < c$.

By definition, $a < b$ and $b < c$ means $b-a$ and $c-b$ are positive. By [P_2], the sum of two positive numbers $(b-a)+(c-b) = c-a$ is positive. Thus, by definition, $a < c$.

21.14. Prove Theorem 21.1: (i) If $a < b$, then $a+c < b+c$.
(ii) If $a < b$ and c is positive, then $ac < bc$.
(iii) If $a < b$ and c is negative, then $ac > bc$.

(i) By definition, $a < b$ means $b - a$ is positive. But

$$(b+c) - (a+c) = b - a$$

Hence $(b+c) - (a+c)$ is positive and so $a+c < b+c$.

(ii) By definition, $a < b$ means $b - a$ is positive. But c is also positive; hence by **[P₂]** the product $c(b-a) = bc - ac$ is positive. Accordingly, $ac < bc$.

(iii) By definition, $a < b$ means $b - a$ is positive. By **[P₁]**, if c is negative then $-c$ is positive; hence by **[P₂]** the product $(b-a)(-c) = ac - bc$ is also positive. Thus, by definition, $bc < ac$ or, equivalently, $ac > bc$.

21.15. Prove Theorem 21.2(iv): $|a+b| \leq |a| + |b|$.

Method 1.
Since $|a| = \pm a$, $-|a| \leq a \leq |a|$; also, $-|b| \leq b \leq |b|$. Then, adding,

$$-(|a| + |b|) \leq a + b \leq |a| + |b|$$

Therefore

$$|a+b| \leq \big||a| + |b|\big| = |a| + |b|$$

Method 2.
Now $ab \leq |ab| = |a|\,|b|$ implies $2ab \leq 2\,|a|\,|b|$, and so

$$(a+b)^2 = a^2 + 2ab + b^2 \leq a^2 + 2\,|a|\,|b| + b^2 = |a|^2 + 2\,|a|\,|b| + |b|^2 = (|a| + |b|)^2$$

But $\sqrt{(a+b)^2} = |a+b|$ and so, by the square root of the above, $|a+b| \leq |a| + |b|$.

21.16. Prove: $|a-b| \leq |a| + |b|$.

Using the result of the preceding problem, we have $|a-b| = |a + (-b)| \leq |a| + |-b| = |a| + |b|$.

21.17. Prove: $|a-c| \leq |a-b| + |b-c|$.

$$|a-c| = |(a-b) + (b-c)| \leq |a-b| + |b-c|$$

21.18. Prove that $\frac{1}{2}$ is a positive number.

By **[P₁]**, either $-\frac{1}{2}$ or $\frac{1}{2}$ is positive. Suppose $-\frac{1}{2}$ is positive and so, by **[P₂]**, the sum $(-\frac{1}{2}) + (-\frac{1}{2}) = -1$ is also positive. But by Example 1.1, the number 1 is positive and not -1. Thus we have a contradiction, and so $\frac{1}{2}$ is positive.

21.19. Prove that the product $a \cdot b$ of a positive number a and a negative number b is negative.

If b is negative then, by **[P₁]**, $-b$ is positive; hence by **[P₂]** the product $a \cdot (-b)$ is also positive. But $a \cdot (-b) = -(a \cdot b)$. Thus $-(a \cdot b)$ is positive and so, by **[P₁]**, $a \cdot b$ is negative.

21.20. Prove that the product $a \cdot b$ of negative numbers a and b is positive.

If a and b are negative then, by **[P₁]**, $-a$ and $-b$ are positive. Hence by **[P₂]**, the product $(-a) \cdot (-b)$ is positive. But $a \cdot b = (-a) \cdot (-b)$, and so $a \cdot b$ is positive.

21.21. Prove: If a and b are positive, then $a < b$ iff $a^2 < b^2$.

Suppose $a < b$. Since a and b are positive, $a^2 < ab$ and $ab < b^2$; hence $a^2 < b^2$.

On the other hand, suppose $a^2 < b^2$. Then $b^2 - a^2 = (b+a)(b-a)$ is positive. Since a and b are positive, the sum $b+a$ is positive; hence $b-a$ is positive or else the product $(b+a)(b-a)$ would be negative. Thus, by definition, $a < b$.

21.22. Prove that the sum of a positive number a and its reciprocal $1/a$ is greater than or equal to 2: if $a > 0$, then $a + 1/a \geqq 2$.

If $a = 1$, then $1/a = 1$ and so $a + 1/a = 1 + 1 = 2$. On the other hand, if $a \neq 1$, then $a - 1 \neq 0$ and so

$$(a-1)^2 > 0 \quad \text{or} \quad a^2 - 2a + 1 > 0 \quad \text{or} \quad a^2 + 1 > 2a$$

Since a is positive, we can divide both sides of the inequality by a to obtain $a + 1/a > 2$.

Supplementary Problems

INEQUALITIES

21.23. Write each statement in notational form:

(i) x is not greater than y.
(ii) r is greater than or equal to t.
(iii) m is not less than n.
(iv) u is less than v.

21.24. Using inequality notation, rewrite each geometric relationship between the given real numbers:

(i) x lies to the right of y.
(ii) r lies to the left of t.
(iii) x lies between 4 and -6.
(iv) y lies between -2 and -5.

21.25. Describe and diagram the following intervals:

(i) $-1 < x \leqq 3$, (ii) $1 \leqq x \leqq 4$, (iii) $-2 \leqq x < 0$, (iv) $x \leqq 1$.

21.26. Solve: (i) $6x + 5 \leqq 2x - 7$, (ii) $2x - 3 > 4x + 5$.

21.27. Solve: (i) $x - 2 < 2x - 6$, (ii) $3x + 2 \geqq 6x + 11$.

21.28. Solve: $\dfrac{x-1}{3} + \dfrac{1}{2} \leqq \dfrac{x}{2} - 1$.

21.29. Solve, i.e. rewrite so that x is alone between the inequality signs:

(i) $-2 \leqq x - 3 \leqq 4$, (ii) $-5 < x + 2 < 1$, (iii) $-12 < 4x < -8$.

21.30. Solve: (i) $4 \leqq -2x \leqq 10$, (ii) $-1 < 2x - 3 < 5$, (iii) $-3 \leqq 5 - 2x \leqq 7$.

ABSOLUTE VALUES

21.31. Evaluate: (i) $|4-7|$, (ii) $|7-4|$, (iii) $|4+7|$, (iv) $|-4-7|$.

21.32. Evaluate: (i) $|-3| - |-5|$, (ii) $|2-3| + |-6|$, (iii) $|-2| + |1-5|$, (iv) $|3-8| - |3-1|$.

21.33. Evaluate: (i) $\big||-3| - |-9|\big|$, (ii) $\big||2-6| - |1-9|\big|$.

21.34. Rewrite without the absolute value sign:
(i) $|x| \leqq 8$, (ii) $|x-3| < 8$, (iii) $|2x+4| \leqq 8$, (iv) $|2-2x| < 8$.

21.35. Rewrite using the absolute value sign:
(i) $-3 < x < 9$, (ii) $2 \leqq x \leqq 8$, (iii) $-7 < x < -1$, (iv) $-1 \leqq 2x+3 \leqq 5$.

THEOREMS ON INEQUALITIES

21.36. Prove: If $a < b$ and c is positive, then $c/a > c/b$.

21.37. Prove: If $a < b$ and $c < d$, then $a+c < b+d$.

21.38. Prove: If a and b are positive, then $\sqrt{ab} \leqq (a+b)/2$.

21.39. Prove: If $\dfrac{a}{b} < \dfrac{c}{d}$ where a, b, c and d are positive, then $\dfrac{a+c}{b+d} < \dfrac{c}{d}$.

21.40. Prove: For any real numbers a, b and c, (i) $2ab \leqq a^2+b^2$ and (ii) $ab+ac+bc \leqq a^2+b^2+c^2$.

THEOREMS ON ABSOLUTE VALUES

21.41. Prove: (i) $|-a| = |a|$, (ii) $a^2 = |a|^2$, (iii) $|a| = \sqrt{a^2}$, (iv) $|x| < a$ iff $-a < x < a$.

21.42. Prove Theorem 21.2(iii): $|ab| = |a| \cdot |b|$.

21.43. Prove Theorem 21.2(v): $|a| - |b| \leqq |a+b|$.

21.44. Prove: $|a| - |b| \leqq |a-b|$.

Answers to Supplementary Problems

21.23. (i) $x \not> y$, (ii) $r \geqq t$, (iii) $m \not< n$, (iv) $u < v$.

21.24. (i) $x > y$ or $y < x$, (ii) $r < t$, (iii) $-6 < x < 4$, (iv) $-5 < y < -2$

21.25.

21.26. (i) $x \leqq -3$, (ii) $x < -4$

21.27. (i) $x > 4$, (ii) $x \leqq -3$

21.28. $x \geqq 7$

21.29. (i) $1 \leqq x \leqq 7$, (ii) $-7 < x < -1$, (iii) $-3 < x < -2$

21.30. (i) $-5 \leqq x \leqq -2$, (ii) $1 < x < 4$, (iii) $-1 \leqq x \leqq 4$

21.31. (i) 3, (ii) 3, (iii) 11, (iv) 11

21.32. (i) -2, (ii) 7, (iii) 6, (iv) 3

21.33. (i) 6, (ii) 4

21.34. (i) $-8 \leqq x \leqq 8$, (ii) $-5 < x < 11$, (iii) $-6 \leqq x \leqq 2$, (iv) $-3 < x < 5$

21.35. (i) $|x-3| < 6$, (ii) $|x-5| \leqq 3$, (iii) $|x+4| < 3$, (iv) $|2x+1| \leqq 3$

Chapter 22

Points, Lines and Hyperplanes

CARTESIAN PLANE

We assume the reader is familiar with the Cartesian plane, denoted by $\mathbf{R}^2$, as shown on the right. Here each point P represents an ordered pair of real numbers (a, b) and vice versa; the vertical line through the point P meets the horizontal axis (x axis) at a, and the horizontal line through P meets the vertical axis (y axis) at b. Thus in discussing the Cartesian plane $\mathbf{R}^2$ we use the words point and ordered pair of real numbers interchangeably.

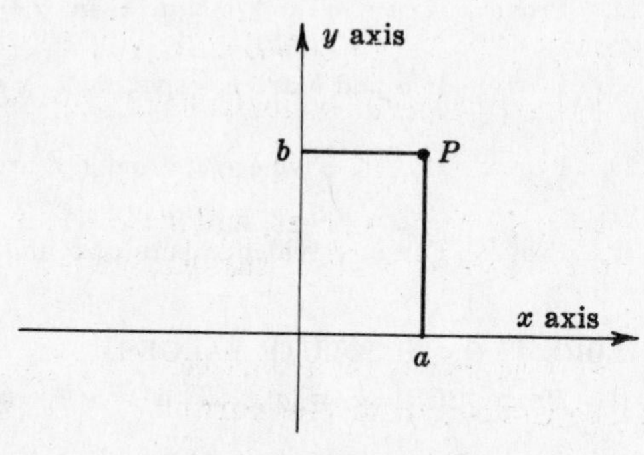

Cartesian plane $\mathbf{R}^2$

DISTANCE BETWEEN POINTS

The distance d between two points $p = (x_1, y_1)$ and $q = (x_2, y_2)$ is given by

$$d = \sqrt{(x_2 - x_1)^2 + (y_2 - y_1)^2}$$

For example, the distance between $p = (3, -1)$ and $q = (7, 2)$ is

$$d = \sqrt{(7-3)^2 + (2-(-1))^2} = \sqrt{16+9} = 5$$

We shall denote the distance between p and q by $d(p, q)$.

INCLINATION AND SLOPE OF A LINE

The inclination θ of a line l is the angle that l makes with the horizontal; that is, $\theta = 0°$ if l is horizontal or θ is the smallest positive angle measured counterclockwise from the positive end of the x axis to the line l. The range of θ is given by $0° \leq \theta < 180°$.

The slope m of a line is defined as the tangent of its angle θ of inclination: $m = \tan\theta$. In particular, the slope of the line passing through two points $p = (x_1, y_1)$ and $q = (x_2, y_2)$ is given by

$$m = \tan\theta = \frac{y_2 - y_1}{x_2 - x_1}$$

The slope of a vertical line, i.e. when $x_2 = x_1$, is not defined.

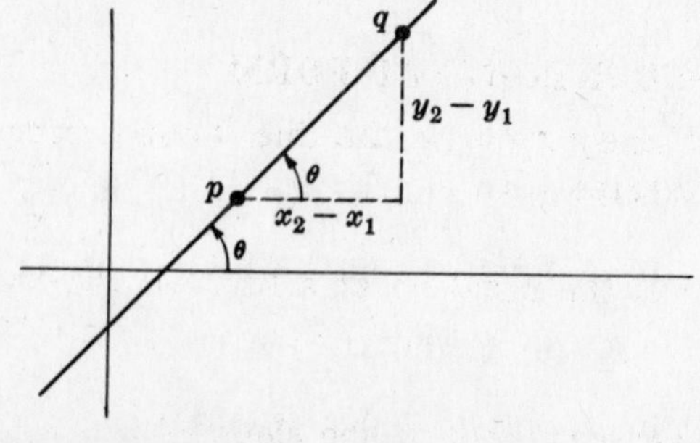

Two distinct lines l_1 and l_2 are *parallel* if and only if their slopes, say m_1 and m_2 respectively, are equal: $m_1 = m_2$. On the other hand, the lines are *perpendicular* if and only if the slope of one is the negative reciprocal of the other: $m_1 = -1/m_2$ or $m_1 m_2 = -1$.

LINES AND LINEAR EQUATIONS

Every line l in the Cartesian plane $\mathbf{R}^2$ can be represented by a linear equation of the form

$$ax + by = c, \qquad \text{where } a \text{ and } b \text{ are not both zero} \tag{1}$$

i.e. each point on l is a solution to (1) and each solution to (1) is a point on l. Conversely, every equation of the form (1) represents a line in the plane. Thus, in discussing the Cartesian plane, we sometimes use the terms line and linear equation synonymously. (See Chapter 11.)

HORIZONTAL AND VERTICAL LINES

The equation of a *horizontal* line, i.e. a line parallel to the x axis, is of the form

$$y = k$$

where k is the point at which the line intersects the y axis. In particular, the equation of the x axis is $y = 0$.

The equation of a *vertical* line, i.e. a line parallel to the y axis, is of the form

$$x = k$$

where k is the point at which the line intersects the x axis. In particular, the equation of the y axis is $x = 0$.

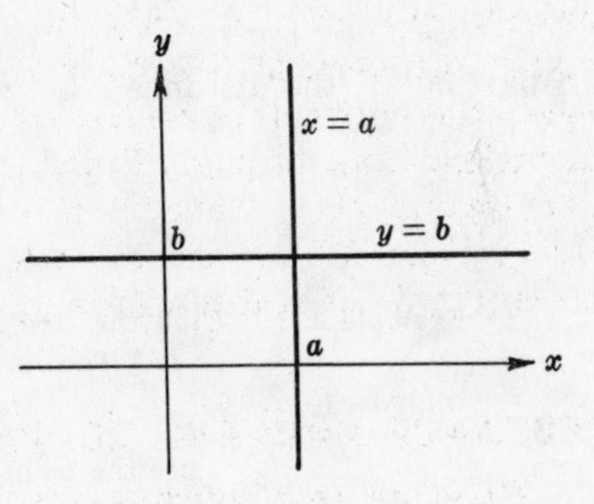

POINT-SLOPE FORM

A line is completely determined once its direction and a point on the line are known. The equation of the line having slope m and passing through the point (x_1, y_1) is

$$y - y_1 = m(x - x_1)$$

Example 1.1: Find the equation of the line having slope 3 and passing through the point $(5, -2)$. Substitute in the above formula to obtain $y + 2 = 3(x - 5)$ or $y = 3x - 17$.

Example 1.2: Find the equation of the line passing through the two points $(3, 2)$ and $(5, -6)$. The slope m of the line is $m = \dfrac{y_2 - y_1}{x_2 - x_1} = \dfrac{-6 - 2}{5 - 3} = -4$. Thus the equation of the line is $y - 2 = -4(x - 3)$ or $4x + y = 14$.

SLOPE-INTERCEPT FORM

The equation of the line having slope m and intersecting the y axis at k, i.e. having y-intercept k, is

$$y = mx + k$$

Thus if the equation of a line l is $ax + by = c$, where $b \neq 0$, then

$$y = -\frac{a}{b}x + \frac{c}{b}$$

and so $m = -a/b$ is the slope of l.

Example 2.1: The slope of the line $y = -2x + 8$ is -2.

Example 2.2: Find the slope m of the line $2x - 3y = 5$. Rewrite the equation in the form $y = \frac{2}{3}x - \frac{5}{3}$; hence $m = \frac{2}{3}$.

PARALLEL LINES. DISTANCE BETWEEN A POINT AND A LINE

Let l be the line $ax + by = c$. The set of all lines parallel to l is represented by the equation

$$ax + by = k \qquad (2)$$

i.e. every value for k determines a line parallel to l and every line parallel to l is of the above form for some k. We call k a *parameter*, and the set (2) a *one parameter family of lines.*

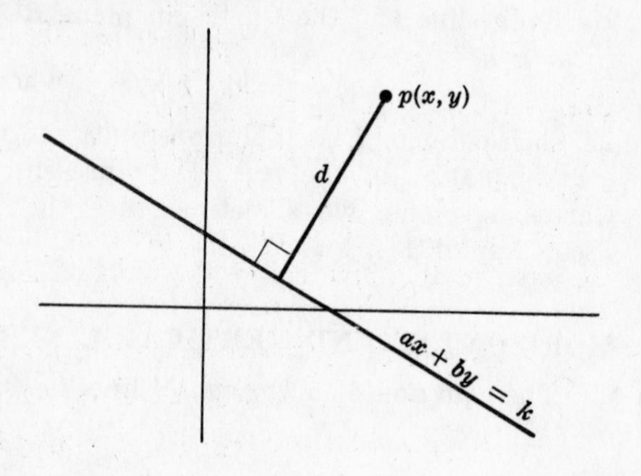

The distance d between a point $p = (x_1, y_1)$ and the line l is given by

$$d = \frac{|ax_1 + by_1 - c|}{\sqrt{a^2 + b^2}}$$

In particular, the distance ρ between the origin $(0, 0)$ and the line l is

$$\rho = \frac{|c|}{\sqrt{a^2 + b^2}}$$

Example 3.1: Find the line l parallel to the line $2x - 5y = 7$ and passing through the point $(3, 4)$. The line l belongs to the family

$$2x - 5y = k$$

Since the point $(3, 4)$ belongs to l it must satisfy the equation of l:

$$2 \cdot 3 - 5 \cdot 4 = k \quad \text{or} \quad k = -14$$

Thus the equation of l is $2x - 5y = -14$.

Example 3.2: Let l be the line $3x + 4y = 2$. The distance d between the point $(2, -6)$ and l is

$$d = \frac{|3 \cdot 2 + 4 \cdot (-6) - 2|}{\sqrt{3^2 + 4^2}} = \frac{|-20|}{5} = 4$$

The distance ρ between the origin and l is $\rho = \dfrac{|2|}{\sqrt{3^2 + 4^2}} = \dfrac{2}{5}$.

EUCLIDEAN m-SPACE

Let u and v be vectors with m components of real numbers:

$$u = (u_1, u_2, \ldots, u_m) \quad \text{and} \quad v = (v_1, v_2, \ldots, v_m)$$

Recall (see Chapter 9) the following operations of vector addition, scalar multiplication, and dot product:

$$u + v = (u_1 + v_1, u_2 + v_2, \ldots, u_m + v_m)$$

$$ku = (ku_1, ku_2, \ldots, ku_m), \qquad \text{for any scalar } k$$

$$u \cdot v = u_1v_1 + u_2v_2 + \cdots + u_mv_m$$

The set of all such vectors with the above operations is called Euclidean m-space or m dimensional Euclidean space or, simply, m-space, and will be denoted by $\mathbf{R}^m$.

The distance between the above vectors u and v, written $d(u, v)$, is defined by

$$d(u, v) = \sqrt{(v_1 - u_1)^2 + (v_2 - u_2)^2 + \cdots + (v_m - u_m)^2} = \sqrt{\sum_{i=1}^{m} (v_i - u_i)^2}$$

The *norm* (or: *length*) of the vector u, written $\|u\|$, is defined by

$$\|u\| = \sqrt{u \cdot u} = \sqrt{u_1^2 + u_2^2 + \cdots + u_m^2}$$

Observe that $d(u, v) = \|u - v\|$.

The vectors u and v are said to be *orthogonal* (or: perpendicular) if their dot product is zero: $u \cdot v = 0$.

The set of vectors of the form $(0, \ldots, 0, u_i, 0, \ldots, 0)$ is called the x_i *axis*. Note that the zero vector, also called the origin, belongs to every axis.

Example 4.1: Consider the following vectors in $\mathbf{R}^4$: $u = (1, -2, 4, 1)$ and $v = (3, 1, -5, 0)$. Then:

$$u + v = (1+3,\ -2+1,\ 4-5,\ 1+0) = (4, -1, -1, 1)$$

$$7u = (7, -14, 28, 7)$$

$$u \cdot v = 1 \cdot 3 + (-2) \cdot 1 + 4 \cdot (-5) + 1 \cdot 0 = 3 - 2 - 20 + 0 = -19$$

$$d(u, v) = \sqrt{(3-1)^2 + (1+2)^2 + (-5-4)^2 + (0-1)^2} = \sqrt{95}$$

$$\|v\| = \sqrt{3^2 + 1^2 + (-5)^2 + 0^2} = \sqrt{9 + 1 + 25} = \sqrt{35}$$

Example 4.2: We frequently identify a point in, say, $\mathbf{R}^2$ with the arrow from the origin to the given point. Then the arrow corresponding to the sum $p + q$ is the diagonal of the parallelogram determined by the arrows corresponding to the points p and q as shown on the right. Furthermore, $\|p\|$, the norm of p, is precisely the length of the arrow to p.

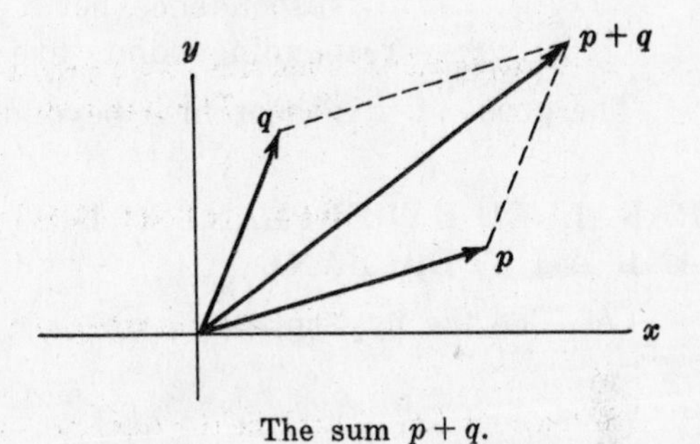

The sum $p + q$.

BOUNDED SETS

A subset A of $\mathbf{R}^m$ is bounded iff there exists a real number $m > 0$ such that $\|u\| < m$ for every point $u \in A$. In the case of the plane $\mathbf{R}^2$, this says that the points of A lie inside of the circle of radius m centered at the origin, as indicated on the right.

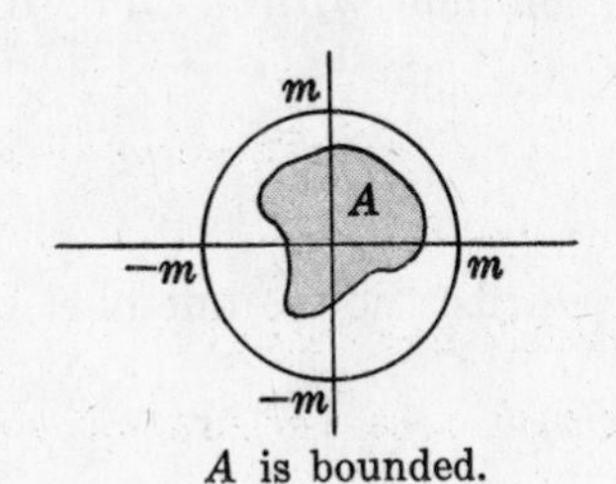

A is bounded.

HYPERPLANES

Consider a linear equation in the m unknowns $x_1, \ldots, x_m$:

$$c_1 x_1 + c_2 x_2 + \cdots + c_m x_m = b \tag{3}$$

Using the notation of the dot product of $\mathbf{R}^m$, the above equation can be written in the form

$$c \cdot x = b$$

where $c = (c_1, c_2, \ldots, c_m)$ and $x = (x_1, x_2, \ldots, x_m)$. Unless otherwise stated, we assume that $c \neq 0$, i.e. that the c_i in (3) are not all zero.

Definition: The set of all points in $\mathbf{R}^m$ which satisfy the equation $c \cdot x = b$, i.e. the set of all solutions of the linear equation (3), is called a *hyperplane*.

Example 5.1: The hyperplanes in the Cartesian plane $\mathbf{R}^2$ are the straight lines as discussed at the beginning of this chapter.

Example 5.2: The hyperplanes in 3-space $\mathbf{R}^3$, i.e. in solid geometry, correspond to the planes as illustrated on the right. In particular, the equation of the xy-plane is $z=0$, the xz-plane is $y=0$, and the yz-plane is $x=0$.

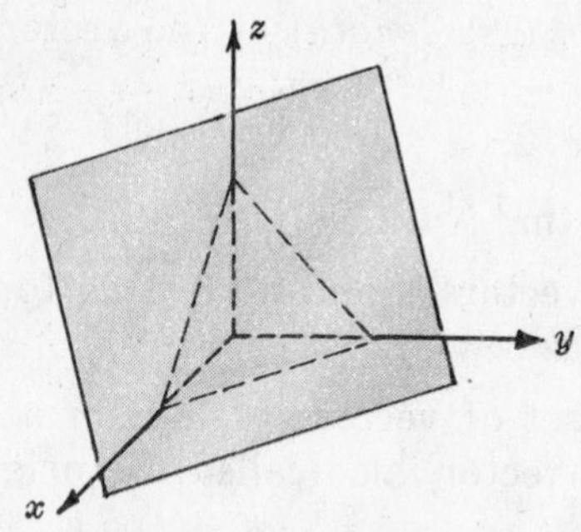

Hyperplane in $\mathbf{R}^3$.

Observe that the "dimension" of a hyperplane in $\mathbf{R}^2$ is one, and in $\mathbf{R}^3$ is two; that is, we can identify the points of a hyperplane in $\mathbf{R}^2$ with the points on the real line $\mathbf{R}$ and we can identify the points on a hyperplane of $\mathbf{R}^3$ with the points in $\mathbf{R}^2$ with the further property that distances between corresponding points are equal. In each case the dimension of the hyperplane is one less than the dimension of the whole space. In fact, this property completely characterizes hyperplanes in general; namely,

Theorem 22.1: A subset A of $\mathbf{R}^m$ is a hyperplane if and only if there exists a one-to-one correspondence between A and $\mathbf{R}^{m-1}$ such that distances between corresponding points are equal.

The proof of this theorem is beyond the scope of this text.

PARALLEL HYPERPLANES. DISTANCE BETWEEN A POINT AND A HYPERPLANE

Let α be the hyperplane $c_1x_1 + c_2x_2 + \cdots + c_mx_m = b$. Then

$$c_1x_1 + c_2x_2 + \cdots + c_mx_m = k$$

is the equation of the family of all hyperplanes parallel to α, i.e. which do not have a point in common with α. The distance d between the point $u = (u_1, u_2, \ldots, u_m)$ and α is given by

$$d = \frac{|c_1u_1 + c_2u_2 + \cdots + c_mu_m - b|}{\sqrt{c_1^2 + c_2^2 + \cdots + c_m^2}}$$

In other words, the distance between u and the hyperplane $c \cdot x = b$ is

$$d = \frac{|c \cdot u - b|}{||c||}$$

In particular, the distance ρ between the origin, i.e. the zero vector, and α is

$$\rho = \frac{|b|}{\sqrt{c_1^2 + c_2^2 + \cdots + c_m^2}} = \frac{|b|}{||c||}$$

Example 6.1: In solid geometry, i.e. $\mathbf{R}^3$, the family of hyperplanes parallel to the xy-plane, i.e. to $z=0$, is represented by $z=k$ where k is the point at which the hyperplane intersects the z axis. (See diagram below.)

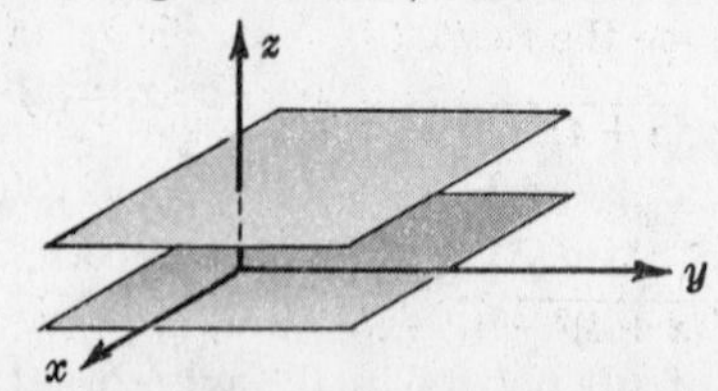

Two members of the family $z=k$.

Example 6.2: Let α be the hyperplane $3x - 4y + z - 2w = 7$ in $\mathbf{R}^4$. The distance d between the point $(1, -4, -2, 3)$ and α is

$$d = \frac{|3 \cdot 1 - 4 \cdot (-4) + 1 \cdot (-2) - 2 \cdot 3 - 7|}{\sqrt{3^2 + (-4)^2 + 1^2 + (-2)^2}} = \frac{4}{\sqrt{30}}$$

The distance ρ between α and the origin is $\rho = \dfrac{|7|}{\sqrt{3^2 + (-4)^2 + 1^2 + (-2)^2}} = \dfrac{7}{\sqrt{30}}$.

Solved Problems

DISTANCE IN THE PLANE

22.1. Find the distance d between (i) $(-3, 5)$ and $(6, 4)$, (ii) $(5, -2)$ and $(-1, -4)$.

In each case use the formula $d = \sqrt{(x_2 - x_1)^2 + (y_2 - y_1)^2}$.

(i) $d = \sqrt{(6+3)^2 + (4-5)^2} = \sqrt{81 + 1} = \sqrt{82}$

(ii) $d = \sqrt{(-1-5)^2 + (-4+2)^2} = \sqrt{36 + 4} = \sqrt{40}$

22.2. Show that the points $p = (-1, 4)$, $q = (2, 1)$ and $r = (3, 5)$ are the vertices of an isosceles triangle.

Compute the distance between each pair of points:

$$d(p, q) = \sqrt{(2+1)^2 + (1-4)^2} = \sqrt{9 + 9} = \sqrt{18}$$

$$d(p, r) = \sqrt{(3+1)^2 + (5-4)^2} = \sqrt{16 + 1} = \sqrt{17}$$

$$d(q, r) = \sqrt{(3-2)^2 + (5-1)^2} = \sqrt{1 + 16} = \sqrt{17}$$

Since $d(p, r) = d(q, r)$, the triangle is isosceles.

22.3. Show that the points $p = (4, -2)$, $q = (-1, 3)$ and $r = (1, 4)$ are the vertices of a right triangle, and find its area.

Compute the distance between each pair of points:

$$d(p, q) = \sqrt{(-1-4)^2 + (3+2)^2} = \sqrt{25 + 25} = \sqrt{50}$$

$$d(p, r) = \sqrt{(1-4)^2 + (4+2)^2} = \sqrt{9 + 36} = \sqrt{45}$$

$$d(q, r) = \sqrt{(1+1)^2 + (4-3)^2} = \sqrt{4 + 1} = \sqrt{5}$$

Since $d(q, r)^2 + d(p, r)^2 = d(p, q)^2$, the triangle is a right triangle.

The area of the right triangle is $A = \frac{1}{2}(\sqrt{45})(\sqrt{5}) = \frac{15}{2}$.

22.4. Find the point equidistant from the points $a = (-4, 3)$, $b = (5, 6)$ and $c = (4, -1)$.

Let $p = (x, y)$ be the required point. Then $d(p, a) = d(p, b) = d(p, c)$. Since $d(p, a) = d(p, b)$,

$$\sqrt{(x+4)^2 + (y-3)^2} = \sqrt{(x-5)^2 + (y-6)^2} \quad \text{or} \quad 3x + y = 6$$

Since $d(p, a) = d(p, c)$,

$$\sqrt{(x+4)^2 + (y-3)^2} = \sqrt{(x-4)^2 + (y+1)^2} \quad \text{or} \quad 2x - y = -1$$

Solve the two linear equations simultaneously to obtain $x = 1$ and $y = 3$. Thus $p = (1, 3)$ is the required point.

22.5. Show that the points $p = (1,-2)$, $q = (3,2)$ and $r = (6,8)$ are collinear, i.e. lie on the same straight line.

Method 1.

Compute the distance between each pair of points:

$$d(p,q) = \sqrt{(3-1)^2 + (2+2)^2} = \sqrt{4+16} = \sqrt{20} = 2\sqrt{5}$$
$$d(q,r) = \sqrt{(6-3)^2 + (8-2)^2} = \sqrt{9+36} = \sqrt{45} = 3\sqrt{5}$$
$$d(p,r) = \sqrt{(6-1)^2 + (8+2)^2} = \sqrt{25+100} = \sqrt{125} = 5\sqrt{5}$$

Since $d(p,r) = d(p,q) + d(q,r)$, the points lie on the same straight line.

Method 2.

Compute the slope m_1 of the line passing through p and q, and the slope m_2 of the line passing through p and r:

$$m_1 = \frac{2-(-2)}{3-1} = \frac{4}{2} = 2, \qquad m_2 = \frac{8-(-2)}{6-1} = \frac{10}{5} = 2$$

Since $m_1 = m_2$, the three points lie on the same line.

LINES

22.6. Plot the following lines:
(i) $x=-4$, $x=-1$, $x=3$ and $x=5$; (ii) $y=4$, $y=1$, $y=-2$, and $y=-3$.

(i) Each line is vertical and meets the x axis at its respective value of x. See diagram below.
(ii) Each line is horizontal and meets the y axis at its respective value of y. See diagram below.

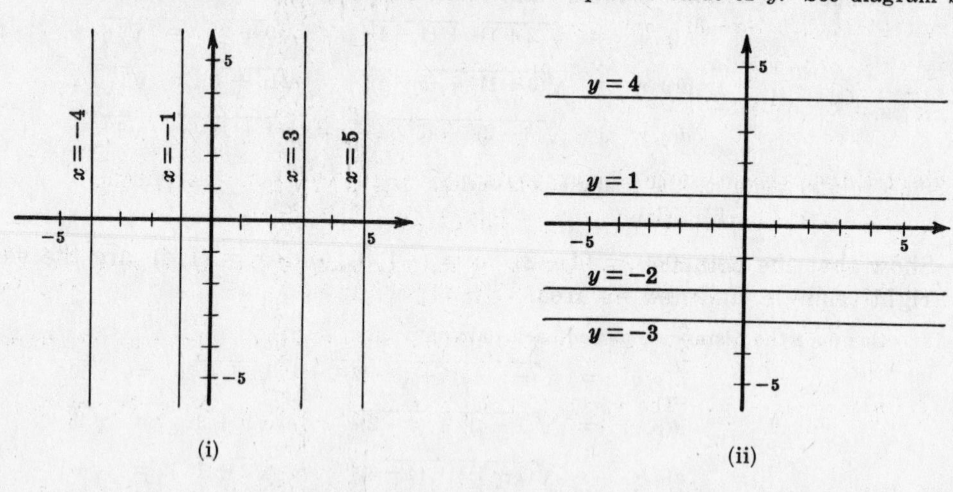

(i) (ii)

22.7. Find the equation of the line (i) parallel to the x axis and passing through the point $(-2,-5)$, (ii) parallel to the y axis and passing through the point $(3,4)$.

(i) If the line is parallel to the x axis, then it is horizontal and its equation is of the form $y=k$. Since it passes through the point $(-2,-5)$, k must equal the y-component of the point. Thus the required equation is $y=-5$.

(ii) If the line is parallel to the y axis, then it is vertical and its equation is of the form $x=k$. Since it passes through the point $(3,4)$, k must equal the x-component of the point. Thus the required equation is $x=3$.

22.8. Plot the following lines: (i) $2x-3y = 6$, (ii) $4x+3y = 6$, (iii) $3x+2y = 0$.

In each case find at least two points on the line, preferably the intercepts, and as a check a third point. The line drawn through these points is the required line.

(i) If $x = 0$, $y = -2$; if $y = 0$, $x = 3$. Also, if $y = 1$, $x = 4.5$. Plot the points $(0, -2)$, $(3, 0)$ and $(4.5, 1)$ and draw the line through the points as in the diagram below.

(ii) If $x = 0$, $y = 2$; if $y = 0$, $x = 3/2 = 1.5$. Also, if $y = -2$, $x = 3$. Plot the points $(0, 2)$, $(1.5, 0)$ and $(3, -2)$ and draw the lines through the points as shown below.

(iii) If $x = 0$, $y = 0$. Note that the line passes through the origin. If $x = 2$, $y = -3$; if $x = -4$, $y = 6$. Plot the points $(0, 0)$, $(2, -3)$ and $(-4, 6)$ and draw the line through the points as shown below.

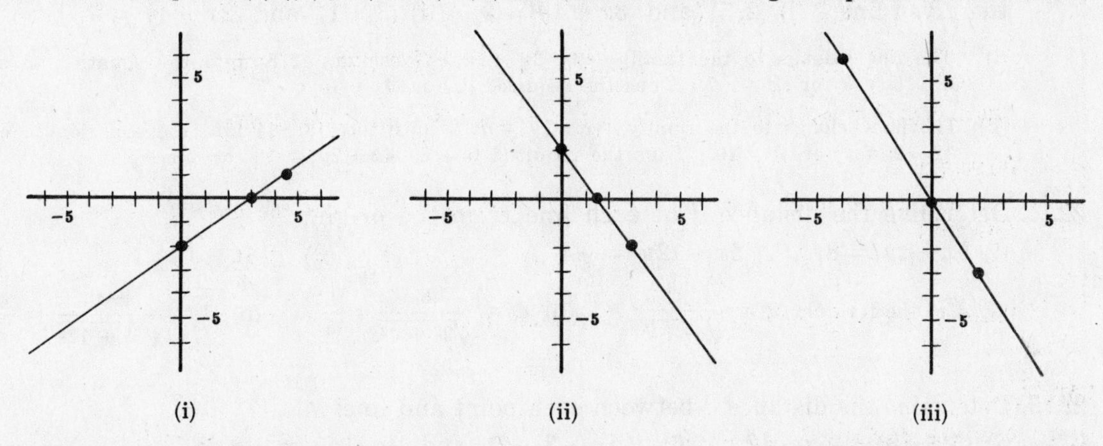

(i) (ii) (iii)

22.9. Find the equation of the line (i) passing through $(3, 2)$ with slope -1, (ii) passing through $(-4, 1)$ with slope $\frac{1}{2}$, (iii) passing through $(-2, -5)$ with slope 2.

In each case use the formula $y - y_1 = m(x - x_1)$ where m is the slope and (x_1, y_1) the given point.

(i) $y - 2 = -1(x - 3)$ or $x + y = 5$

(ii) $y - 1 = \frac{1}{2}(x + 4)$ or $x - 2y = -6$

(iii) $y + 5 = 2(x + 2)$ or $2x - y = 1$

22.10. Find the equation of the line passing through each pair of points: (i) $(1, 3)$ and $(4, -5)$; (ii) $(-2, -6)$ and $(3, 1)$.

In each case, first find the slope m of the line using the formula $m = \dfrac{y_2 - y_1}{x_2 - x_1}$ and then find the equation of the line using the point-slope formula.

(i) $m = \dfrac{-5 - 3}{4 - 1} = -\dfrac{8}{3}$. The equation of the line is $y - 3 = -\frac{8}{3}(x - 1)$ or $8x + 3y = 17$.

(ii) $m = \dfrac{1 + 6}{3 + 2} = \dfrac{7}{5}$. The equation of the line is $y - 1 = \frac{7}{5}(x - 3)$ or $7x - 5y = 16$.

22.11. Find the equation of the line passing through each point and parallel to the given line: (i) $(3, -2)$ and $5x + 4y = -5$; (ii) $(1, 6)$ and $3x - 7y = 4$.

(i) The line belongs to the family $5x + 4y = k$. Substitute $(3, -2)$ into the equation to obtain $5 \cdot 3 + 4 \cdot (-2) = k$ or $k = 7$. Thus the required line is $5x + 4y = 7$.

(ii) The line belongs to the family $3x - 7y = k$. Substitute $(1, 6)$ into the equation to obtain $3 \cdot 1 - 7 \cdot 6 = k$ or $k = -39$. Thus the required line is $3x - 7y = -39$.

22.12. Let $ax + by = c$ be the equation of a line l. Show that

$$bx - ay = k \tag{1}$$

is the equation of the family of lines perpendicular to l.

If $b = 0$, then l is vertical and (*1*) is the equation of the family of horizontal lines; hence (*1*) is the family of lines perpendicular to l.

On the other hand, if $a = 0$ then l is horizontal and (1) is the equation of the family of vertical lines and so is the family of lines perpendicular to l.

Lastly, suppose $b \neq 0$ and $a \neq 0$. Then the slope of l is $-a/b$ and (1) is the equation of the family of lines with slope b/a. But $(-a/b)(b/a) = -1$; hence, again, (1) is the family of lines perpendicular to l.

22.13. Find the equation of the line passing through the given point and perpendicular to the given line: (i) $(2, 5)$ and $3x + 4y = 9$; (ii) $(3, -1)$ and $2x - 4y = 7$.

(i) The line belongs to the family $4x - 3y = k$. Substitute $(2, 5)$ into the equation to obtain $8 - 15 = k$ or $k = -7$. Thus the required line is $4x - 3y = -7$.

(ii) The line belongs to the family $4x + 2y = k$. Substitute $(3, -1)$ into the equation to obtain $12 - 2 = k$ or $k = 10$. Thus the required line is $4x + 2y = 10$ or $2x + y = 5$.

22.14. Determine the distance d of each line from the origin: (i) $3x - 4y = 8$, (ii) $5x + 12y = -4$.

Use the formula $d = \dfrac{|c|}{\sqrt{a^2 + b^2}}$. (i) $d = \dfrac{|8|}{\sqrt{3^2 + 4^2}} = \dfrac{8}{5}$. (ii) $d = \dfrac{|-4|}{\sqrt{5^2 + 12^2}} = \dfrac{4}{13}$.

22.15. Determine the distance d between each point and line: (i) $(5, 3)$ and $5x - 12y = 6$; (ii) $(-3, -2)$ and $4x - 5y = -8$.

Use the formula $d = \dfrac{|ax_1 + by_1 - c|}{\sqrt{a^2 + b^2}}$.

(i) $d = \dfrac{|5 \cdot 5 - 12 \cdot 3 - 6|}{\sqrt{5^2 + 12^2}} = \dfrac{17}{13}$. (ii) $d = \dfrac{|4 \cdot (-3) - 5 \cdot (-2) + 8|}{\sqrt{4^2 + 5^2}} = \dfrac{6}{\sqrt{41}}$.

EUCLIDEAN *m*-SPACE

22.16. Find the distance $d(u, v)$ and the dot product $u \cdot v$ for each pair of vectors:
(i) $u = (3, -2, 0, 5)$ and $v = (6, 2, -3, -4)$;
(ii) $u = (2, 3, -5, -7, 1)$ and $v = (5, 3, -2, -4, -1)$;
(iii) $u = (4, -1, 6, 5)$ and $v = (1, -4, 2, 0, 7)$.

Use the formulas $d(u, v) = \sqrt{(v_1 - u_1)^2 + \cdots + (v_m - u_m)^2}$ and $u \cdot v = u_1 v_1 + \cdots + u_m v_m$.

(i) $d(u, v) = \sqrt{(6-3)^2 + (2+2)^2 + (-3-0)^2 + (-4-5)^2} = \sqrt{115}$

$u \cdot v = 3 \cdot 6 + (-2) \cdot 2 + 0 \cdot (-3) + 5 \cdot (-4) = 18 - 4 + 0 - 20 = -6$

(ii) $d(u, v) = \sqrt{(5-2)^2 + (3-3)^2 + (-2+5)^2 + (-4+7)^2 + (-1-1)^2} = \sqrt{31}$

$u \cdot v = 10 + 9 + 10 + 28 - 1 = 56$

(iii) The distance $d(u, v)$ and the dot product $u \cdot v$ are not defined since the vectors belong to different spaces: $u \in \mathbf{R}^4$ and $v \in \mathbf{R}^5$.

22.17. Find the norm of each vector: $u = (2, -1, 3, 5)$; $v = (3, -2, -1)$.

In each case use the formula $||u|| = \sqrt{u_1^2 + u_2^2 + \cdots + u_m^2}$.

$||u|| = \sqrt{2^2 + (-1)^2 + 3^2 + 5^2} = \sqrt{39}$. $||v|| = \sqrt{3^2 + (-2)^2 + (-1)^2} = \sqrt{14}$.

22.18. Let $u = (1, -3, 2, 4)$ and $v = (3, 5, -1, -2)$. Find:
(i) $u + v$; (ii) $5u$; (iii) $2u - 3v$; (iv) $u \cdot v$; (v) $||u||$ and $||v||$.

(i) $u + v = (1+3, -3+5, 2-1, 4-2) = (4, 2, 1, 2)$

(ii) $5u = (5 \cdot 1, 5 \cdot (-3), 5 \cdot 2, 5 \cdot 4) = (5, -15, 10, 20)$

(iii) $2u - 3v = (2, -6, 4, 8) + (-9, -15, 3, 6) = (-7, -21, 7, 14)$

(iv) $u \cdot v = 3 - 15 - 2 - 8 = -22$

(v) $||u|| = \sqrt{1 + 9 + 4 + 16} = \sqrt{30}$. $||v|| = \sqrt{9 + 25 + 1 + 4} = \sqrt{39}$.

22.19. For each pair of vectors, determine k so that the vectors are orthogonal: (i) $(3, -2, k, 5)$ and $(2k, 5, -3, -1)$; (ii) $(2, 3k, -4, 1, 5)$ and $(6, -1, 3, 7, 2k)$.

In each case, set the product of the vectors equal to zero and solve for k.

(i) $3 \cdot 2k + (-2) \cdot 5 + k \cdot (-3) + 5 \cdot (-1) = 0$ or $6k - 10 - 3k - 5 = 0$ or $k = 5$

(ii) $2 \cdot 6 + 3k \cdot (-1) + (-4) \cdot 3 + 7 + 5 \cdot 2k = 0$ or $12 - 3k - 12 + 7 + 10k = 0$ or $k = -1$

22.20. Determine which of the following subsets of $\mathbf{R}^m$ are bounded: (i) A consists of all probability vectors; (ii) B consists of the points on the x_1 axis; (iii) C is finite; (iv) D consists of all points with positive components.

Recall that $X \subset \mathbf{R}^m$ is bounded if there exists a real number $m > 0$ such that $||u|| \leq m$ for every $u \in X$.

(i) A is bounded since $||u|| \leq 1$ for every probability vector u.

(ii) B is not bounded; for if m is any positive real number, then $(m+1, 0, 0, \ldots, 0)$ belongs to B and has norm greater than m.

(iii) C is bounded. Choose m to be the maximum norm of the vectors in C; then m satisfies the required property.

(iv) D is not bounded.

HYPERPLANES

22.21. Find the distance d of each hyperplane from the origin: (i) $2x - 3y + 5z = 8$; (ii) $2x + 7y - z + 2w = -4$.

In each case use the formula $d = \frac{|b|}{||c||}$ where the hyperplane is $c \cdot x = b$.

(i) $d = \frac{|8|}{\sqrt{2^2 + (-3)^2 + 5^2}} = \frac{8}{\sqrt{38}}$. (ii) $d = \frac{|-4|}{\sqrt{4 + 49 + 1 + 4}} = \frac{4}{\sqrt{58}}$.

22.22. Determine the distance d between each point and hyperplane: (i) $(4, 0, -2, -3)$ and $x - 6y + 5z + 2w = 3$; (ii) $(2, -1, 3, 7, -2, -4)$ and $3x_1 - 5x_2 - 2x_3 + 4x_4 + x_5 - 3x_6 = -9$.

In each case use the formula $d = \frac{|c \cdot u - b|}{||c||}$ where u is the point and $c \cdot x = b$ the hyperplane.

(i) $d = \frac{|4 + 0 - 10 - 6 - 3|}{\sqrt{1 + 36 + 25 + 4}} = \frac{15}{\sqrt{66}}$. (ii) $d = \frac{|6 + 5 - 6 + 28 - 2 + 12 + 9|}{\sqrt{9 + 25 + 4 + 16 + 1 + 9}} = \frac{52}{\sqrt{64}} = \frac{13}{2}$.

22.23. Find the equation of the hyperplane passing through the given point and parallel to the given hyperplane:
(i) $(3, -2, 1, -4)$ and $2x + 5y - 6z - 2w = 8$;
(ii) $(1, -2, 3, 5)$ and $4x - 5y + 2z + w = 11$.

(i) The hyperplane belongs to the family $2x + 5y - 6z - 2w = k$. Substitute $(3, -2, 1, -4)$ into the equation to obtain $6 - 10 - 6 + 8 = k$ or $k = -2$. The required hyperplane is $2x + 5y - 6z - 2w = -2$.

(ii) The hyperplane belongs to the family $4x - 5y + 2z + w = k$. Substitute $(1, -2, 3, 5)$ into the equation to obtain $4 + 10 + 6 + 5 = k$ or $k = 25$. The required hyperplane is $4x - 5y + 2z + w = 25$.

22.24. Show that the equation of the hyperplane α in $\mathbf{R}^m$ which intersects the x_i axis at $a_i \neq 0$ is

$$\frac{x_1}{a_1} + \frac{x_2}{a_2} + \cdots + \frac{x_m}{a_m} = 1$$

Let $c_1x_1 + c_2x_2 + \cdots + c_mx_m = b$ be the equation of α. Substituting each x_i-intercept $(0, \ldots, 0, a_i, 0, \ldots, 0)$ into the equation, we obtain the homogeneous system

$$\begin{aligned} a_1c_1 \qquad\qquad\qquad - b &= 0 \\ a_2c_2 \qquad\quad - b &= 0 \\ \cdots\cdots\cdots\cdots\cdots\cdots & \\ a_mc_m - b &= 0 \end{aligned}$$

where the c_i and b are the unknowns. The system is in echelon form and we can arbitrarily assign a value to b. Set $b = 1$; then $c_i = 1/a_i$. Thus the equation of α is as claimed.

22.25. Find the equation of the hyperplane in $\mathbf{R}^4$ which intersects the axis at $-2, 4, 3$ and -6, respectively.

By the preceding problem, the equation is

$$\frac{x_1}{-2} + \frac{x_2}{4} + \frac{x_3}{3} + \frac{x_4}{-6} = 1 \quad \text{or} \quad 6x_1 - 3x_2 - 4x_3 + 2x_4 = -12$$

22.26. Find the equation of the hyperplane α in $\mathbf{R}^4$ which passes through the points $(1, 3, -4, 2)$, $(0, 1, 6, -3)$, $(0, 0, 2, 5)$ and $(0, 0, 3, 7)$.

Let $ax + by + cz + dw = e$ be the equation of α. Substituting each point into the equation, we obtain the following homogeneous system of linear equation where a, b, c, d and e are the unknowns:

$$\begin{cases} a + 3b - 4c + 2d - e = 0 \\ \qquad b + 6c - 3d - e = 0 \\ \qquad\qquad 2c + 5d - e = 0 \\ \qquad\qquad 3c + 7d - e = 0 \end{cases} \quad \text{or} \quad \begin{cases} a + 3b - 4c + 2d - e = 0 \\ \qquad b + 6c - 3d - e = 0 \\ \qquad\qquad 2c + 5d - e = 0 \\ \qquad\qquad\qquad d - e = 0 \end{cases}$$

The system has been reduced to echelon form on the right. There is only one free variable, i.e. we can arbitrarily assign a value to e. Set $e = 1$. Then $d = 1$ by equation four, $c = -2$ by equation three, $b = 16$ by equation two, and $a = -41$ by equation one. Thus the required equation is $-41x + 16y - 2z + w = 1$.

Remark: The hyperplane α is uniquely determined since there is only one free variable in the homogeneous system. In other words, every value of e determines a different equation but all these equations are multiples of each other and so they determine the same hyperplane α.

22.27. Show that there is more than one hyperplane in $\mathbf{R}^4$ which passes through the points $(1, 6, -1, 4)$, $(0, 1, 2, 2)$, $(0, 1, 1, 6)$ and $(0, 1, 3, -2)$.

Let $ax + by + cz + dw = e$ be the general equation of the hyperplanes passing through the given points. Substitute each point into the general equation and then reduce the system of equations to echelon form as follows:

$$\begin{aligned} a + 6b - c + 4d - e &= 0 \\ b + 2c + 2d - e &= 0 \\ b + c + 6d - e &= 0 \\ b + 3c - 2d - e &= 0 \end{aligned} \quad \text{or} \quad \begin{aligned} a + 6b - c + 4d - e &= 0 \\ b + 2c + 2d - e &= 0 \\ c - 4d \qquad &= 0 \\ c - 4d \qquad &= 0 \end{aligned} \quad \text{or} \quad \begin{aligned} a + 6b - c + 4d - e &= 0 \\ b + 2c + 2d - e &= 0 \\ c - 4d \qquad &= 0 \end{aligned}$$

The system has two free variables, e and d; that is, we can arbitrarily assign values to both e and d and obtain an equation of a hyperplane passing through the given points. Accordingly, all the equations will not be multiples of each other and so there is more than one hyperplane passing through the given points.

22.28. **Show that there is at least one hyperplane in $\mathbf{R}^m$ passing through any m arbitrary points.**

Substituting each of the points into the equation $c_1x_1 + c_2x_2 + \cdots + c_mx_m = b$, we obtain a system of m homogeneous linear equations in the $m+1$ unknowns, $c_1, \ldots, c_m$ and b. Thus the system has at least one free variable and so has a non-zero solution. The non-zero solution determines a hyperplane passing through the given points.

22.29. **Show that any $r < m$ arbitrary points in $\mathbf{R}^m$ do not uniquely determine a hyperplane, i.e. that there is more than one hyperplane passing through the r points.**

Substituting each of the r points into the equation $c_1x_1 + c_2x_2 + \cdots + c_mx_m = b$, we obtain a system of r homogeneous equations in the $m+1$ unknowns, $c_1, \ldots, c_m$ and b. Since $r < m$, the system has at least two free variables. Accordingly, all the solutions of the system will not be multiples of each other and so they determine more than one hyperplane.

22.30. **Show that the hyperplanes $x+2y+3z = 1$, $2x-y-z = 3$ and $3x+2y+2z = 1$ in $\mathbf{R}^3$ intersect in a unique point. Find the intersection.**

Compute the determinant D of the matrix of coefficients of the three equations:

$$D = \begin{vmatrix} 1 & 2 & 3 \\ 2 & -1 & -1 \\ 3 & 2 & 2 \end{vmatrix} = 1(-2+2) - 2(4+3) + 3(4+3) = 0 - 14 + 21 = 7$$

Since $D \neq 0$, the system of 3 equations in 3 unknowns has a unique solution.

To find the solution, reduce the system to echelon form:

$$\begin{cases} x + 2y + 3z = 1 \\ 2x - y - z = 3 \\ 3x + 2y + 2z = 1 \end{cases} \quad \text{or} \quad \begin{cases} x + 2y + 3z = 1 \\ 5y + 7z = -1 \\ 4y + 7z = 2 \end{cases} \quad \text{or} \quad \begin{cases} x + 2y + 3z = 1 \\ 5y + 7z = -1 \\ y = -3 \end{cases}$$

Substitute in the second equation to obtain $z = 2$, and then substitute in the first equation to find $x = 1$. Thus the hyperplanes intersect in the unique point $(1, -3, 2)$.

22.31. **Determine whether the given hyperplanes in $\mathbf{R}^3$ intersect in a unique point:**

(i) $x+3y-2z = 1$, $2x-y+z = 2$ and $4x+5y-3z = 4$;

(ii) $x+y+z = 1$, $2x+y-z = 6$ and $3x-3y+z = -5$.

In each case, compute the determinant D of the matrix of coefficients:

$$\text{(i)} \quad D = \begin{vmatrix} 1 & 3 & -2 \\ 2 & -1 & 1 \\ 4 & 5 & -3 \end{vmatrix} = 1(3-5) - 3(-6-4) - 2(10+4) = -2 + 30 - 28 = 0$$

Since $D = 0$, the hyperplanes do not intersect in a unique point.

$$\text{(ii)} \quad D = \begin{vmatrix} 1 & 1 & 1 \\ 2 & 1 & -1 \\ 3 & -3 & 1 \end{vmatrix} = 1(1-3) - 1(2+3) + 1(-6-3) = -2 - 5 - 9 = -16$$

Since $D \neq 0$, the hyperplanes intersect in a unique point.

22.32. Show that m arbitrary hyperplanes $u_1 \cdot x = b_1,\ u_2 \cdot x = b_2,\ \ldots,\ u_m \cdot x = b_m$ in $\mathbf{R}^m$ intersect in a unique point if and only if the m vectors $u_1, u_2, \ldots, u_m$ are (linearly) independent.

The system of m linear equations in m unknowns has a unique solution if and only if the determinant of the matrix A of coefficients is not zero. However, the determinant of A is not zero if and only if the rows of A are independent vectors. The result now follows from the fact that the rows of A are the vectors u_i.

MISCELLANEOUS PROBLEMS

22.33. Prove the Cauchy-Schwarz Inequality: For any pair of vectors $u = (u_1, \ldots, u_m)$ and $v = (v_1, \ldots, v_m)$ in $\mathbf{R}^m$,

$$|u \cdot v| \;\leq\; \sum_{i=1}^{m} |u_i v_i| \;\leq\; ||u||\,||v||$$

If $u = 0$ or $v = 0$, then the inequality reduces to $0 \leq 0 \leq 0$ and is therefore true. Thus we need only consider the case in which $u \neq 0$ and $v \neq 0$, i.e. in which $||u|| \neq 0$ and $||v|| \neq 0$. Furthermore,

$$|u \cdot v| \;=\; |u_1v_1 + \cdots + u_mv_m| \;\leq\; |u_1v_1| + \cdots + |u_mv_m| \;=\; \sum_{i=1}^{m} |u_iv_i|$$

Thus we need only prove the second inequality.

Now for any real numbers $x, y \in \mathbf{R}$, $0 \leq (x-y)^2 = x^2 - 2xy + y^2$ or, equivalently,

$$2xy \;\leq\; x^2 + y^2 \tag{1}$$

Set $x = |u_i|/||u||$ and $y = |v_i|/||v||$ in (1) to obtain, for any i,

$$2\,\frac{|u_i|}{||u||}\,\frac{|v_i|}{||v||} \;\leq\; \frac{|u_i|^2}{||u||^2} + \frac{|v_i|^2}{||v||^2} \tag{2}$$

But, by definition of the norm of a vector, $||u|| = \sum u_i^2 = \sum |u_i|^2$ and $||v|| = \sum v_i^2 = \sum |v_i|^2$. Thus summing (2) with respect to i and using $|u_iv_i| = |u_i|\,|v_i|$, we have

$$2\,\frac{\sum |u_iv_i|}{||u||\,||v||} \;\leq\; \frac{\sum |u_i|^2}{||u||^2} + \frac{\sum |v_i|^2}{||v||^2} \;=\; \frac{||u||^2}{||u||^2} + \frac{||v||^2}{||v||^2} \;=\; 2$$

that is,

$$\frac{\sum |u_iv_i|}{||u||\,||v||} \;\leq\; 1$$

Multiplying both sides by $||u||\,||v||$, we obtain the required inequality.

22.34. Prove Minkowski's Inequality: For any pair of vectors $u = (u_1, \ldots, u_m)$ and $v = (v_1, \ldots, v_m)$ in $\mathbf{R}^m$, $||u+v|| \leq ||u|| + ||v||$.

If $||u+v|| = 0$, the inequality clearly holds. Thus we need only consider the case in which $||u+v|| \neq 0$.

Now $|u_i + v_i| \leq |u_i| + |v_i|$ for any real numbers $u_i, v_i \in \mathbf{R}$. Hence

$$\begin{aligned} ||u+v||^2 &= \sum (u_i+v_i)^2 = \sum |u_i+v_i|^2 = \sum |u_i+v_i|\,|u_i+v_i| \\ &\leq \sum |u_i+v_i|\,(|u_i|+|v_i|) = \sum |u_i+v_i|\,|u_i| + \sum |u_i+v_i|\,|v_i| \end{aligned}$$

But by the Cauchy-Schwarz Inequality (see preceding problem),

$$\sum |u_i+v_i|\,|u_i| \;\leq\; ||u+v||\,||u|| \qquad \text{and} \qquad \sum |u_i+v_i|\,|v_i| \;\leq\; ||u+v||\,||v||$$

Thus

$$||u+v||^2 \;\leq\; ||u+v||\,||u|| + ||u+v||\,||v|| \;=\; ||u+v||\,(||u|| + ||v||)$$

Dividing by $||u+v||$, we obtain the required inequality.

Supplementary Problems

DISTANCE BETWEEN POINTS

22.35. Compute the distance between each pair of points: (i) $(5,-1)$ and $(2,6)$; (ii) $(2,2)$ and $(5,-5)$; (iii) $(4,0)$ and $(-2,-7)$; (iv) $(-3,-5)$ and $(4,-1)$.

22.36. Determine k so that the distance between $(1,3)$ and $(2k,7)$ is 5.

22.37. Find the point equidistant from $(1,2)$, $(-1,4)$ and $(5,3)$.

22.38. Determine k so that the points $(2,1)$, $(-2,k)$ and $(6,4)$ are collinear, i.e. belong to the same straight line.

22.39. Show that the points $(6,5)$, $(2,-5)$ and $(-1,2)$ are the vertices of a right triangle, and find its area.

LINES

22.40. Plot the following lines: (i) $x=-5$, $x=-2$, $x=1$, $x=4$; (ii) $y=-4$, $y=-1$, $y=2$, $y=6$.

22.41. Find the equation of the line:
(i) parallel to the x axis and passing through the point $(3,-4)$;
(ii) parallel to the y axis and passing through the point $(-2,1)$;
(iii) horizontal and passing through the point $(5,-7)$;
(iv) vertical and passing through the point $(6,-3)$.

22.42. Plot the following lines: (i) $3x-5y=15$ and $2x+3y=12$; (ii) $4x-3y=-6$ and $5x+2y=8$; (iii) $2x+5y=0$ and $4x-3y=0$.

22.43. Find the slope of each line: (i) $y=-2x+8$; (ii) $2x-5y=14$; (iii) $x=3y-7$; (iv) $x=-2$; (v) $y=7$.

22.44. Find the equation of the line:
(i) passing through $(3,-2)$ with slope $\frac{1}{4}$;
(ii) passing through $(-5,0)$ and $(3,5)$;
(iii) passing through $(1,-3)$ and parallel to $2x-3y=7$;
(iv) passing through $(2,-1)$ and perpendicular to $3x+4y=6$.

22.45. Determine the distance of each line from the origin:
(i) $4x+3y=-8$; (ii) $2x+5y=3$; (iii) $12x-5y=26$.

22.46. Determine the distance between each point and line:
(i) $(2,-1)$ and $6x-8y=15$; (ii) $(3,1)$ and $x-3y=-6$; (iii) $(-2,4)$ and $12x+5y=3$.

EUCLIDEAN *m*-SPACE

22.47. Let $u=(1,-2,5)$ and $v=(3,1,-2)$. Find:
(i) $u+v$; (ii) $2u-5v$; (iii) $u\cdot v$; (iv) $||u||$; (v) $d(u,v)$.

22.48. Let $u=(2,1,-3,0,4)$ and $v=(5,-3,-1,2,7)$. Find:
(i) $u+v$; (ii) $3u-2v$; (iii) $u\cdot v$; (iv) $||v||$; (v) $d(u,v)$, i.e. $||u-v||$.

22.49. For each pair of vectors determine the value of k so that the vectors are orthogonal:
(i) $(5,k,-4,2)$ and $(1,-3,2,2k)$; (ii) $(1,7,k+2,-2)$ and $(3,k,-3,k)$.

22.50. Determine the value of k so that the distance between $(2,k,1,-4)$ and $(3,-1,6,-3)$ is 6.

22.51. Show that a hyperplane is not bounded.

22.52. Show that a set A is bounded if and only if there exists a real number $m^*>0$ such that $-m^*\leq u_i\leq m^*$ for every component of any vector $u\in A$.

HYPERPLANES

22.53. Calculate the distance of each hyperplane from the origin, i.e. the zero vector:
(i) $2x-y+3z+4w=7$; (ii) $4x-12y-3z=-10$; (iii) $3x+2y+2z-w=-6$.

22.54. Calculate the distance between the point and the hyperplane:
(i) $(2,-3,1)$ and $3x+4y-12z = 8$; (ii) $(5,-2,1,-3)$ and $x+7y-5z+5w = 1$; (iii) $(4,1,-2)$ and $x+8y-4z = -7$.

22.55. Find the equation of the hyperplane in $\mathbf{R}^3$ which:
(i) contains $(2,-3,-5)$ and is parallel to the xz plane;
(ii) intersects the x, y and z axes at 2, -3 and 4 respectively;
(iii) contains $(1,-2,2)$, $(0,1,3)$ and $(0,2,-1)$;
(iv) contains $(1,-5,4)$ and is parallel to $3x-7y-6z = 3$.

22.56. Find the equation of the hyperplane in $\mathbf{R}^4$ which:
(i) contains the x_1, x_2 and x_4 axes;
(ii) contains $(1,5,-4,3)$ and is parallel to $x_1-3x_2-4x_3-2x_4 = -1$;
(iii) contains $(1,1,1,1)$, $(0,1,-3,1)$, $(0,0,1,4)$ and $(0,0,2,-1)$;
(iv) intersects the axes at $3,-1,2$, and 6, respectively.

22.57. Determine whether the given hyperplanes in $\mathbf{R}^3$ intersect in a unique point:

(i) $\begin{cases} x+3y+5z = 1 \\ 5x+y+z = 2 \\ 2x-y-2z = 7 \end{cases}$; (ii) $\begin{cases} x+2y+z = 1 \\ 3x-2y-z = 2 \\ 5x+2y+z = 4 \end{cases}$; (iii) $\begin{cases} x-4y-2z = 1 \\ 2x+2y+z = 2 \\ 3x-y-z = 0 \end{cases}$.

22.58. Show that the intersection of $r<m$ hyperplanes in $\mathbf{R}^m$ is either empty or contains an infinite number of points.

Answers to Supplementary Problems

22.35. (i) $\sqrt{58}$, (ii) $\sqrt{58}$, (iii) $\sqrt{85}$, (iv) $\sqrt{65}$

22.36. $k=2$ or $k=-1$

22.37. (2.3, 5.3)

22.38. $k=-2$

22.39. 29

22.41. (i) $y=-4$, (ii) $x=-2$, (iii) $y=-7$, (iv) $x=6$

22.43. (i) -2, (ii) 2/5, (iii) 1/3, (iv) not defined, (v) 0

22.44. (i) $x-4y = 11$, (ii) $5x-8y = -25$, (iii) $2x-3y = 11$, (iv) $4x-3y = 11$

22.45. (i) 8/5, (ii) $3/\sqrt{29}$, (iii) 2

22.46. (i) $\frac{1}{2}$, (ii) $6/\sqrt{10}$, (iii) 7/13

22.47. (i) $u+v = (4,-1,3)$; (ii) $2u-5v = (-13,-9,20)$; (iii) $u \cdot v = -9$; (iv) $\|u\| = \sqrt{30}$; (v) $d(u,v) = \sqrt{62}$

22.48. (i) $u+v = (7,-2,-4,2,11)$; (ii) $3u-2v = (-4,9,-7,-4,-2)$; (iii) $u \cdot v = 38$; (iv) $\|v\| = \sqrt{88} = 2\sqrt{22}$; (v) $d(u,v) = \sqrt{42}$

22.49. (i) $k=3$; (ii) $k=3/2$

22.50. $k=2$ or $k=-4$

22.53. (i) $7/\sqrt{30}$, (ii) 10/13, (iii) $6/\sqrt{18} = 2/\sqrt{2}$

22.54. (i) 2, (ii) 3, (iii) 3

22.55. (i) $y=-3$; (ii) $6x-4y+3z = 12$; (iii) $13x+4y+z = 7$; (iv) $3x-7y-6z = 14$

22.56. (i) $x_3 = 0$; (ii) $x_1-3x_2-4x_3-2x_4 = -4$; (iii) $20x_1-23x_2-5x_3-x_4 = -9$; (iv) $2x_1-6x_2+3x_3+x_4 = 6$

22.57. (i) No. (ii) No. (iii) Yes.

Chapter 23

Convex Sets and Linear Inequalities

LINE SEGMENTS AND CONVEX SETS

The line segment joining two points $P = (a_1, a_2, \ldots, a_m)$ and $Q = (b_1, b_2, \ldots, b_m)$ in $\mathbf{R}^m$, denoted by $\overline{PQ}$, is the set of points

$$tP + (1-t)Q = (ta_1 + (1-t)b_1,\ ta_2 + (1-t)b_2,\ \ldots,\ ta_m + (1-t)b_m), \qquad \text{where } 0 \leq t \leq 1$$

A subset X of $\mathbf{R}^m$ is convex iff the line segment joining any two points P and Q in X is contained in X: $P, Q \in X$ implies $\overline{PQ} \subset X$.

Example 1.1.

The rectangular and elliptical areas below are each convex since the line segment joining any two points P and Q, in either figure, is contained in the figure. On the other hand, the U-shaped area to the right below is not convex, since the line segment joining the points P and Q also has points lying outside of the figure.

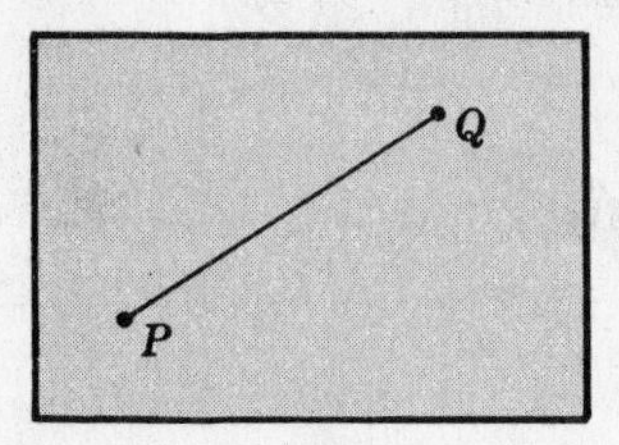

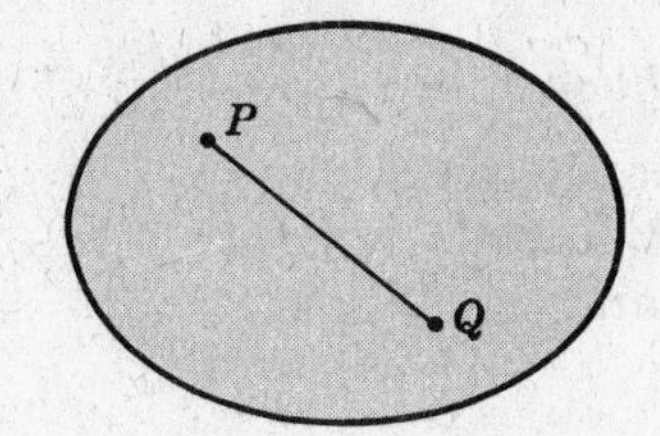

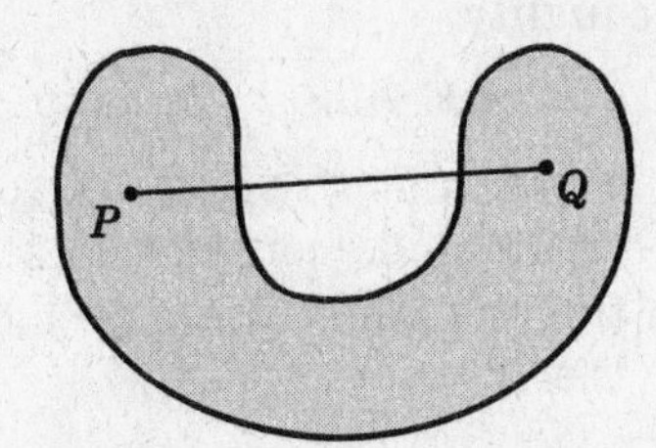

Example 1.2.

Let X and Y be any two convex sets in $\mathbf{R}^m$. We show that the intersection $X \cap Y$ is also convex. For let P and Q be any two points in $X \cap Y$; then $P, Q \in X$ and $P, Q \in Y$. Since X and Y are each convex, the line segment $\overline{PQ}$ is contained in both X and Y: $\overline{PQ} \subset X$ and $\overline{PQ} \subset Y$. Hence $\overline{PQ}$ is contained in the intersection: $\overline{PQ} \subset X \cap Y$. Thus $X \cap Y$ is convex.

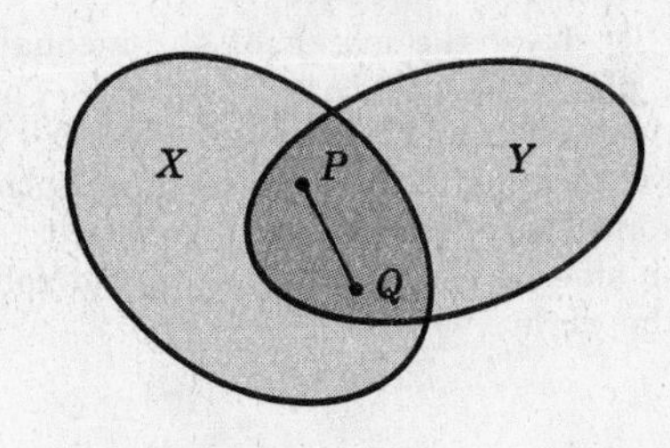

In fact,

Theorem 23.1: The intersection of any number of convex sets is convex.

We next state a theorem about line segments which will be used throughout.

Theorem 23.2: Let f be a linear function defined over a line segment $\overline{PQ}$. Then for every point $A \in \overline{PQ}$, either

$$\text{(i) } f(P) \leq f(A) \leq f(Q) \quad \text{or} \quad \text{(ii) } f(Q) \leq f(A) \leq f(P)$$

In particular, if $f(P) = f(Q)$ then f has the same value at every point in $\overline{PQ}$.

In other words, the value of a linear function f at any point of a line segment lies between the values of f at the end points.

Remark: A function f defined over $\mathbf{R}^m$ is said to be *linear* if it is of the form

$$f = f(x_1, x_2, \ldots, x_m) = c_1x_1 + c_2x_2 + \cdots + c_mx_m$$

LINEAR INEQUALITIES

The linear equation $a_1x_1 + a_2x_2 + \cdots + a_mx_m = b$ partitions $\mathbf{R}^m$ into 3 subsets:

(i) those points which satisfy the given equation;

(ii) those points which satisfy the inequality $a_1x_1 + a_2x_2 + \cdots + a_mx_m < b$;

(iii) and those points which satisfy the inequality $a_1x_1 + a_2x_2 + \cdots + a_mx_m > b$.

The first set, as defined previously, is a hyperplane; the other two sets are called *open half spaces.*

Accordingly, the *solution set* or, simply, *solution* (or: graph) of the linear inequality

$$a_1x_1 + a_2x_2 + \cdots + a_mx_m \leqq b \qquad (\text{or: } a_1x_1 + a_2x_2 + \cdots + a_mx_m \geqq b)$$

consists of the points of the hyperplane $a_1x_1 + a_2x_2 + \cdots + a_mx_m = b$, called the *bounding hyperplane,* and of one of the open half spaces; such a set is termed a *closed half space* or, simply, a *half space.*

An important property of half spaces follows.

Theorem 23.3: Half spaces are convex.

In particular, the solution of the linear inequality

$$ax + by \leqq c \qquad (\text{or: } ax + by \geqq c)$$

is called a *half plane.* Geometrically speaking, the half plane consists of all of the points lying on the bounding line $ax + by = c$ and on one side of the line.

Example 2.1.

To draw the graph of the inequality $2x - 3y \leqq -6$, first plot the line $2x - 3y = -6$. Then choose an arbitrary point that doesn't belong to the line, say $(0, 0)$, and test it to see if it belongs to the solution set. Since $(0, 0)$ does not satisfy the given inequality, the points on the other side of the line belong to the solution set as shown on the right.

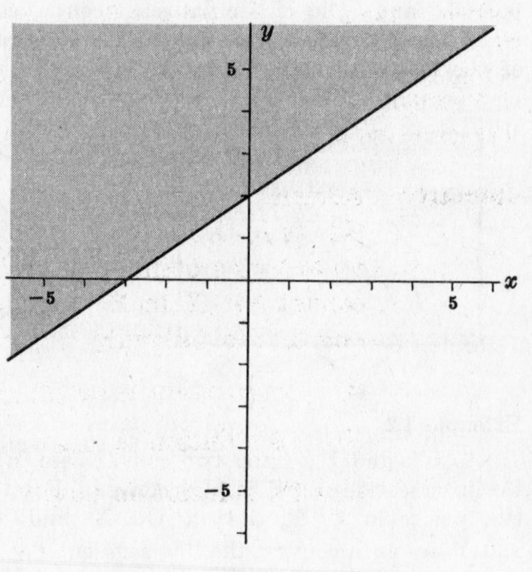

Graph of $2x - 3y \leqq -6$

POLYHEDRAL CONVEX SETS AND EXTREME POINTS

The intersection X of a finite number of (closed) half spaces in $\mathbf{R}^m$ is called a *polyhedral convex set.* A point $P \in X$ is called an *extreme point* or *corner point* of X if P is the intersection of m of the bounding hyperplanes determining X. Observe that Theorem 23.1 and Theorem 23.3 justify the use of the term convex; that is, half spaces are convex and so their intersection is also convex.

Theorem 23.4: The solution of a system of linear inequalities

$$\begin{array}{ccccccccc} a_{11}x_1 & + & a_{12}x_2 & + & \cdots & + & a_{1m}x_m & \leqq & b_1 \\ a_{21}x_1 & + & a_{22}x_2 & + & \cdots & + & a_{2m}x_m & \leqq & b_2 \\ \multicolumn{9}{c}{\cdots\cdots\cdots\cdots\cdots\cdots\cdots\cdots\cdots\cdots} \\ a_{r1}x_1 & + & a_{r2}x_2 & + & \cdots & + & a_{rm}x_m & \leqq & b_r \end{array}$$

is a polyhedral convex set.

The theorem follows from the fact that the solution to each inequality is a half space and that the solution of the system is the intersection of the individual solutions.

Remark: Since the sense of an inequality can be changed by multiplying the inequality by, say, -1, there is no loss in generality in assuming the inequalities in Theorem 23.4 are of the same given sense.

Example 3.1.

The bounded polyhedral convex set in the figure on the right is the solution of the following system of five linear inequalities:

$$\begin{aligned} x+y+z &\leqq 2 \\ x-y &\geqq 0 \\ z &\leqq 1 \\ x &\geqq 0 \\ z &\geqq 0 \end{aligned}$$

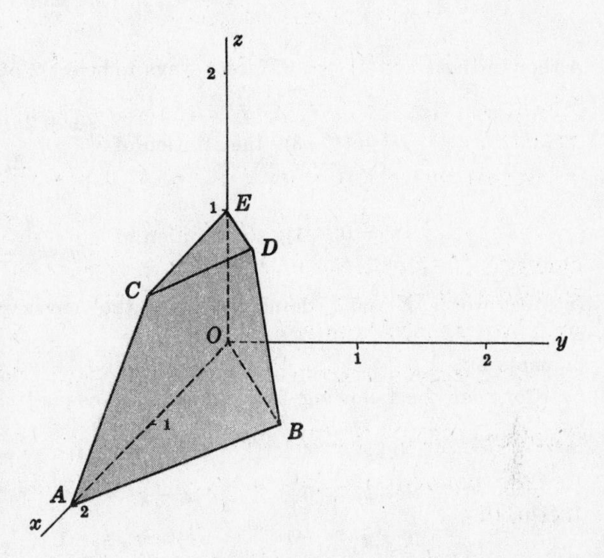

The points A, B, C, D, E and O in the diagram are the corner points, i.e. extreme points of the solution; each point is the intersection of three of the bounding planes. For example, D is the intersection of the planes $x+y+z=2$, $x-y=0$ and $z=1$; solving the three equations simultaneously, we obtain $D=(\frac{1}{2},\frac{1}{2},1)$. Similarly we can obtain $A=(2,0,0)$, $B=(1,1,0)$, $C=(1,0,1)$, $E=(0,0,1)$ and $O=(0,0,0)$.

Remark: Although there are $\binom{5}{3}=10$ different ways to select 3 of the 5 given bounding planes in the above example, four of the ten sets of equations will not determine an extreme point. In general, a set of m bounding hyperplanes of a polyhedral convex set X in $\mathbf{R}^m$ does not determine an extreme point of X if and only if one of the following holds:

(i) its solution is empty, i.e. the equations are inconsistent;

(ii) its solution is infinite, i.e. the equations are dependent;

(iii) its solution is unique but does not belong to X.

POLYGONAL CONVEX SETS AND CONVEX POLYGONS

In the special case of the plane $\mathbf{R}^2$, a polyhedral convex set X is called a *polygonal convex set* and, if it is bounded, a *convex polygon*. Furthermore, a point $P \in X$ is an extreme point of X iff it is the intersection of two of the bounding lines of X.

Example 4.1.

The convex polygon shaded in the adjoining diagram is the solution to the following system of four linear inequalities in two unknowns:

$$\begin{aligned} x+2y &\leqq 4 \\ x-\ y &\leqq 4 \\ x &\geqq 1 \\ y &\geqq -1 \end{aligned}$$

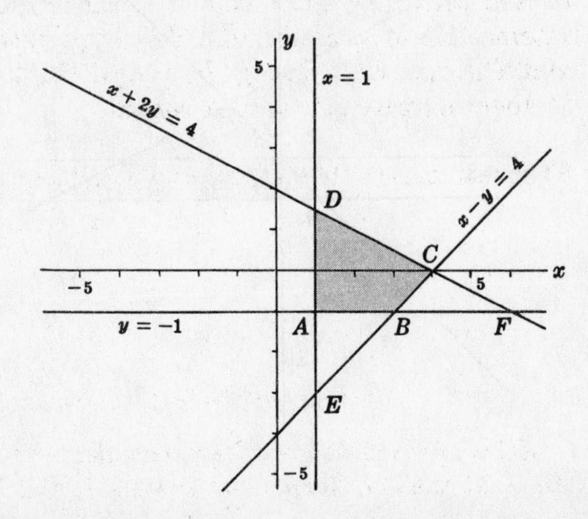

There are four extreme points:

$A=(1,-1)$, the solution of $\begin{cases} x = 1 \\ y = -1 \end{cases}$

$B=(3,-1)$, the solution of $\begin{cases} x-y = 4 \\ y = -1 \end{cases}$

$$C = (4,0), \text{ the solution of } \begin{cases} x + 2y = 4 \\ x - y = 4 \end{cases}$$

$$D = (1, \tfrac{3}{2}), \text{ the solution of } \begin{cases} x + 2y = 4 \\ x = 1 \end{cases}$$

Although there are $\binom{4}{2} = 6$ different ways to select 2 of the 4 bounding lines,

$$E = (1,-3), \text{ the solution of } \begin{cases} x + 2y = 4 \\ x = 1 \end{cases}, \text{ does not satisfy } x - y \leqq 4$$

$$F = (6,-1), \text{ the solution of } \begin{cases} x - y = 4 \\ y = -1 \end{cases}, \text{ does not satisfy } y \geqq -1$$

In other words, E and F do not belong to the convex polygon.

Example 4.2.

Consider the following four systems of inequalities:

(i)	(ii)	(iii)	(iv)
$x + 2y \leqq 4$	$x + 2y \leqq 4$	$x + 2y \leqq 4$	$x + 2y \geqq 4$
$x - y \geqq 0$	$x - y \leqq 0$	$x - y \leqq 0$	$x - y \leqq 0$
$y \geqq -1$	$y \geqq -1$	$y \leqq -1$	$y \leqq -1$

In each case the bounding lines of the given half planes are $x + 2y = 4$, $x - y = 0$ and $y = -1$. There are $\binom{3}{2} = 3$ different ways to select 2 of the 3 bounding lines. Furthermore,

$$A = (-1,-1), \text{ is the solution of } \begin{cases} x - y = 0 \\ y = -1 \end{cases}$$

$$B = (6,-1), \text{ is the solution of } \begin{cases} x + 2y = 4 \\ y = -1 \end{cases}$$

$$C = (\tfrac{4}{3}, \tfrac{4}{3}), \text{ is the solution of } \begin{cases} x + 2y = 4 \\ x - y = 0 \end{cases}$$

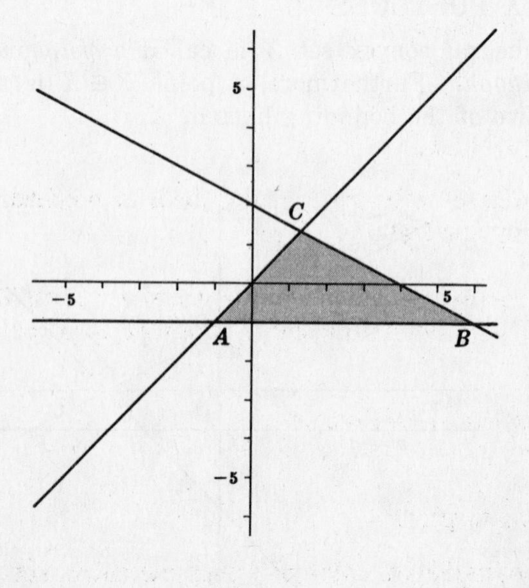

(i)

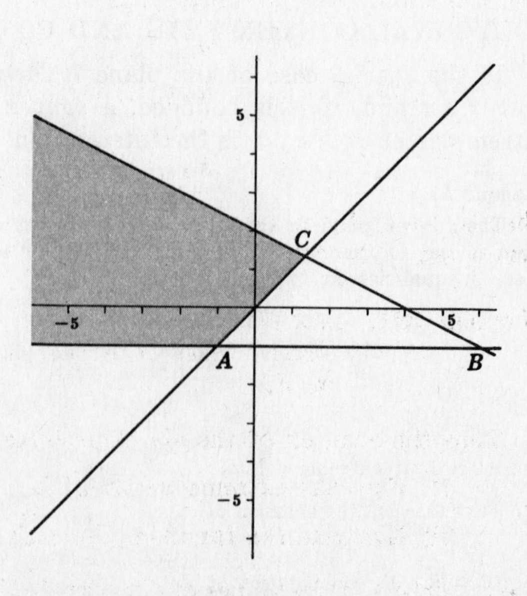

(ii)

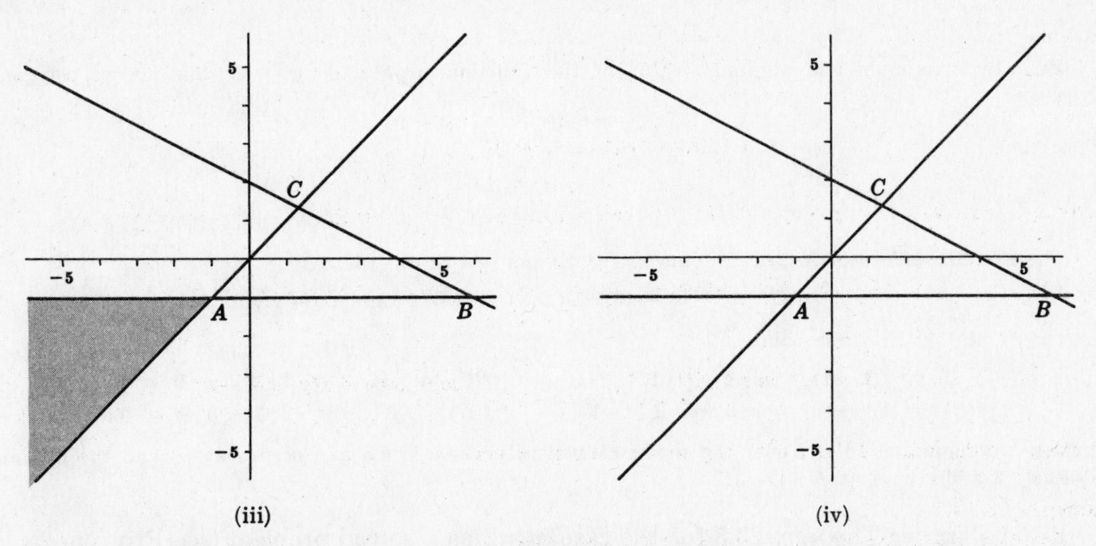

(iii) (iv)

We now discuss the solution of each system which is sketched above.

(i) Here the solutions are bounded and hence called a convex polygon. All three points A, B and C are its extreme points.

(ii) Here the solutions are not bounded. Only the points A and C are its extreme points.

(iii) Here the solutions are the same as the solutions to $\begin{cases} x - y \leqq 0 \\ y \leqq -1 \end{cases}$, that is, without the inequality $x + 2y \leqq 4$. In this case we say that the inequality $x + 2y \leqq 4$ is *redundant* (or: *superfluous*). The point A is the only extreme point.

(iv) Here the solution is empty, and we say that the system is *inconsistent*. There are no extreme points.

LINEAR FUNCTIONS ON POLYHEDRAL CONVEX SETS

We seek the maximum or minimum value of a linear function

$$f = c_1x_1 + c_2x_2 + \cdots + c_mx_m$$

where the unknowns are subject to the following constraints:

$$\begin{array}{l} a_{11}x_1 + a_{12}x_2 + \cdots + a_{1m}x_m \leqq b_1 \\ a_{21}x_1 + a_{22}x_2 + \cdots + a_{2m}x_m \leqq b_2 \\ \cdots\cdots\cdots\cdots\cdots\cdots\cdots\cdots \\ a_{r1}x_1 + a_{r2}x_2 + \cdots + a_{rm}x_m \leqq b_r \end{array}$$

That is, we seek the maximum or minimum value of a linear function over a polyhedral convex set. The following fundamental theorem applies:

Theorem 23.5: Let f be a linear function defined over a bounded polyhedral set X in $\mathbf{R}^m$. Then f takes on its maximum (and minimum) value at an extreme point of X.

Thus the solution of the above problem consists of two steps:

(i) Find the extreme points of X.

(ii) Evaluate the function f at each extreme point.

The maximum of the values of f in step (ii) is the maximum value of f over the entire set X, and the minimum value of f in step (ii) is the minimum value of f over X.

Example 5.1.

Find the maximum and minimum value of the function $f = 2x + 5y$ over the convex polygon X given by

$$x + 2y \leqq 4$$
$$x - y \leqq 4$$
$$x \geqq 1$$
$$y \geqq -1$$

The extreme points of X are as follows (see Example 4.1):

$$A = (1,-1), \quad B = (3,-1), \quad C = (4,0) \quad \text{and} \quad D = (1,\tfrac{3}{2})$$

Evaluate f at each of these points:

$$f(A) = f(1,-1) = 2\cdot 1 - 5\cdot 1 = -3 \qquad f(C) = f(4,0) = 2\cdot 4 + 5\cdot 0 = 8$$
$$f(B) = f(3,-1) = 2\cdot 3 - 5\cdot 1 = 1 \qquad f(D) = f(1,\tfrac{3}{2}) = 2\cdot 1 + 5\cdot\tfrac{3}{2} = 9.5$$

Thus the maximum value of f over the entire convex polygon X is 9.5 and occurs at D, and the minimum value is -3 and occurs at A.

We shall prove Theorem 23.5 for the case $m = 2$ as a solved problem (see Problem 23.17). However, it is instructive to show here how this result follows from geometric considerations in the case of the preceding example.

We can consider the equation $f = 2x + 5y$ as a one parameter family of lines in the plane $\mathbf{R}^2$ (see adjoining diagram). We seek the maximum (or: minimum) value of f with the condition that the line $f = 2x + 5y$ intersects the given convex polygon X. As indicated by the diagram, starting with lines of the family with large values of f, the family will first intersect X at the point D. Similarly, starting with lines having small values of f, the family will first intersect X at the point A. In other words, the linear function $f = 2x + 5y$ will take on its maximum and minimum values over X at the points D and A, respectively.

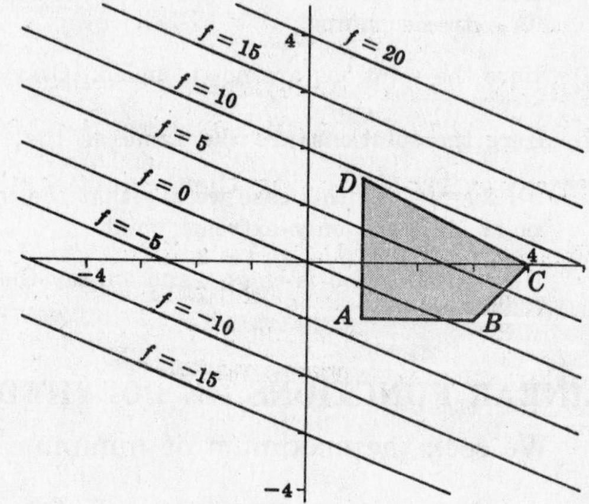

The family $f = 2x + 5y$ and X are plotted.

Example 5.2.

Find the maximum value of the function $f = x + 2y + 8z$ over the polyhedral convex set X given by

$$x + y + z \leqq 2$$
$$x - y \geqq 0$$
$$z \leqq 1$$
$$x \geqq 0$$
$$z \geqq 0$$

The extreme points of X are (see Example 3.1, Page 283).

$$A = (2,0,0), \quad B = (1,1,0), \quad C = (1,0,1), \quad D = (\tfrac{1}{2},\tfrac{1}{2},1), \quad E = (0,0,1), \quad O = (0,0,0)$$

Evaluate f at each of the six extreme points:

$$f(A) = 2, \quad f(B) = 3, \quad f(C) = 9, \quad f(D) = 9.5, \quad f(E) = 8, \quad f(O) = 0$$

Thus, by Theorem 23.5, the maximum value of f over the entire polyhedral convex set X occurs at D and is 9.5.

Consider now a function f of the form

$$f = c_1x_1 + c_2x_2 + \cdots + c_mx_m + d$$

over a polyhedral convex set X in $\mathbf{R}^m$. The function f differs from the linear function

$$g = c_1x_1 + c_2x_2 + \cdots + c_mx_m$$

at every point of X by the fixed amount d; hence f takes on its maximum (and minimum) value at the same point as g. In other words,

Theorem 23.6: A function of the form

$$f = c_1x_1 + c_2x_2 + \cdots + c_mx_m + d$$

defined over a bounded polyhedral set X in $\mathbf{R}^m$ takes on its maximum (and minimum) value at an extreme point of X.

Example 5.3.

Find the minimum value of the function

$$f = 2x - 4y + 3z + 2$$

over the bounded polyhedral convex set X of the preceding example.

Evaluate f at each of the six extreme points of X:

$f(A) = f(2,0,0) = 4+0+0+2 = 6$ $\qquad$ $f(D) = f(\frac{1}{2},\frac{1}{2},1) = 1-2+3+2 = 4$

$f(B) = f(1,1,0) = 2-4+0+2 = 0$ $\qquad$ $f(E) = f(0,0,1) = 0+0+3+2 = 5$

$f(C) = f(1,0,1) = 2+0+3+2 = 7$ $\qquad$ $f(O) = f(0,0,0) = 0+0+0+2 = 2$

Thus the minimum value of f over X occurs at the point B and is 0.

Remark: If a polyhedral convex set X is not bounded, then a linear function f may or may not take on a maximum (and minimum) value.

However, the following theorem always applies (see Problem 23.9):

Theorem 23.7: Let f be a linear function defined over a polyhedral convex set X. Then (i) f will take on a maximum (minimum) value over X if and only if the values of f are bounded above (below) and (ii) the maximum (minimum) value will always occur at an extreme point of X.

Solved Problems

LINEAR INEQUALITIES IN TWO UNKNOWNS

23.1. Graph each inequality:

(i) $2x + y \leq -4$,

(ii) $4x - 3y \geq -6$,

(iii) $2x - 3y \leq 0$.

In each case, first plot the line bounding the half plane by finding at least two points on the line; then choose a point which doesn't belong to the line and test it in the given inequality.

(i) On the line $2x + y = -4$: if $x = 0$, $y = -4$; if $y = 0$, $x = -2$. Draw the line through $(0, -4)$ and $(-2, 0)$. Test, say, the origin $(0, 0)$. Since $(0, 0)$ does not satisfy $2x + y \leq -4$, i.e. $0 \not\leq -4$, shade the side of the line which does not contain the origin.

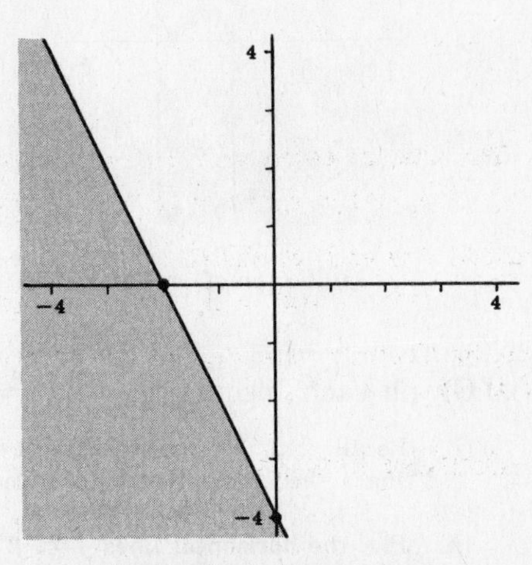

(i) Graph of $2x + y \leq -4$.

(ii) On the line $4x - 3y = -6$: if $x = 0$, $y = 2$; if $y = 0$, $x = -1.5$. Draw the line through $(0, 2)$ and $(-1.5, 0)$. Test the origin $(0, 0)$. Since $(0, 0)$ satisfies the given inequality $4x - 3y \geq -6$, shade the side of the line containing the origin.

(iii) On the line $2x - 3y = 0$: if $x = 0$, $y = 0$; if $x = 3$, $y = 2$. Draw the line through $(0, 0)$ and $(3, 2)$. Since the origin belongs to the line, test the point $(2, 0)$. Since $(2, 0)$ does not satisfy the given inequality $2x - 3y \leq 0$, shade the other side of the line.

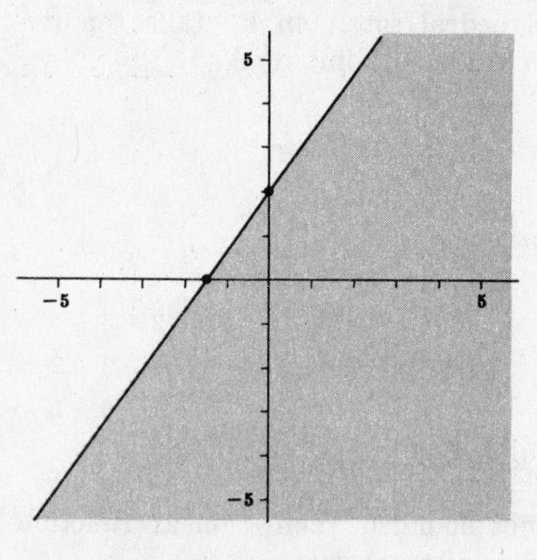

(ii) Graph of $4x - 3y \geq -6$.

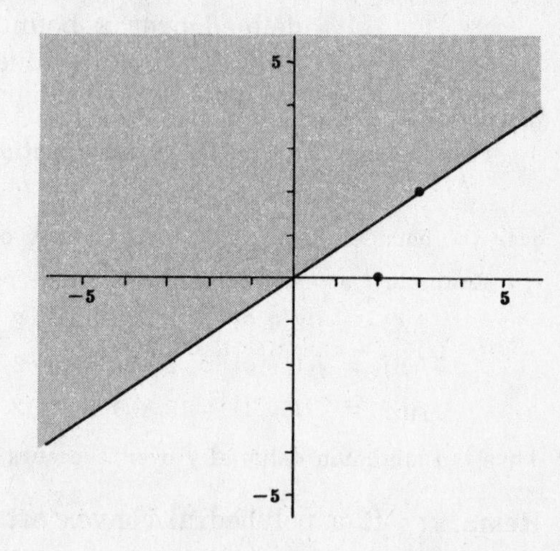

(iii) Graph of $2x - 3y \leq 0$.

23.2. Graph each inequality: (i) $x \leq 2$, (ii) $x \geq 1$, (iii) $y \geq -1$.

(i) First plot the line $x = 2$, the vertical line meeting the x axis at 2. Then shade the area to the left of the line since the x-coordinates of these points are less than 2.

(ii) First plot the line $x = 1$, the vertical line meeting the x axis at 1. Then shade the area to the right of this line since the x-coordinates of these points are greater than 1.

(iii) First plot the line $y = -1$, the horizontal line meeting the y axis at -1. Then shade the area above the line since the y-coordinate of these points are greater than -1.

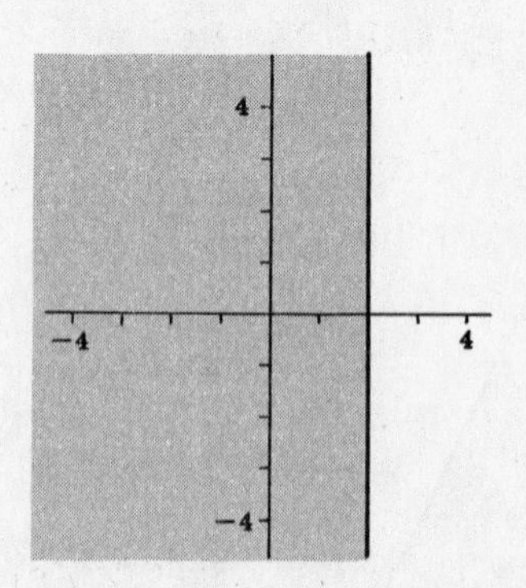

(i) Graph of $x \leq 2$.

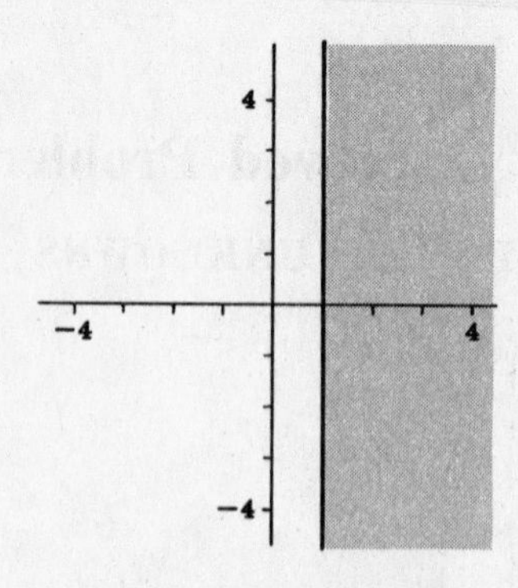

(ii) Graph of $x \geq 1$.

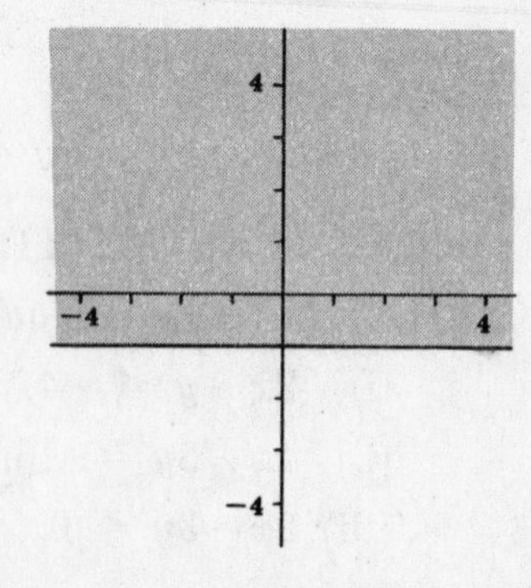

(iii) Graph of $y \geq -1$.

23.3. Graph each inequality: (i) $1 \leq x \leq 4$, (ii) $-2 \leq y \leq 1$.

(i) Recall that $1 \leq x \leq 4$ means $1 \leq x$ and $x \leq 4$. Plot the vertical lines $x = 1$ and $x = 4$; then shade the area between the two lines as shown in Fig. (i) below.

(ii) Plot the horizontal lines $y = -2$ and $y = 1$; then shade the area between the lines as shown in Fig. (ii) below.

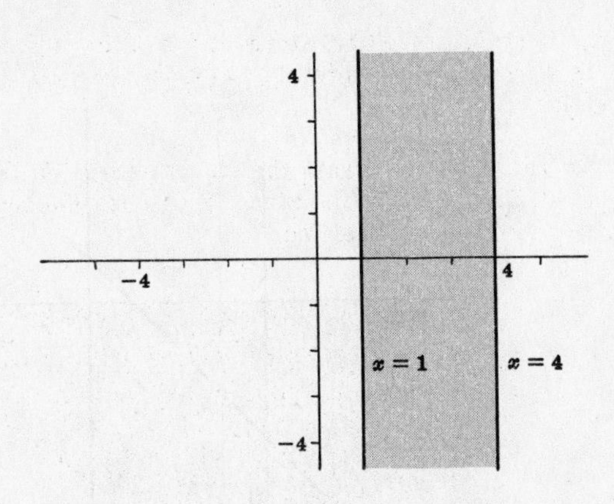

(i) Graph of $1 \leqq x \leqq 4$.

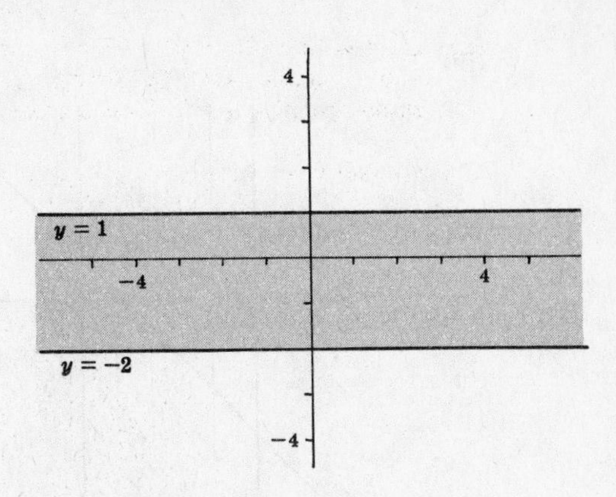

(ii) Graph of $-2 \leqq y \leqq 1$.

23.4. Plot the solution of each pair of inequalities:

(i) $2x - 3y \leqq 6$, $2x + y \leqq 4$ (ii) $2x + 5y \leqq 10$, $x \geqq 1$ (iii) $-1 \leqq x \leqq 3$, $x - y \leqq 2$

In each case, plot each half-plane with different slanted lines; the cross-hatched area is the required solution.

(i)

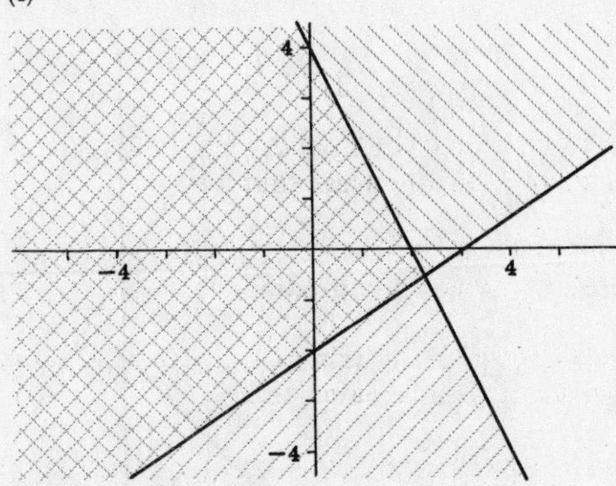

Graphs of $2x - 3y \leqq 6$ and $2x + y \leqq 4$.

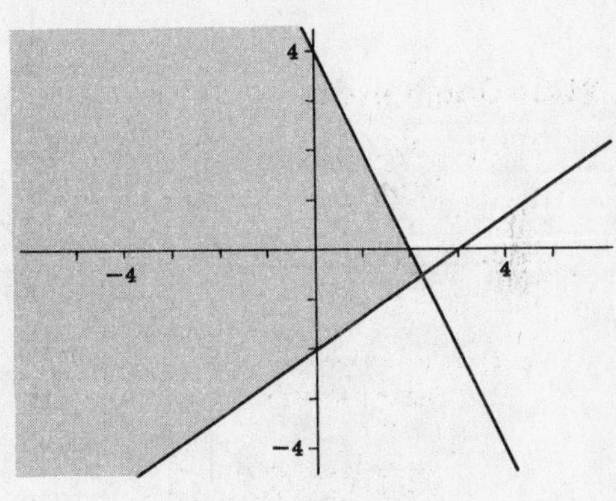

Solution of $\begin{cases} 2x - 3y \leqq 6 \\ 2x + y \leqq 4 \end{cases}$

(ii)

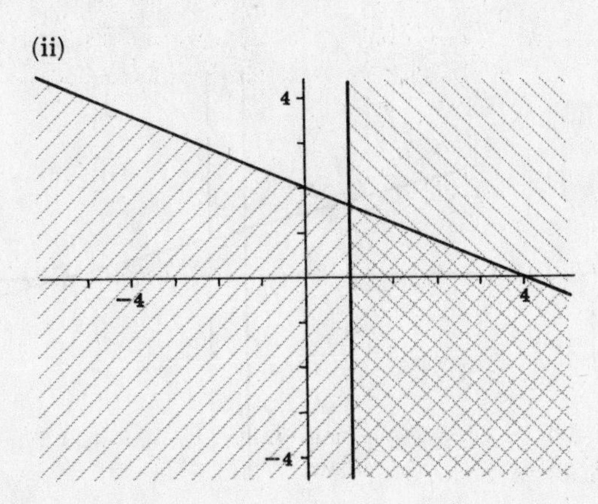

Graphs of $2x + 5y \leqq 10$ and $x \geqq 1$.

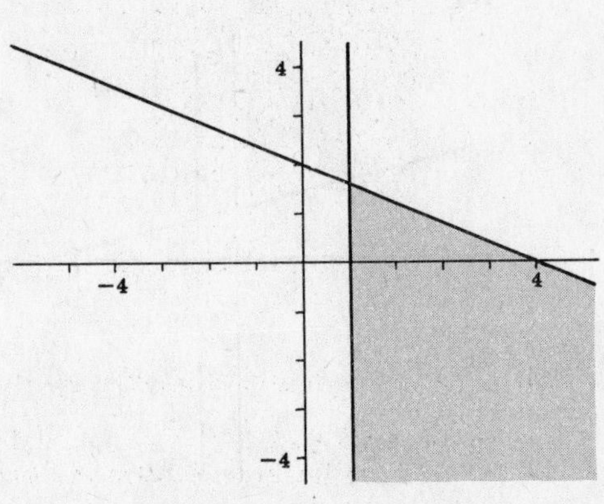

Solution of $\begin{cases} 2z + 5y \leqq 10 \\ x \geqq 1 \end{cases}$

(iii)

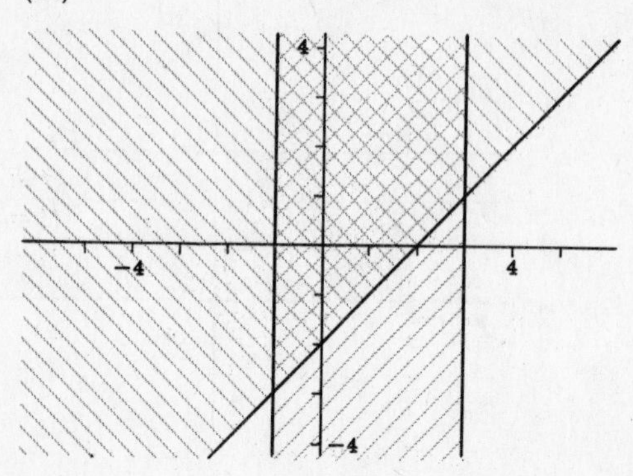

Graphs of $-1 \leqq x \leqq 3$ and $x - y \leqq 2$.

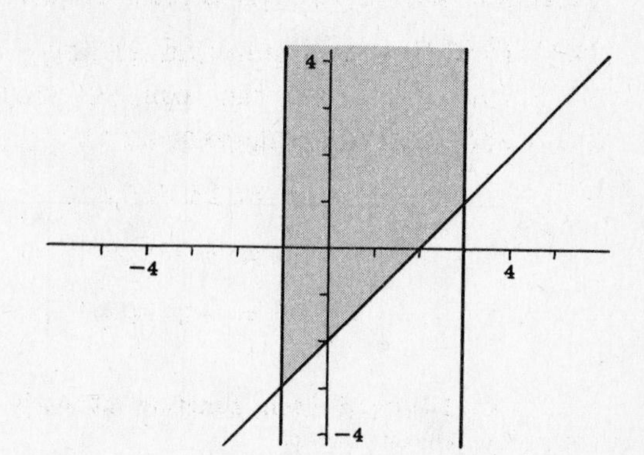

Solution of $\begin{cases} -1 \leqq x \leqq 3 \\ x - y \leqq 2 \end{cases}$

23.5. Plot the solution of each system of inequalities:

(i) $3x + 2y \leqq 6$, $x - y \leqq 4$, $x \leqq 2$ (ii) $x + 4y \leqq 4$, $x \geqq -1$, $y \geqq 0$

(i)

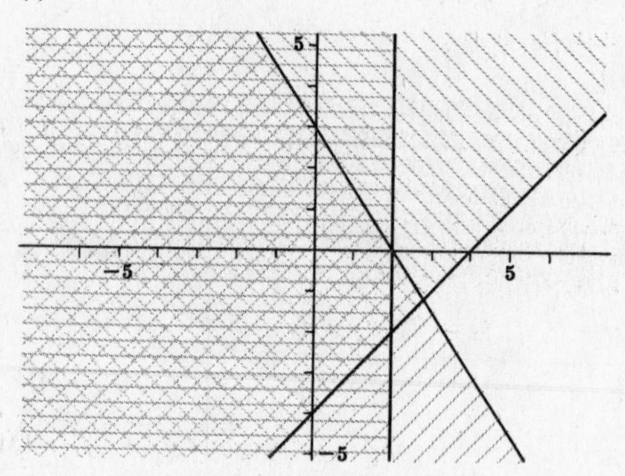

Each half plane is plotted.

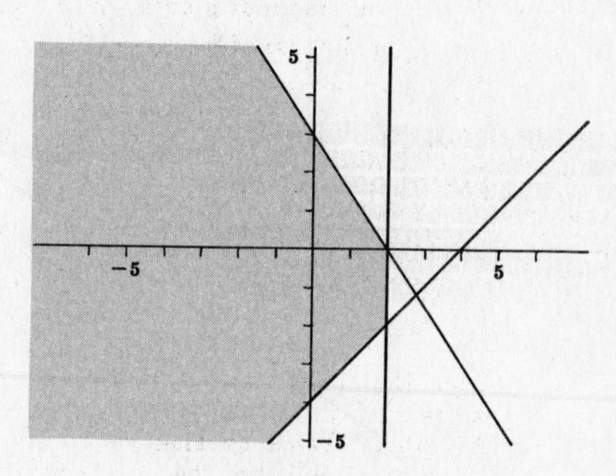

The required solution.

Observe that the solution of the system is not bounded.

(ii)

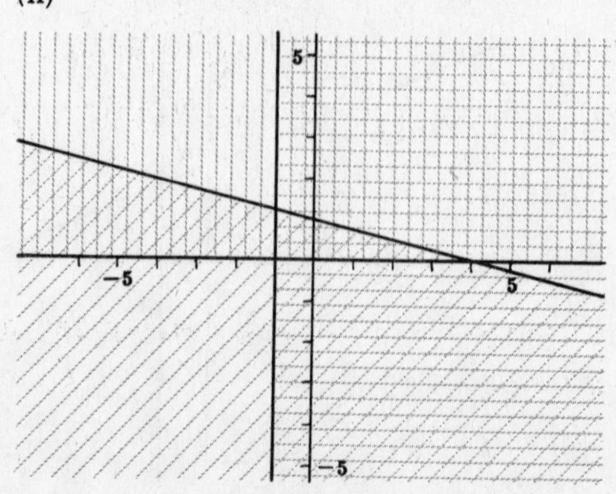

Each half plane is plotted.

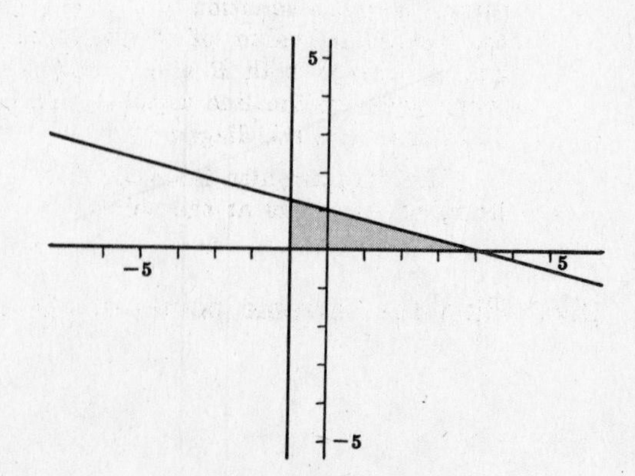

The required solution.

Here the solution of the system is bounded.

LINEAR FUNCTIONS OVER POLYHEDRAL CONVEX SETS

23.6. Find the maximum and minimum of each function over the convex polygon X shown in the diagram on the right:

$$f = 2x + 5y$$
$$g = x - 2y$$
$$h = -3x + y$$

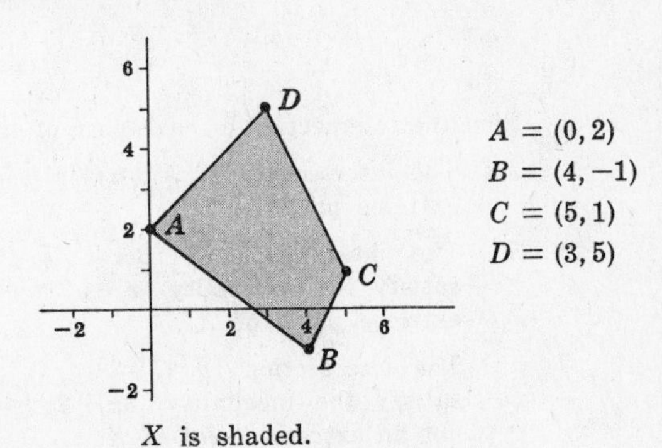

X is shaded.

Evaluate each function at each extreme, i.e. corner point:

$f(A) = f(0,2) = 0 + 10 = 10$ $\qquad g(A) = 0 - 4 = -4$ $\qquad h(A) = 0 + 2 = 2$

$f(B) = f(4,-1) = 8 - 5 = 3$ $\qquad g(B) = 4 + 2 = 6$ $\qquad h(B) = -12 - 1 = -13$

$f(C) = f(5,1) = 10 + 5 = 15$ $\qquad g(C) = 5 - 2 = 3$ $\qquad h(C) = -15 + 1 = -14$

$f(D) = f(3,5) = 6 + 25 = 31$ $\qquad g(D) = 3 - 10 = -7$ $\qquad h(D) = -9 + 5 = -4$

Thus: maximum of f over X is 31 and occurs at the point D
maximum of g over X is 6 and occurs at the point B
maximum of h over X is 2 and occurs at the point A

and minimum of f over X is 3 and occurs at the point B
minimum of g over X is -7 and occurs at the point D
minimum of h over X is -14 and occurs at the point C.

23.7. Find the maximum of the function $f = 2x - y$ over the convex polygon X of the preceding example.

Evaluate f at each of the four extreme points of X:

$$f(A) = f(0,2) = -2 \qquad f(C) = f(5,1) = 9$$
$$f(B) = f(4,-1) = 9 \qquad f(D) = f(3,5) = 1$$

Thus the maximum of f over X is 9 and occurs at two of the corner points, $B = (4,-1)$ and $C = (5,1)$; hence by Theorem 23.2 it will occur at every point on the line segment joining B and C.

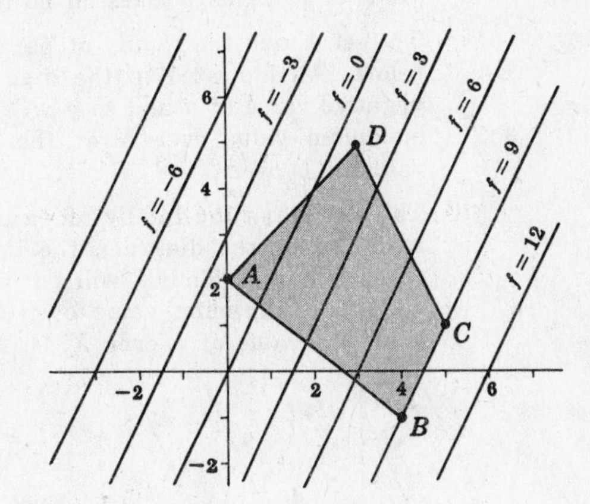

The above can be interpreted geometrically; the member of the family of parallel lines determined by the equation $f = 2x - y$ which has the maximum value of f and intersects X, passes through both B and C and so contains every point on the line segment joining B and C, as shown in the diagram.

The diagram also indicates that f takes on its minimum value at the point A.

23.8. Find the extreme points of the polygonal convex set X determined by the system

$$3x + 2y \leq 6$$
$$x - y \geq 0$$
$$x \leq 4$$

Method 1 (Algebraic).

There are $\binom{3}{2} = 3$ ways to select 2 of the 3 bounding lines:

$$\begin{array}{ccc} 3x + 2y = 6 & 3x + 2y = 6 & x - y = 0 \\ x - y = 0 & x = 4 & x = 4 \\ \text{(i)} & \text{(ii)} & \text{(iii)} \end{array}$$

Find the intersection of each pair of lines and test the point in the other inequality.

(i) The intersection is $A = (6/5, 6/5)$ and it does satisfy the inequality $x \leq 4$; hence A is an extreme point of X.

(ii) The intersection is $B = (4, -3)$ and it does satisfy the inequality $x - y \geq 0$; hence B is an extreme point of X.

(iii) The intersection is $C = (4, 4)$, but C does not satisfy the inequality $3x + 2y \leq 6$; hence C is not an extreme point of X.

Method 2 (Geometric).

Draw the polygonal convex set X as shown on the right. Then A and B are corner points of X, but C is not. Find A by finding the intersection of the two lines $3x + 2y = 6$ and $x - y = 0$ which meet at A; and find B by determining the intersection of the two lines $3x + 2y = 6$ and $x = 4$ which meet at B: $A = (6/5, 6/5)$ and $B = (4, -3)$.

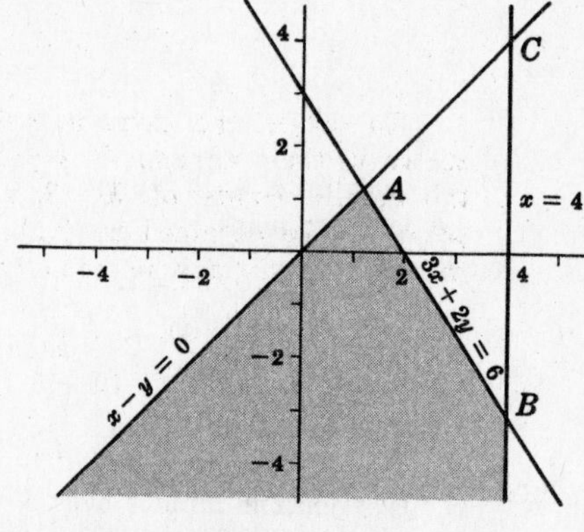

X is shaded.

23.9. Consider the following three functions over the polygonal convex set X of the preceding problem:

$$f = 2x - y; \quad g = x + y; \quad h = x - 2y$$

Show that:

(i) f takes on neither a maximum nor a minimum value over X;

(ii) g takes on a maximum but no minimum value over X;

(iii) h takes on a minimum but no maximum value over X.

(i) The set X and the family of parallel lines determined by the equation $f = 2x - y$ are plotted below. As indicated by the diagram, the lines $f = 2x - y$ will each intersect X for every value of f. Thus f takes on no maximum value and no minimum value over X.

(ii) The set X and the family of parallel lines determined by the equation $g = x + y$ are plotted below. As indicated in the diagram, the lines $g = x + y$ will each intersect X for every negative value of g and so g will not take on a minimum value over X. But g will take on a maximum value over X at the point A. Thus $g(A) = g(6/5, 6/5) = 2.4$ is the maximum value of g over X.

(iii) The set X and the family of parallel lines determined by $h = x - 2y$ are plotted below. As indicated by the diagram, the lines $h = x - 2y$ will each intersect X for all values of h greater than 4. Thus h will not take on a maximum value over X. On the other hand, h will take on a minimum value over X at the point A; hence $h(A) = h(6/5, 6/5) = -1.2$ is the minimum value of h over X.

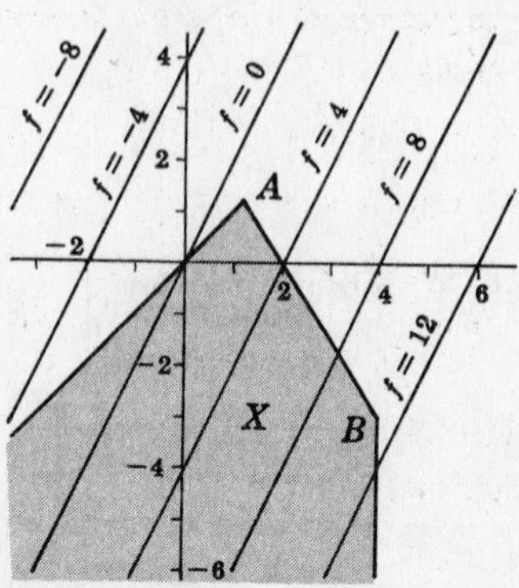

X and the family $f = 2x - y$.

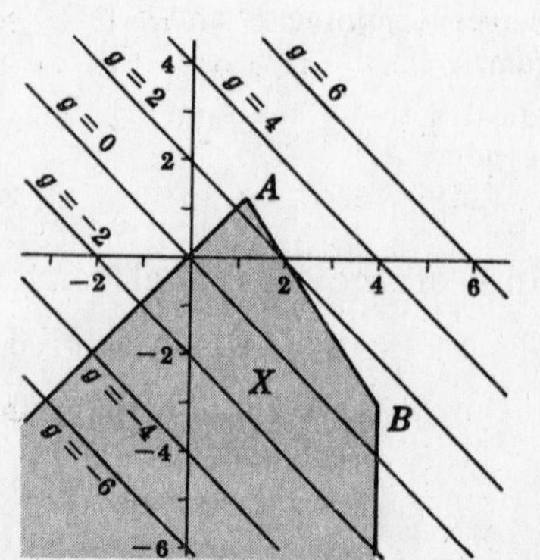

X and the family $g = x + y$.

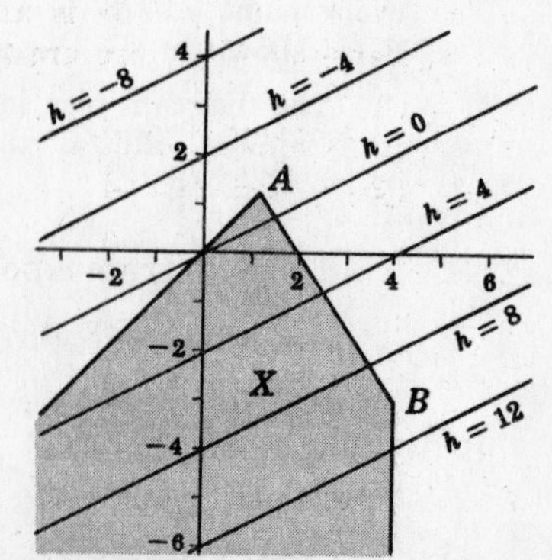

X and the family $h = x - 2y$.

23.10. Find the maximum and minimum of the function $f = x - 2y + 4$ over the convex polygon X given by

$$x + y \leq 4$$
$$x + 2y \geq -2$$
$$x - y \geq -2$$
$$x \leq 3$$

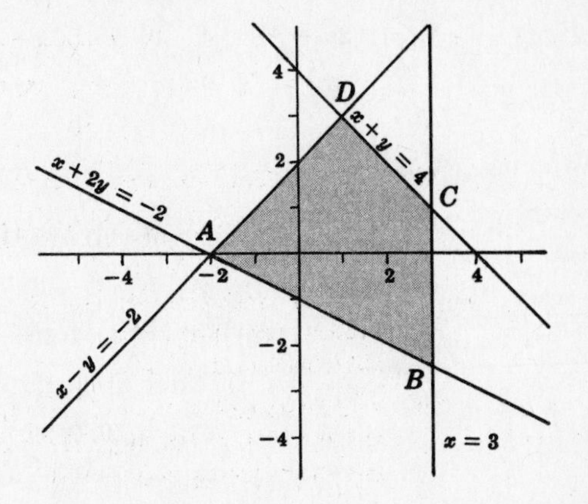

To obtain the extreme points of X, plot X as shown in the diagram. The points A, B, C and D are the corner points of X. The coordinates of each point can be obtained by finding the intersection of the appropriate pair of bounding lines:

$A = (-2, 0)$ is the solution of $\begin{cases} x + 2y = -2 \\ x - y = -2 \end{cases}$

$B = (3, -\frac{5}{2})$ is the solution of $\begin{cases} x + 2y = -2 \\ x = 3 \end{cases}$

$C = (3, 1)$ is the solution of $\begin{cases} x + y = 4 \\ x = 3 \end{cases}$

$D = (1, 3)$ is the solution of $\begin{cases} x + y = 4 \\ x - y = -2 \end{cases}$

Next evaluate the function f at each extreme point:

$$f(A) = -2 + 0 + 4 = 2 \qquad f(C) = 3 - 2 + 4 = 5$$
$$f(B) = 3 + 5 + 4 = 12 \qquad f(D) = 1 - 6 + 4 = -1$$

Thus the maximum of f over X is 12 and occurs at B, and the minimum is -1 and occurs at D.

23.11. Find the extreme points of the polyhedral convex set X given by

$$x + y + z \leq 3$$
$$2x + 2y + z \leq 4$$
$$x - y \leq 0$$
$$x \geq 0$$
$$z \geq 0$$

A point $p \in \mathbf{R}^3$ is an extreme point of X iff it is the intersection of three of the bounding planes of X. There are $\binom{5}{3} = 10$ ways to choose 3 of the 5 bounding planes:

$$\begin{cases} x + y + z = 3 \\ 2x + 2y + z = 4 \\ x - y = 0 \end{cases} \quad \text{(i)} \qquad \begin{cases} x + y + z = 3 \\ 2x + 2y + z = 4 \\ x = 0 \end{cases} \quad \text{(ii)} \qquad \begin{cases} x + y + z = 3 \\ 2x + 2y + z = 4 \\ z = 0 \end{cases} \quad \text{(iii)} \qquad \begin{cases} x + y + z = 3 \\ x - y = 0 \\ x = 0 \end{cases} \quad \text{(iv)}$$

$$\begin{cases} x + y + z = 3 \\ x - y = 0 \\ z = 0 \end{cases} \quad \text{(v)} \qquad \begin{cases} x + y + z = 3 \\ x = 0 \\ z = 0 \end{cases} \quad \text{(vi)} \qquad \begin{cases} 2x + 2y + z = 4 \\ x - y = 0 \\ x = 0 \end{cases} \quad \text{(vii)}$$

$$\begin{cases} 2x + 2y + z = 4 \\ x - y = 0 \\ z = 0 \end{cases} \qquad \begin{cases} 2x + 2y + z = 4 \\ x = 0 \\ z = 0 \end{cases} \qquad \begin{cases} x - y = 0 \\ x = 0 \\ z = 0 \end{cases}$$

(viii) (ix) (x)

Each system except (iii) yields a unique solution:

(i) $A_1 = (\frac{1}{2}, \frac{1}{2}, 2)$

(ii) $A_2 = (0, 1, 2)$

(iii) No solution, i.e. system is inconsistent.

(iv) $A_4 = (0, 0, 3)$

(v) $A_5 = (\frac{3}{2}, \frac{3}{2}, 0)$

(vi) $A_6 = (0, 3, 0)$

(vii) $A_7 = (0, 0, 4)$

(viii) $A_8 = (1, 1, 0)$

(ix) $A_9 = (0, 2, 0)$

(x) $A_{10} = (0, 0, 0)$

To find the extreme points of X, test each of the above points in the given inequalities. The following are not extreme points of X:

$A_5 = (\frac{3}{2}, \frac{3}{2}, 0)$ does not satisfy $2x + 2y + z \leq 4$

$A_6 = (0, 3, 0)$ does not satisfy $2x + 2y + z \leq 4$

$A_7 = (0, 0, 4)$ does not satisfy $x + y + z \leq 3$

The other six points do satisfy the given inequalities and are the extreme points of X. The adjacent diagram shows the position of these points. Note that we obtained the extreme points purely algebraically.

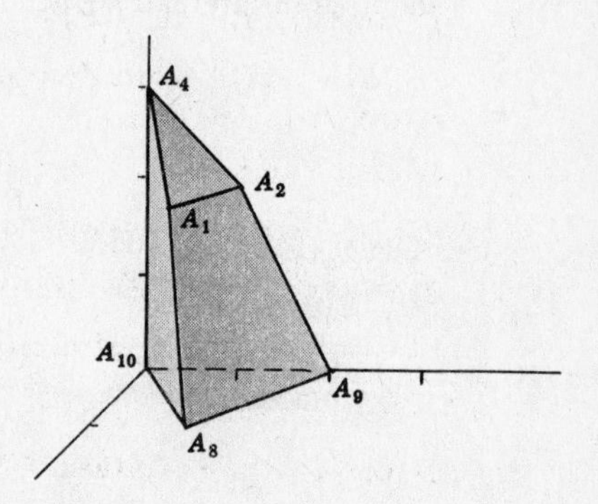

23.12. Find the maximum and minimum values of the function $f = 4x - 2y - z$ over the bounded polyhedral convex set X of the preceding problem.

Evaluate f at each of the six extreme points of X:

$f(A_1) = f(\frac{1}{2}, \frac{1}{2}, 2) = 2 - 1 - 2 = -1$

$f(A_2) = f(0,1,2) = 0 - 2 - 2 = -4$

$f(A_4) = f(0,0,3) = 0 + 0 - 3 = -3$

$f(A_8) = f(1,1,0) = 4 - 2 + 0 = 2$

$f(A_9) = f(0,2,0) = 0 - 4 + 0 = -4$

$f(A_{10}) = f(0,0,0) = 0 + 0 + 0 = 0$

The maximum value of f over X is 2 and occurs at the point A_8. The minimum value of f over X is -4 and occurs at the extreme points A_2 and A_9 and so also occurs at each point on the line segment joining A_2 to A_9.

MISCELLANEOUS PROBLEMS

23.13. Prove: Let f be a linear function defined over $\mathbf{R}^m$:

$$f(x_1, x_2, \ldots, x_m) = c_1x_1 + c_2x_2 + \cdots + c_mx_m$$

Then for any vectors $P, Q \in \mathbf{R}^m$ and any real number k,
(i) $f(P+Q) = f(P) + f(Q)$, (ii) $f(kP) = k\,f(P)$.

Suppose $P = (p_1, p_2, \ldots, p_m)$ and $Q = (q_1, q_2, \ldots, q_m)$; then

$$P + Q = (p_1 + q_1,\ p_2 + q_2,\ \ldots,\ p_m + q_m) \quad \text{and} \quad kP = (kp_1, kp_2, \ldots, kp_m)$$

Hence

$$\begin{aligned} f(P+Q) &= c_1(p_1+q_1) + c_2(p_2+q_2) + \cdots + c_m(p_m+q_m) \\ &= c_1p_1 + c_1q_1 + c_2p_2 + c_2q_2 + \cdots + c_mp_m + c_mq_m \\ &= (c_1p_1 + c_2p_2 + \cdots + c_mp_m) + (c_1q_1 + c_2q_2 + \cdots + c_mq_m) = f(P) + f(Q) \end{aligned}$$

and

$$f(kP) = c_1kp_1 + c_2kp_2 + \cdots + c_mkp_m = k(c_1p_1 + c_2p_2 + \cdots + c_mp_m) = k\,f(P)$$

23.14. Prove: Let f be a linear function over $\mathbf{R}^m$ with $f(P) \leq f(Q)$. Then for $0 \leq t \leq 1$, $f(P) \leq f(tP + (1-t)Q) \leq f(Q)$.

Since t and $(1-t)$ are each non-negative, $f(P) \leq f(Q)$ implies

$$t\,f(P) \leq t\,f(Q) \quad \text{and} \quad (1-t)\,f(P) \leq (1-t)\,f(Q)$$

Hence, using the result of the preceding problem, we have

$$f(P) = t\,f(P) + (1-t)\,f(P) \leq t\,f(P) + (1-t)\,f(Q) = f(tP + (1-t)Q)$$

and

$$f(tP + (1-t)Q) = t\,f(P) + (1-t)\,f(Q) \leq t\,f(Q) + (1-t)\,f(Q) = f(Q)$$

23.15. Prove Theorem 23.2: Let f be a linear function defined over a line segment $\overline{PQ}$ in $\mathbf{R}^m$. Then for every point $A \in \overline{PQ}$, either (i) $f(P) \leq f(A) \leq f(Q)$ or (ii) $f(Q) \leq f(A) \leq f(P)$. In particular, if $f(P) = f(Q)$ then f has the same value at every point in $\overline{PQ}$.

The proof follows from the preceding problem and the fact that $A = tP + (1-t)Q$ for some t between 0 and 1.

23.16. Prove Theorem 23.3: Let X be a half space $a_1x_1 + \cdots + a_mx_m \leq b$. Then X is convex: $P, Q \in X$ implies $\overline{PQ} \subset X$.

Now X consists of all the points $A \in \mathbf{R}^m$ such that

$$f(A) \leq b \quad \text{where} \quad f = a_1x_1 + a_2x_2 + \cdots + a_mx_m$$

Hence if $P, Q \in X$, then $f(P) \leq b$ and $f(Q) \leq b$. Thus, by Theorem 23.2, if $A \in \overline{PQ}$ then either

$$f(P) \leq f(A) \leq f(Q) \leq b \quad \text{or} \quad f(Q) \leq f(A) \leq f(P) \leq b$$

In either case $f(A) \leq b$ and so $\overline{PQ} \subset X$.

23.17. Prove Theorem 23.5 (for the case $m = 2$): Let f be a linear function defined over a convex polygon X. Then f takes on its maximum (and minimum) value at a corner point of X.

Suppose that f takes on its largest corner value at the corner point P. Let A be any other point in X. We want to prove that $f(A) \leq f(P)$.

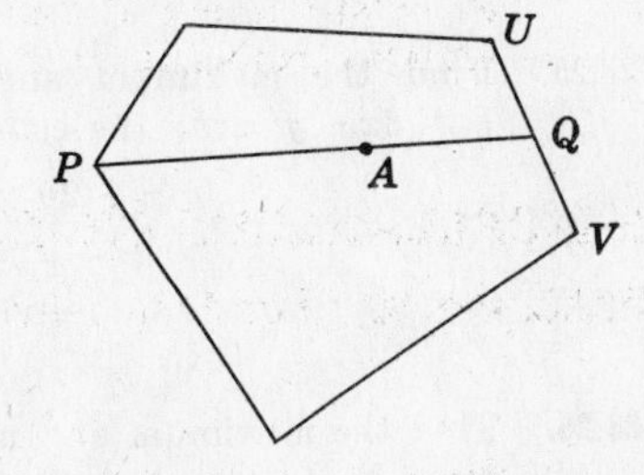

Since X is bounded, the line passing through the points P and A will intersect a bounding line of X at a point Q such that A lies between P and Q: $A \in \overline{PQ}$. Furthermore, Q must lie on the line segment joining two corners, say, U and V: $Q \in \overline{UV}$.

Suppose $f(P) < f(A)$; then by Theorem 23.2, $f(P) < f(A) \leq f(Q)$. That is, $f(A)$ lies between $f(P)$ and $f(Q)$. On the other hand, by Theorem 23.2, either $f(Q) \leq f(U)$ or $f(Q) \leq f(V)$. Combining the various inequalities, we finally have

$$f(P) < f(A) \leq f(Q) \leq f(U) \quad \text{or} \quad f(P) < f(A) \leq f(Q) \leq f(V)$$

However, both cases contradict the assumption that $f(P)$ was the largest corner value. Accordingly, $f(P) < f(A)$ is impossible and so $f(A) \leq f(P)$.

23.18. Prove Theorem 23.5: Let f be a linear function defined over a bounded polyhedral convex set X in $\mathbf{R}^m$. Then f takes on its maximum (and minimum) value at an extreme point of X.

The proof is by induction on m. If $m = 1$, then X is a line segment and by Theorem 23.2 f takes on its maximum (and minimum) value at an extreme point, i.e. an end point.

We assume the theorem holds for a polygonal convex set in $\mathbf{R}^{m-1}$. Let $A \in X$. Since X is bounded, the point A belongs to some line segment $\overline{PQ}$ where P and Q each belongs to a bounding half space of X. By Theorem 23.2, $f(A)$ must lie between $f(P)$ and $f(Q)$ and so f must take on its maximum (and minimum) value at a point on a bounding half space.

Now every point $B \in X$ on a bounding half space of X belongs to the intersection Y of X and a hyperplane H. However, Y is a bounded polyhedral convex set of H and H is isomorphic to $\mathbf{R}^{m-1}$. Thus by induction f takes on its maximum value at an extreme point of Y which is also an extreme point of X.

Supplementary Problems

LINEAR INEQUALITIES IN TWO UNKNOWNS

23.19. Graph each inequality: (i) $x - 2y \leqq 4$, (ii) $3x + y \leqq 6$, (iii) $2x + 5y \geqq 0$, (iv) $3x - 5y \leqq -15$.

23.20. Graph each inequality: (i) $x \leqq 3$, (ii) $x \geqq -4$, (iii) $y \leqq 2$, (iv) $y \geqq -3$.

23.21. Graph each inequality: (i) $-2 \leqq x \leqq 2$, (ii) $0 \leqq y \leqq 3$.

23.22. Plot the solution to each system:

(i) $\begin{aligned} 2x - y &\leqq 4 \\ x &\leqq 1 \end{aligned}$ (ii) $\begin{aligned} x + y &\leqq 5 \\ x - 2y &\geqq 2 \end{aligned}$ (iii) $\begin{aligned} 2x + 3y &\leqq 6 \\ -1 \leqq y &\leqq 2 \end{aligned}$ (iv) $\begin{aligned} -1 \leqq x &\leqq 4 \\ -2 \leqq y &\leqq 1 \end{aligned}$

23.23. Plot the solution to each system:

(i) $\begin{aligned} x - y &\leqq 2 \\ 2x + y &\geqq -2 \\ y &\leqq 1 \end{aligned}$ (ii) $\begin{aligned} 3x + 2y &\leqq 6 \\ x - y &\geqq 1 \\ x &\geqq -1 \end{aligned}$ (iii) $\begin{aligned} x - 2y &\geqq -4 \\ -2 \leqq x &\leqq 1 \\ y &\geqq -3 \end{aligned}$

LINEAR FUNCTIONS OVER POLYHEDRAL CONVEX SETS

23.24. Find the maximum and minimum values of each function over the convex polygon X shown in the diagram on the right: (i) $f = 3x + 2y$; (ii) $g = -x + 2y - 4$; (iii) $h = x - 2y$.

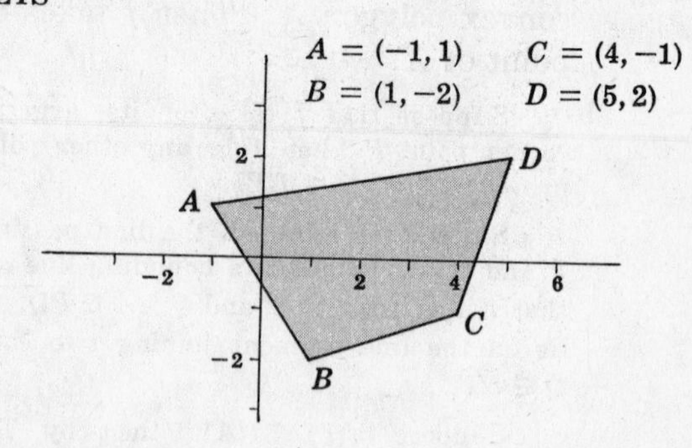

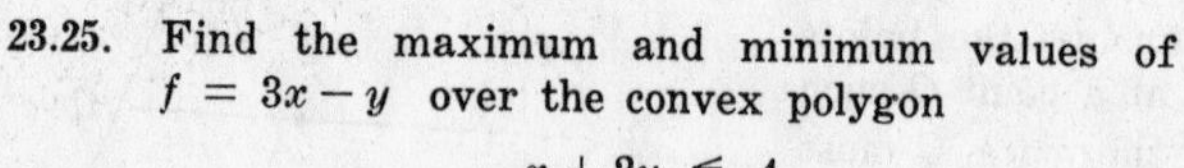

23.25. Find the maximum and minimum values of $f = 3x - y$ over the convex polygon

$$\begin{aligned} x + 2y &\leqq 4 \\ x - y &\leqq 1 \\ x &\geqq -1 \end{aligned}$$

23.26. Find the maximum and minimum values of the function $f = 2x + y + 3$ over the convex polygon given by

$$\begin{aligned} x - y &\geqq 0 \\ 3x + y &\geqq 3 \\ x &\leqq 4 \end{aligned}$$

23.27. Draw the convex polygon X given by

$$\begin{aligned} x + y &\leqq 3 \\ x - y &\geqq -3 \\ y &\geqq 0 \\ x &\geqq -1 \\ x &\leqq 2 \end{aligned}$$

Find the corner points of X and the maximum and minimum values of the function $f = 2x - y$ over X.

23.28. Draw the convex polygon X given by

$$\begin{aligned} x + 2y &\leqq 6 \\ x - y &\geqq 0 \\ x &\leqq 4 \\ y &\geqq 0 \end{aligned}$$

Find the corner points of X and the maximum and minimum values of the function $f = 2x - 3y$ over X.

23.29. Find the extreme points of the following bounded polyhedral convex set X in $\mathbf{R}^3$:

$$\begin{aligned} 3x + 3y + z &\leqq 6 \\ x - y &\geqq 0 \\ x &\geqq 0 \\ y &\geqq 0 \\ z &\geqq 0 \\ z &\leqq 3 \end{aligned}$$

Find the maximum and minimum values of the function $f = x + 2y - z$ over X.

23.30. Draw the polygonal convex set X given by

$$\begin{aligned} x + 2y &\leqq 4 \\ x - y &\geqq -2 \\ x &\leqq 2 \end{aligned}$$

Find the maximum and minimum values (if they exist) of each function over X:

$$f = 2x + y, \quad g = 2x - y, \quad h = x - 3y$$

Answers to Supplementary Problems

23.19.

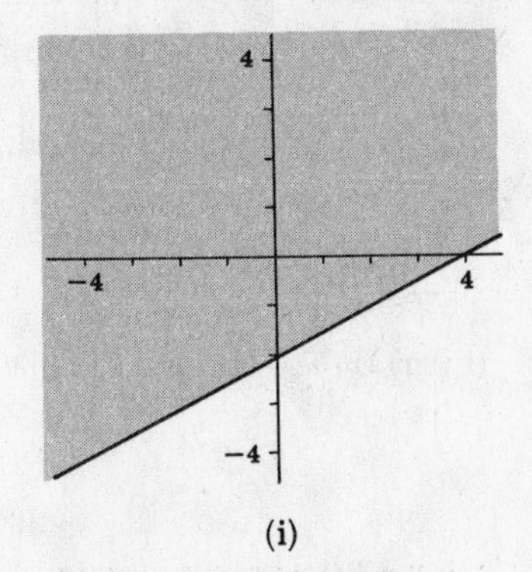

(i)

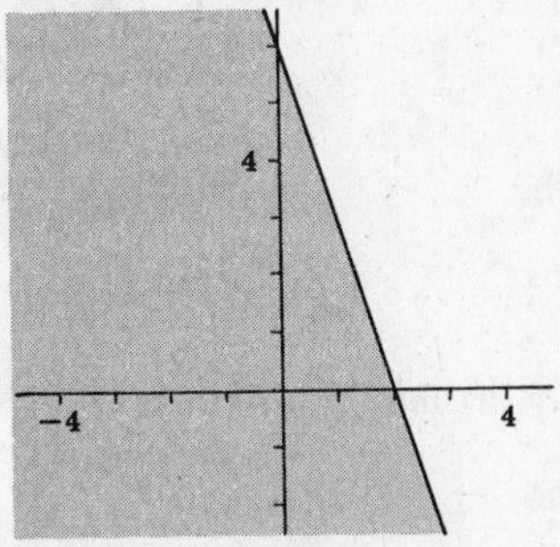

(ii)

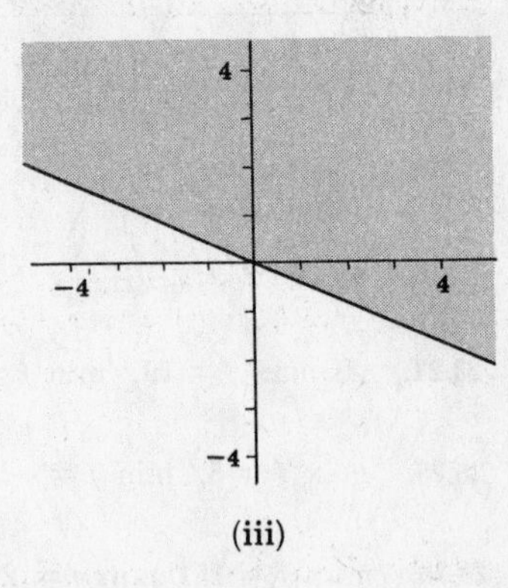

(iii)

23.20.

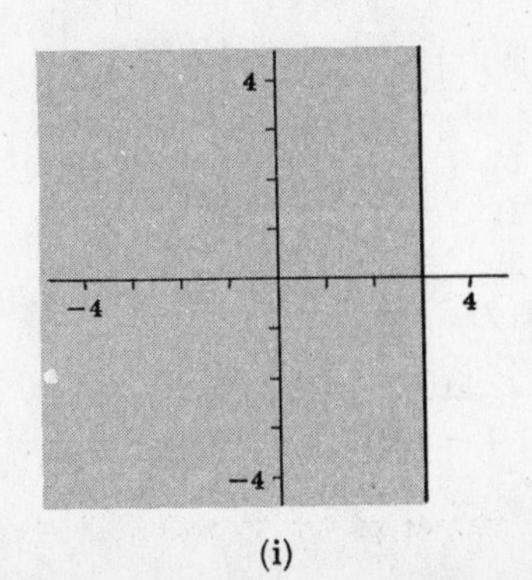

(i)

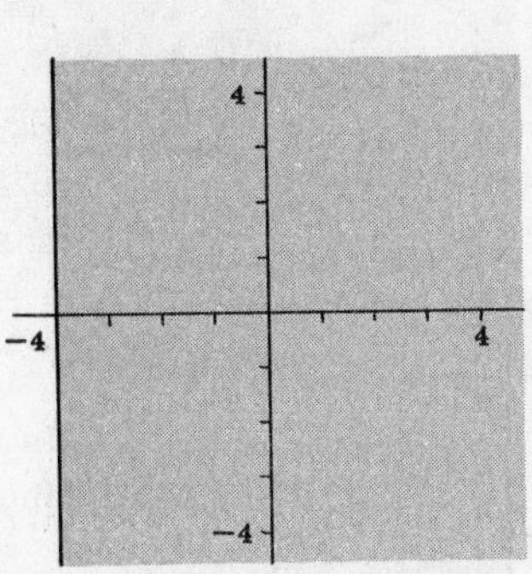

(ii)

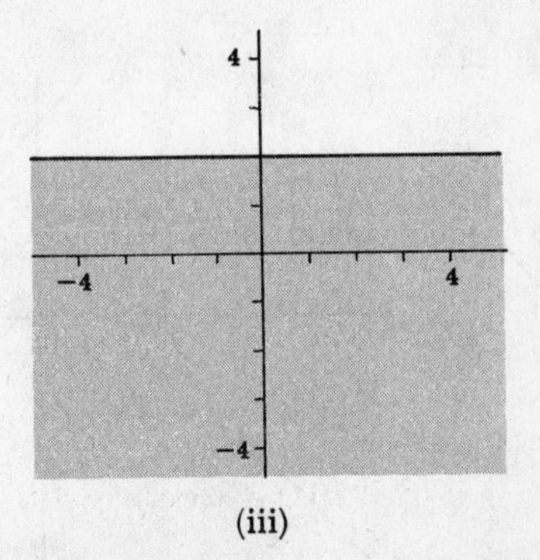

(iii)

23.21.

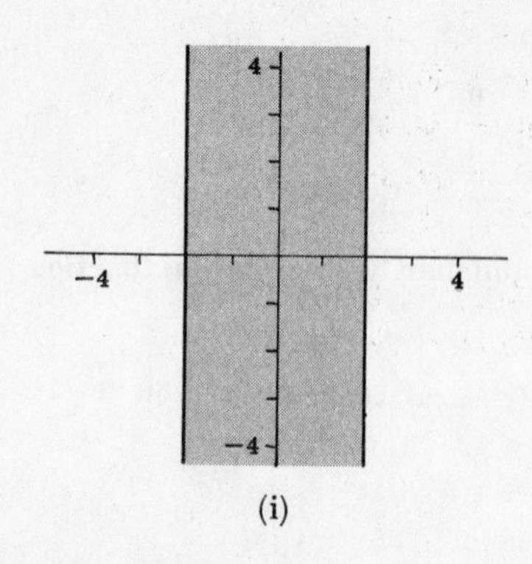
(i)

(ii)

23.22.

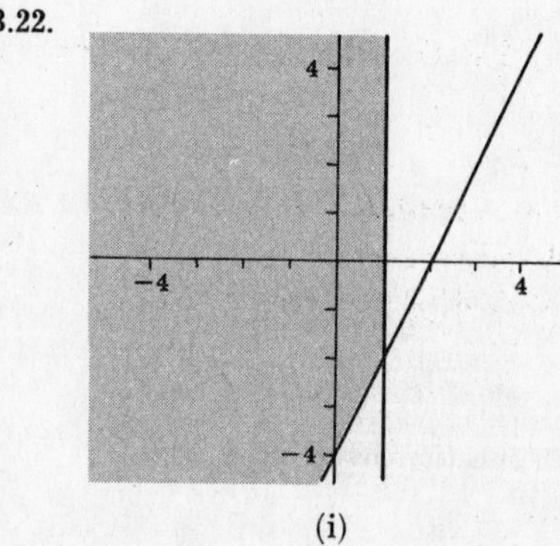
(i)

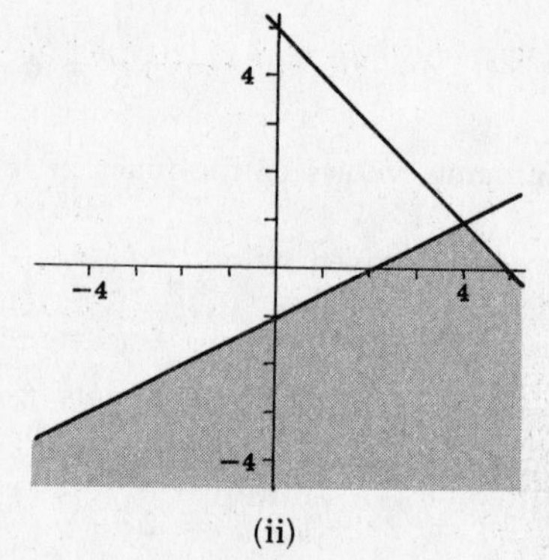
(ii)

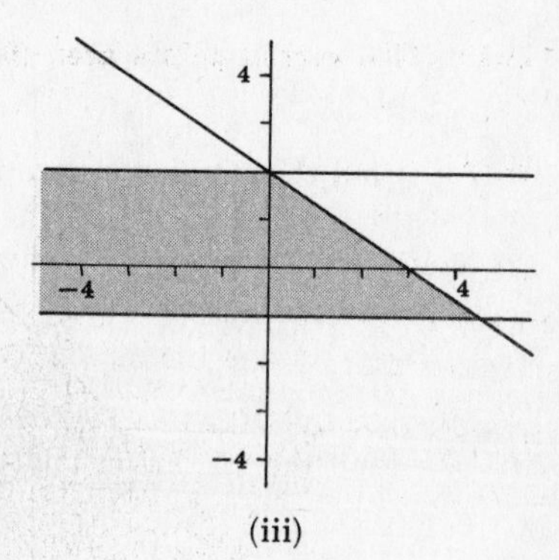
(iii)

23.23.

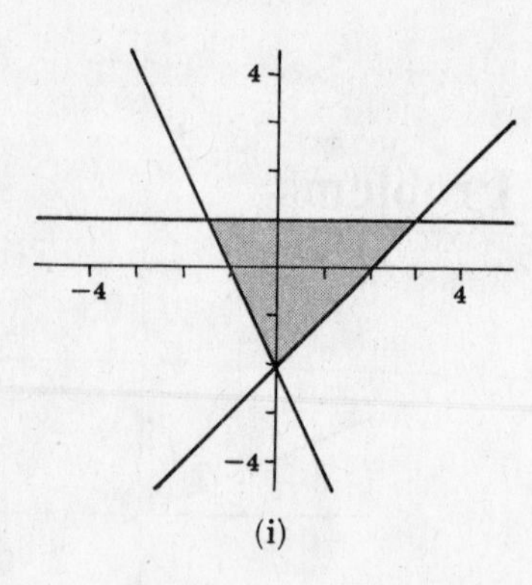
(i)

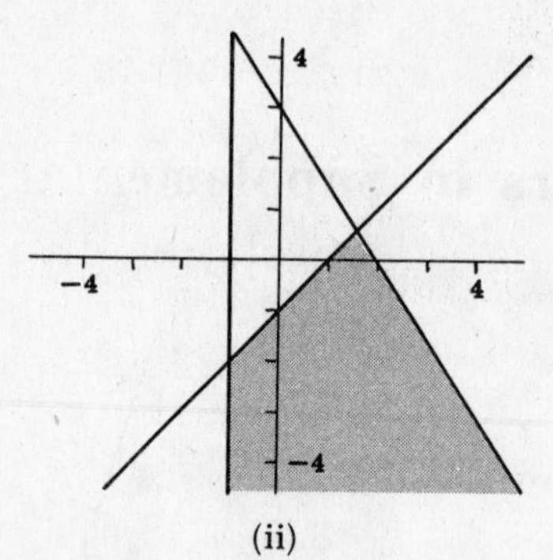
(ii)

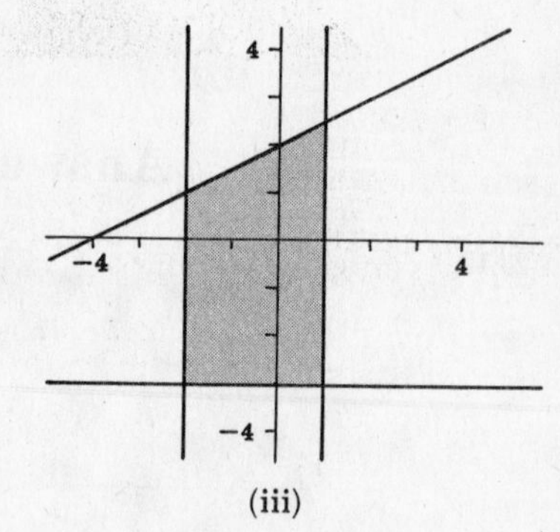
(iii)

23.24. (i) max $f = 19$, min $f = -1$; (ii) max $g = -1$, min $g = -10$; (iii) max $h = 6$, min $h = -3$

23.25. max $f = 5$, min $f = -5.5$

23.26. max $f = 15$, min $f = 2$

23.27.

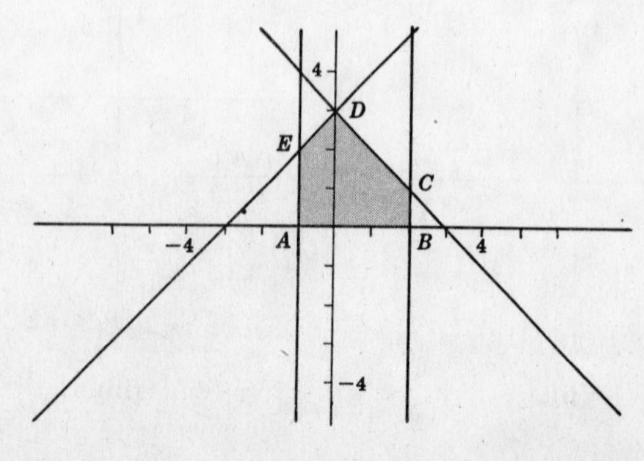

$A = (-1, 0)$
$B = (2, 0)$
$C = (2, 1)$
$D = (0, 3)$
$E = (-1, 2)$

Max $f = 4$, at B.
Min $f = -4$, at E.

23.28.

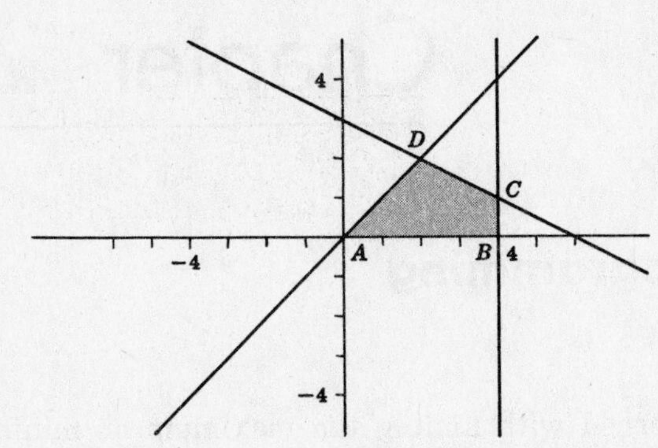

$A = (0, 0)$
$B = (4, 0)$
$C = (4, 1)$
$D = (2, 2)$

Max $f = 8$, at B.
Min $f = -2$, at D.

23.29. The corner points are: $(0, 0, 0)$, $(2, 0, 0)$, $(1, 1, 0)$, $(0, 0, 3)$, $(1, 0, 3)$, $(\frac{1}{2}, \frac{1}{2}, 3)$. Max $f = 4$, min $f = -3$.

23.30.

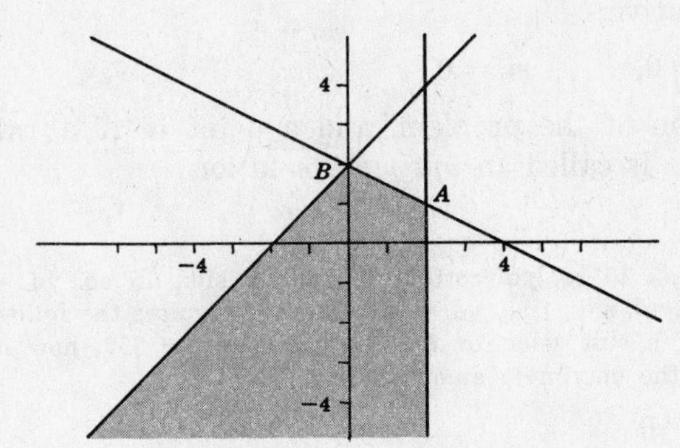

Max $f = 5$ at A; min f does not exist.
Max g and min g do not exist.
Min $h = -6$ at B; max h does not exist.

Chapter 24

Linear Programming

LINEAR PROGRAMMING PROBLEMS

Linear programming problems are concerned with finding the maximum or minimum of a linear function

$$f = c_1x_1 + c_2x_2 + \cdots + c_mx_m$$

called the *objective function*, over a polyhedral convex set X which includes the condition that the unknowns (variables) are non-negative:

$$x_1 \geq 0, \; x_2 \geq 0, \; \ldots, \; x_m \geq 0$$

Each point in X is called a *feasible* solution of the problem, and a point in X at which f takes on its maximum (or minimum) value is called an *optimum* solution.

Example 1.1. (Maximum Problem.)

A tailor has the following materials available: 16 sq. yd. cotton, 11 sq. yd. silk, 15 sq. yd. wool. A suit requires the following: 2 sq. yd. cotton, 1 sq. yd. silk, 1 sq. yd. wool. A gown requires the following: 1 sq. yd. cotton, 2 sq. yd. silk, 3 sq. yd. wool. If a suit sells for \$30 and a gown for \$50, how many of each garment should the tailor make to obtain the maximum amount of money?

The following table summarizes the above facts:

	Suit	Gown	Available
Cotton	2	1	16
Silk	1	2	11
Wool	1	3	15
Price	\$30	\$50	

Let x be the number of suits the tailor makes and y the number of gowns. We seek to maximize

$$f = 30x + 50y$$

subject to $\quad 2x + y \leq 16, \quad x + 2y \leq 11, \quad x + 3y \leq 15, \quad x \geq 0, \; y \geq 0$

Observe that the inequality $2x + y \leq 16$ corresponds to the fact that 2 sq. yd. of cotton is required for each suit and 1 sq. yd. for each gown, but only 16 sq. yd. is available. The inequalities $x + 2y \leq 11$ and $x + 3y \leq 15$ correspond to the analogous facts about the silk and wool.

Fig. 24-1 is the graph of the convex polygon X given by the above inequalities. The corner points of X are

$$O = (0,0), \; A = (8,0), \; B = (7,2), \; C = (3,4), \; D = (0,5)$$

The values of the objective function $f = 30x + 50y$ at the corner points are

$$f(0) = 0, \; f(A) = 240, \; f(B) = 310, \; f(C) = 290, \; f(D) = 250$$

Thus the maximum of f over X is 310 and occurs at the point $B = (7,2)$. In other words, the maximum amount of money the tailor can get is \$310, and this will happen if he makes 7 suits and 2 gowns.

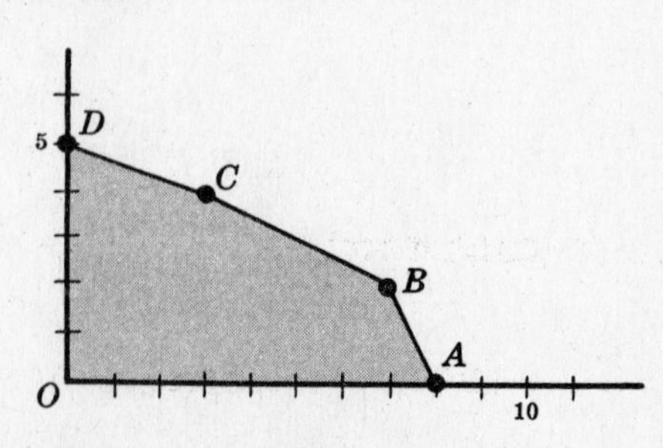

Fig. 24-1

Example 1.2. (Minimum Problem.)

A person requires 10, 12 and 12 units of chemicals A, B and C, respectively, for his garden. A liquid product contains 5, 2 and 1 units of A, B and C, respectively, per jar; and a dry product contains 1, 2 and 4 units of A, B and C, respectively, per carton. If the liquid product sells for \$3 per jar and the dry product sells for \$2 per carton, how many of each should he purchase to minimize the cost and meet the requirements?

The following table summarizes the above facts:

	Units per jar	Units per carton	Units needed
Chemical A	5	1	10
Chemical B	2	2	12
Chemical C	1	4	12
Cost	\$3	\$2	

Let x denote the number of jars bought and y the number of cartons. We seek to minimize the function

$$g = 3x + 2y$$

subject to the conditions

$$5x + y \geq 10, \quad 2x + 2y \geq 12, \quad x + 4y \geq 12$$

and the fact that x and y are non-negative: $x \geq 0$, $y \geq 0$.

Fig. 24-2 is the graph of the polygonal convex set T of this minimum linear programming problem. The corner points of T are

$$P = (0, 10), \quad Q = (1, 5), \quad R = (4, 2), \quad S = (12, 0)$$

Although T is not bounded, the objective function $g = 3x + 2y$ is never negative over T and so takes on a minimum value at a corner point of T. Since

$$g(P) = 20, \quad g(Q) = 13, \quad g(R) = 16, \quad g(S) = 36$$

the minimum value of g over T is 13 and occurs at $Q = (1, 5)$. In other words, he should buy 1 jar and 5 cartons.

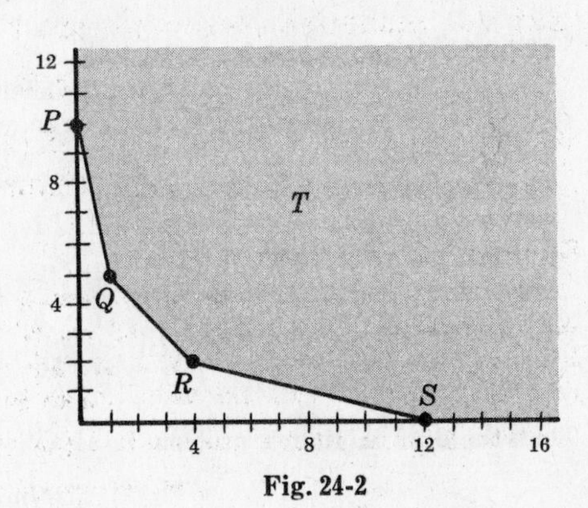

Fig. 24-2

DUAL PROBLEMS

Every maximum linear programming problem has associated with it a corresponding minimum problem called its *dual* and, conversely, every minimum linear programming problem has associated with it a corresponding maximum problem called its *dual*. The original problem is termed the *primal* problem. To be more specific, we define each of the following problems to be the dual of the other:

(i) Maximum problem:

$$\text{Maximize } f = c_1x_1 + c_2x_2 + \cdots + c_mx_m$$

$$\begin{aligned} \text{subject to } \quad & a_{11}x_1 + a_{12}x_2 + \cdots + a_{1m}x_m \leq b_1 \\ & a_{21}x_1 + a_{22}x_2 + \cdots + a_{2m}x_m \leq b_2 \\ & \cdots\cdots\cdots\cdots\cdots\cdots\cdots\cdots \\ & a_{k1}x_1 + a_{k2}x_2 + \cdots + a_{km}x_m \leq b_k \\ \text{and } \quad & x_1 \geq 0, \; x_2 \geq 0, \; \ldots, \; x_m \geq 0. \end{aligned}$$

(ii) Minimum problem:

$$\text{Minimize } g = b_1w_1 + b_2w_2 + \cdots + b_kw_k$$

$$\begin{aligned} \text{subject to } \quad & a_{11}w_1 + a_{21}w_2 + \cdots + a_{k1}w_k \geq c_1 \\ & a_{12}w_1 + a_{22}w_2 + \cdots + a_{k2}w_k \geq c_2 \\ & \cdots\cdots\cdots\cdots\cdots\cdots\cdots\cdots \\ & a_{1m}w_1 + a_{2m}w_2 + \cdots + a_{km}w_k \geq c_m \\ \text{and } \quad & w_1 \geq 0, \; w_2 \geq 0, \; \ldots, \; w_k \geq 0. \end{aligned}$$

$$(1)$$

Let us write the coefficients of the above inequalities in the form of a matrix as follows with the coefficients of the objective function as the last row:

(i) Maximum problem:

$$\begin{pmatrix} a_{11} & a_{12} & \cdots & a_{1m} & b_1 \\ a_{21} & a_{22} & \cdots & a_{2m} & b_2 \\ \cdots & \cdots & \cdots & \cdots & \cdots \\ a_{k1} & a_{k2} & \cdots & a_{km} & b_k \\ c_1 & c_2 & \cdots & c_k & * \end{pmatrix}$$

(ii) Minimum problem:

$$\begin{pmatrix} a_{11} & a_{21} & \cdots & a_{k1} & c_1 \\ a_{12} & a_{22} & \cdots & a_{k2} & c_2 \\ \cdots & \cdots & \cdots & \cdots & \cdots \\ a_{1m} & a_{2m} & \cdots & a_{km} & c_m \\ b_1 & b_2 & \cdots & b_k & * \end{pmatrix}$$

In each case the matrix of coefficients of the dual problem can be obtained by taking the transpose of the matrix of coefficients of the primal problem. (Note that there is no last entry in the last row and last column of either matrix.)

Example 2.1.

Find the dual of the maximum problem of Example 1.1. Write the matrix of coefficients of the given (primal) problem and then take its transpose:

Primal problem

$$\begin{pmatrix} 2 & 1 & 16 \\ 1 & 2 & 11 \\ 1 & 3 & 15 \\ 30 & 50 & * \end{pmatrix}$$

Dual problem

$$\begin{pmatrix} 2 & 1 & 1 & 30 \\ 1 & 2 & 3 & 50 \\ 16 & 11 & 15 & * \end{pmatrix}$$

Thus the dual minimum problem is as follows:

$$\begin{aligned} &\text{Minimize} \quad g = 16u + 11v + 15w \\ &\text{subject to} \quad 2u + v + w \geqq 30 \\ &\qquad\qquad\quad u + 2v + 3w \geqq 50 \\ &\text{and} \qquad\quad u \geqq 0,\ v \geqq 0,\ w \geqq 0. \end{aligned}$$

Example 2.2.

Find the dual of the minimum problem of Example 1.2. Write the matrix of coefficients of the given (primal) problem and then take its transpose:

Primal problem

$$\begin{pmatrix} 5 & 1 & 10 \\ 2 & 2 & 12 \\ 1 & 4 & 12 \\ 3 & 2 & * \end{pmatrix}$$

Dual problem

$$\begin{pmatrix} 5 & 2 & 1 & 3 \\ 1 & 2 & 4 & 2 \\ 10 & 12 & 12 & * \end{pmatrix}$$

Thus the dual maximum problem is as follows:

$$\begin{aligned} &\text{Maximize} \quad f = 10u + 12v + 12w \\ &\text{subject to} \quad 5u + 2v + w \leqq 3 \\ &\qquad\qquad\quad u + 2v + 4w \leqq 2 \\ &\text{and} \qquad\quad u \geqq 0,\ v \geqq 0,\ w \geqq 0. \end{aligned}$$

In general, a linear programming problem need not have an optimum solution. However, the following theorem, which is the main result in the theory of linear programming, applies:

Duality Theorem: The objective function f of a maximum linear programming problem takes on a maximum value if and only if the objective function g of the corresponding dual problem takes on a minimum value and, in this case, $\max f = \min g$. Furthermore, if P and Q are feasible solutions such that $f(P) = g(Q)$, then P and Q are optimum solutions of their corresponding problems.

The proof of this theorem lies beyond the scope of this text.

Example 2.3.

Consider the maximum problem of Example 2.2. The graph of the polyhedral convex set X given by the inequalities of that problem appears on the right. The corner points are

$$P = (10/19, 0, 7/19)$$
$$Q = (3/5, 0, 0)$$
$$R = (1/4, 7/8, 0)$$
$$S = (0, 1, 0)$$
$$T = (0, 0, \tfrac{1}{2})$$
$$O = (0, 0, 0)$$

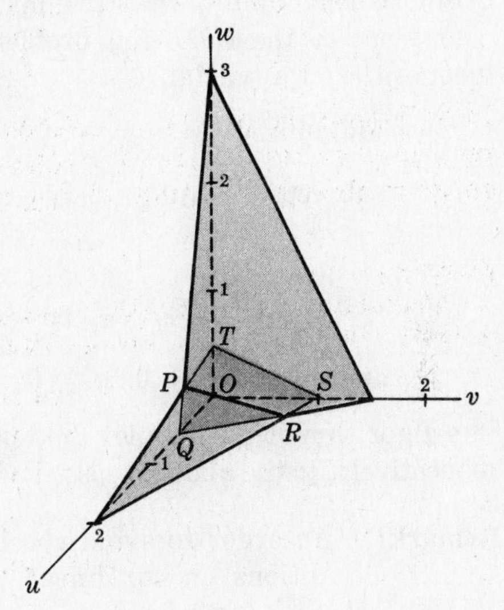

The objective function $f = 10u + 12u + 12w$ takes on its maximum value at the point R and $\max f = 13$. On the other hand, $\min g = 13$ for the corresponding minimum problem appearing in Example 1.2, as predicted by the Duality Theorem.

MATRIX NOTATION

Let the matrix A, the column vectors b and x, and the row vectors c and w be defined as follows:

$$A = \begin{pmatrix} a_{11} & a_{12} & \cdots & a_{1m} \\ a_{21} & a_{22} & \cdots & a_{2m} \\ \cdots & \cdots & \cdots & \cdots \\ a_{k1} & a_{k2} & \cdots & a_{km} \end{pmatrix}, \quad b = \begin{pmatrix} b_1 \\ b_2 \\ \cdot \\ b_k \end{pmatrix}, \quad x = \begin{pmatrix} x_1 \\ x_2 \\ \cdot \\ x_m \end{pmatrix}, \quad \begin{array}{l} c = (c_1, c_2, \ldots, c_m) \\ w = (w_1, w_2, \ldots, w_k) \end{array}$$

Then the maximum and minimum linear programming problems (*1*) in the preceding section can be rewritten in the following short compact form:

(i) Maximum problem:		(ii) Minimum problem:		
Maximize	$f = cx$	Minimize	$g = wb$	
subject to	$Ax \leq b$	subject to	$wA \geq c$	(*1*)
and	$x \geq 0$.	and	$w \geq 0$.	

Remark: If u and v are vectors, then $u \leq v$ shall mean that each component of u is less than or equal to the corresponding component of v.

INTRODUCTION TO THE SIMPLEX METHOD

Let X be the polyhedral convex set, i.e. set of feasible solutions, of the above maximum linear programming problem. An optimum solution can be found by evaluating the objective function f at each corner point of X. However, if the number of unknowns and inequalities is large, then it can be a long and tedious task to find all the corners of X. An alternate way of finding an optimum solution is the following:

(1) Select a corner of X as a starting point.

(2) Choose an edge through this corner such that f increases in value along the edge.

(3) Proceed along the chosen edge to the next corner.

(4) Repeat steps (2) and (3) until an optimum solution is obtained, i.e. a corner is reached such that f does not increase in value along any of its edges.

The above process is accomplished mathematically by the so-called simplex method. Observe, first of all, that the maximum linear programming problem (*1*), Page 301, is equivalent to the following problem which contains only linear equalities and not linear inequalities; the variables $x_{m+1}, \ldots, x_{m+k}$ are called "slack" variables:

$$\text{Maximize } f = c_1x_1 + c_2x_2 + \cdots + c_mx_m + 0 \cdot x_{m+1} + 0 \cdot x_{m+2} + \cdots + 0 \cdot x_{m+k}$$

$$\begin{array}{lllr} \text{subject to} & a_{11}x_1 + a_{12}x_2 + \cdots + a_{1m}x_m + x_{m+1} & = b_1 & \\ & a_{21}x_1 + a_{22}x_2 + \cdots + a_{2m}x_m \qquad + x_{m+2} & = b_2 & \\ & \cdots\cdots\cdots\cdots\cdots\cdots\cdots\cdots\cdots\cdots\cdots\cdots & & \\ & a_{k1}x_1 + a_{k2}x_2 + \cdots + a_{km}x_m \qquad\qquad + x_{m+k} & = b_k & (2) \\ \text{and} & x_1 \geqq 0,\ x_2 \geqq 0,\ \ldots,\ x_m \geqq 0,\ x_{m+1} \geqq 0,\ \ldots,\ x_{m+k} \geqq 0. & & \end{array}$$

The algorithm of the simplex method consists of four steps (see Page 307) which correspond, respectively, to the above steps. For the sake of clarity we shall explain each step separately.

Remark: In order to avoid special cases, we shall assume the following additional conditions on our linear programming problems:

(i) *Positive condition*: Each component of b is non-negative ($b \geqq 0$) and each row of A contains at least one positive entry.

(ii) *Non-degeneracy condition*: If r is the rank of A, i.e. the number of independent rows (or columns) of A, then c is not a linear combination of less than r rows of A and b is not a linear combination of less than r columns of A.

INITIAL SIMPLEX TABLEAU

The initial simplex tableau of the maximum linear programming problem (*1*) or the equivalent problem (*2*) is as follows:

P_1	P_2	$\cdots$	P_m					
a_{11}	a_{12}	$\cdots$	a_{1m}	1	0	$\cdots$	0	b_1
a_{21}	a_{22}	$\cdots$	a_{2m}	0	1	$\cdots$	0	b_2
$\cdots$	$\cdots$	$\cdots$	$\cdots$	$\cdots$	$\cdots$	$\cdots$	$\cdots$	$\cdot$
a_{k1}	a_{k2}	$\cdots$	a_{km}	0	0	$\cdots$	1	b_k
$-c_1$	$-c_2$	$\cdots$	$-c_k$	0	0	$\cdots$	0	0

indicators

Observe that the rows of the tableau, excluding the last, are the coefficients of the linear equalities in (*2*). The columns of the coefficients of the non-slack variables are labeled P_i, respectively; their significance will appear later. The entries in the last row, excluding the last entry, are called *indicators* and are the negatives of the coefficients of the objective function f in (*2*). The entry in the lower right hand corner is zero; it corresponds to the value of the objective function f at the origin.

Remark: Although the simplex method is stated in terms of the maximum linear programming problem, the method will give the optimum solution to both the maximum problem and to its dual minimum problem.

Example 3.1.

Maximize $f = 30x + 50y$

subject to
$$2x + y \leqq 16$$
$$x + 2y \leqq 11$$
$$x + 3y \leqq 15$$
and $x \geqq 0,\ y \geqq 0$

Initial simplex tableau:

P_1	P_2				
2	1	1	0	0	16
1	2	0	1	0	11
1	3	0	0	1	15
−30	−50	0	0	0	0

Example 3.2.

Maximize $f = 10u + 12v + 12w$

subject to
$$5u + 2v + w \leqq 3$$
$$u + 2v + 4w \leqq 2$$
and $u \geqq 0,\ v \geqq 0,\ w \geqq 0$

Initial simplex tableau:

P_1	P_2	P_3			
5	2	1	1	0	3
1	2	4	0	1	2
−10	−12	−12	0	0	0

Example 3.3.

Maximize $f = x + 2y + 8z$

subject to
$$z \leqq 1$$
$$x + y + z \leqq 2$$
$$-x + y \leqq 0$$
and $x \geqq 0,\ y \geqq 0,\ z \geqq 0$

Initial simplex tableau:

P_1	P_2	P_3				
0	0	1	1	0	0	1
1	1	1	0	1	0	2
−1	1	0	0	0	1	0
−1	−2	−8	0	0	0	0

PIVOT ENTRY OF A SIMPLEX TABLEAU

A pivot entry or simply a pivot of a simplex tableau is obtained as follows:

(i) Arbitrarily select a column which contains a negative indicator.

(ii) Divide each positive entry in this column *into* the corresponding element in the last column.

The entry which yields the smallest quotient in step (ii) is the pivot; its column is called the pivotal column and its row the pivotal row. If the tableau has no negative indicator, then it is a *terminal tableau* and has no pivot. Properly interpreted, the terminal tableau yields a solution to our problem. The procedure which follows represents a method of passing to such a terminal configuration.

Example 4.1.

Consider the following simplex tableaux:

P_1	P_2				
1	2	1	0	0	12
(2)	−1	0	1	0	3
3	(1)	0	0	1	5
−30	−50	0	0	0	0

(i)

P_1	P_2	P_3				
0	0	1	1	0	0	1
1	(1)	0	−1	1	0	2
0	2	0	−1	1	1	5
0	−1	0	7	1	0	16

(ii)

P_1	P_2				
0	0	1	−5	(3)	6
1	0	0	3	−2	3
0	1	0	−1	1	4
0	0	0	40	−10	290

(iii)

P_1	P_2	P_3				
0	0	1	1	0	0	1
0	0	0	$-\frac{1}{2}$	$\frac{1}{2}$	0	$\frac{1}{2}$
0	1	0	$-\frac{1}{2}$	$\frac{1}{2}$	$\frac{1}{2}$	$\frac{1}{2}$
0	0	0	13	3	3	19

(iv)

(i) Either the first column or the second column can be chosen as the pivotal column. In the first case 2 is the pivot since 3/2 is less than 12/1 and 5/3. In the second case 1 is the pivot since 5/1 is less than 12/2. We need not consider the quotient 3/(−1) since −1 is not positive.

(ii) Here the second column must be chosen as the pivotal column since it is the only one with a negative indicator. Furthermore, 1 is the pivot since 2/1 is less than 5/2.

(iii) The fifth column is the pivotal column and 3 is the pivot since 6/3 is less than 4/1.

(iv) There is no pivotal column and no pivot since there is no negative indicator. This tableau is a terminal tableau.

CALCULATING THE NEW SIMPLEX TABLEAU

Let D be a simplex tableau with rows R_i whose pivot appears in the rth row and sth column; say D is the matrix (d_{ij}) and so d_{rs} is the pivot. A new tableau D^* with rows R_i^* is calculated from D (using appropriate "elementary row operations") so that 1 appears in the original pivot position and zeros elsewhere in that column. This is accomplished as follows:

(i) First divide each entry of the pivotal row R_r of D by the pivot d_{rs} to obtain the corresponding row R_r^* of D^*:

$$R_r^* \;=\; \frac{1}{d_{rs}}R_r$$

If the pivotal column of D is labeled P_i, then this row of D^* is labeled (or relabeled) P_i.

(ii) Every other row R_i^* of D^* is obtained by adding the appropriate multiple of the above row R_r^* to the row R_i of D:

$$R_i^* \;=\; -d_{is}R_r^* + R_i$$

If one of these rows of D is labeled P_i, then the corresponding row of D^* is given the same label; otherwise the row is not labeled.

Example 5.1.

Consider the following tableau with a pivot entry circled:

	P_1	P_2				
R_1:	1	2	1	0	0	22
R_2:	(2)	−1	0	1	0	4
R_3:	4	1	0	0	1	9
R_4:	−8	−6	0	0	0	0

We remark that the circled 2 is a pivot entry since it appears in a column with a negative indicator and 4/2 is less than both 22/1 and 9/4. We now calculate the new tableau step by step:

(i) Divide each entry of the pivotal row R_2 by the pivot 2 to obtain the new row R_2^*; label R_2^* with the label of the pivotal column, i.e. P_1:

R_1^*:							
R_2^*:	P_1	1	$-\frac{1}{2}$	0	$\frac{1}{2}$	0	2
R_3^*:							
R_4^*:							

(ii) R_1^* is obtained by multiplying the above row R_2^* by -1 and adding it to R_1: $R_1^* = -R_2^* + R_1$; R_3^* is obtained by multiplying R_2^* by -4 and adding it to R_3: $R_3^* = -4R_2^* + R_3$; R_4^* is obtained by multiplying R_2^* by 8 and adding it to R_4: $R_4^* = 8R_2^* + R_4$. These rows do not change labels. Thus the completed new tableau is

	P_1	P_2				
	0	$\frac{5}{2}$	1	$-\frac{1}{2}$	0	20
P_1	1	$-\frac{1}{2}$	0	$\frac{1}{2}$	0	2
	0	3	0	-2	1	1
	0	-10	0	4	0	16

The intermediate tableau appearing in the preceding example was illustrated separately from the completed tableau only for the sake of clarity; in general, we calculate the new tableau directly from the given tableau. We remark that if we choose the second column as the pivotal column and so obtain a different pivot, then we would calculate a different new tableau. This corresponds to the fact that in moving from corner to corner there may be more than one edge on which the objective function f increases in value; hence there is more than one way to finally reach the optimum solution.

The simplex method begins with the initial tableau and then calculates new tableaux one after the other until a terminal tableau is obtained. We remark that in each step at most one row, the pivotal row, changes its label; the columns keep the same label throughout.

Example 5.2.

Consider the following tableau with a pivot entry circled:

	P_1	P_2				
	0	3	1	$-\frac{1}{2}$	0	20
P_1	1	$-\frac{1}{2}$	0	$\frac{1}{2}$	0	2
	0	(2)	0	2	1	1
	0	-10	0	4	0	16

(*)

Divide each entry of the pivotal row by the pivot 2 to obtain the row

$$0 \quad 1 \quad 0 \quad 1 \quad \tfrac{1}{2} \quad \tfrac{1}{2} \qquad (**)$$

The new tableau is

	P_1	P_2				
	0	0	1	$-\frac{7}{2}$	$-\frac{3}{2}$	$\frac{37}{2}$
P_1	1	0	0	1	$\frac{1}{4}$	$\frac{9}{4}$
P_2	0	1	0	1	$\frac{1}{2}$	$\frac{1}{2}$
	0	0	0	14	5	21

It is calculated as follows:

(i) the first row is obtained by multiplying (**) by -3 and adding the result to the first row of (*);

(ii) the second row is obtained by multiplying (**) by $\frac{1}{2}$ and adding the result to the second row of (*); it retains the label P_1;

(iii) the third row is (**); it is labeled P_2, the label of the pivotal column of (*);

(iv) the fourth row is obtained by multiplying (**) by 10 and adding the result to the fourth row of (*).

Note that the new tableau is a terminal tableau, i.e. it has no negative indicators.

INTERPRETING THE TERMINAL TABLEAU

Suppose after performing the algorithm of the simplex method on the linear programming problem (*1*), Page 301, we obtain the following terminal tableau:

	P_1	P_2	$\cdots$	P_m					
some labeled with P_i	*	*	...	*	*	*	...	*	b_1^*
	*	*	...	*	*	*	...	*	b_2^*
	.....				.....				.
	*	*	...	*	*	*	...	*	b_k^*
	*	*	...	*	q_1	q_2	$\cdots$	q_k	v

Then: (i) v, the entry in the lower right hand corner of the tableau, is the maximum value of f (and so by the duality theorem is the minimum value of g: $\max f = \min g = v$);

(ii) the point $Q = (q_1, q_2, \ldots, q_k)$ is the optimum solution of the minimum problem; its components are the last entries in the columns corresponding to the slack variables;

(iii) the point $P = (p_1, p_2, \ldots, p_m)$, where

$$p_i = \begin{cases} \text{the last entry of a row labeled } P_i \\ 0 \text{ if no row is labeled } P_i \end{cases}$$

is the optimum solution of the maximum problem.

Thus a necessary condition (and so a check) that the calculations are correct is that $f(P) = g(Q) = v$.

Example 6.1.

Find the optimum solution $Q = (q_1, \ldots, q_k)$ of the minimum problem, the optimum solution $P = (p_1, \ldots, p_m)$ of the maximum problem, and $v = \max f = \min g$ according to each of the following terminal tableaux:

	P_1	P_2	P_3				
P_3	*	*	*	*	*	*	6
	*	*	*	*	*	*	1
P_2	*	*	*	*	*	*	2
	*	*	*	3	4	5	19

(i)

	P_1	P_2	P_3			
P_2	*	*	*	*	*	7
	*	*	*	*	*	4
	*	*	*	2	5	11

(ii)

(i) $Q = (3, 4, 5)$, $P = (0, 2, 6)$, $v = \max f = \min g = 19$. Note that the first component of P is 0 since no row is labeled P_1.

(ii) $Q = (2, 5)$, $P = (0, 7, 0)$, $v = \max f = \min g = 11$. Note that the first and third components of P are each 0 since no row is labeled P_1 or P_3.

ALGORITHM OF THE SIMPLEX METHOD

The algorithm of the simplex method consists of four steps which correspond, respectively, to the steps on Page 304:

(1) Construct the initial tableau.

(2) Find and circle a pivot entry in the tableau.

(3) Calculate the new tableau from the given tableau and circled pivot.

(4) Repeat steps (2) and (3) until a terminal tableau is obtained.

The following examples illustrate the use of this algorithm.

Example 7.1.

Applying the simplex method to Example 1.1, Page 300, we obtain the following sequence of tableaux:

	P_1	P_2				
	2	1	1	0	0	16
	1	2	0	1	0	11
	1	3	0	0	1	15
	−30	−50	0	0	0	0

(i)

	P_1	P_2				
	$\frac{5}{3}$	0	1	0	$-\frac{1}{3}$	11
	$\frac{1}{3}$	0	0	1	$-\frac{2}{3}$	1
P_2	$\frac{1}{3}$	1	0	0	$\frac{1}{3}$	5
	$-\frac{40}{3}$	0	0	0	$\frac{50}{3}$	250

(ii)

	P_1	P_2				
	0	0	1	−5	3	6
P_1	1	0	0	3	−2	3
P_2	0	1	0	−1	1	4
	0	0	0	40	−10	290

(iii)

	P_1	P_2				
	0	0	$\frac{1}{3}$	$-\frac{5}{3}$	1	2
P_1	1	0	$\frac{2}{3}$	$-\frac{1}{3}$	0	7
P_2	0	1	$-\frac{1}{3}$	$\frac{2}{3}$	0	2
	0	0	$\frac{10}{3}$	$\frac{70}{3}$	0	310

(iv)

Thus, by the terminal tableau (iv), the optimum solution is $P = (7, 2)$ since 7 and 2 are the last entries in the rows labeled P_1 and P_2 respectively, and the maximum value of f is 310. This agrees with the previous solution of the problem. Observe that the numbers 0, 250, 290 and 310 which appear respectively in the lower right hand corners of the tableaux correspond precisely to the values of the objective function f at the points O, D, C and B in Fig. 24-1.

Example 7.2.

We solve the problem in Example 1.2, Page 301, by the simplex method; the first tableau is the initial tableau of the dual maximum problem:

	P_1	P_2	P_3			
	5	2	1	1	0	3
	1	2	4	0	1	2
	−10	−12	−12	0	0	0

(i)

	P_1	P_2	P_3			
	4	0	−3	1	−1	1
P_2	$\frac{1}{2}$	1	2	0	$\frac{1}{2}$	1
	−4	0	12	0	6	12

(ii)

	P_1	P_2	P_3			
P_1	1	0	$-\frac{3}{4}$	$\frac{1}{4}$	$-\frac{1}{4}$	$\frac{1}{4}$
P_2	0	1	$\frac{19}{8}$	$-\frac{1}{8}$	$\frac{5}{8}$	$\frac{7}{8}$
	0	0	9	1	5	13

(iii)

The optimum solution of the given (minimum) problem is, by the terminal tableau (iii), the point $Q = (1, 5)$ and the minimum value of the objective function g is 13. This agrees with the previous solution.

Remark: The rows and columns of the tableaux in the simplex method are labeled so that we can obtain the optimum solution of the maximum problem from the terminal tableau. However, if only the optimum solution of the minimum problem is required, then we can dispense with the labels. For example, the labels in the preceding example were not needed.

Solved Problems

DUALITY

24.1. Find the dual of each problem.

(i) Maximize $f = 3x + 5y + 2z$

subject to
$$2x - y + 3z \leqq 6$$
$$x + 2y + 4z \leqq 8$$
and $x \geqq 0,\ y \geqq 0,\ z \geqq 0.$

(ii) Minimize $g = 2x + 6y + 7z$

subject to
$$x + 2y + 5z \geqq 4$$
$$2x - y + 2z \geqq 3$$
$$3x + 5y + z \geqq 1$$
and $x \geqq 0,\ y \geqq 0,\ z \geqq 0.$

In each case form the matrix of coefficients of the given problem with the objective function as the last row, and then form the transpose matrix.

(i) Coefficient matrix:
$$\begin{pmatrix} 2 & -1 & 3 & 6 \\ 1 & 2 & 4 & 8 \\ 3 & 5 & 2 & * \end{pmatrix}$$
Transpose:
$$\begin{pmatrix} 2 & 1 & 3 \\ -1 & 2 & 5 \\ 3 & 4 & 2 \\ 6 & 8 & * \end{pmatrix}$$
Dual problem:

Minimize $g = 6u + 8v$

subject to
$$2u + v \geqq 3$$
$$-u + 2v \geqq 5$$
$$3u + 4v \geqq 2$$
and $u \geqq 0,\ v \geqq 0.$

(ii) Coefficient matrix:
$$\begin{pmatrix} 1 & 2 & 5 & 4 \\ 2 & -1 & 2 & 3 \\ 3 & 5 & 1 & 1 \\ 2 & 6 & 7 & * \end{pmatrix}$$
Transpose:
$$\begin{pmatrix} 1 & 2 & 3 & 2 \\ 2 & -1 & 5 & 6 \\ 5 & 2 & 1 & 7 \\ 4 & 3 & 1 & * \end{pmatrix}$$
Dual problem:

Maximize $f = 4u + 3v + w$

subject to
$$u + 2v + 3w \leqq 2$$
$$2u - v + 5w \leqq 6$$
$$5u + 2v + w \leqq 7$$
and $u \geqq 0,\ v \geqq 0,\ w \geqq 0.$

24.2. Find the dual of each problem.

(i) Maximize $f = 5x - y + 2z$

subject to
$$7x + 2y - z \leqq 8$$
$$x - 3y - 2z \geqq 4$$
$$3x - y + 6z \leqq 5$$
and $x \geqq 0,\ y \geqq 0,\ z \geqq 0.$

(ii) Minimize $g = 2r + 5s + t$

subject to
$$9r - 3s + 5t \geqq 7$$
$$6r - 4s - 3t \leqq 2$$
and $r \geqq 0,\ s \geqq 0,\ t \geqq 0.$

(i) First multiply the second inequality by -1 to change its sense to $\leqq$.

Coefficient matrix:
$$\begin{pmatrix} 7 & 2 & -1 & 8 \\ -1 & 3 & 2 & -4 \\ 3 & -1 & 6 & 5 \\ 5 & -1 & 2 & * \end{pmatrix}$$
Transpose:
$$\begin{pmatrix} 7 & -1 & 3 & 5 \\ 2 & 3 & -1 & -1 \\ -1 & 2 & 6 & 2 \\ 8 & -4 & 5 & * \end{pmatrix}$$
Dual problem:

Minimize $g = 8u - 4v + 5w$

subject to
$$7u - v + 3w \geqq 5$$
$$2u + 3v - w \geqq -1$$
$$-u + 2v + 6w \geqq 2$$
and $u \geqq 0,\ v \geqq 0,\ w \geqq 0.$

(ii) First multiply the second inequality by -1 to change its sense to $\geqq$.

Coefficient matrix:
$$\begin{pmatrix} 9 & -3 & 5 & 7 \\ -6 & 4 & 3 & -2 \\ 2 & 5 & 1 & * \end{pmatrix}$$
Transpose:
$$\begin{pmatrix} 9 & -6 & 2 \\ -3 & 4 & 5 \\ 5 & 3 & 1 \\ 7 & -2 & * \end{pmatrix}$$
Dual problem:

Maximize $f = 7x - 2y$

subject to
$$9x - 6y \leqq 2$$
$$-3x + 4y \leqq 5$$
$$5x + 3y \leqq 1$$
and $x \geqq 0,\ y \geqq 0.$

SIMPLEX METHOD

24.3. Find the initial tableau of each problem.

(i) Maximize $f = 5x + 2y - z$

subject to
$$x - 3y + 4z \leqq 8$$
$$3x + 2y + 6z \leqq 5$$
$$2x + 7y - z \leqq 9$$
and $x \geqq 0,\ y \geqq 0,\ z \geqq 0.$

(ii) Minimize $g = 2r + 5s + t$

subject to
$$9r + 2s + 4t \geqq 6$$
$$6r + 3s - 2t \geqq 8$$
and $r \geqq 0,\ s \geqq 0,\ t \geqq 0.$

(i) The initial tableau is of the form

A	I	b
$-c$	0	0

with the columns of A labeled P_i.

P_1	P_2	P_3				
1	−3	4	1	0	0	8
3	2	6	0	1	0	5
2	7	−1	0	0	1	9
−5	−2	1	0	0	0	0

(ii) Find the dual maximum problem and then construct the initial tableau as in (i).

P_1	P_2				
9	6	1	0	0	2
2	3	0	1	0	5
4	−2	0	0	1	1
−6	−8	0	0	0	0

24.4. Find a pivot entry of each of the initial tableaux in the preceding problem.

In each case, (1) select any column with a negative indicator as the pivotal column, (2) divide each positive entry in this column into the corresponding entry in the last column, and (3) choose as the pivot that entry which yields the smallest quotient in (2).

(i) We can choose either the first or second column as the pivotal column.

(*a*) If the first column is chosen, form the quotients 8/1, 5/3 and 9/2. Since 5/3 is the smallest number of the three, the entry 3 in the second row is the pivot.

(*b*) If the second column is chosen, form the quotients 5/2 and 9/7. Since 9/7 is the smaller of the two numbers, the entry 7 in the third row is the pivot. Note that the quotient 8/(−3) is not considered since −3 is not a positive number.

(ii) We can choose either the first or the second column as the pivotal column. In the first column, 9 is the pivot since 2/9 is the smallest of the numbers 2/9, 5/2 and 1/4. In the second column, 6 is the pivot since 2/6 is the smaller of the numbers 2/6 and 5/3.

24.5. Using the pivot circled in the following tableau, calculate the new tableau and then find a pivot entry of it.

P_1	P_2				
2	1	1	0	0	7
(4)	−1	0	1	0	4
4	3	0	0	1	5
−8	−5	0	0	0	0

Divide each entry in the pivotal row, i.e. the second row, by the pivot 4 to obtain the row

$$1 \quad -\tfrac{1}{4} \quad 0 \quad \tfrac{1}{4} \quad 0 \quad 1 \tag{1}$$

Multiply (1) by -2 and add it to the first row to obtain

$$0 \quad \tfrac{3}{2} \quad 1 \quad -\tfrac{1}{2} \quad 0 \quad 5$$

Multiply (1) by -4 and add it to the third row to obtain

$$0 \quad 4 \quad 0 \quad -1 \quad 1 \quad 1$$

Multiply (1) by 8 and add it to the last row of the tableau to obtain

$$0 \quad -7 \quad 0 \quad 2 \quad 0 \quad 8$$

Thus the new tableau is

	P_1	P_2				
	0	$\frac{3}{2}$	1	$-\frac{1}{2}$	0	5
P_1	1	$-\frac{1}{4}$	0	$\frac{1}{4}$	0	1
	0	(4)	0	-1	1	1
	0	-7	0	2	0	8

Note that (1), the row corresponding to the original pivotal row, is labeled P_1 which is the label of the pivotal column. Note that 1 now appears in the original pivot position and zeros elsewhere in the column.

We must choose the second column as the pivotal column since it is the only column with a negative indicator. Since 1/4 is less than $5/(3/2) = 10/3$, the 4 is the pivot.

24.6. Using the tableau and the pivot obtained in the preceding problem, calculate the new tableau.

Divide each entry in the pivotal row, i.e. the third row, by the pivot 4 to obtain the row

$$0 \quad 1 \quad 0 \quad -\tfrac{1}{4} \quad \tfrac{1}{4} \quad \tfrac{1}{4} \tag{1}$$

Multiply (1) by $-3/2$ and add the result to the first row to obtain

$$0 \quad 0 \quad 1 \quad -\tfrac{1}{8} \quad -\tfrac{3}{8} \quad \tfrac{37}{8}$$

Multiply (1) by $\frac{1}{4}$ and add the result to the second row to obtain

$$1 \quad 0 \quad 0 \quad \tfrac{3}{16} \quad \tfrac{1}{16} \quad \tfrac{17}{16}$$

Multiply (1) by 7 and add the result to the last row to obtain

$$0 \quad 0 \quad 0 \quad \tfrac{1}{4} \quad \tfrac{7}{4} \quad \tfrac{39}{4}$$

Thus the calculated tableau is

	P_1	P_2				
	0	0	1	$-\frac{1}{8}$	$-\frac{3}{8}$	$\frac{37}{8}$
P_1	1	0	0	$\frac{3}{16}$	$\frac{1}{16}$	$\frac{17}{16}$
P_2	0	1	0	$-\frac{1}{4}$	$\frac{1}{4}$	$\frac{1}{4}$
	0	0	0	$\frac{1}{4}$	$\frac{7}{4}$	$\frac{39}{4}$

where the third row is now labeled P_2, the label of the preceding pivotal column.

Observe that this tableau is a terminal tableau; the optimum solution of the minimum problem is $Q = (0, 1/4, 7/4)$, the optimum solution of the maximum problem is $P = (17/16, 1/4)$, and $\max f = \min g = 39/4$.

24.7. **Using the pivot circled in the following tableau, calculate the new tableau and then find a pivot entry of it.**

P_1	P_2				
2	3	1	0	0	7
3	8	0	1	0	11
(1)	2	0	0	1	3
−2	−10	0	0	0	0

Divide each entry in the pivotal row, i.e. the third row, by the pivot 1 to obtain the (same) row

$$1 \quad 2 \quad 0 \quad 0 \quad 1 \quad 3 \tag{1}$$

Multiply (*1*) by −2 and add the result to the first row to obtain

$$0 \quad -1 \quad 1 \quad 0 \quad -2 \quad 1$$

Multiply (*1*) by −3 and add the result to the second row to obtain

$$0 \quad 2 \quad 0 \quad 1 \quad -3 \quad 2$$

Multiply (*1*) by 2 and add the result to the last row to obtain

$$0 \quad -6 \quad 0 \quad 0 \quad 2 \quad 6$$

Thus the new tableau is

	P_1	P_2				
	0	−1	1	0	−2	1
	0	(2)	0	1	−3	2
P_1	1	2	0	0	1	3
	0	−6	0	0	2	6

where the third row is labeled P_1, the label of the original pivotal column.

The second column of this tableau has the only negative indicator and so must be chosen as the pivotal column. Since 2/2 is less than 3/2, the 2 in the second row is the pivot.

24.8. **Using the tableau and the pivot obtained in the preceding problem, calculate the new tableau.**

Divide each entry of the pivotal row, i.e. the second row, by the pivot 2 to obtain the row

$$0 \quad 1 \quad 0 \quad \tfrac{1}{2} \quad -\tfrac{3}{2} \quad 1 \tag{1}$$

Then the new tableau is

	P_1	P_2				
	0	0	1	$\frac{1}{2}$	$-\frac{7}{2}$	2
P_2	0	1	0	$\frac{1}{2}$	$-\frac{3}{2}$	1
P_1	1	0	0	−1	4	1
	0	0	0	3	−7	12

where:

(i) the first row is 1 times (*1*) added to the preceding first row;

(ii) the second row is the row (*1*);

(iii) the third row is −2 times (*1*) added to the preceding third row;

(iv) the last row is 6 times (*1*) added to the preceding last row;

and the second row is labeled P_2, the label of the preceding pivotal column.

Note that this tableau is not yet a terminal tableau since the fifth column has a negative indicator.

24.9. In each of the following terminal tableaux, find the optimum solution $Q = (q_1, \ldots, q_k)$ of the minimum problem, the optimum solution $P = (p_1, \ldots, p_m)$ of the maximum problem, and $v = \max f = \min g$.

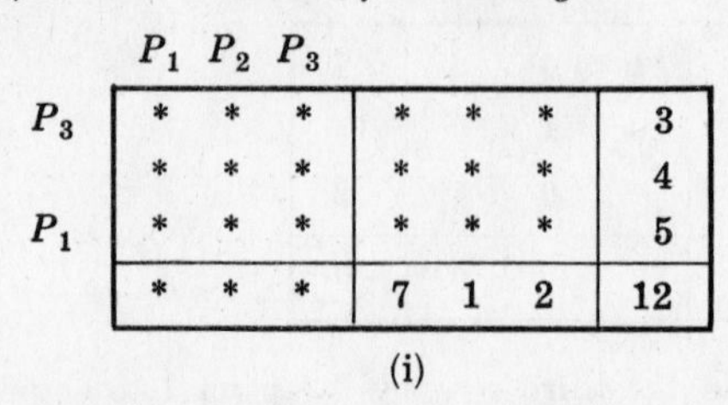

	P_1	P_2	P_3				
P_3	*	*	*	*	*	*	3
	*	*	*	*	*	*	4
P_1	*	*	*	*	*	*	5
	*	*	*	7	1	2	12

(i)

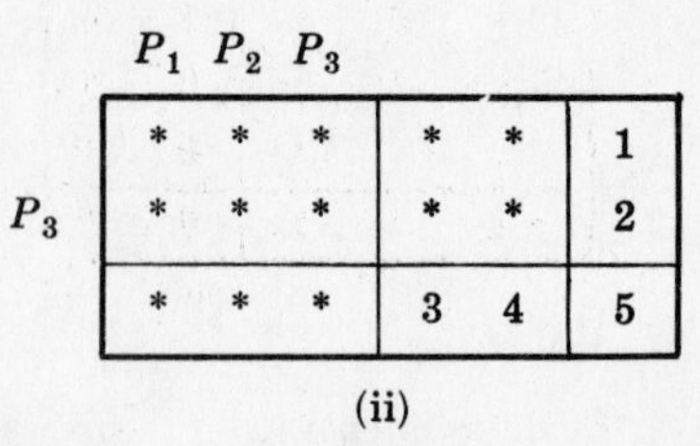

	P_1	P_2	P_3			
	*	*	*	*	*	1
P_3	*	*	*	*	*	2
	*	*	*	3	4	5

(ii)

Recall that: (1) the components of Q are the last entries, respectively, in the columns corresponding to the slack variables; (2) the component p_i of P is 0 if no row is labeled P_i or the last entry in a row labeled P_i; (3) v is the entry in the lower right hand corner of the tableau.

(i) $Q = (7, 1, 2)$, $P = (5, 0, 3)$, $v = 12$. Note that the second component of P is 0 since no row is labeled P_2.

(ii) $Q = (3, 4)$, $P = (0, 0, 2)$, $v = 5$. Note that the first two components of P are 0 since no row is labeled P_1 or P_2.

24.10. Solve the following linear programming problem by the simplex method.

$$\text{Maximize } f = 4x - 2y - z$$

$$\text{subject to} \quad \begin{aligned} x + y + z &\leqq 3 \\ 2x + 2y + z &\leqq 4 \\ x - y &\leqq 0 \end{aligned}$$

$$\text{and} \quad x \geqq 0,\ y \geqq 0,\ z \geqq 0.$$

We obtain the following sequence of tableaux:

	P_1	P_2	P_3				
	1	1	1	1	0	0	3
	2	2	1	0	1	0	4
	(1)	−1	0	0	0	1	0
	−4	2	1	0	0	0	0

(i)

	P_1	P_2	P_3				
	0	2	1	1	0	−1	3
	0	(4)	1	0	1	−2	4
P_1	1	−1	0	0	0	1	0
	0	−2	1	0	0	4	0

(ii)

	P_1	P_2	P_3				
	0	0	$\frac{1}{2}$	1	$-\frac{1}{2}$	0	1
P_2	0	1	$\frac{1}{4}$	0	$\frac{1}{4}$	$-\frac{1}{2}$	1
P_1	1	0	$\frac{1}{4}$	0	$\frac{1}{4}$	$\frac{1}{2}$	1
	0	0	$\frac{3}{2}$	0	$\frac{1}{2}$	3	2

(iii)

Thus, by the terminal tableau, the maximum of f is 2 and occurs at the point $P = (1, 1, 0)$. (See Problem 23.11, Page 293.)

24.11. Solve the following linear programming problem by the simplex method.

$$\text{Maximize } f = 2x - 3y + z$$

$$\text{subject to} \quad \begin{aligned} 3x + 6y + z &\leqq 6 \\ 4x + 2y + z &\leqq 4 \\ x - y + z &\leqq 3 \end{aligned}$$

$$\text{and} \quad x \geqq 0,\ y \geqq 0,\ z \geqq 0.$$

We obtain the following sequence of tableaux:

	P_1	P_2	P_3				
	3	6	1	1	0	0	6
	4	2	1	0	1	0	4
	1	−1	(1)	0	0	1	3
	−2	3	−1	0	0	0	0

(i)

	P_1	P_2	P_3				
	2	7	0	1	0	−1	3
	(3)	3	0	0	1	−1	1
P_3	1	−1	1	0	0	1	3
	−1	2	0	0	0	1	3

(ii)

	P_1	P_2	P_3				
	0	5	0	1	$-\frac{2}{3}$	$-\frac{1}{3}$	$\frac{7}{3}$
P_1	1	1	0	0	$\frac{1}{3}$	$-\frac{1}{3}$	$\frac{1}{3}$
P_3	0	−2	1	0	$-\frac{1}{3}$	$\frac{4}{3}$	$\frac{8}{3}$
	0	3	0	0	$\frac{1}{3}$	$\frac{2}{3}$	$\frac{10}{3}$

(iii)

Thus, by the terminal tableau, the maximum value of f is 10/3 and occurs at the point $P = (1/3, 0, 8/3)$.

24.12. Solve the following linear programming problem by the simplex method.

$$\text{Minimize } g = 6x + 5y + 2z$$

$$\text{subject to } \quad x + 3y + 2z \geqq 5$$
$$2x + 2y + z \geqq 2$$
$$4x - 2y + 3z \geqq -1$$
$$\text{and } \quad x \geqq 0,\ y \geqq 0,\ z \geqq 0.$$

We obtain the following sequence of tableaux:

	P_1	P_2	P_3				
	1	2	4	1	0	0	6
	3	2	−2	0	1	0	5
	2	(1)	3	0	0	1	2
	−5	−2	1	0	0	0	0

(i)

	P_1	P_2	P_3				
	−3	0	−2	1	0	−2	2
	−1	0	−8	0	1	−2	1
P_2	(2)	1	3	0	0	1	2
	−1	0	7	0	0	2	4

(ii)

	P_1	P_2	P_3				
	0	$\frac{3}{2}$	$\frac{5}{2}$	1	0	$-\frac{1}{2}$	5
	0	$\frac{1}{2}$	$-\frac{13}{2}$	0	1	$-\frac{3}{2}$	2
P_1	1	$\frac{1}{2}$	$\frac{3}{2}$	0	0	$\frac{1}{2}$	1
	0	$\frac{1}{2}$	$\frac{17}{2}$	0	0	$\frac{5}{2}$	5

(iii)

Note first that (i) is the tableau of the dual maximum problem. Note also that the third row of (iii) is relabeled P_1, the label of the pivotal column of the preceding tableau (ii). By the terminal tableau, the minimum value of g is 5 and occurs at the point $Q = (0, 0, 5/2)$.

LINEAR PROGRAMMING WORD PROBLEMS

24.13. A tailor has 80 sq. yd. of cotton material and 120 sq. yd. of wool material. A suit requires 1 sq. yd. of cotton and 3 sq. yd. of wool, and a dress requires 2 sq. yd. of each. How many of each garment should the tailor make to maximize his income if

(i) a suit and dress each sells for \$30, (ii) a suit sells for \$40 and a dress for \$20, (iii) a suit sells for \$30 and a dress for \$20?

Construct the following table which summarizes the given facts:

	Suit	Dress	Available
Cotton	1	2	80
Wool	3	2	120

Let x denote the number of suits the tailor makes and y the number of dresses. Then x and y are subject to the conditions

$$x + 2y \leqq 80, \quad 3x + 2y \leqq 120, \quad x \geqq 0, \quad y \geqq 0$$

The convex polygon consisting of the feasible solutions is drawn on the right; its corner points are

$$O = (0,0), \; A = (0,40), \; B = (20,30), \; C = (40,0)$$

Here B is the intersection of $x + 2y = 80$ and $3x + 2y = 120$.

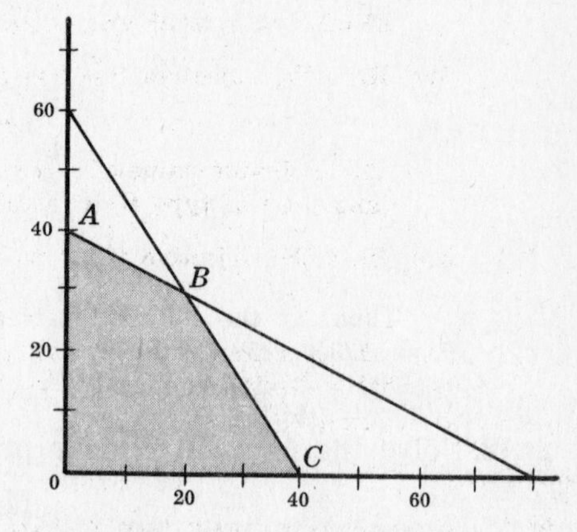

(i) Here the objective function is $f = 30x + 30y$. Evaluate f at the corner points:

$$f(O) = f(0,0) = 0 \qquad f(A) = f(0,40) = 1200$$
$$f(B) = f(20,30) = 1500 \qquad f(C) = f(40,0) = 1200$$

The maximum value of f occurs at $B = (20,30)$. Thus he should make 20 suits and 30 dresses, for which he will receive \$1500.

(ii) Here the objective function is $g = 40x + 20y$. Evaluate g at the corner points:

$$g(O) = 0, \; g(A) = 800, \; g(B) = 1400, \; g(C) = 1600$$

The maximum value of g occurs at $C = (40,0)$. Thus he should make 40 suits (and no dresses), for which he will receive \$1600.

(iii) Here the objective function is $h = 30x + 20y$. Evaluate h at the corner points:

$$h(O) = 0, \; h(A) = 800, \; h(B) = 1200, \; h(C) = 1200$$

The maximum value of h occurs at both $B = (20,30)$ and $C = (40,0)$. Thus the tailor should make either 20 suits and 30 dresses or only 40 suits; in either case he will receive \$1200. (Actually, any point on the line segment joining B and C is an optimum solution.)

24.14. A truck rental company has two types of trucks; type A has 20 ft^3 of refrigerated space and 40 ft^3 of non-refrigerated, and type B has 30 ft^3 each of refrigerated and non-refrigerated space. A food plant must ship 900 ft^3 of refrigerated produce and 1200 ft^3 of non-refrigerated produce. How many trucks of each type should the plant rent so as to minimize cost if (i) truck A rents for 30¢ per mile and B for 40¢ per mile, (ii) truck A rents for 30¢ per mile and B for 50¢ per mile, (iii) truck A rents for 20¢ per mile and B for 30¢ per mile?

Construct the following table which summarizes the given facts:

	Type A	Type B	Requirements
Refrigerated	20	30	900
Non-refrigerated	40	30	1200

Let x denote the number of trucks of type A and y the number of type B that the plant rents. Then x and y are subject to the following conditions:

$$20x + 30y \geqq 900, \quad 40x + 30y \geqq 1200, \quad x \geqq 0, \quad y \geqq 0$$

The set of feasible solutions is drawn on the right; its corner points are

$$P = (0,40),\quad Q = (15,20),\quad R = (45,0)$$

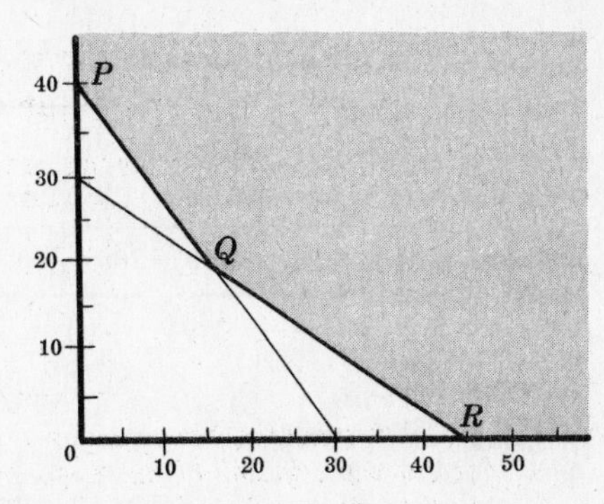

(i) Here the objective function to minimize is $f = 30x + 40y$. Evaluate f at the corner points:

$$f(P) = f(0,40) = 1600,\quad f(Q) = f(15,20) = 1250,$$
$$f(R) = f(45,0) = 1350$$

The minimum value of f occurs at $Q = (15,20)$. Thus the plant should rent 15 trucks of type A and 20 of B at a total cost of \$12.50 per mile.

(ii) Here the objective function is $g = 30x + 50y$. Evaluate g at the corner points:

$$g(P) = 2000,\quad g(Q) = 1450,\quad g(R) = 1350$$

The minimum value of g occurs at $R = (45,0)$. Thus the plant should rent 45 trucks of type A (and none of type B) at a total cost of \$13.50 per mile.

(iii) Here the objective function is $h = 20x + 30y$. Evaluate h at the corner points:

$$h(P) = 1200,\quad h(Q) = 900,\quad h(R) = 900$$

The minimum value of h occurs at both $Q = (15,20)$ and $R = (45,0)$. Thus the plant can rent either 15 trucks of type A and 20 of B or only 45 of type A at a total cost of \$9.00 per mile.

24.15. A company owns two mines: mine A produces 1 ton of high grade ore, 3 tons of medium grade ore and 5 tons of low grade ore each day; and mine B produces 2 tons of each of the three grades of ore each day. The company needs 80 tons of high grade ore, 160 tons of medium grade ore and 200 tons of low grade ore. How many days should each mine be operated if it costs \$200 per day to work each mine?

Construct the following table which summarizes the given facts:

	Mine A	Mine B	Requirements
High grade	1	2	80
Medium grade	3	2	160
Low grade	5	2	200

Let x denote the number of days mine A is open and y the number of days B is open. Then x and y are subject to the following conditions:

$$x + 2y \geq 80,\quad 3x + 2y \geq 160,\quad 5x + 2y \geq 200,\quad x \geq 0,\quad y \geq 0$$

We wish to minimize the function $g = 200x + 200y$ subject to the above conditions.

Method 1.

Draw the set of feasible solutions as on the right. The corner points are: $p = (0,100)$, $Q = (20,50)$, $R = (40,20)$, $S = (80,0)$.

Evaluate the objective function $g = 200x + 200y$ at the corner points: $g(P) = 20{,}000$, $g(Q) = 14{,}000$, $g(R) = 12{,}000$, $g(S) = 16{,}000$. The minimum value of g occurs at $R = (40,20)$. Thus the company should keep mine A open 40 days and mine B open 20 days for a total cost of \$12,000.

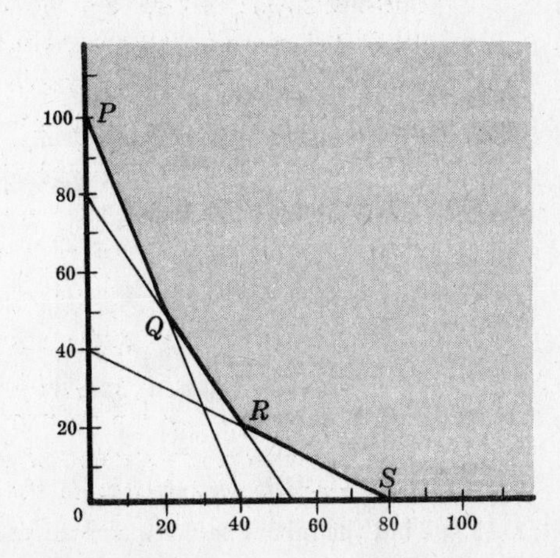

Method 2. Consider the dual problem:

Maximize $f = 80u + 160v + 200w$

subject to
$$u + 3v + 5w \leq 200$$
$$2u + 2v + 2w \leq 200$$

and $u \geq 0,\ v \geq 0,\ w \geq 0.$

Using the simplex method we obtain the following sequence of tableaux:

(i)

1	3	5	1	0	200
(2)	2	2	0	1	200
−80	−160	−200	0	0	0

(ii)

0	(2)	4	1	$-\frac{1}{2}$	100
1	1	1	0	$\frac{1}{2}$	100
0	−80	−120	0	40	8,000

(iii)

0	1	2	$\frac{1}{2}$	$-\frac{1}{4}$	50
1	0	−1	$-\frac{1}{2}$	$\frac{3}{4}$	50
0	0	40	40	20	12,000

Thus the minimum value of g is 12,000 and occurs at the point (40, 20). (Note that labels were not necessary since we were only solving the minimum problem.)

24.16. Suppose, in the preceding problem, the production cost at mine A is \$250 per day and at mine B is \$150 per day. How should the production schedule be changed?

In this case, the objective function to minimize is $g = 250x + 150y$ subject to the same set of feasible solutions, i.e. conditions. Evaluate g at the corner points of the set of feasible solutions:

$$g(P) = g(0,100) = 15{,}000, \quad g(Q) = g(20,50) = 12{,}500, \quad g(R) = g(40,20) = 13{,}000,$$
$$g(S) = g(80,0) = 20{,}000$$

The minimum value of g occurs at $Q = (20, 50)$. Thus the company should now keep mine A open 20 days and mine B open 50 days for a total cost of \$12,500.

Supplementary Problems

DUALITY

24.17. Find the dual of each problem.

(i) Maximize $f = 5x - 2y + 3z$

subject to $3x - 4y + 2z \leqq 9$

$2x + 7y + 5z \leqq 11$

and $x \geqq 0,\ y \geqq 0,\ z \geqq 0.$

(ii) Minimize $g = 12x + 3y + 7z$

subject to $6x - 2y + 5z \geqq 3$

$2x + 3y - 4z \geqq -2$

$3x + 9y + z \geqq 8$

and $x \geqq 0,\ y \geqq 0,\ z \geqq 0.$

24.18. Find the dual of each problem.

(i) Maximize $f = 3x + 2y - 2z$

subject to $4x + 3y - 2z \leqq 11$

$5x - 2y + 7z \geqq 2$

$x + y + 3z \leqq 15$

and $x \geqq 0,\ y \geqq 0,\ z \geqq 0.$

(ii) Minimize $g = r + 7s + 2t$

subject to $7r + 3s + 5t \geqq 9$

$4r + 8s - t \leqq 10$

and $r \geqq 0,\ s \geqq 0,\ t \geqq 0.$

SIMPLEX METHOD

24.19. Find the initial tableau and all possible pivot entries for each of the problems given in Problem 24.17.

24.20. Calculate the new tableau from each tableau and circled pivot.

(i)

P_1	P_2	P_3				
2	3	4	1	0	0	8
5	−5	2	0	1	0	3
1	(2)	6	0	0	1	4
−3	−4	2	0	0	0	0

(ii)

	P_1	P_2	P_3			
	0	1	4	1	−2	3
P_1	1	(2)	1	0	1	4
	0	−3	4	0	3	12

24.21. In each of the following terminal tableaux, find the optimum solution $Q = (q_1, \ldots, q_k)$ of the minimum problem, the optimum solution $P = (p_1, \ldots, p_m)$ of the maximum problem, and $v = \max f = \min g$.

(i)

	P_1	P_2	P_3				
P_2	*	*	*	*	*	*	$\frac{1}{4}$
	*	*	*	*	*	*	2
P_1	*	*	*	*	*	*	$\frac{1}{2}$
	*	*	*	3	1	5	9

(ii)

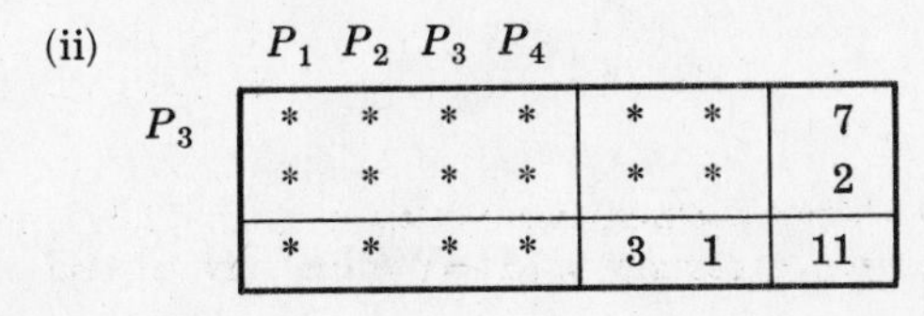

	P_1	P_2	P_3	P_4			
P_3	*	*	*	*	*	*	7
	*	*	*	*	*	*	2
	*	*	*	*	3	1	11

24.22. Solve by the simplex method: Maximize $f = 9x + 2y + 5z$

subject to
$$2x + 3y - 5z \leqq 12$$
$$2x - y + 3z \leqq 3$$
$$3x + y - 2z \leqq 2$$
and $x \geqq 0,\ y \geqq 0,\ z \geqq 0.$

24.23. Solve by the simplex method: Minimize $g = 4x + 5y + z$

subject to
$$x + y + z \geqq 3$$
$$2x + 2y - z \geqq 2$$
$$x + 2y \geqq 4$$
and $x \geqq 0,\ y \geqq 0,\ z \geqq 0.$

LINEAR PROGRAMMING WORD PROBLEMS

24.24. A carpenter has 90, 80 and 50 running feet of plywood, pine and birch, respectively. Product A requires 2, 1 and 1 running feet of plywood, pine and birch respectively; and product B requires 1, 2 and 1 running feet of plywood, pine and birch respectively. (i) If A sells for \$12 and B sells for \$10, how many of each should he make to obtain the maximum gross income? (ii) If B sells for \$12 and A sells for \$10, how many of each should he make?

24.25. Suppose that 8, 12 and 9 units of proteins, carbohydrates and fats respectively are the minimum weekly requirements for a person. Food A contains 2, 6 and 1 units of proteins, carbohydrates and fats respectively per pound; and food B contains 1, 1 and 3 units respectively per pound. (i) If A costs 85¢ per pound and B costs 40¢ per pound, how many pounds of each should he buy per week to minimize his cost and still meet his minimum requirements? (ii) If the price of A drops to 75¢ per pound while B still sells at 40¢ per pound, how many pounds of each should he then buy?

24.26. A baker has 150, 90 and 150 units of ingredients A, B and C respectively. A loaf of bread requires 1, 1 and 2 units of A, B and C respectively; and a cake requires 5, 2 and 1 units of A, B and C respectively. (i) If a loaf of bread sells for 35¢ and a cake for 80¢, how many of each should he bake and what would his gross income be? (ii) If the price of bread is increased to 45¢ per loaf and a cake still sells for 80¢, how many of each should he then bake and what would be the gross income?

24.27. A manufacturer has 240, 370 and 180 pounds of wood, plastic and steel respectively. Product A requires 1, 3 and 2 pounds of wood, plastic and steel respectively; and product B requires 3, 4 and 1 pounds of wood, plastic and steel respectively. (i) If A sells for \$4 and B for \$6, how many of each should be made to obtain the maximum gross income? (ii) If the price of B drops to \$5 and A remains at \$4, how many of each should then be made to obtain the maximum gross income?

24.28. An oil company requires 9,000, 12,000 and 26,000 barrels of high, medium, and low grade oil, respectively. It owns two oil refineries, say A and B. A produces 100, 300 and 400 barrels of high, medium and low grade oil respectively per day; and B produces 200, 100 and 300 barrels of high, medium and low grade oil respectively per day. (i) If each refinery costs \$200 per day to operate, how many days should each be run to minimize the cost and meet the requirements? (ii) If the cost at A rises to \$300 per day and B remains at \$200 per day, how many days should each be operated?

MISCELLANEOUS PROBLEMS

24.29. Solve Problem 24.16 by the simplex method.

24.30. Show that the maximum linear programming problem (*1*), Page 301, is equivalent to the problem (*2*), Page 304.

Answers to Supplementary Problems

24.17. (i) Minimize $g = 9u + 11v$

subject to
$$3u + 2v \geqq 5$$
$$-4u + 7v \geqq -2$$
$$2u + 5v \geqq 3$$
and $u \geqq 0,\ v \geqq 0.$

(ii) Maximize $f = 3u - 2v + 8w$

subject to
$$6u + 2v + 3w \leqq 12$$
$$-2u + 3v + 9w \leqq 3$$
$$5u - 4v + w \leqq 7$$
and $u \geqq 0,\ v \geqq 0,\ w \geqq 0.$

24.18. (i) Minimize $g = 11u - 2v + 15w$

subject to
$$4u - 5v + w \geqq 3$$
$$3u + 2v + w \geqq 2$$
$$-2u - 7v + 3w \geqq -2$$
and $u \geqq 0,\ v \geqq 0,\ w \geqq 0.$

(ii) Maximize $f = 9x - 10y$

subject to
$$7x - 4y \leqq 1$$
$$3x - 8y \leqq 7$$
$$5x + y \leqq 2$$
and $x \geqq 0,\ y \geqq 0.$

24.19. (i)

P_1	P_2	P_3			
(3)	−4	2	1	0	9
2	7	(5)	0	1	11
−5	2	−3	0	0	0

(ii)

P_1	P_2	P_3				
6	2	3	1	0	0	12
−2	3	(9)	0	1	0	3
(5)	−4	1	0	0	1	7
−3	2	−8	0	0	0	0

24.20. (i)

	P_1	P_2	P_3				
	$\frac{1}{2}$	0	−5	1	0	$-\frac{3}{2}$	2
	$\frac{15}{2}$	0	17	0	1	$\frac{5}{2}$	13
P_2	$\frac{1}{2}$	1	3	0	0	$\frac{1}{2}$	2
	−1	0	14	0	0	2	8

(ii)

	P_1	P_2	P_3			
	$-\frac{1}{2}$	0	$\frac{7}{2}$	1	$-\frac{5}{2}$	1
P_2	$\frac{1}{2}$	1	$\frac{1}{2}$	0	$\frac{1}{2}$	2
	$\frac{3}{2}$	0	$\frac{11}{2}$	0	$\frac{9}{2}$	18

24.21. (i) $Q = (3,1,5)$, $P = (\frac{1}{2}, \frac{1}{4}, 0)$, $v = 9$. (ii) $Q = (3,1)$, $P = (0,0,7,0)$, $v = 11$.

24.22. The maximum value of f is 49 and occurs at $P = (0,12,5)$.

24.23. The minimum value of g is 11 and occurs at the point $Q = (0,2,1)$.

24.24. (i) 40 of A and 10 of B for \$580. (ii) 20 of A and 30 of B for \$560.

24.25. (i) 1 of A and 6 of B for \$3.25. (ii) 3 of A and 2 of B for \$3.05.

24.26. (i) 50 loaves of bread and 20 cakes for \$33.50. (ii) 70 loaves of bread and 10 cakes for \$39.50.

24.27. (i) 30 of A and 70 of B for \$540. (ii) 70 of A and 40 of B for \$480.

24.28. (i) A run 50 days and B run 20 days. (ii) A run 20 days and B run 60 days.

Chapter 25

Theory of Games

INTRODUCTION TO MATRIX GAMES

Every matrix A defines a game as follows:

(1) There are two players, one called R and the other C.

(2) A *play* of the game consists of R's choosing a row of A and, simultaneously, C's choosing a column of A.

(3) After each play of the game, R receives from C an amount equal to the entry in the chosen row and column, a negative entry denoting a payment from R to C.

Example 1.1.

Consider the following matrix games:

(i)

3	−1	4
−1	1	−2

(ii)

0	−5
1	2

In the first game, if R consistently plays the first row, hoping to win an amount 3 or 4, then C can play the second column and so win an amount 1. However, if C consistently plays the second column then R can play the second row and so win an amount 1. Thus we see that if either player should consistently play a particular row or column, the other player can take advantage of that fact.

Now in the second game, player R can guarantee winning 1 or more by consistently playing row 2. Player C can then minimize his loss by playing column 1. Thus in this game it is "best" if R plays a given row and C plays a given column.

On the other hand, suppose a competitive situation is as follows:

(i) there are two people (or: organizations, companies, countries, . . .);

(ii) each person has a finite number of courses of action;

(iii) the two people choose a course of action simultaneously;

(iv) for each pair (i, j) of choices, there is a payment b_{ij} from one given player to the other.

Then the above competitive situation is equivalent to the matrix game determined by the matrix $B = (b_{ij})$.

Example 1.2.

Each of the two players R and C simultaneously shows 1 or 2 fingers. If the sum of the fingers shown is even, R wins the sum from C; if the sum is odd, R loses the sum to C. The matrix of this game is as follows:

		Player C: 1	2
Player R	1	2	−3
	2	−3	4

We shall show later (Example 3.1) that this game is "favorable" to player C.

We remark that matrix games and the above competitive games which can be represented as matrix games are termed *two-person zero sum games.* The zero sum corresponds to the fact that the sum of the winnings and losses of the players after each play is zero. There also exists a theory for n-person games and non-zero sum games but that lies beyond the scope of this text.

STRATEGIES

By a *strategy* for R in a matrix game, we mean a decision by R to play the various rows with a given probability distribution, say to play row 1 with probability p_1, to play row 2 with probability p_2, ... This strategy for R is formally denoted by the probability vector $p = (p_1, p_2, \ldots, p_k)$. For example, if the matrix game has two rows and R tosses a coin to decide which row to play, then his strategy is the probability vector $p = (\frac{1}{2}, \frac{1}{2})$.

Similarly, by a *strategy* for C we mean a decision by C to play the various columns with a given probability distribution, say to play column 1 with probability q_1, to play column 2 with probability q_2, ... This strategy for C is formally denoted by the probability vector $q = (q_1, q_2, \ldots, q_m)$.

A strategy which contains a 1 as a component and 0 everywhere else, i.e. where R decides to play consistently a given row or C decides to play consistently a given column, is called a *pure strategy*; otherwise it is called a *mixed strategy.*

OPTIMUM STRATEGIES AND THE VALUE OF A GAME

For simplicity, we shall first consider a 2×2 matrix game, say

$$A = \begin{array}{|c|c|} \hline a & b \\ \hline c & d \\ \hline \end{array}$$

Suppose player R adopts the strategy $p = (p_1, p_2)$ and player C adopts the strategy $q = (q_1, q_2)$. Then R plays row 1 with probability p_1 and C plays column 1 with probability q_1, and so the entry a will occur with probability p_1q_1. Similarly, b will occur with probability p_1q_2, c with probability p_2q_1 and d with probability p_2q_2. Hence the expected winning (see Page 220) of R is

$$\begin{aligned} E(p,q) &= p_1q_1a + p_1q_2b + p_2q_1c + p_2q_2d = (p_1, p_2)\begin{pmatrix} a & b \\ c & d \end{pmatrix}\begin{pmatrix} q_1 \\ q_2 \end{pmatrix} = p\,A\,q^t \\ &= (q_1, q_2)\begin{pmatrix} a & c \\ b & d \end{pmatrix}\begin{pmatrix} p_1 \\ p_2 \end{pmatrix} = q\,A^t\,p^t \end{aligned}$$

where q^t, A^t and p^t denote the transpose of q, A and p, respectively.

Similarly, if A is any $k \times m$ matrix game and player R adopts the strategy $p = (p_1, \ldots, p_k)$ and C adopts the strategy $q = (q_1, \ldots, q_m)$, then the expected winnings of R is given by

$$E(p,q) = p\,A\,q^t = q\,A^t\,p^t$$

Now suppose that v_1 is the maximum number for which there is a strategy p^0 for R such that the expected winnings of R is at least v_1 for any strategy adopted by C:

$$E(p^0, q) \geq v_1, \quad \text{for every strategy } q \text{ of } C \qquad (1)$$

Since v_1 is the maximum of such numbers, the strategy p^0 is in a certain sense a "best" strategy that R can adopt; we call p^0 an *optimum strategy* for R. We show in Problem 25.9, Page 233, that (*1*) is equivalent to the condition

$$p^0 A \geqq (v_1, v_1, \ldots, v_1)$$

On the other hand, suppose v_2 is the minimum number for which there is a strategy q^0 for player C such that the expected loss of C is not greater than v_2 for any strategy adopted by R:

$$E(p, q^0) \leqq v_2, \quad \text{for every strategy } p \text{ for } R \tag{2}$$

Then the strategy q^0 is in a certain sense a "best" strategy that C can adopt; we call q^0 an *optimum strategy* for C. Similarly, (*2*) can be shown to be equivalent to the condition

$$q^0 A^t \leqq (v_2, v_2, \ldots, v_2)$$

The fundamental theorem on game theory by von Neumann states that v_1 and v_2 exist for any matrix game and are equal to each other: $v = v_1 = v_2$. This number v is called the *value* of the game, and the game is said to be *fair* if $v = 0$. We observe that (*1*) and (*2*) imply that $v_1 \leqq E(p^0, q^0) \leqq v_2$; hence by von Neumann's theorem we have the equality relation

$$E(p^0, q^0) = p^0 A q^{0t} = q^0 A^t p^{0t} = v \tag{3}$$

We summarize and formally define the terms introduced above:

Definition: In a matrix game A, a strategy p^0 for player R is an *optimum strategy*, a strategy q^0 for player C is an *optimum strategy*, and a number v is the *value* of the game if p^0, q^0 and v are related as follows:

$$\text{(i)} \;\; p^0 A \geqq (v, v, \ldots, v); \qquad \text{(ii)} \;\; q^0 A^t \leqq (v, v, \ldots, v)$$

If $v = 0$, the game is said to be *fair*.

We point out that there may be more than one pair of strategies for which (i) and (ii) hold; that is, a player may have more than one optimum strategy.

We remark that the *solution* of a matrix game means finding optimum strategies for the players and the value of the game. Before we give the solution to an arbitrary matrix game by means of the simplex method developed in the preceding chapter, we shall discuss two special cases which occur frequently and which offer simple solutions: strictly determined games and 2×2 games.

STRICTLY DETERMINED GAMES

A matrix game is called *strictly determined* if the matrix has an entry which is a minimum in its row and a maximum in its column; such an entry is called a "saddle point". The following theorem applies.

Theorem 25.1: Let v be a saddle point of a strictly determined game. Then an optimum strategy for R is always to play the row containing v, an optimum strategy for C is always to play the column containing v, and v is the value of the game.

Thus in a strictly determined game a pure strategy for each player is an optimum strategy.

Example 2.1.

Consider the following game:

$$A = \begin{array}{|c|c|c|c|} \hline 3 & 0 & -4 & -3 \\ \hline 2 & 3 & 1 & 2 \\ \hline -4 & 2 & -1 & 3 \\ \hline \end{array}$$

If we circle the minimum entry in each row and put a square about the maximum entry in each column, we obtain

$$A \;=\; \begin{array}{|c|c|c|c|}\hline \boxed{3} & 0 & \textcircled{-4} & -3 \\ \hline 2 & \boxed{3} & \boxed{\textcircled{1}} & 2 \\ \hline \textcircled{-4} & 2 & -1 & \boxed{3} \\ \hline \end{array}$$

The point 1 is both circled and "boxed" and so is a saddle point of A. Thus $v = 1$ is the value of the game, and optimum strategies p^0 for R and q^0 for C are as follows:

$$p^0 \;=\; (0, 1, 0) \qquad \text{and} \qquad q^0 \;=\; (0, 0, 1, 0)$$

That is, R consistently plays row 2, and C consistently plays column 3.

We now show that p^0, q^0 and v do satisfy the required properties to be optimum strategies and the value of the game:

(i) $$p^0 A \;=\; (0,1,0)\begin{pmatrix} 3 & 0 & -4 & -3 \\ 2 & 3 & 1 & 2 \\ -4 & 2 & -1 & 3 \end{pmatrix} \;=\; (2,3,1,2) \;\geq\; (1,1,1,1) \;=\; (v,v,v,v)$$

(ii) $$q^0 A^t \;=\; (0,0,1,0)\begin{pmatrix} 3 & 2 & -4 \\ 0 & 3 & 2 \\ -4 & 1 & -1 \\ -3 & 2 & 3 \end{pmatrix} \;=\; (-4,1,-1) \;\leq\; (1,1,1) \;=\; (v,v,v)$$

We remark that the proof of Theorem 25.1 is similar to the proof above for the special case.

2×2 MATRIX GAMES

Consider the matrix game $\quad A \;=\; \begin{array}{|c|c|}\hline a & b \\ \hline c & d \\ \hline \end{array}$

If A is strictly determined, then the solution is indicated in the preceding section. Thus we need only consider the case in which A is non-strictly determined. We first state a useful criterion to determine whether or not a 2×2 matrix game is strictly determined.

Lemma 25.2: The above matrix game A is non-strictly determined if and only if each of the entries on one of the diagonals is greater than each of the entries on the other diagonal:

(i) $a, d > b$ and $a, d > c$; or (ii) $b, c > a$ and $b, c > d$.

The following theorem applies.

Theorem 25.3: Suppose the above matrix game A is non-strictly determined. Then $p^0 = (x_1, x_2)$ is an optimum strategy for player R, $q^0 = (y_1, y_2)$ is an optimum strategy for player C, and v is the value of the game where

$$x_1 \;=\; \frac{d-c}{a+d-b-c}, \quad x_2 \;=\; \frac{a-b}{a+d-b-c}$$

$$y_1 \;=\; \frac{d-b}{a+d-b-c}, \quad y_2 \;=\; \frac{a-c}{a+d-b-c}$$

and $$v \;=\; \frac{ad-bc}{a+d-b-c}$$

Note that the five quotients above have the same denominator, and that the numerator of the formula for v is the determinant of the matrix A.

Example 3.1.

Consider the following matrix game (see Example 1.2):

2	−3
−3	4

Using the above formula, we can obtain the value v of the game:

$$v = \frac{2 \cdot 4 - (-3) \cdot (-3)}{2+4+3+3} = -\frac{1}{12}$$

Thus the game is not fair and is in favor of the column player C. Optimum strategies p^0 for R and q^0 for C are as follows: $p^0 = (\frac{7}{12}, \frac{5}{12})$ and $q^0 = (\frac{7}{12}, \frac{5}{12})$.

RECESSIVE ROWS AND COLUMNS

Let A be a matrix game. Suppose A contains a row r_i such that $r_i \leqq r_j$ for some other row r_j. Recall that $r_i \leqq r_j$ means that every entry of r_i is less than or equal to the corresponding entry of r_j. Then r_i is called a *recessive row* and r_j is said to dominate it. Clearly, player R would always rather play row r_j than row r_i since he is guaranteed to win the same or a greater amount in every possible play of the game. Accordingly, the recessive row r_i can be omitted from the game.

On the other hand, suppose A contains a column c_i such that $c_i \geqq c_j$ for some other column c_j. Then c_i is called a *recessive column* and c_j is said to dominate it. For analogous reasons player C would always rather play column c_j than column c_i. Hence the recessive column c_i can be omitted from the game.

We emphasize that a recessive row contains numbers smaller than those of another row, whereas a recessive column contains numbers larger than those of another column.

Example 4.1.

Consider the game $A =$

−5	−3	1
2	−1	2
−2	3	4

. Note that $(-5,-3,1) \leqq (2,-1,2)$, i.e. every entry in the first row is $\leqq$ the corresponding entry in the second row. Thus the first row is recessive and can be omitted from the game and the game may be reduced to

2	−1	2
−2	3	4

Now observe that the third column is recessive since each entry is $\geqq$ the corresponding entry in the second column. Thus the game may be reduced to the 2×2 game

$A^* =$

2	−1
−2	3

The solution to A^* can be found by using the formulas in Theorem 25.3 and is

$$v = \tfrac{4}{8} = \tfrac{1}{2}; \quad p_0^* = (\tfrac{5}{8}, \tfrac{3}{8}) \quad \text{and} \quad q_0^* = (\tfrac{1}{2}, \tfrac{1}{2})$$

Thus the solution to the original game A is

$$v = \tfrac{1}{2}; \quad p_0 = (0, \tfrac{5}{8}, \tfrac{3}{8}) \quad \text{and} \quad q_0 = (\tfrac{1}{2}, \tfrac{1}{2}, 0)$$

SOLUTION OF A MATRIX GAME BY THE SIMPLEX METHOD

Suppose we are given a matrix game A. We assume that A is non-strictly determined and does not contain any recessive rows or columns. We obtain the solution to A as follows.

(1) Add a sufficiently large number k to every entry of A to form, say, the following matrix game which has only positive entries:

$$A^* = \begin{pmatrix} a_{11} & a_{12} & \cdots & a_{1m} \\ a_{21} & a_{22} & \cdots & a_{2m} \\ \cdots & \cdots & \cdots & \cdots \\ a_{k1} & a_{k2} & \cdots & a_{km} \end{pmatrix}$$

(The purpose of this step is to guarantee that the value of the new matrix game is positive.)

(2) Form the following initial tableau:

P_1	P_2	$\cdots$	P_m					
a_{11}	a_{12}	$\cdots$	a_{1m}	1	0	$\cdots$	0	1
a_{21}	a_{22}	$\cdots$	a_{2m}	0	1	$\cdots$	0	1
$\cdots$	$\cdots$	$\cdots$	$\cdots$	$\cdots$	$\cdots$	$\cdots$	$\cdots$	.
a_{k1}	a_{k2}	$\cdots$	a_{km}	0	0	$\cdots$	1	1
-1	-1	$\cdots$	-1	0	0	$\cdots$	0	0

This tableau corresponds to the linear programming problem:

$$\text{Maximize } f = x_1 + x_2 + \cdots + x_m$$

$$\begin{aligned} \text{subject to} \quad & a_{11}x_1 + a_{12}x_2 + \cdots + a_{1m}x_m \leqq 1 \\ & a_{21}x_1 + a_{22}x_2 + \cdots + a_{2m}x_m \leqq 1 \\ & \cdots\cdots\cdots\cdots\cdots\cdots \\ & a_{k1}x_1 + a_{k2}x_2 + \cdots + a_{km}x_m \leqq 1 \\ \text{and} \quad & x_1 \geqq 0,\ x_2 \geqq 0,\ \ldots,\ x_m \geqq 0 \end{aligned}$$

(3) Solve the above linear programming problem by the simplex method. Let P be the optimum solution to the maximum problem, Q the optimum solution to the dual minimum problem, and $v^* = f(P) = g(Q)$ the entry in the lower right hand corner of the terminal tableau. Set

$$\text{(i)}\quad p^0 = \frac{1}{v^*}Q \qquad \text{(ii)}\quad q^0 = \frac{1}{v^*}P \qquad \text{(iii)}\quad v = \frac{1}{v^*} - k$$

Then p^0 is an optimum strategy for player R in game A, q^0 is an optimum strategy for player C, and v is the value of the game.

We remark that the games A and A^* have the same optimum strategies for their respective players and that their values differ by the added constant k; that is, p^0 is an optimum strategy for player R in game A^*, q^0 is an optimum strategy for player C in game A^*, and $v + k = 1/v^*$ is the value of the matrix game A^*.

Example 5.1.

Consider the matrix game

$$A = \begin{array}{|c|c|c|} \hline 3 & -1 & 0 \\ \hline -2 & 1 & -1 \\ \hline \end{array}$$

Add 3 to each entry of A to form the matrix game

6	2	3
1	4	2

The initial simplex tableau and successive tableaux are as follows:

(i)

P_1	P_2	P_3			
6	2	$\textcircled{3}$	1	0	1
1	4	2	0	1	1
−1	−1	−1	0	0	0

(ii)

	P_1	P_2	P_3			
P_3	2	$\frac{2}{3}$	1	$\frac{1}{3}$	0	$\frac{1}{3}$
	−3	$\textcircled{\frac{8}{3}}$	0	$-\frac{2}{3}$	1	$\frac{1}{3}$
	1	$-\frac{1}{3}$	0	$\frac{1}{3}$	0	$\frac{1}{3}$

(iii)

	P_1	P_2	P_3			
P_3	$\frac{11}{4}$	0	1	$\frac{1}{2}$	$-\frac{1}{4}$	$\frac{1}{4}$
P_2	$-\frac{9}{8}$	1	0	$-\frac{1}{4}$	$\frac{3}{8}$	$\frac{1}{8}$
	$\frac{5}{8}$	0	0	$\frac{1}{4}$	$\frac{1}{8}$	$\frac{3}{8}$

Thus $P = (0, \frac{1}{8}, \frac{1}{4})$, $Q = (\frac{1}{4}, \frac{1}{8})$ and $v^* = \frac{3}{8}$; hence $1/v^* = 8/3$. Then for the original game A,

$p^0 = (8/3)Q = (2/3, 1/3)$ is an optimum strategy for player R,

$q^0 = (8/3)P = (0, 1/3, 2/3)$ is an optimum strategy for player C,

$v = 8/3 - 3 = -1/3$ is the value of the game.

Observe that the game is favorable to the column player C.

Example 5.2.

Consider the matrix game A shown below:

A =

2	1	−1
−1	2	−1
−1	−1	0

; A^* =

4	3	1
1	4	1
1	1	2

Adding 2 to each entry of A, we obtain the matrix game A^* on the right.

The initial tableau and the successive tableaux are as follows:

(i)

P_1	P_2	P_3				
4	3	1	1	0	0	1
1	4	1	0	1	0	1
1	1	$\textcircled{2}$	0	0	1	1
−1	−1	−1	0	0	0	0

(ii)

	P_1	P_2	P_3				
	$\textcircled{\frac{7}{2}}$	$\frac{5}{2}$	0	1	0	$-\frac{1}{2}$	$\frac{1}{2}$
	$\frac{1}{2}$	$\frac{7}{2}$	0	0	1	$-\frac{1}{2}$	$\frac{1}{2}$
P_3	$\frac{1}{2}$	$\frac{1}{2}$	1	0	0	$\frac{1}{2}$	$\frac{1}{2}$
	$-\frac{1}{2}$	$-\frac{1}{2}$	0	0	0	$\frac{1}{2}$	$\frac{1}{2}$

(iii)

	P_1	P_2	P_3				
P_1	1	$\frac{5}{7}$	0	$\frac{2}{7}$	0	$-\frac{1}{7}$	$\frac{1}{7}$
	0	$\textcircled{\frac{22}{7}}$	0	$-\frac{1}{7}$	1	$-\frac{3}{7}$	$\frac{3}{7}$
P_3	0	$\frac{1}{7}$	1	$-\frac{1}{7}$	0	$\frac{4}{7}$	$\frac{3}{7}$
	0	$-\frac{1}{7}$	0	$\frac{1}{7}$	0	$\frac{3}{7}$	$\frac{4}{7}$

(iv)

	P_1	P_2	P_3				
P_1	1	0	0	$\frac{7}{22}$	$-\frac{5}{22}$	$-\frac{1}{22}$	$\frac{1}{22}$
P_2	0	1	0	$-\frac{1}{22}$	$\frac{7}{22}$	$-\frac{3}{22}$	$\frac{3}{22}$
P_3	0	0	1	$-\frac{3}{22}$	$-\frac{1}{22}$	$\frac{13}{22}$	$\frac{9}{22}$
	0	0	0	$\frac{3}{22}$	$\frac{1}{22}$	$\frac{9}{22}$	$\frac{13}{22}$

Thus $P = (1/22, 3/22, 9/22)$, $Q = (3/22, 1/22, 9/22)$ and $v^* = 13/22$; hence $1/v^* = 22/13$. Accordingly, for the original game A,

$$p^0 = (22/13)Q = (3/13, 1/13, 9/13) \quad \text{is an optimum strategy for } R,$$
$$q^0 = (22/13)P = (1/13, 3/13, 9/13) \quad \text{is an optimum strategy for } C,$$
$$v = 22/13 - 2 = -4/13 \quad \text{is the value of the game.}$$

$2 \times m$ AND $m \times 2$ MATRIX GAMES

Consider the $2 \times m$ matrix game

$$A = \begin{array}{|c|c|c|c|}\hline a_1 & a_2 & \cdots & a_m \\ \hline b_1 & b_2 & \cdots & b_m \\ \hline \end{array}$$

Suppose we apply the simplex method to the above game. Note that the optimum solution of the corresponding maximum linear programming problem will contain at most two non-zero components; hence so will an optimum strategy for the column player C. In other words, the $2 \times m$ matrix game A can be reduced to a 2×2 game A^*.

In the next example we show a way of solving such a $2 \times m$ game A and, in particular, how to reduce the game to a 2×2 game A^*. A similar method holds for any $m \times 2$ matrix game.

Example 6.1.

Consider the following matrix game (see Example 5.1):

$$A = \begin{pmatrix} 3 & -1 & 0 \\ -2 & 1 & -1 \end{pmatrix}$$

Let $p^0 = (x, 1-x)$ denote an optimum strategy for R and let v be the value of the game. Then

$$p^0 A = (x, 1-x)\begin{pmatrix} 3 & -1 & 0 \\ -2 & 1 & -1 \end{pmatrix} \geqq (v, v, v)$$

or

$$\begin{cases} 3x - 2(1-x) \geqq v \\ -x + (1-x) \geqq v \\ 0 - (1-x) \geqq v \end{cases} \quad \text{or} \quad \begin{cases} v \leqq 5x - 2 \\ v \leqq -2x + 1 \\ v \leqq x - 1 \end{cases}$$

Let us plot the three linear inequalities in v and x on a coordinate plane where $0 \leqq x \leqq 1$, as shown in the diagram. The shaded region denotes all possible values of v; however, R chooses his optimum strategy so that v is a maximum. This occurs at the point W which is the intersection of $v = -2x + 1$ and $v = x - 1$; that is, where $x = \frac{2}{3}$ and $v = -\frac{1}{3}$. Thus the value of the game is $-\frac{1}{3}$ and the optimum strategy for R is $(\frac{2}{3}, \frac{1}{3})$.

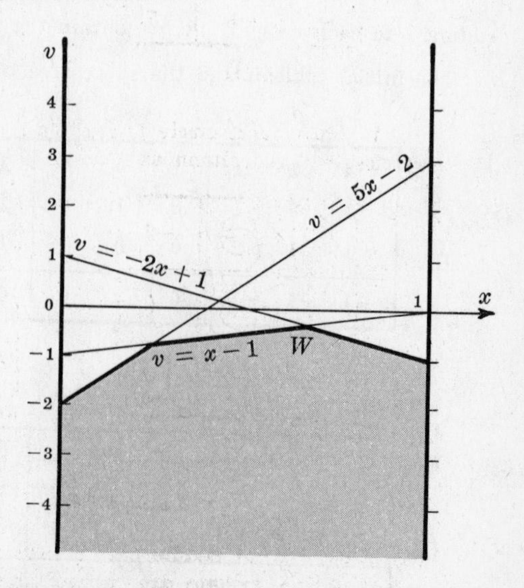

To find an optimum strategy for C, we can set $q^0 = (y, z, 1-y-z)$ and use the condition $q^0 A^t \geqq (v, v, v) = (-\frac{1}{3}, -\frac{1}{3}, -\frac{1}{3})$. However, a simpler way is as follows. Observe that the two lines intersecting at W are determined by only the second and third columns of A. Omit the other column and reduce the game to the 2×2 matrix game

$$A^* = \begin{array}{|c|c|}\hline -1 & 0 \\ \hline 1 & -1 \\ \hline \end{array}$$

By Theorem 25.3, the solution to A^* is

$$v' = -\tfrac{1}{3}, \quad p^{0'} = (\tfrac{2}{3}, \tfrac{1}{3}) \quad \text{and} \quad q^{0'} = (\tfrac{1}{3}, \tfrac{2}{3})$$

An optimum strategy for the column player C in the matrix game A is then $q^0 = (0, \frac{1}{3}, \frac{2}{3})$. (Note that $v = v'$ and $p^0 = p^{0'}$, as expected.)

Remark: In the diagram above, the three bounding lines intersect the line $x = 0$ at -2, 1 and -1 respectively, and the line $x = 1$ at 3, -1 and 0, respectively. Observe that these numbers correspond to the components of the rows of the original matrix A.

SUMMARY

In finding a solution to a matrix game, the reader should follow the following steps:

(1) Test to see if the game is strictly determined.

(2) Eliminate all recessive rows and columns.

(3) In the case of a 2×2 game, use the formulas in Theorem 25.3.

(4) In the case of a $2 \times m$ game or an $m \times 2$ game, reduce the game to a 2×2 game as in Example 6.1.

(5) Use the simplex method in all other cases.

Solved Problems

MATRIX GAMES

25.1. Which of the following games are strictly determined? For the strictly determined games, find the value v of the game and find an optimum strategy p^0 for the row player R and an optimum strategy q^0 for the column player C.

6	−4	−2
1	−1	3
−8	−3	7

(i)

3	−1	4	−2
0	3	−1	6
3	−3	5	1

(ii)

1	2	0	3
2	3	2	5
3	0	1	4

(iii)

In each case, circle the minimum entries in each row and put a box around the maximum entries in each column as follows:

6	−4	−2
1	−1	3
−8	−3	7

(i)

3	−1	4	−2
0	3	−1	6
3	−3	5	1

(ii)

1	2	0	3
2	3	2	5
3	0	1	4

(iii)

(i) The -1 in row 2 and column 2 is a saddle point; hence $v = -1$, $p^0 = (0, 1, 0)$, $q^0 = (0, 1, 0)$.

(ii) There is no saddle point; hence the game is non-strictly determined. Note that both 3's in the first column are "boxed", since both are maximum entries in the column.

(iii) The 2 in row 2 and column 3 is a saddle point; hence $v = 2$, $p^0 = (0, 1, 0)$, $q^0 = (0, 0, 1, 0)$.

25.2. Find the solution to each of the following 2×2 games.

(i)

2	−1
0	2

(ii)

1	−1
3	−2

(iii)

1	−1
−3	5

(iv)

−3	5
2	−3

If the game is non-strictly determined, use the following formulas of Theorem 25.3, where $p^0 = (x_1, x_2)$ and $q^0 = (y_1, y_2)$:

$$x_1 = \frac{d-c}{a+d-b-c}, \quad x_2 = \frac{a-b}{a+d-b-c}, \quad y_1 = \frac{d-b}{a+d-b-c},$$

$$y_2 = \frac{a-c}{a+d-b-c}, \quad v = \frac{ad-bc}{a+d-b-c}$$

(i) The game is non-strictly determined. First compute $a+d-b-c = 2+2-0+1 = 5$. Then

$$x_1 = \frac{2-0}{5} = \frac{2}{5}, \quad x_2 = \frac{2-(-1)}{5} = \frac{3}{5}, \quad y_1 = \frac{2-(-1)}{5} = \frac{3}{5}, \quad y_2 = \frac{2-0}{5} = \frac{2}{5}$$

That is, $(\frac{2}{5}, \frac{3}{5})$ is an optimum strategy for R and $(\frac{3}{5}, \frac{2}{5})$ is an optimum strategy for C. The value of the game is $v = \frac{2 \cdot 2 - 0 \cdot (-1)}{5} = \frac{4}{5}$. Observe that the game is not fair and is in favor of the row player R.

(ii) This game is strictly determined with saddle point -1. Thus $v = -1$; and $p^0 = (1, 0)$ and $q^0 = (0, 1)$, i.e. R should always play row 1 and C should always play column 2.

(iii) The game is non-strictly determined. First compute $a+d-b-c = 1+5+3+1 = 10$. Then $x_1 = 8/10 = .8$, $x_2 = .2$, $y_1 = .6$, $y_2 = .4$. That is, $(.8, .2)$ is an optimum strategy for R and $(.6, .4)$ is an optimum strategy for C. The value of the game is $v = \frac{1 \cdot 5 - (-3) \cdot (-1)}{10} = .2$. The game is not fair and, since v is positive, the game is in favor of the row player R.

(iv) The game is non-strictly determined. First compute $a+d-b-c = -3-3-2-5 = -13$. Then $x_1 = 5/13$, $x_2 = 8/13$, $y_1 = 8/13$, $y_2 = 5/13$. Thus $(\frac{5}{13}, \frac{8}{13})$ is an optimum strategy for R and $(\frac{8}{13}, \frac{5}{13})$ is an optimum strategy for C. The value of the game is $v = \frac{(-3) \cdot (-3) - 2 \cdot 5}{-13} = \frac{1}{13}$. The game is not fair and is in favor of the row player R.

25.3. Find a solution to the matrix game

1	−3	−2
0	−4	2
−5	2	3

.

First circle each row minimum and box each column maximum:

[1]	(−3)	−2
0	(−4)	2
(−5)	[2]	[3]

Observe that there is no saddle point and so the game is non-strictly determined. We now test for recessive rows and columns. Note that the third column is recessive since each entry is larger than the corresponding entry in the second column; hence the third column can be omitted to obtain the game

1	−3
0	−4
−5	2

Note that the second row is recessive since each entry is smaller than the corresponding entry in the first row. Thus the second row can be omitted to obtain the game

1	−3
−5	2

Using the formulas in Theorem 25.3, the solution to the above 2×2 game is

$$v' = -13/11, \quad p^{0'} = (\tfrac{7}{11}, \tfrac{4}{11}), \quad q^{0'} = (\tfrac{5}{11}, \tfrac{6}{11})$$

Thus a solution to the original game is

$$v = -13/11, \quad p^0 = (\tfrac{7}{11}, 0, \tfrac{4}{11}), \quad q^0 = (\tfrac{5}{11}, \tfrac{6}{11}, 0)$$

25.4. Find a solution to the matrix game $A =$

−1	5	1	−2
1	−3	−2	5

Set $p^0 = (x, 1-x)$. Then $p^0A \geqq (v, v, v, v)$ is equivalent to

$$(x, 1-x)\begin{pmatrix} -1 & 5 & 1 & -2 \\ 1 & -3 & -2 & 5 \end{pmatrix} \geqq (v, v, v, v)$$

or

$$\begin{aligned} -2x + 1 &\geqq v \\ 8x - 3 &\geqq v \\ 3x - 2 &\geqq v \\ -7x + 5 &\geqq v \end{aligned}$$

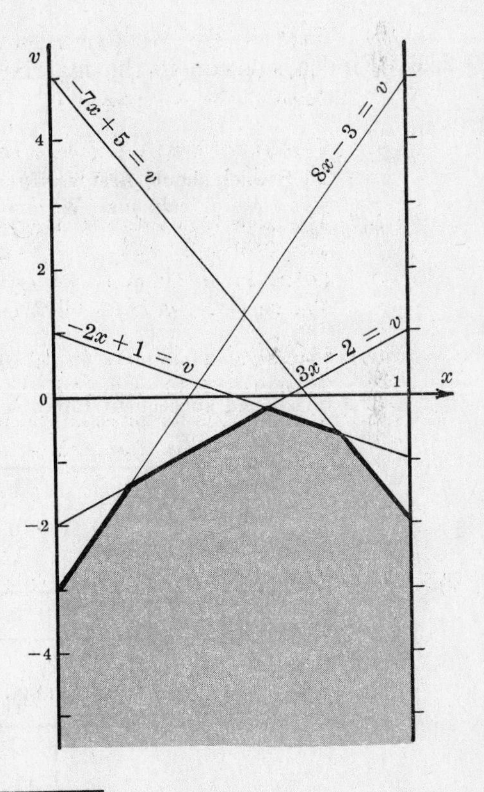

Plot the above four linear inequalities as shown in the diagram. Note that the maximum v satisfying the inequalities occurs at the intersection of the lines $-2x+1 = v$ and $3x-2 = v$, that is, at the point $x = \frac{3}{5}$ and $v = -\frac{1}{5}$. Thus $v = -\frac{1}{5}$ is the value of the game, and $p^0 = (\frac{3}{5}, \frac{2}{5})$ is an optimum strategy for player R.

Since the two lines determining v come from the first and third columns of the matrix game A, omit the other columns to reduce A to the 2×2 matrix game $A^* =$

−1	1
1	−2

. A solution of A^*, by Theorem 23.5, is

$$v' = -\tfrac{1}{5}, \quad p^{0'} = (\tfrac{3}{5}, \tfrac{2}{5}), \quad q^{0'} = (\tfrac{3}{5}, \tfrac{2}{5})$$

Thus $q^0 = (\frac{3}{5}, 0, \frac{2}{5}, 0)$ is an optimum strategy for player C in game A. (Note that $v = v'$ and $p^0 = p^{0'}$, as expected.)

25.5. Find a solution to the matrix game $A =$

3	−1
−2	1
2	0

Set $q^0 = (x, 1-x)$. Then $q^0A^t \leqq (v, v, v)$ is equivalent to

$$(x, 1-x)\begin{pmatrix} 3 & -2 & 2 \\ -1 & 1 & 0 \end{pmatrix} \leqq (v, v, v)$$

or

$$\begin{aligned} 4x - 1 &\leqq v \\ -3x + 1 &\leqq v \\ 2x &\leqq v \end{aligned}$$

Plot the above three inequalities (where $0 \leq x \leq 1$) as shown in the diagram. We seek the *minimum* v satisfying the given inequalities; note that it occurs at the intersection of the lines $-3x+1 = v$ and $2x = v$, i.e. at the point $x = \frac{1}{5}$ and $v = \frac{2}{5}$. Thus $v = \frac{2}{5}$ is the value of the game and $q^0 = (\frac{1}{5}, \frac{4}{5})$ is an optimum strategy for player C.

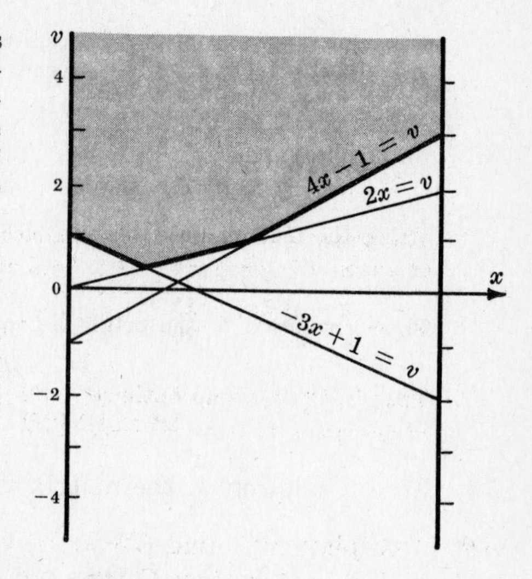

Since the two lines determining v come from the second and third rows of A, omit the other row to obtain the 2×2 matrix game $A^* = \begin{pmatrix} -2 & 1 \\ 2 & 0 \end{pmatrix}$. A solution of A^*, by Theorem 23.5, is

$$v' = \tfrac{2}{5}, \quad p^{0\prime} = (\tfrac{2}{5}, \tfrac{3}{5}), \quad q^{0\prime} = (\tfrac{1}{5}, \tfrac{4}{5})$$

Thus $p^0 = (0, \frac{2}{5}, \frac{3}{5})$ is an optimum strategy for player R. (Note that $v = v'$ and $q^0 = q^{0\prime}$, as expected.)

25.6. Find a solution to the matrix game $\begin{pmatrix} 1 & -2 & 2 \\ -3 & 0 & -1 \\ 1 & -1 & 0 \end{pmatrix}$.

The reader should first verify that the game is non-strictly determined and that there are no recessive rows or columns. We find a solution by the simplex method. First add 4 to each entry to obtain the game

$$\begin{pmatrix} 5 & 2 & 6 \\ 1 & 4 & 3 \\ 5 & 3 & 4 \end{pmatrix}$$

The initial and subsequent tableaux follow:

	P_1	P_2	P_3				
	5	2	6	1	0	0	1
	1	(4)	3	0	1	0	1
	5	3	4	0	0	1	1
	−1	−1	−1	0	0	0	0

(i)

	P_1	P_2	P_3				
	$\frac{9}{2}$	0	$\frac{9}{2}$	1	$-\frac{1}{2}$	0	$\frac{1}{2}$
P_2	$\frac{1}{4}$	1	$\frac{3}{4}$	0	$\frac{1}{4}$	0	$\frac{1}{4}$
	($\frac{17}{4}$)	0	$\frac{7}{4}$	0	$-\frac{3}{4}$	1	$\frac{1}{4}$
	$-\frac{3}{4}$	0	$-\frac{1}{4}$	0	$\frac{1}{4}$	0	$\frac{1}{4}$

(ii)

	P_1	P_2	P_3				
	*	*	*	*	*	*	*
P_2	*	*	*	*	*	*	$\frac{4}{17}$
P_1	1	0	$\frac{7}{17}$	0	$-\frac{3}{17}$	$\frac{4}{17}$	$\frac{1}{17}$
	0	0	$\frac{1}{17}$	0	$\frac{2}{17}$	$\frac{3}{17}$	$\frac{5}{17}$

(iii)

(*Remark*: Since the last row in (iii) has no negative indicators, the tableau is a terminal one and so all of its entries need not be computed.)

Thus $P = (\frac{1}{17}, \frac{4}{17}, 0)$, $Q = (0, \frac{2}{17}, \frac{3}{17})$ and $v^* = \frac{5}{17}$; hence $1/v^* = \frac{17}{5}$. Accordingly, $p^0 = (\frac{17}{5})Q = (0, \frac{2}{5}, \frac{3}{5})$, $q^0 = (\frac{17}{5})P = (\frac{1}{5}, \frac{4}{5}, 0)$ and $v = \frac{17}{5} - 4 = -\frac{3}{5}$.

25.7. Two players R and C simultaneously show 2 or 3 fingers. If the sum of the fingers shown is even, then R wins the sum from C; if the sum is odd, then R loses the sum to C. Find optimum strategies for the players and to whom the game is favorable.

The matrix of the game is

	2	3
2	4	−5
3	−5	6

The game is non-strictly determined; hence the formulas in Theorem 25.3, Page 234, apply. First compute $a+d-b-c = 4+6+5+5 = 20$. Then

$$x_1 = \frac{6-(-5)}{20} = \frac{11}{20}, \quad x_2 = \frac{4-(-5)}{20} = \frac{9}{20}, \quad y_1 = \frac{4-(-5)}{20} = \frac{9}{20}, \quad y_2 = \frac{6-(-5)}{20} = \frac{11}{20}$$

That is, $(\frac{11}{20}, \frac{9}{20})$ is an optimum strategy for R and $(\frac{9}{20}, \frac{11}{20})$ is an optimum strategy for C. The value of the game is $v = \frac{4\cdot 6-(-5)\cdot(-5)}{20} = -\frac{1}{20}$; and, since it is negative, the game is favorable to C.

25.8. Two players R and C match pennies. If the coins match, then R wins; if the coins do not match then C wins. Determine optimum strategies for the players and the value of the game.

The matrix of the game is

	H	T
H	1	−1
T	−1	1

The game is non-strictly determined. Using the formulas in Theorem 25.3, Page 234, we obtain $x_1 = x_2 = y_1 = y_2 = \frac{1}{2}$. Thus $(\frac{1}{2}, \frac{1}{2})$ is an optimum strategy for both R and C. The value of the game is $v = \frac{1\cdot 1-(-1)\cdot(-1)}{4} = 0$; hence the game is fair.

THEOREMS

25.9. Let p^0 be a given strategy for player R in a matrix game A. Show that the following conditions are equivalent:

(i) $p^0 A q^t \geqq v$, for every strategy q for C; (ii) $p^0 A \geqq (v, v, \ldots, v)$.

Assume that (ii) holds. Then, for any strategy q for C,

$$p^0 A q^t \geqq (v, v, \ldots, v)\, q^t = v$$

On the other hand, assume that (i) holds and that $p^0 A = (a_1, a_2, \ldots, a_k)$. Choosing the pure strategy $q = (1, 0, \ldots, 0)$, we have

$$p^0 A q^t = (a_1, a_2, \ldots, a_k)\, q^t = a_1 \geqq v$$

Similarly, $a_2 \geqq v, \ldots, a_k \geqq v$. In other words, $p^0 A \geqq (v, v, \ldots, v)$.

25.10. Let $p = (a_1, a_2, \ldots, a_k)$ and $p^* = (b_1, b_2, \ldots, b_k)$ be probability vectors. Show that each point on the line segment joining p and p^*, i.e. $tp + (1-t)p^*$ where $0 \leqq t \leqq 1$, is also a probability vector.

Observe that

$$tp + (1-t)p^* = (ta_1 + (1-t)b_1,\ ta_2 + (1-t)b_2,\ \ldots,\ ta_k + (1-t)b_k)$$

Since t, $1-t$, the a_i and the b_i are all non-negative, $ta_i + (1-t)b_i$ is also non-negative for every i. Furthermore,

$$\begin{aligned} & ta_1 + (1-t)b_1 + ta_2 + (1-t)b_2 + \cdots + ta_k + (1-t)b_k \\ &= ta_1 + ta_2 + \cdots + ta_k + (1-t)b_1 + (1-t)b_2 + \cdots + (1-t)b_k \\ &= t(a_1 + a_2 + \cdots + a_k) + (1-t)(b_1 + b_2 + \cdots + b_k) = t + (1-t) = 1 \end{aligned}$$

25.11. Let p^0 and p^* be optimum strategies for player R in a matrix game A with value v. Show that $p' = tp^0 + (1-t)p^*$, where $0 \leq t \leq 1$, is also an optimum strategy for R.

By the preceding problem, p' is a probability vector; hence we need only show that $p'A \geq (v, v, \ldots, v)$. We are given that

$$p^0A \geq (v, v, \ldots, v) \quad \text{and} \quad p^*A \geq (v, v, \ldots, v)$$

Thus

$$\begin{aligned} p'A = (tp^0 + (1-t)p^*)A &= tp^0A + (1-t)p^*A \\ &\geq t(v, v, \ldots, v) + (1-t)(v, v, \ldots, v) \\ &= (v, v, \ldots, v) \end{aligned}$$

25.12. Show that if a and b are saddle points of a matrix A, then $a = b$.

If a and b are in the same row, then each is a minimum of the row and so $a = b$. Similarly if a and b are in the same column, then a and b are maxima of the column and so $a = b$.

Now assume that a and b appear in different columns and different rows as illustrated below:

$$\begin{pmatrix} * & * & * & * & * & * & * & * \\ * & * & \boxed{a} & * & * & \boxed{d} & * & * \\ * & * & * & * & * & * & * & * \\ * & * & \boxed{c} & * & * & \boxed{b} & * & * \\ * & * & * & * & * & * & * & * \end{pmatrix}$$

Since a is a saddle point, $a \leq d$ and $c \leq a$; and since b is a saddle point, $b \leq c$ and $d \leq b$. Thus

$$a \leq d \leq b \leq c \leq a$$

Accordingly, the equality relation holds above: $a = d = b = c$.

25.13. Show that the game $A = \begin{array}{|c|c|} \hline a & a \\ \hline b & c \\ \hline \end{array}$ is strictly determined.

Observe that each a is a minimum of the first row. If $a \geq b$ then the first a is a saddle point, and if $a \geq c$ then the second a is a saddle point.

Thus we are left with the case that $b > a$ and $c > a$, i.e. where b and c are maxima of their respective columns. Now if $b \leq c$ then b is a saddle point, and if $c \leq b$ then c is a saddle point.

Accordingly, in all cases A has a saddle point and so A is strictly determined.

25.14. Prove Lemma 25.2: The matrix game $A = \begin{array}{|c|c|} \hline a & b \\ \hline c & d \\ \hline \end{array}$ is non-strictly determined if and only if each of the entries on one of the diagonals is greater than each of the entries on the other diagonal:

(i) $a, d > b$ and $a, d > c$; or (ii) $b, c > a$ and $b, c > d$.

If (i) or (ii) holds, then there is no saddle point for A and so A is non-strictly determined.

Now suppose A is non-strictly determined. By the preceding problem, $a \neq b$. We consider two cases.

Case (1): $a < b$. If $a \geq c$ then a is a saddle point; hence $a < c$. We need only show that $d < b$ and $d < c$. Suppose $d \geq b$; then $d > c$ or else d is a saddle point. But $a < c$ and $d > c$ implies c is a saddle point. Accordingly, $d \geq b$ leads to a contradiction, and so $d < b$. Similarly, $d < c$.

Case (2): $a > b$. The proof is similar to the proof in Case (1).

25.15. Prove Theorem 25.3: Suppose the matrix game $A = \begin{array}{|c|c|}\hline a & b \\ \hline c & d \\ \hline\end{array}$ is not strictly determined. Then

$$p^0 = \left(\frac{d-c}{a+d-b-c}, \frac{a-b}{a+d-b-c}\right), \quad q^0 = \left(\frac{d-b}{a+d-b-c}, \frac{a-c}{a+d-b-c}\right),$$

$$v = \frac{ad-bc}{a+d-b-c}$$

are respectively an optimum strategy for R, an optimum strategy for C and the value of the game.

Set $p^0 = (x, 1-x)$ and $q^0 = (y, 1-y)$. Then $p^0A \geqq (v, v)$ and $q^0A^t \leqq (v, v)$ is equivalent to

$$(x, 1-x)\begin{pmatrix} a & b \\ c & d \end{pmatrix} \geqq (v, v) \quad \text{and} \quad (y, 1-y)\begin{pmatrix} a & c \\ b & d \end{pmatrix} \leqq (v, v)$$

or

$$\begin{aligned}(a-c)x + c &\geqq v \\ (b-d)x + d &\geqq v\end{aligned} \quad \text{and} \quad \begin{aligned}(a-b)y + b &\leqq v \\ (c-d)y + d &\leqq v\end{aligned}$$

Solving the above inequalities as equations for the unknowns x, y and v, we obtain the required result.

We remark that $a+b-c-d \neq 0$ by the preceding problem, i.e. by Lemma 25.2. We also note that in the case of a 2×2 non-strictly determined matrix game A, we have the following stronger result: $p^0A = (v, v)$ and $q^0A^t = (v, v)$.

25.16. Let A be a matrix game with optimum strategy p^0 for R, optimum strategy q^0 for C, and value v; and let k be a positive number. Show that for the game kA, p^0 is an optimum strategy for R, q^0 is an optimum strategy for C, and kv is the value of the game.

We are given that $p^0A \geqq (v, v, \ldots, v)$ and $q^0A \leqq (v, v, \ldots, v)$. Then

$$p^0(kA) = k(p^0A) \geqq k(v, v, \ldots, v) = (kv, kv, \ldots, kv)$$

$$q^0(kA)^t = q^0kA^t = k(q^0A^t) \leqq k(v, v, \ldots, v) = (kv, kv, \ldots, kv)$$

which was to be shown.

Supplementary Problems

MATRIX GAMES

25.17. Find a solution to each game, i.e. an optimum strategy p^0 for R, an optimum strategy q^0 for C, and the value of the game.

(i)

3	−3
2	1

(ii)

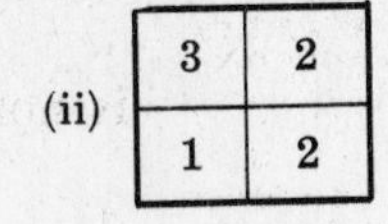

3	2
1	2

(iii)

1	2
1	3

25.18. Find a solution to each game.

(i)

2	−1
−4	3

(ii)

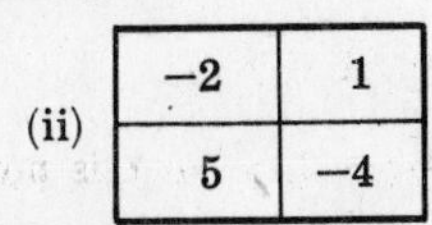

−2	1
5	−4

(iii)

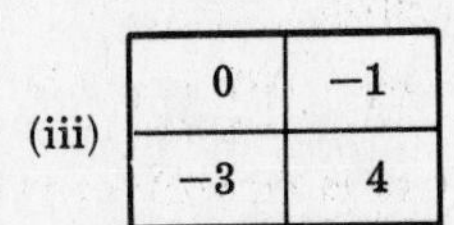

0	−1
−3	4

25.19. Find a solution to each game.

(i)

2	3	−1
−3	−1	2

(ii)

1	−1	3
−1	4	−2

(iii)

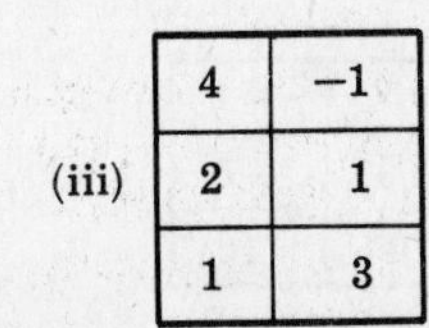

4	−1
2	1
1	3

25.20. Find a solution to each game.

(i)

2	−4	−3
−1	−2	1
0	1	1

(ii)

−2	1	2
3	−1	−2
−1	1	3

25.21. Consider the game

4	1
3	a

. Find a solution to the game if (i) $a < 1$, (ii) $1 < a < 3$, (iii) $a > 3$.

25.22. Each of two players R and C has a dime and a quarter. They each show a coin simultaneously. If the coins are the same, R wins C's coin; if the coins are different, C wins R's coin. Represent the game as a matrix game and find a solution.

25.23. Each of two players R and C has a penny, nickel and dime. They each show a coin simultaneously. If the total amount of money shown is even, R wins C's coin; if it is odd, C wins R's coin. Represent the game as a matrix and find a solution.

THEOREMS

25.24. Suppose every entry in a matrix game A is increased by an amount k. Show that the value of the game also increases by k, but that the optimum strategies remain the same.

25.25. Show that if every entry in a matrix game is positive, then the value of the game is positive.

Answers to Supplementary Problems

25.17. (i) $p^0 = (0,1)$, $q^0 = (0,1)$, $v = 1$; (ii) $p^0 = (1,0)$, $q^0 = (0,1)$, $v = 2$; (iii) p^0 can be any strategy, $q^0 = (1,0)$, $v = 1$.

25.18. (i) $p^0 = (.7, .3)$, $q^0 = (.4, .6)$, $v = .2$; (ii) $p^0 = (3/4, 1/4)$, $q^0 = (5/12, 7/12)$, $v = 1/4$; (iii) $p^0 = (7/8, 1/8)$, $q^0 = (5/8, 3/8)$, $v = 3/8$.

25.19. (i) $p^0 = (5/8, 3/8)$, $q^0 = (3/8, 0, 5/8)$, $v = 1/8$;
(ii) $p^0 = (5/7, 2/7)$, $q^0 = (5/7, 2/7, 0)$, $v = 3/7$;
(iii) $p^0 = (2/7, 0, 5/7)$, $q^0 = (4/7, 3/7)$, $v = 13/7$.

25.20. (i) $p^0 = (1/7, 0, 6/7)$, $q^0 = (5/7, 2/7, 0)$, $v = 2/7$;
(ii) $p^0 = (0, 1/3, 2/3)$, $q^0 = (1/3, 2/3, 0)$, $v = 1/3$.

25.21. (i) $p^0 = (1,0)$, $q^0 = (0,1)$, $v = 1$;
(ii) $p^0 = (0,1)$, $q^0 = (0,1)$, $v = a$;
(iii) $p^0 = \left(\frac{a-3}{a}, \frac{3}{a}\right)$, $q^0 = \left(\frac{a-1}{a}, \frac{1}{a}\right)$, $v = (4a-3)/a$.

25.22.

10	−10
−25	25

; $p^0 = (5/7, 2/7)$, $q^0 = (\frac{1}{2}, \frac{1}{2})$, $v = 0$.

25.23.

	1	5	10
1	1	5	−1
5	1	5	−5
10	−10	−10	10

; $p^0 = (10/11, 0, 1/11)$, $q^0 = (1/2, 0, 1/2)$, $v = 0$.

INDEX